T0140369

Microlocal Analysis, Sharp Spectral Asymptotics and Applications II

Victor Ivrii

Microlocal Analysis, Sharp Spectral Asymptotics and Applications II

Functional Methods and Eigenvalue Asymptotics

Springer

Victor Ivrii
Department of Mathematics
University of Toronto
Toronto, ON, Canada

ISBN 978-3-030-30543-7 ISBN 978-3-030-30541-3 (eBook)
https://doi.org/10.1007/978-3-030-30541-3

Mathematics Subject Classification (2010): 35P20, 35S05, 35S30, 81V70

This Springer imprint is published by the registered company Springer Nature Switzerland AG
The registered company address is: Gewerbestrasse 11, 6330 Cham, Switzerland

Preface

The Problem of the *Spectral Asymptotics*, in particular the problem of the *Asymptotic Distribution of the Eigenvalues*, is one of the central problems in the *Spectral Theory of Partial Differential Operators*; moreover, it is very important for the *General Theory of Partial Differential Operators*.

I started working in this domain in 1979 after R. Seeley [1] justified a remainder estimate of the same order as the then hypothetical second term for the Laplacian in domains with boundary, and M. Shubin and B. M. Levitan suggested me to try to prove Weyl's conjecture. During the past almost 40 years I have not left the topic, although I had such intentions in 1985, when the methods I invented seemed to fail to provide the further progress and only a couple of not very exciting problems remained to be solved. However, at that time I made the step toward local semiclassical spectral asymptotics and rescaling, and new much wider horizons opened.

So I can say that this book is the result of 40 years of work in the Theory of Spectral Asymptotics and related domains of Microlocal Analysis and Mathematical Physics (I started analysis of *Propagation of singularities* (which plays the crucial role in my approach to the spectral asymptotics) in 1975).

This monograph consists of five volumes. This Volume II concludes the general theory. It consists of two parts. In the first one we develop methods of combining local asymptotics derived in Volume I, with estimates of the eigenvalue counting functions in the small domains (singular zone) derived by methods of functional analysis. In the second part we derive eigenvalue asymptotics which either follow directly from the general theory, or require applications of the developed methods (if the operator has singularities or degenerations, strong enough to affect the asymptotics).

Victor Ivrii,
Toronto, June 10, 2019.

Contents

Introduction

Let us discuss in details the current volume, concluding the general theory.

Part IV. Estimates of the Spectrum.

Chapter 9. Estimates of the Negative Spectrum. We assume here that the operator is semibounded from below. Then under assumption $\rho(y)\gamma(y) \ge h$ everywhere the partition-rescaling technique results in

$$|\mathsf{N}^-(0) - \mathcal{N}^-(0)| \le CR_1 \tag{0.1}$$

with the same (Weyl or non-Weyl) principal part $\mathcal{N}^-(0)$ as before and with some remainder estimate R_1. However, there are many important cases in which this condition $\rho(y)\gamma(y) \ge h$ is fulfilled only in a subdomain X' of the domain X (as we mentioned we call X' semiclassical or regular zone); then summing with respect to a partition of unity in X' we obtain estimates only for $\mathcal{E}_\psi(0)$ (but not for $\mathsf{N}^-(0)$) with an admissible function ψ such that $\operatorname{supp}(\psi) \subset X'$ and $\psi = 1$ outside of $X'_{3/2} = \{y\colon \rho(y)\gamma(y) \le \frac{3}{2}h\}$. Thus, in order to obtain an estimate for $\mathsf{N}^-(0) = \mathcal{E}_1(0)$, we need to derive an estimate for $\mathcal{E}_{1-\psi}(0)$.

On the other hand, if we had $X = X'' \supset \{y\colon \rho(y)\gamma(y) \le 2h\}$, then we very often would have some estimates for $\mathsf{N}^-(0)$ of a completely different (variational) origin which have the form $\mathsf{N}^-(0) \le CR_2$.

In this book we almost exclusively apply the Rozenblioum variational estimate, but one can use other estimates of this type. It appears that in the general case one can prove these estimates together and derive the estimate

$$|\mathsf{N}^-(0) - \mathcal{N}^-(0)| \le C(R_1 + R_2 + R_{12}), \tag{0.2}$$

where $\mathcal{N}^-(0)$ and R_1 (associated with $\mathcal{E}_\psi(0)$) are exactly the same as before in the zone X', the estimate R_2 is derived for a self-adjoint operator A'' defined in the domain X'' where A'' coincides with A in the zone $\{y\colon \rho(y)\gamma(y) \le \frac{7}{4}h\}$ (and the boundary conditions also coincide at the common part of the boundaries of X and $\partial X''$), and R_{12} is an additional term, depending on the auxiliary construction.

The proper construction using the *Birman-Schwinger principle* with an appropriate auxiliary function makes this term less than $C(R_1 + R_2)$. This construction is done in Chapter 9.

First of all, in *Section 9.1* the principal theorems are formulated and proven.

Next, in *Section 9.2* we illustrate these theorems in the simpler cases and in *Section 9.3* we reformulate them for quadratic forms. Here we have two important ideas:

(a) *rescaling with multiplication and a change of parameters* (we mentioned this idea in the discussion of Section 5.3) and

(b) *combining the asymptotics derived in Parts II and III with estimates of the variational origin which makes it possible to consider a domain containing not only a semiclassical zone X' but also a singular zone X''.*

This task is done in this chapter for semi-bounded operators.

Chapter 10. Estimates of the Spectrum in the Interval. Here we consider the non-semibounded case and our analysis here is essentially parallel to the analysis of Chapter 9; a very important role is played by the following form of the *Birman-Schwinger principle:*

$$\text{(0.3)}\quad \mathsf{N}\bigl(A;[\lambda_1,\lambda_2)\bigr) = \mathsf{N}\bigl(A-\bar{t}J;[\lambda_1,\lambda_2)\bigr) - \mathsf{N}\bigl(-J^{-1}(A-\lambda I);[0,\bar{t})\bigr)\Bigr|_{\lambda=\lambda_1}^{\lambda=\lambda_2},s$$

which applies (under some hypotheses) in the case of a positively defined matrix-valued function J and for $\bar{t} > 0$ [1], where $\mathsf{N}(A;[\lambda_1,\lambda_2))$ is the number of eigenvalues of the operator A in the interval $[\lambda_1,\lambda_2)$ [2] and similarly $\mathsf{N}(-J^{-1}(A-\lambda I);[0,\bar{t}))$ is the number of eigenvalues of the spectral problem

$$\text{(0.4)}\qquad \bigl((A-\lambda I)+tJ\bigr)u = 0$$

[1] A similar form applies to the case $\bar{t} < 0$.

[2] Provided $\mathsf{Spec}_{\mathsf{ess}}(A)\cap[\lambda_1,\lambda_2)=\emptyset$; otherwise $\mathsf{N}(A;[\lambda_1,\lambda_2))=\infty$.

in the interval $[0, \bar{t})$.

One can iterate this formula, at each step replacing the operator A by $A - t_1J_1 - \dots - t_{k-1}J_{k-1}$, the number $\bar{t}$ by t_k and the function J by J_k with $k = 1, \dots, n$; here $t_1, \dots, t_n$ *are not necessarily of the same sign*. As a result we obtain the following more complicated formula

$$\begin{aligned}\mathsf{N}(A;[\lambda_1,\lambda_2)) =& \mathsf{N}(A - t_1J_1 - \dots - t_{k-1}J_{k-1};[\lambda_1,\lambda_2)) - \\ & \mathsf{N}(-J_1^{-1}(A-\lambda I);[0,t_1))\Big|_{\lambda=\lambda_1}^{\lambda=\lambda_2} - \dots \\ & \mathsf{N}(-J_n^{-1}(A-\lambda I - t_1J_1 - \dots - t_{n-1}J_{n-1});[0,t_n))\Big|_{\lambda=\lambda_1}^{\lambda=\lambda_2},\end{aligned}$$

to the right-hand expression of which we apply arguments, similar to those discussed above (in the discussion of Chapter 9)[3].

It should be noted that the reader who is interested only in semibounded operators has no need to read Chapter 10. It should also be noted that Chapters 9 and 10 would be absolutely trivial if we assumed that the singular zone is empty; however in this case we would obtain a much more restricted series of results in Chapters 11, 23 and 24 and we would be ready for Chapter 12 neither technically nor psychologically.

Part V. Asymptotics of Spectra

In Part V we return to spectral asymptotics, now based on Parts II and III and IV.

Chapter 11. Weyl Asymptotics of Spectra. The essence of Chapter 11 is very simple: *we substitute an operator A depending on certain parameters into the assertions of Chapters 9 and 10 (and even Chapters 4–8 after partition of unity, rescaling, multiplication and change of parameters), check that the conditions of these assertions are uniformly fulfilled, and as a result we obtain asymptotics with respect to these parameters.*

The arguments of this Chapter are in terms of "arithmetics", dynamical systems theory (when we check conditions associated with the Hamiltonian trajectories) and rather simple functional analysis (when we check conditions associated with the singular zone).

[3] Actually, we multiply the last equality by $\varphi_1(t_1)\cdots\varphi_n(t_n)$ with appropriate functions φ_k and integrate on $(t_1, \dots, t_n) \in \mathbb{R}^n$ (this rather technical trick is useful).

So *having started with (micro)local semiclassical asymptotics we come to asymptotics with respect to quite different parameters (e.g., the spectral parameter)*. Actually, semiclassical asymptotics are also treated but our results are not the same as before and one should distinguish the actual semiclassical parameter h and the effective semiclassical parameter $h_{\mathsf{eff}} = h/(\rho\gamma)$. The latter depends on the point and arises always, no matter what parameters were in the original problem.

In *Section 11.1* we treat semiclassical asymptotics again, but now we are interested in $\mathsf{N}_h(\lambda_1, \lambda_2)$ and in situations when there are singularities in the domain and coefficients, exits to infinity of different shapes (or even the domain $\mathbb{R}^d$) and other non-regular phenomena.

In Sections 11.2–11.5 we investigate *asymptotics with respect to a large spectral parameter*. Namely, in *Sections 11.2* and *11.3* we consider cases in which a domain may have singularities on the boundary (edges, vertices, conical points, spikes, etc.), cusps (i.e., exits to infinity) which are not too massive and also cases in which coefficients may have singularities which are not too strong (a good example is the Schrödinger operator with potential similar to $\pm|x|^{2m}$ near the origin with $m \in (-1, 0)$).

Further, in *Section 11.4* stronger singularities of coefficients are allowed but a *coercivity* condition is assumed (a good example is the Schrödinger operator with potential similar to $+|x|^{2m}$ near the origin with $m \le -1$).

Furthermore, in *Section 11.5* we consider the situation in which a domain is "essentially unbounded" or even coincides with $\mathbb{R}^d$; e.g., we consider the Schrödinger operator in $\mathbb{R}^d$ with potential similar to $+|x|^{2m}$ near infinity with $m > 0$. We are able to consider certain situations when the potential does not tend to $+\infty$ along some directions (along these directions, $V(x) = 0$ or even $V(x)$ tends to $-\infty$, for example) but the arising "canyons" are assumed to be narrow and shallow enough to provide the standard asymptotics.

Next, in *Section 11.6* we analyze eigenvalues tending to -0 which is the bottom of the essential spectrum; e.g., we consider the Schrödinger operator in $\mathbb{R}^d$ with potential similar to $-|x|^{2m}$ near infinity with $m \in (-1, 0)$; for this operator the essential spectrum is $[0, +\infty)$ and $(-\infty, 0)$ contains the discrete spectrum tending to -0, the asymptotics of which is the object of our interest.

Next, in *Section 11.7* we consider asymptotics with several parameters: when we have different semiclassical parameters with respect to different variables or when we have semiclassical and spectral parameters at the same time.

It should be noted, that in all Sections 11.1–11.7 there are three cases: when singularities are rather week and affect neither principal part nor remainder estimate, when singularities are too strong and we cannot derive any meaningful results this way, and the intermediate case, when we derive asymptotics with the same principal part albeit with deteriorated remainder estimate. Two latter cases are partially covered in Chapter 12.

In rather different *Section 11.8* the distribution of the spectra of negative-order pseudodifferential operators on compact closed manifolds is derived.

Chapter 12. Miscellaneous Asymptotics of Spectra. The character of this chapter is quite different: we improve and generalize some results of Chapter 11, but the results of Chapters 4–8 are insufficient for our purposes in this case and we are forced to repeat some arguments of Chapter 4 and even those of Chapter 2 in certain specific situations. Basically we apply the ideas discussed above. The asymptotics obtained are more or less non-Weyl (or, more precisely, "partially Weyl").

In *Section 12.2* elliptic operators in domains with *thick cusps* are discussed: a cusp is thick if the area of its boundary is infinite, and the standard Weyl asymptotics with the remainder estimate $O\bigl(\tau^{\frac{1}{m}(d-1)}\bigr)$ fails to hold. Moreover, if the volume of the cusp is infinite, then $\mathsf{N}(\tau) = O\bigl(\tau^{\frac{d}{m}}\bigr)$ is no longer true[4)].

For simplicity let us consider the Dirichlet Laplacian in a domain which for $|x'| > c$ coincides with $\{(x', x'') : x'' \in f(x')\Omega\}$ with $\Omega \Subset \mathbb{R}^{d''}$. Here Ω and $f(x')$ describe the shape of the cusp's cross section and its thickness. Making a change of variables $(x', x'') \mapsto (x', x''/f(x'))$, which transforms the cusp into the cylinder $\mathbb{R}^{d'} \times \Omega$, one can consider the transformed operator as a Schrödinger operator in $\mathbb{R}^{d'}$ with operator-valued coefficients.

The auxiliary space is $\mathbb{H} = \mathscr{L}^2(\Omega)$. Applying the results of Chapter 4 concerning operators with operator-valued symbols we obtain non-Weyl asymptotics. However, this asymptotics is not very effective in the sense that the principal part contains a term

$$(2\pi)^{-d'} \iint n(x', \xi'; \lambda)(1 - \psi_0(x'))\, dx' d\xi' \tag{0.5}$$

4) This is correct as long as we consider $\mathsf{N}(\lambda)$; for its Riesz means $\mathsf{N}_\vartheta(\lambda)$ the definition of the thickness depends on ϑ.

(with $\psi_0 \in \mathscr{C}_0^\infty$), thus depending on the eigenvalue counting function $\mathsf{n}(x', \xi')$ for the family of the auxiliary operators in $\mathbb{H}$ while we would like to have just one such operator.

Observe that one can study the propagation of singularities for the transformed operator, treated as an operator with an operator-valued symbol. We then obtain some results for the original operator as well. Namely, we obtain that for a not very remote part of the cusp in Weyl asymptotics one can take take much larger T (depending on τ, $|x'|$) and drastically improve the remainder estimate. These asymptotics are not very effective either because in addition to the Weyl principal part they contain plenty of extra terms.

On the other hand, under a stabilization condition in the remote part $\{x\colon |x'| \ge r\}$ of the cusp ($r = r(\tau) \to \infty$ as $\tau \to +\infty$) the Schrödinger operator is rather close to $\Delta' + |x'|^{2m}\Delta''$ provided $f(x') = |x'|^{-m}$; for the latter operator we can express the right-hand expression of (0.5) through eigenvalue counting function $\mathsf{n}(.)$ for the Dirichlet Laplacian Δ'' in Ω. Then for the Schrödinger operator which we really got we have an asymptotics which is not very effective either; its main part depends only on $\mathsf{n}(.)$ but one needs to add plenty of extra terms.

So, we get two different asymptotics, both of which are not effective. However, combining them we obtain much more effective asymptotics because in the result most of these extra terms turn out to be superficial since they strongly depend on parameter $r(\tau)$, which we can select almost arbitrarily; comparing these terms for different choices of this parameter we can conclude that they must vanish. In the best possible case one has only one auxiliary operator instead of a family. Thus, the ideas are basically the same as in many forthcoming sections and chapters:

We take two different approaches (on one hand, we apply the theory of operators with operator-valued symbols to derive spectral asymptotics, and on the other, we apply the operator-valued theory only in order to get better propagation results) and then we combine the asymptotics obtained.

In *Section 12.3* we consider the class of elliptic operators in $\mathbb{R}^d$. A good example of such operators is the Schrödinger operator with potential V, failing to tend to $+\infty$ along directions $x \in \Lambda$, where Λ is a conical set of the "bad" directions. So, the domain $\{x\colon V(x) \le \tau\}$ contains cusps and if the "canyons" are not narrow and shallow enough the Weyl asymptotics with the standard remainder estimate fails.

Let us assume that Λ is a smooth d'-dimensional manifold and let us transform it to $\mathbb{R}^{d'}$. Then the ideas of Section 12.2 yield results of the same type (heuristically one can say that we apply the results of the Section 12.2 but the cusp depends on τ and widens when τ increases).

In *Section 12.4* we generalize the results of Section 11.6 to singular potentials, decaying at infinity in the case when singularity is strong enough; again non-Weyl correction terms appear; moreover, if singularity is strong enough the principal term in the asymptotics is non-Weyl.

In *Section 12.5 maximally hypoelliptic operators* on compact manifolds are treated. We assume that the operator uniformly degenerates on the symplectic conic manifold Λ. Microlocally $\Lambda = \{(x,\xi)\colon x'' = \xi'' = 0\}$. The same basic ideas as above are applied. We treat the operator in question as $A(x', D')$ in $\mathscr{L}^2(\mathbb{R}^{d'}, \mathbb{H})$ with $\mathbb{H} = \mathscr{L}^2(\mathbb{R}^{d''})$. Instead of the remote part of the cusp we have now a tight (depending on the spectral parameter) vicinity of Λ: $\{(x,\xi)\colon |x'| + |\xi'| \le \varepsilon(\tau)\}$ with $\varepsilon(\tau) \to +0$ as $\tau \to +\infty$.

In *Section 12.6* we consider semiclassical asymptotics of the spectral Riesz means for operators with singularities at disjoint points. In this case the asymptotics of $\mathsf{N}^-(h)$ are Weyl as are the asymptotics of $\mathsf{N}^-_\vartheta(h)$ with $\vartheta < \bar{\vartheta}$ where ϑ is the order of the Riesz mean. So, we are interested in the case $\vartheta \ge \bar{\vartheta}$. One very early idea was to use the equality

$$\operatorname{Tr}\bigl(E^1_\vartheta(\tau) - E^0_\vartheta(\tau)\bigr) = \mathsf{const} \int_0^1 \operatorname{Tr}(A^1 - A^0) E^t_{\vartheta-1}(\tau)\, d\tau, \tag{0.6}$$

where $A^t = tA^1 + (1-t)A^0$ and the superscript $E^t(\tau)$ is related to A^t and $E_\vartheta(\tau) := \Gamma(\vartheta+1)^{-1} \operatorname{Tr}\bigl((\tau - A)^\vartheta_+\bigr)$. Then taking appropriate partitions and applying the standard Weyl approximation to some elements and this formula (or its more sophisticated and general form) to the other elements we obtain the final answer as a sum of the Weyl expression and the term(s) associated with the model operator A^0. In comparison with purely Weyl asymptotics correction terms appear. They are similar to *Scott correction term* for $-|x|^{-1}$-like potential in dimension 3 for $\vartheta = 1$.

We pick A^0 as an operator with positively homogeneous coefficients and get a more or less simple expression for the non-Weyl term(s). Actually this idea is adjusted in different ways, one of which is an investigation of the very long range (in the local scale) propagation of singularities which allows to improve significantly the remainder estimate.

In *Section 12.7* we apply the same idea as in Section 12.2 to the Neumann Laplacian in domain with cusps. However situation is very different now because the Neumann Laplacian in Ω of the cusp has the lowest eigenvalue 0 and the Schrödinger operator described there breaks into direct sum of two: one with the potential growing as in Section 12.2 in $\mathscr{L}^2(\mathbb{R}^{d'}, \mathbb{H}\ominus\mathbb{K})$ and a special one in $\mathscr{L}^2(\mathbb{R}^{d'}, \mathbb{K})$ coinciding here with $\mathsf{const}|\nabla \log f(x')|^2$. Here $\mathbb{K} = \mathbb{C}^1$ (or $\mathbb{C}^p$ in more general examples). This term is due to the the change of the density when we change variables.

As a result, unless the cusp is really very thin, the Neumann Laplacian has an essential spectrum $[0, \infty)$ or $[c_*, \infty)$. However if $|\log f| \asymp |x'|^m$, $|\nabla \log f| \asymp |x'|^{m-1}$ with $m > 1$, then the essential spectrum is empty and we derive sharp spectral asymptotics for $\mathsf{N}(\tau)$ which may contain non-Weyl correction or even principal term.

Finally, in *Section 12.8* we consider periodic operator A^0; its spectrum consists of *spectral bands* separated by *spectral gaps*. Let λ belong to a spectral gap (or its border). Consider perturbed operator $A^t := A^0 + tW$ with W non-negative and decaying fast enough at infinity. This operator can have discrete eigenvalues inside spectral gaps. We study asymptotics for the number $N(\tau)$ of such eigenvalues crossing λ as t changes from 0 to τ and $\tau \to \pm\infty$.

Part IV

Estimates of the Spectrum

Chapter 9

Estimates of the Negative Spectrum

This short chapter plays a crucial role in the whole book bridging Parts I, II and Part V. Namely here we combine the results of Parts II, III devoted to local semiclassical spectral asymptotics with the variational estimates of Rozenblioum type (upper estimates for $\mathsf{N}^-(A)$) in order to derive upper and lower estimates for $\mathsf{N}^-(A) = \mathsf{N}(-\infty, A)$ where $\mathsf{N}^-(A)$ is the number of negative eigenvalues of the self-adjoint operator A (counting their multiplicities) provided $(-\infty, 0)$ does not intersect with the essential spectrum of A and $\mathsf{N}^-(A) = \infty$ otherwise.

More precisely, the idea is that the geometric domain X consists of two overlapping zones: the *regular zone* where local semiclassical spectral asymptotics are applied and the *singular zone* where formally variational estimates are applied. We use only the regular zone and semiclassical asymptotics there to derive a lower estimate of $\mathsf{N}^-(A)$. However we use both the regular zone with its semiclassical asymptotics and the singular zone with variational estimates to derive upper estimates of $\mathsf{N}^-(A)$. Both these estimate have the same main part derived somewhere in Parts II and III.

On the *inner boundary* of the singular zone we can introduce boundary conditions of our choice and Dirichlet boundary conditions are usually the best. In the justification of this procedure (decoupling of asymptotics) the results of Chapters 2 and 3 concerning the finite speed of propagation of singularities in the intersection of these zones play an important role.

In Section 9.1 we formulate the main statements, discuss and prove them. Actually there is just one main Theorem 9.1.7 and its sidekick Theorem 9.1.13

V. Ivrii, *Microlocal Analysis, Sharp Spectral Asymptotics and Applications II*, https://doi.org/10.1007/978-3-030-30541-3_9

for spectral Riesz means.

Section 9.2 is devoted to the series of initial examples; each of them could be formulated as a theorem.

In Section 9.3 we transfer the results of Section 9.2 to quadratic forms.

In Appendix 9.A we survey variational estimates and discuss some their consequences.

Later (in Part 4) we treat the case when the operator A depends on parameter(s) and we derive asymptotics of $\mathsf{N}^-(A)$ with respect to the parameters. These parameters may be of different nature but even if one of them is a semiclassical parameter h it is not the same as used here effective semiclassical parameter h which depends on partition element in the regular zone.

Surely, the results of Parts II and III provide some of the asymptotics of Part V but without the additional analysis of this part we cannot cover many important cases.

9.1 Main Theorem

9.1.1 Discussion

As a preliminary example let us consider operator $\Delta-\lambda$ in the domain having singularities of different types in the domain on Figure 9.1. This domain has singularities of the type "angle" or "spike" and of the type "cusp". Later we consider many other examples of singularities of the boundary (including the case when the whole boundary itself is a singularity) or/and of the operator.

Consider *spatial scaling function* $\gamma(x)$ such that ∂X intersected with $B(\bar{x},\gamma(\bar{x}))$ after scaling $x\mapsto (x-\bar{x})\gamma(\bar{x})^{-1}$ is uniformly smooth. After this scaling and division by λ we have Schrödinger operator $h^2\Delta-1$ with $h=\lambda^{-\frac{1}{2}}\gamma(\bar{x})^{-1}$.

Therefore contribution of $B(\bar{x},\gamma(\bar{x}))$ to the remainder does not exceed $Ch^{1-d}=C\lambda^{\frac{1}{2}(d-1)}\gamma(\bar{x})^{d-1}$ as long as $h\le 1$. So we can introduce a *regular zone* $X_{\mathsf{reg}}=\{x:\ \gamma(x)\ge\lambda^{-\frac{1}{2}}\}$ and its contribution to the remainder does not exceed

$$C\lambda^{\frac{1}{2}(d-1)}\int_{X_{\mathsf{reg}}}\gamma(x)^{-1}\,dx \tag{9.1.1}$$

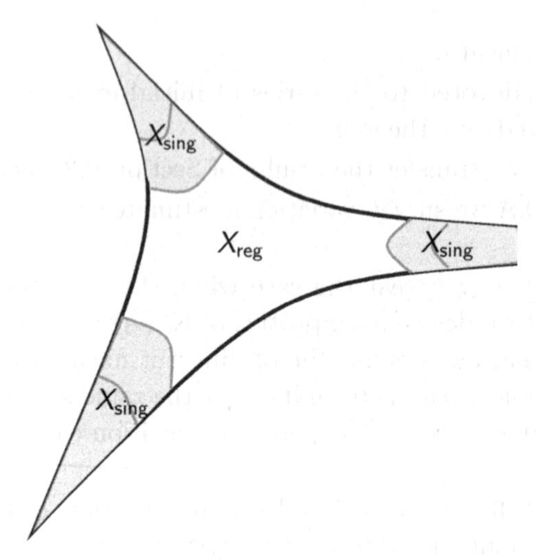

Figure 9.1: Black lines bound domain X, regular zone X_{reg} is bounded by gray and black lines, singular zone X_{sing} is shAded (both zones overlap).

while the principal part of asymptotics is given by the Weyl formula (with the integration over X_{reg})

$$\mathsf{const}\cdot\lambda^{\frac{1}{2}d}\int_{X_{\mathsf{reg}}} dx \tag{9.1.2}$$

with the standard constant.

Therefore we need to consider zone $X'_{\mathsf{sing}}=\{x:\ \gamma(x)\le 2\lambda^{-\frac{1}{2}}\}$ which is part of the *singular zone* $X_{\mathsf{sing}}=\{x:\ \gamma(x)\le 3\lambda^{-\frac{1}{2}}\}$. These zones shrink as $\lambda\to\infty$.

Assume that operator Δ was defined with the Dirichlet boundary conditions on ∂X. Let us formally add Dirichlet boundary conditions on the *inner boundary of singular zone* $\partial X_{\mathsf{sing}}\setminus\partial X$ (i.e. thick lines on Figure 9.1). In this case contribution of X_{sing} and X'_{sing} to asymptotics does not exceed as $d\ge 3$

$$C\lambda^{\frac{1}{2}d}\int_{X_{\mathsf{sing}}} dx. \tag{9.1.3}$$

So,

(9.1.4) The main part of asymptotics is (9.1.2) while the remainder estimate is given by the sum of (9.1.1) and (9.1.3).

Remark 9.1.1. (i) Obviously, the are no Dirichlet boundary condition on part of $\partial X_{\mathsf{sing}}$ and therefore this claim so far is justified only in the very special case when X is the union of infinitely many connected components, some of them completely belonging to X_{reg} and the rest to X_{sing}. However as we mentioned we will justify it.

(ii) The main part of the asymptotics is given by cut Weyl formula (9.1.2). Sure, we can always complete it, making integral over X since integral over X_{sing} is part of the error estimate. However, in the certain cases (f.e. in domains with cusps of the infinite volume) we will be able to get better estimate than (9.1.3).

(iii) Under some assumptions we are able to use sharp asymptotics[1)] in X_{reg} thus replacing (9.1.1) by a smaller expression; however in certain general cases we do not have even $O(h^{1-d})$ estimate thus replacing (9.1.1) by a larger expression.

(iv) Neumann boundary problem could be treated under certain assumptions as well but those are really restrictive assumptions due to very special nature of the Neumann boundary condition.

(v) In the discussed case the meaning of the *reference theorem* was obvious; in the general case we can use different reference theorems in the different parts of the regular zone.

In what follows by the *reference theorem* we mean the local semiclassical asymptotics we used.

9.1.2 Main Theorem Setup

Scaling Functions

Let us cover our domain X into two overlapping subdomains X_{reg} and X_{sing} in which specific conditions are assumed to be fulfilled.

Let us assume that

$$X \subset \mathbb{R}^d/(\bar{\gamma}_1\mathbb{Z} \times \bar{\gamma}_2\mathbb{Z} \times \ldots \times \bar{\gamma}_d\mathbb{Z}), \qquad 0 < \bar{\gamma}_j \le \infty \quad (j = 1, \ldots, d) \tag{9.1.5}$$

1) With remainder estimate $o(h^{1-d})$ or better.

and $X = X_{\mathsf{reg}} \cup X_{\mathsf{sing}}$.

Remark 9.1.2. Surely one can treat more general manifolds and in specific cases we will do it.

Let us assume that on X_{reg} a function $\gamma = \gamma(x)$ is given such that

$$0 < \gamma(x) \le \min_j \bar{\gamma}_j, \quad |\nabla \gamma(x)| \le \frac{1}{2}. \tag{9.1.6$_1$}$$

We say that $\gamma(x)$ is a *spatial scaling function.* We also assume that on X_{reg} a *momentum scaling function* $\rho(x)$[2] is given such that[3]

$$x \in \mathcal{B}(y, \gamma(y), X) \implies \rho_k(x) > 0, \quad \epsilon_0 \le \frac{\rho_k(x)}{\rho_k(y)} \le c, \tag{9.1.6$_2$}$$
$$|\nabla \rho_k(x)| \le c\rho_k(y)\gamma^{-1}(y)$$

and also scaling functions ϱ and ϱ_0 are given satisfying

$$x \in \mathcal{B}(y, \gamma(y), X) \implies \varrho_j(x) > 0, \quad \epsilon_0 \le \frac{\varrho_j(x)}{\varrho_j(y)} \le c, \tag{9.1.6$_3$}$$
$$|\nabla \varrho_j(x)| \le c\varrho_j(y)\gamma^{-1}(y)$$

where here and below

(9.1.7) $\mathcal{B}(y, \gamma(y), X)$ denotes a connected component of $B(y, \gamma(y)) \cap \bar{X}$ containing y.

Remark 9.1.3. Obviously

$$\mathcal{B}(y, \gamma(y), X) \supset \{x \in \bar{X} \colon \mathsf{dist}(x, y) < \gamma(y)\}$$

where $\mathsf{dist}(x, y) = \mathsf{dist}_X(x, y)$ is a *path-distance* in X.

Further, let us assume that

$$\rho(y)\gamma(y) \ge 1 \qquad \forall y \in X_{\mathsf{reg}}. \tag{9.1.6$_4$}$$

Finally, X_{reg} and X_{sing} be properly overlapping: let $X_{\mathsf{reg},0} \subset X_{\mathsf{reg}}$ and

$$X = X_{\mathsf{sing}} \cup X_{\mathsf{reg},0}; \quad y \in X_{\mathsf{reg},0} \implies \mathcal{B}(y, \gamma(y), X) \subset X_{\mathsf{reg}}. \tag{9.1.8}$$

[2] In principle one can take this function also depending on ξ.

[3] Later other scaling functions ρ_k and ϱ_j may be introduced which should all satisfy condition (9.1.6)$_2$.

Definition 9.1.4. In what follows X_{reg} is the *zone of semiclassical asymptotics* or the *regular zone* and X_{sing} is the *zone of variational estimates* or the *singular zone.*

Operators

Let us consider the operator A defined on X_{reg} by the formula

$$A = \sum_{\alpha:|\alpha|\le m} a_\alpha(x) D^\alpha \tag{9.1.9}$$

with boundary conditions

$$B_j u = \sum_{\alpha:|\alpha|\le m_j} b_{j\alpha}(x) D^\alpha u = 0 \quad (j = 1,\dots,\mu) \qquad \text{on} \quad \partial X\cap \bar{X}_{\mathsf{reg}} \tag{9.1.10}$$

where a_α, $b_{j\alpha}$ are $\mathsf{D}\times\mathsf{D}$ matrices and $m_j\le m-1$.

That means that A is an operator in $\mathscr{L}^2(X,\mathbb{H})$, $\mathbb{H}=\mathbb{C}^{\mathsf{D}}$ [4] with domain $\mathfrak{D}(A)$ but $Au(x)$ is given by (9.1.9) for $u\in\mathfrak{D}(A)$, $x\in X_{\mathsf{reg}}$. Moreover, it means that if $u\in\mathscr{H}^m(X,\mathbb{H})$ is compactly supported in X_{reg} then $u\in\mathfrak{D}(A)$ if and only if $Bu|_{\partial X}=0$.

In our analysis we replace the operator A by the operator

$$\boldsymbol{A} = J^{-1}A \tag{9.1.11}$$

in $\mathcal{H}$ where

(9.1.12) J is a Hermitian positive definite matrix depending on x

and here and below $\mathcal{H}=\mathcal{H}_J$ is a Hilbert space with inner product $(u,v)_{\mathcal{H}}=(Ju,v)$ and $(.,.)$ is the standard $\mathscr{L}^2(X,\mathbb{H})$-inner product.

We assume that

(9.1.13) $\boldsymbol{A}$ is a self-adjoint operator in $\mathcal{H}$.

Our goal is to derive upper and lower estimates for

$$\mathsf{N}^-(\boldsymbol{A}) = \int \operatorname{tr}\bigl(e(x,x,-\infty,0)\bigr)\,dx \tag{9.1.14}$$

where $e(.,.,.,.)$ is the Schwartz kernel of the spectral projector $E(\tau)$ of $\boldsymbol{A}$.

---

[4] One can also treat the case when $\mathbb{H}$ is an infinite dimensional auxiliary Hilbert space.

Remark 9.1.5. Let us note that

(i) $\boldsymbol{A}$ is unitarily equivalent to the operator $J^{-\frac{1}{2}}AJ^{-\frac{1}{2}}$ in $\mathcal{H}=\mathscr{L}^2(X,\mathbb{H})$. In what follows we use the notation $\boldsymbol{A}$ for the operator $J^{-\frac{1}{2}}AJ^{-\frac{1}{2}}$ as well. We hope that there will be no confusion.

(ii) Usually $\mathsf{N}^-(\boldsymbol{A})=\mathsf{N}^-(A)$. It is the case, f.e. if smooth functions u satisfying boundary conditions and such that $\|J\|$ and $\|J^{-1}\|$ are bounded by $C(u)$ on $\operatorname{supp}(u)$ are dense in $\mathcal{H}$. We will discuss this with more details in Section 9.3.

(iii) Equality $\mathsf{N}^-(A)=\mathsf{N}^-(\boldsymbol{A})$ is an example of the *Birman-Schwinger* principle (see Subsection 10.1.2 for more details).

Let us describe assumptions to A, B and J.

Regular Zone

First of all, let us describe assumptions in regular zone X_{reg}. Namely, assume that uniformly with respect to $y\in X_{\mathsf{reg}}$ assumptions (9.1.15)–(9.1.18) are fulfilled:

(9.1.15) If $y\in X_{\mathsf{reg}}$ then either $B(y,\gamma(y))\subset X$ or there exist a number $k=k(y)\in\{1,\dots,d\}$, $\varsigma=\pm 1$ and a real-valued function $\phi(x_{\hat{k}})$ such that $|\nabla\phi|\le c$ and

$$\mathcal{B}(y,\gamma(y),X)=\{\varsigma(x_k-\phi(x_{\hat{k}}))>0\}\cap B(y,\gamma(y)) \tag{9.1.16}$$

where $x_{\hat{k}}=(x_1,\dots,x_{k-1},x_{k+1},\dots,x_d)$;

(9.1.17) $X_{\mathsf{reg}}=\bigcup_l X_{\mathsf{reg}(l)}$ and if $y\in X_{\mathsf{reg}(l)}$ then rescaling[5] $\mathcal{B}(y,\gamma(y),X)$ by $x\mapsto(x-y)\gamma^{-1}$, $\xi\mapsto\xi\rho^{-1}$, $\tau\mapsto\tau\varrho^{-1}\varrho_0$ and multiplying $\boldsymbol{A}$ by $\varrho^{-1}\varrho_0$ we find ourselves in the framework of one of the theorems of Chapters 4–8 which providing estimate

$$\mathsf{R}_{h,\psi}:=|\int \operatorname{tr}\bigl(e(x,x,0)\bigr)\psi(x)\,dx-\mathcal{N}_{h,\psi}|\le h^{-d}\vartheta_l(h) \tag{9.1.18}$$

---

[5] All scaling functions are calculated at y.

where $\mathcal{N}_\psi$ denotes main part of semiclassical asymptotics, $h=(\rho\gamma)^{-1}$ is an *effective semiclassical parameter* and $\psi\in\mathscr{C}^\infty(\bar{X})$ supported in $\mathcal{B}(y,\frac{1}{2}\gamma(y))$ such that

$$|D^\beta\psi|\le c_\beta\gamma^{-|\beta|}\qquad\forall\beta:|\beta|\le K \tag{9.1.19}$$

and in (9.1.19) we already scaled back to original coordinates.

Later we list the most important cases. One can prove easily

(9.1.20) There exist real-valued functions $\psi_\nu\in\mathscr{C}^\infty(\bar{X})$, supported in $\mathcal{B}(y_\nu,\frac{1}{2}\gamma(y_\nu),X)$ such that $\psi_{\mathsf{reg}}:=\sum_\nu\psi_\nu^2$ equals 1 in $X_{\mathsf{reg},\frac{7}{16}}$ and no more than $L(d)$ of different balls $\mathcal{B}(y_\nu,\frac{3}{4}\gamma(y_\nu),X)$ can have non-empty intersection.

Here and below γ'-vicinity of set M is a set $\bigcup_{y\in M}\mathcal{B}(y,\gamma'(y),X)$ and $X_{\mathsf{reg},\sigma}$ denotes $\sigma\gamma$-vicinity of $X_{\mathsf{reg},0}$, $0\le\sigma\le 1$. Condition (9.1.8) means that $X_{\mathsf{reg}}\supset X_{\mathsf{reg},1}$.

Then we instantly arrive to

Proposition 9.1.6. *Let all the above assumptions be fulfilled and let ψ satisfy (9.1.19) and be supported in $X_{\mathsf{reg},\frac{1}{2}}$. Then*

$$\mathsf{R}_\psi\le R_1:=C_0\sum_l\int_{X_{\mathsf{reg}(l)}}\rho^d\vartheta_l\bigl((\rho\gamma)^{-1}\bigr)\,dx \tag{9.1.21}$$

with $C_0=C_0(d,c,m)$.

Singular Zone

To estimate $\mathsf{N}^-(\boldsymbol{A})$ from above we need to consider $\psi=\psi_{\mathsf{sing}}$ satisfying (9.1.19) and supported in $\bar{X}\setminus\bar{X}_{\mathsf{reg},\frac{6}{16}}$.

Let us assume that

(9.1.22) There exist an operator $\boldsymbol{A}'$ in $\mathscr{L}^2(X_{\mathsf{sing}},\mathbb{H})$ and boundary operators B'_j coinciding with $\boldsymbol{A}$ and B_j respectively on functions supported in $\bar{X}_{\mathsf{sing}}\setminus\bar{X}_{\mathsf{reg},\frac{1}{4}}$ such that under the boundary conditions $B'_ju|_{\partial X_{\mathsf{sing}}}=0$ $(j=1,\dots,\mu)$ this operator $\boldsymbol{A}'$ is self-adjoint and the estimate

$$\mathsf{N}^-(\boldsymbol{A}'-t)\le R_0(t+1)^p\qquad\forall t\ge 0 \tag{9.1.23}$$

holds.

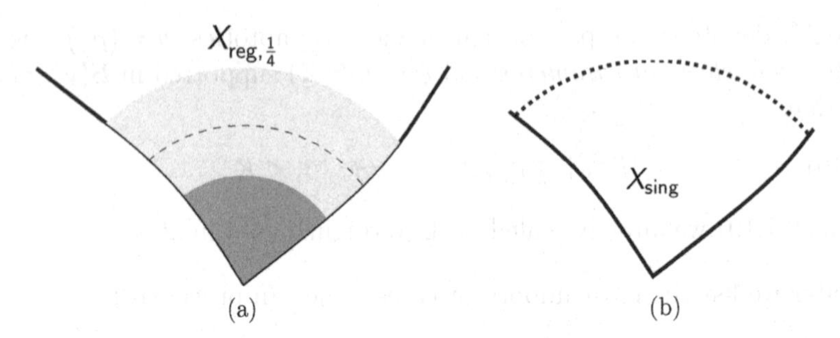

Figure 9.2: On (a) dark shaded zone denotes $\mathsf{supp}(\psi_{\mathsf{sing}})$, together with light shaded zone making X_{sing}; dashed line is the border of $X_{\mathsf{reg},\frac{1}{4}}$. On (b) X_{sing} is considered alone; dotted line is $\partial X_{\mathsf{sing}} \setminus \partial X$.

Finally, let us assume that

$$\varrho_0\rho\gamma \ge \epsilon_0\varrho \quad \text{in} \quad \bigl(X_{\mathsf{reg},\frac{1}{4}} \setminus X_{\mathsf{reg,ell}}\bigr) \cap X_{\mathsf{sing}} \cap \{\rho\gamma \ge c\}, \tag{9.1.24}$$

$$\epsilon_0\varrho(\rho\gamma)^{-p} \le \varrho_0 \le c\varrho(\rho\gamma)^{p} \quad \text{in} \quad X_{\mathsf{reg}} \cap X_{\mathsf{sing}} \tag{9.1.25}$$

where $X_{\mathsf{reg,ell}}$ denotes a part of X_{reg} where $\{A, B\}$ is elliptic in the semiclassical sense (after described rescaling).

The first assumption is assumed to provide the following statement:

(9.1.26) As $0 \le \sigma_1 < \sigma \le 1$ singularities starting at $X \setminus X_{\mathsf{reg},\sigma}$ reach $X_{\mathsf{reg},\sigma_1}$ only after a time no less than $\epsilon(\sigma - \sigma_1)$.

In fact, as $\rho\gamma \le c$ the propagation of singularities is absent because $h = (\rho\gamma)^{-1} \asymp 1$ and hence the notion of the singularity is empty. Further, there is no propagation in the elliptic zone either.

On the other hand, the results of Chapters 2 and 3 obviously yield that the spatial velocity of propagation of singularities for the operator $D_t - \boldsymbol{A}$ in $X_{\mathsf{reg},\frac{1}{4}} \cap X_{\mathsf{sing}}$ does not exceed $C_0\varrho\varrho_0^{-1}\rho^{-1}$ while the distance between $y \in X \setminus X_{\mathsf{reg},\sigma}$ and $x \in X_{\mathsf{reg},\sigma_1}$ is not less than $(\sigma - \sigma_1)\gamma(x)$. Therefore the time is not less than $\epsilon_2(\sigma - \sigma_1)\varrho^{-1}\varrho_0\rho\gamma$.

Meanwhile assumption (9.1.25) guarantees that $\varrho\varrho_0^{-1}$ is neither too large nor too small in this zone.

Finally, assume that

(9.1.27) In $X_{\mathsf{reg}}\cap X_{\mathsf{sing}}$ after rescaling $\mathcal{B}(y,\gamma(y),X)$ by $x\mapsto(x-y)\gamma^{-1}$, $\xi\mapsto\xi\rho^{-1}$, $\tau\mapsto\tau\varrho^{-1}\varrho_0$ and multiplying $\boldsymbol{A}$ by $\varrho^{-1}\varrho_0$ we find that problem $\{A,B\}$ is uniformly elliptic and positive in the classical sense[6)].

Our major result is the following

Theorem 9.1.7. *Let conditions* (9.1.5)–(9.1.13) *and* (9.1.15)–(9.1.18) *be fulfilled.*

(i) Then estimate

$$\mathsf{N}^-(\boldsymbol{A})\ge\mathcal{N}-C_0R_1 \tag{9.1.28}$$

holds with $\mathcal{N}=\mathcal{N}_{1,\psi}$ *defined as a sum of main parts of asymptotics in* (9.1.17)–(9.1.18) *and* R_1 *defined by* (9.1.21) *as a sum of remainder estimates there.*

(ii) Moreover, if conditions (9.1.22)–(9.1.25) *and* (9.1.27) *are fulfilled then estimate*

$$\mathsf{N}^-(\boldsymbol{A})\le\mathcal{N}+C_0R_1+C_1R_0+CR_2 \tag{9.1.29}$$

holds with R_0 *appearing in* (9.1.23) *and*

$$R_2=\int_{X_{\mathsf{reg},\frac{1}{4}}\cap X_{\mathsf{sing}}}\rho^{d-s}\gamma^{-s}\,dx, \tag{9.1.30}$$

$C_0=C_0(d,c)$, $C_1=C_1(d,p,c)$, $C=C(d,\mathsf{D},c,s)$.

9.1.3 Proof of the Main Theorem

Let us note first that the lower estimate (9.1.23) for $\mathsf{N}^-(\boldsymbol{A})$ follows immediately from Proposition 9.1.6 since $\mathsf{N}^-(\boldsymbol{A})\ge\mathsf{Tr}(E(0)\psi_{\mathsf{reg}})$ as $0\le\psi_{\mathsf{reg}}\le1$. Further note that the upper estimate for $\mathsf{N}^-(\boldsymbol{A})$ also follows from Proposition 9.1.6 in the special case when $X_{\mathsf{sing}}=\emptyset$. Meanwhile in the general case we obtained an upper estimate for $\mathsf{Tr}(E(0)\psi_{\mathsf{reg}})$ provided

[6)] Which means that selecting only homogeneous main parts of A, B_j with coefficients fixed in y we get a problem in the whole or half-space which is elliptic uniformly with respect to y and non-negative.

(9.1.31) $\operatorname{supp}(\psi_{\mathsf{reg}}) \subset X_{\mathsf{reg},\frac{1}{4}}$, $\psi_{\mathsf{reg}} = 1$ in $\operatorname{supp}(\psi) \subset X_{\mathsf{reg},\frac{1}{3}}$ and ψ_{reg} satisfies (9.1.19).

To derive an upper estimate for $\mathsf{N}^-(\boldsymbol{A})$ we need an upper estimate for $\operatorname{Tr}(E(0)(1-\psi_{\mathsf{reg}}))$. Let $U(x,y,t)$ be the Schwartz kernel of the operator $\exp(it\boldsymbol{A})$; recall that $\boldsymbol{A} = J^{-\frac{1}{2}}AJ^{-\frac{1}{2}}$. We need some additional analysis in the zone $X_{\mathsf{reg,ell}} \cap X_{\mathsf{sing}} \cap \{\rho\gamma \ge c\}$.

Proposition 9.1.8. *Let conditions* (9.1.5)–(9.1.13), (9.1.15)–(9.1.18), (9.1.24)–(9.1.25) *and* (9.1.27) *be fulfilled.*

Let $\epsilon_0 > 0$ *and* $T_0 = T_0(d,c,\epsilon_0) > 0$ *be small enough constants. Then the following estimates hold uniformly with respect to* $x \in X_{\mathsf{reg},\frac{2}{7}}$:

$$\|D_x^\alpha F_{t\to h^{-1}\tau}\chi_T(t)U^\pm\psi'(y)\| \le C_1'h^s(|\tau|+1)^{-s}\gamma(x)^{-\frac{d}{2}-|\alpha|}\qquad \forall\tau\le c\qquad\forall\alpha:|\alpha|\le s, \tag{9.1.32}$$

and

$$\|D_x^\alpha F_{t\to h^{-1}\tau}\chi_T(t)U^\pm\psi'(y)\| \le C_1'h^{-M}(|\tau|+1)^{M}\gamma(x)^{-\frac{d}{2}-|\alpha|}\qquad \forall\tau\ge c\qquad\forall\alpha:|\alpha|\le s \tag{9.1.33}$$

where $\mathscr{L}^2$*-norms with respect to* y *are calculated in the left-hand expression and here and below* $\psi' \in \mathscr{L}^\infty$ *is an arbitrary function such that*

$$|\psi'|\le 1,\ \operatorname{supp}(\psi')\cap\mathcal{B}(x,\epsilon_0\gamma(x),X)=\emptyset \tag{9.1.34}$$

with $M = M(d,m,s)$, $C_1' = C_1'(d,m,s,c)$ *and* $h := \bigl(\rho(x)\gamma(x)\bigr)^{-1}$.

Proof. (a) Let us consider point $\bar{x}$ instead of x and operator $P = hD_t - \zeta\boldsymbol{A}$ with $h = (\bar\rho\bar\gamma)^{-1}$, $\zeta = h\bar\varrho^{-1}\bar\varrho_0$ where $\bar\gamma = \gamma(\bar x)$ etc. Without any loss of the generality one can assume that $\bar\gamma = 1$ (otherwise we rescale).

Consider $f \in \mathscr{L}^2(X,\mathbb{H})$ with $\operatorname{supp} f \cap \mathcal{B}(\bar x,\epsilon_0\bar\gamma,X)$. Let us set

$$u(x,t) = \int U(x,y,t)(y)f(y)\,dy. \tag{9.1.35}$$

Then

$$Pu^\pm = 0 \quad\text{in}\quad \mathcal{B}(\bar x,\epsilon_0\bar\gamma,X), \tag{9.1.36}$$

$$B_ju^\pm = 0 \quad\text{at}\quad \partial X\cap\mathcal{B}(\bar x,\bar\gamma,X)\qquad (j=1,\dots,\mu). \tag{9.1.37}$$

(b) Let us first assume that $\zeta \le c$. Then one can apply the theorem about the finite speed of propagation of singularities[7] and therefore the following estimate holds:

$$\|F_{t\to h^{-1}\tau}\chi_T(t)u^{\pm}\| \le C_1' h^s \|f\| \qquad \forall x \in \mathcal{B}(\bar{x}, \frac{1}{2}\epsilon_0\bar{\gamma}, X) \qquad \forall \tau : |\tau| \le c \tag{9.1.38}$$

where we have also used the estimate

$$\|u^{\pm}\| \le \|f\| \tag{9.1.39}$$

which follows from the fact that the operator $\exp(it\boldsymbol{A})$ is unitary in $\mathcal{H}_0$; here $\mathscr{L}^2(X, \mathbb{H})$-norms (with respect to x) are calculated.

(c) Now let us apply the following *additional arguments:* the classical ellipticity of the rescaled problem as stated in (9.1.27), the inequality $\zeta \ge h^p$ due to (9.1.25) and the standard embedding inequality; then we obtain estimate

$$|D^\alpha F_{t\to h^{-1}\tau}\chi_T(t)u^{\pm}| \le C_1' h^s \|f\| \qquad \forall x \in \mathcal{B}(\bar{x}, \frac{1}{2}\epsilon_0\bar{\gamma}, X) \quad \forall \alpha : |\alpha| \le s \quad \forall \tau : |\tau| \le c. \tag{9.1.40}$$

Furthermore, the same *additional arguments* and the classical ellipticity and positivity of the rescaled problem as stated in (9.1.27) easily yield estimate

$$|D^\alpha F_{t\to h^{-1}\tau}\chi_T(t)u^{\pm}| \le C_1' h^s (|\tau|+1)^{-s} \|f\| \qquad \forall x \in \mathcal{B}(\bar{x}, \frac{1}{2}\epsilon_0\bar{\gamma}, X) \quad \forall \alpha : |\alpha| \le s \quad \forall \tau \le -C_0. \tag{9.1.41}$$

(d) On the other hand, let $\zeta \ge c$. Then as $\rho\gamma \ge c$ the problem is elliptic in the semiclassical sense in $\mathcal{B}(\bar{x}, \epsilon_0\bar{\gamma})$ due to assumption (9.1.24) and the semiclassical elliptic arguments and the above *additional arguments* yield estimate (9.1.41) for all $\tau \le \epsilon_2\zeta$. Therefore estimate (9.1.41) remains true in this case as well.

---

[7] Let us recall that we assumed not just estimate in (9.1.18) but also being in the frameworks of one of the theorems of Parts II or III in X_{reg} and it implies the finite speed of propagation and all other results involved.

(e) Finally, the *additional arguments* alone easily yield the estimate

$$|D^\alpha F_{t\to h^{-1}\tau}\chi_T(t)u^\pm| \le C_1' h^{-M}(|\tau|+1)^M \|f\| \qquad \forall x\in\mathcal{B}(\bar{x},\frac{1}{2}\epsilon_0\bar{\gamma},X)\quad \forall\alpha:|\alpha|\le s\quad\forall\tau. \tag{9.1.42}$$

Since $f\in\mathscr{L}^2(X,\mathbb{H})$ is an arbitrary function such that $\operatorname{supp}(f)\cap\mathcal{B}(\bar{x},\epsilon_0\bar{\gamma},X)=\emptyset$, estimates (9.1.40)–(9.1.42) yield estimates (9.1.32) and (9.1.33). □

Remark 9.1.9. The same proof implies that estimates (9.1.32) and (9.1.33) remain true for $U'^\pm(x,y,t)$ where $U'(x,y,t)$ is the Schwartz kernel of the operator $\exp(it\boldsymbol{A}')$ as long as $x\in X\setminus X_{\mathsf{reg},\frac{1}{4}}$.

Indeed, in $X\setminus X_{\mathsf{reg},\frac{1}{4}}$ operator $\boldsymbol{A}'$ coincides with $\boldsymbol{A}$.

Proposition 9.1.10. *In the framework of Proposition 9.1.8 estimates*

$$\|F_{t\to\tau}\chi_T(t)\int \psi_{\mathsf{sing}}(x)\bigl(U^\pm-U'^\pm\bigr)(x,x,t)\,dx\|\le C_1'(|\tau|+1)^{-s}R_2 \tag{9.1.43}$$

and

$$\|F_{t\to\tau}\psi_{\mathsf{sing}}\chi_T(t)\bigl(\exp(it\boldsymbol{A})-\exp(it\boldsymbol{A}')\bigr)\psi_{\mathsf{sing}}\|_1\le C'(|\tau|+1)^{-s}R_2 \tag{9.1.44}$$

hold for all $\tau\le\frac{1}{2}c$. Recall that $\|.\|_1$ means the trace norm and C_1' does not depend on D *but C' does.*

Proof. Let ψ satisfy (9.1.19), $\psi=1$ in $\bar{X}\setminus\bar{X}_{\mathsf{reg},\frac{6}{16}}$ and $\operatorname{supp}(\psi)\subset\bar{X}\setminus\bar{X}_{\mathsf{reg},\frac{5}{16}}$. Let us recall that $\operatorname{supp}(\psi_{\mathsf{sing}})\subset\bar{X}\setminus\bar{X}_{\mathsf{reg},\frac{7}{16}}$.

Let us consider

$$V^\pm(x,y,t)=\psi(x)(U^\pm-U'^\pm)(x,y,t)\psi'(y) \tag{9.1.45}$$

where here and below ψ' is a cut-off function similar to ψ_{sing}. Then

$$PV^\pm=F^\pm,\quad B_kV^\pm=G_k^\pm$$

with $P = D_t - \boldsymbol{A}$, $F^\pm$ and $G_k^\pm$ supported in $\mathcal{X} := \bar{X}_{\mathsf{reg},\frac{6}{16}} \setminus \bar{X}_{\mathsf{reg},\frac{5}{16}}$ and therefore by the Duhamel principle

$$V^+(x,y,t) = \int_0^t dt' \Bigl(\int U^+(x,z,t')F^+(z,y,t-t')\,dz + \sum_{1\le k\le \mu} \int_{\partial X} \bigl(\beta_k(z,D_z)U^+(x,z,t')\bigr)\, G_k^+(z,y,t-t')\,dz'\Bigr) \tag{9.1.46}$$

where β_k are differential operators on z.

Therefore $V^+ = \sum_n V_n^+$ where V_n^+ are given by formula (9.1.46) with additional factors $\psi_n^2(z)$ in the integrands and $\{\psi_n^2\}$ is a γ-admissible partition of unity in $\mathcal{X}$.

Then Proposition 9.1.8 and Remark 9.1.9 yield that $\mathscr{L}^2$-norms with respect to y of $F_{t\to h^{-1}\tau}\chi_T(t)F^+$ and $F_{t\to h^{-1}\tau}\chi_T(t)G_k^+$ do not exceed

$$C_0 h^{-M}\bigl(|\tau|+1\bigr)^M \bar{\gamma}^{-\frac{d}{2}}$$

and

$$C_0 h^{-M}\bar{\varrho}_k\bigl(|\tau|+1\bigr)^M \bar{\gamma}^{-\frac{d}{2}}$$

respectively for all $z \in \mathsf{supp}(\psi_n)$ where $h = \bar{\rho}^{-1}\bar{\gamma}^{-1}$ and $\bar{\rho}, \bar{\gamma}, \bar{\varrho}_k$ are ρ, γ, ϱ_k calculated at some point of $\mathsf{supp}(\psi_n)$; all of these and the following estimates hold for all n.

Further, Proposition 9.1.8 and Remark 9.1.9 yield that the $\mathscr{L}^2$-norms with respect to x of

$$\psi'(x)F_{t\to h^{-1}\tau}\chi_T(t)U^+(x,z,t)$$

and

$$\psi'(x)F_{t\to h^{-1}\tau}\chi_T(t)B_k'(z,D_z)U^+(x,z,t)\psi_n(z)$$

do not exceed

$$C_0 h^{-M}\bigl(|\tau|+1\bigr)^M \bar{\gamma}^{-\frac{d}{2}}$$

and

$$C_0 h^{-M}\bar{\varrho}_k^{-1}\bigl(|\tau|+1\bigr)^M \bar{\gamma}^{1-\frac{d}{2}}$$

respectively. Recall that $z \in \mathsf{supp}(\psi_n)$ and

$$U(x,y,t) = U^\dagger(y,x,-t) \tag{9.1.47}$$

because $\boldsymbol{A}$ is a self-adjoint operator in $\mathcal{H}_0$.

Furthermore, Proposition 9.1.8 and Remark 9.1.9 also imply that for $T \le \frac{1}{2}c$ these four estimates hold with the exponent M replaced by $-s$ in the right-hand expressions and with constant C_0 replaced by C'_1. Since $\dim \mathbb{H} = \mathsf{D} < \infty$ all these inequalities hold (with constants depending also on D) not only $\mathscr{L}^2$-norms with respect to x or y but also for Hilbert-Schmidt norms of the corresponding operators.

Therefore we conclude that

(9.1.48) Let $W_n^+(x,y,t)$ be given by formula (9.1.46) with the additional factor $\psi_n^2(z)\chi_T(t')\chi_T(t-t')$ in the integrand. Then the trace norm of operator with the Schwartz kernel $F_{t\to h^{-1}\tau} W_n^+\psi'(x)\psi'(y)$ does not exceed

$$
\begin{cases} C'\bar{\gamma}^{-d}h^s(|\tau|+1)^{-s} & \text{as } T\le c,\\ C\bar{\gamma}^{-d}h^{-M}(|\tau|+1)^{M} & \text{as } T\ge c. \end{cases} \tag{9.1.49}
$$

Therefore

(9.1.50) The trace norm of the operator with the Schwartz kernel

$$
F_{t\to h^{-1}\tau}\chi_{\frac{1}{2}T}(t) W_n^+\psi'(x)\psi'(y)
$$

does not exceed $C'\bar{\gamma}^{-d}h^s(|\tau|+1)^{-s}$ as $T\le\frac{1}{2}c$.

Let us replace in the previous analysis and definitions function χ by $\bar{\chi}$ equal to 1 on $\operatorname{supp}(\chi)$. Then $\chi_T(t)\bigl(V_n^+ - W_n^+\bigr) = 0$ and therefore one can replace W_n^+ by V_n^+ in the statement (9.1.50).

Surely one can replace $h^{-1}\tau$ by τ in this statement as well. Replacing h by $\rho^{-1}\gamma^{-1}$ and summing with respect to n we obtain estimate (9.1.44); (9.1.43) follows from it. Finally, these estimates for sign "+" imply themselves for sign "−". □

Proof of Theorem 9.1.7. Let Π be an arbitrary r-dimensional self-adjoint projector in $\mathcal{H}_0$. Then (9.1.44) yields that

$$
|F_{t\to\tau}\chi_T(t)\operatorname{Tr}\Bigl(\Pi\psi_{\mathsf{sing}}\bigl(\exp(it\boldsymbol{A}) - \exp(it\boldsymbol{A}')\bigr)\psi_{\mathsf{sing}}\Bigr)| \le \\ C'R_2\bigl(|\tau|+1\bigr)^{-s} \tag{9.1.51}
$$

where here and below we assume that $\tau \le \frac{1}{2}c$.

Observe that the left-hand expression of (9.1.51) equals the left-hand expression of

$$|\int \widehat{\chi_T}(\tau-\lambda)\, d_\lambda \operatorname{Tr}\Bigl(\Pi\psi_{\mathsf{sing}}\bigl(E(\lambda)-E'(\lambda)\bigr)\psi_{\mathsf{sing}}\Bigr)| \le C'R_2\bigl(|\tau|+1\bigr)^{-s}$$

where $E'(\lambda)$ is the spectral projector of $\boldsymbol{A}'$. The latter estimate holds ince the trace norm of Π does not exceed Cr. Therefore

$$|\int \widehat{\chi_T}(\tau-\lambda)\, d_\lambda \operatorname{Tr}\Bigl(\Pi\psi_{\mathsf{sing}}E(\lambda)\psi_{\mathsf{sing}}\Bigr)| \le \\ |\int \widehat{\chi_T}(\tau-\lambda)\, d_\lambda \operatorname{Tr}\Bigl(\Pi\psi_{\mathsf{sing}}E'(\lambda)\psi_{\mathsf{sing}}\Bigr)| + C'R_2\bigl(|\tau|+1\bigr)^{-s}.$$

Let us take a Hörmander function χ (and so χ_T is of the same type). Then the standard Tauberian arguments yield that

$$\operatorname{Tr}\Bigl(\Pi\psi_{\mathsf{sing}}\bigl(E(\tau)-E(\tau-1)\bigr)\psi_{\mathsf{sing}}\Bigr) \le \\ C\int \widehat{\chi_T}(\tau-\lambda)\operatorname{Tr}\Bigl(\Pi\psi_{\mathsf{sing}}\, d_\lambda E'(\lambda)\psi_{\mathsf{sing}}\Bigr) + C'R_2\bigl(|\tau|+1\bigr)^{-s}.$$

The right-hand expression of this estimate does not decrease if we replace Π by I and ψ_{sing} by 1. Since $\operatorname{Tr}(E'(\lambda)) = \mathsf{N}^-(\boldsymbol{A}'-\lambda)$ we conclude that

$$\operatorname{Tr}\Bigl(\Pi\psi_{\mathsf{sing}}\bigl(E(\tau)-E(\tau-1)\bigr)\psi_{\mathsf{sing}}\Bigr) \le \\ C\int \widehat{\chi_T}(\tau-\lambda)\, d_\lambda \mathsf{N}^-(\boldsymbol{A}'-\lambda) + C'R_2\bigl(|\tau|+1\bigr)^{-s}.$$

Since this estimate is true for all the self-adjoint finite-dimensional projectors in $\mathcal{H}_0$ it is true for Π replaced by I:

$$\operatorname{Tr}\Bigl(\psi_{\mathsf{sing}}\bigl(E(\tau)-E(\tau-1)\bigr)\psi_{\mathsf{sing}}\Bigr) \le \\ C\int \widehat{\chi_T}(\tau-\lambda)\, d_\lambda \mathsf{N}^-(\boldsymbol{A}'-\lambda) + C'R_2\bigl(|\tau|+1\bigr)^{-s}. \tag{9.1.52}$$

Thus we got rid of projector Π which we used only to justify the Fourier transform and the replacement of the left-hand expression in (9.1.51).

Picking up $\tau = -n$ in (9.1.52), after summation with respect to $n = 0, 1, 2, \ldots$ we obtain estimate

$$\operatorname{Tr}\Bigl(\psi_{\mathsf{sing}} E(0)\psi_{\mathsf{sing}}\Bigr)\Bigr) \le C\sum_{n\ge 0}\int \widehat{\chi_T}(-n-\lambda)\,d_\lambda \mathsf{N}^-(\boldsymbol{A}'-\lambda) + C'R_2. \tag{9.1.53}$$

Let us note that $\sum_{n\ge 0}\widehat{\chi_T}(-n-\lambda)$ does not exceed C_1 and $C_1'(|\lambda|+1)^{-s}$ for $\lambda \ge 0$ and $\lambda \le 0$ respectively.

Therefore the first term in the right-hand expression of (9.1.53) does not exceed

$$C_1'\int_0^\infty (|\lambda|+1)^{-s}\,d_\lambda \mathsf{N}^-(\boldsymbol{A}'-\lambda) + C_1\int_{-\infty}^0 d_\lambda \mathsf{N}^-(\boldsymbol{A}'-\lambda) \le \\ C_1'\int_0^\infty (|\lambda|+1)^{-s}\,d_\lambda \mathsf{N}^-(\boldsymbol{A}'-\lambda) + C_1\mathsf{N}^-(\boldsymbol{A}')$$

where we have integrated by parts. Then condition (9.1.23) yields that the first term in the right-hand expression in (9.1.53) does not exceed C_1R_0 and therefore

$$\operatorname{Tr}\bigl(\psi_{\mathsf{sing}} E(0)\psi_{\mathsf{sing}}\bigr) \le C_1R_0 + C'R_2. \tag{9.1.54}$$

Namely this estimate was lacking for the proof of Theorem 9.1.7. □

The following problem does not seem to be too hard but still worth publishing:

Problem 9.1.11. As $\mathbb{H}$ is infinite-dimensional space prove the similar results involving auxiliary operators B and B_1 (not to be confused with boundary operators) and Hilbert scale $\mathbb{H}_j$ (see Chapters 2 and 4 for details.

9.1.4 Riesz Means Estimates

Let us instead of $\mathsf{N}^-(A)$ consider Riesz means, namely

$$\mathsf{N}^-_\omega(A) = \int \tau_+^{\omega-1}\mathsf{N}^-(A-\tau)\,d\tau, \qquad \omega > 0. \tag{9.1.55}$$

Then one can apply local asymptotics for Riesz means. However, it would be more convenient for us to write $\mathsf{N}^-_\omega(A)$ in the form

$$\mathsf{N}^-_\omega(A) = \int \tau_+^{\omega-1}\mathsf{N}^-(J^{-\frac12}(A-\tau)J^{-\frac12})\,d\tau \tag{9.1.56}$$

with the same function J as before in order to get better estimates in the singular zone. In this case one can apply the results of Section 6.5.

As before contribution of the partition element ψ of the the regular zone to the main part of the asymptotics is given by $\mathcal{N}_{\psi,\omega}$ prescribed by Section 6.5 while its contribution to the remainder does not exceed $h^{-d}\varrho^{\omega}\nu_{l,\omega}(h)$ as this element is supported in $X_{\mathsf{reg},(l)}$. So the total main part of asymptotics is $\mathcal{N}_{\psi_{\mathsf{reg}},\omega}$ while contribution of the regular zone to the remainder is

$$\mathsf{R}_{\psi,\omega} \le R_{1,\omega} := C\sum_l \int_{X_{\mathsf{reg}(l)}} \rho^d \varrho^{\omega}\vartheta_{l,\omega}\bigl((\rho\gamma)^{-1}\bigr)\,dx. \tag{9.1.57}$$

As in (9.1.21) the standard case was h^{1-d} but now it is $h^{1-d+\omega}\varrho^{\omega}$; so the standard remainder estimate is

$$C\int_{X_{\mathsf{reg}}} \rho^{d-1-\omega}\gamma^{-1-\omega}\varrho^{\omega}\,dx. \tag{9.1.58}$$

Remark 9.1.12. One can easily prove that application of (9.1.56) instead of (9.1.55) does not lead to a different local expression.

Further, the same arguments as in the proof of Theorem 9.1.7 imply

Theorem 9.1.13. *Let conditions* (9.1.5)–(9.1.13) *and* (9.1.15)–(9.1.18) *be fulfilled.*

(i) Then estimate

$$\mathsf{N}^-_{\omega}(A) \ge \mathcal{N}_{\omega} - C_0 R_{1,\omega} \tag{9.1.59}$$

holds.

(ii) Moreover, if conditions

$$\mathsf{N}^-_{\omega}(A' - t) \le R_{0,\omega}(t+1)^p \qquad \forall t\ge 0, \tag{9.1.60}$$

(9.1.24), (9.1.25) *and* (9.1.27) *are fulfilled then the estimate*

$$\mathsf{N}^-_{\omega}(A) \le \mathcal{N}_{\omega} + C_0 R_{1,\omega} + C_1 R_{0,\omega} + C' R_{2,\omega} \tag{9.1.61}$$

holds with

$$R_{2,\omega} = \int_{X_{\mathsf{reg},\frac{1}{4}}\cap X_{\mathsf{sing}}} \rho^{d-s}\gamma^{-s}\varrho^{\omega}\,dx, \tag{9.1.62}$$

$C_0 = C_0(d, c, \omega)$, $C_1 = C_1(d, c, p, \omega)$, $C' = C'(d, \mathsf{D}, c, M, \omega)$.

Proof. Easy details are left to the reader. □

Problem 9.1.14. Derive similar results for averaging functions φ satisfying (4.2.73)–(4.2.75) rather than $\varphi(\tau)=\tau_+^{\omega}$.

9.2 Examples to Main Theorems

In this section we consider some rather general examples of contributions of X_{reg} and X_{sing} to the remainder.

9.2.1 Regular Zone: Standard Remainder Estimate

Let us consider the first regular zone. In virtue of Parts II and III one can set

$$\mathcal{N}=\mathcal{N}^{\mathsf{W}}:=(2\pi)^{-d}\iint \boldsymbol{n}^-(x,\xi)\,dxd\xi \tag{9.2.1}$$

where $\boldsymbol{n}^-(x,\xi)$ is a number of negative eigenvalues of $a(x,\xi)$ and

$$R_1=C\int_{X_{\mathsf{reg}}\setminus X_{\mathsf{reg,ell}}}\rho^{d-1}\gamma^{-1}\,dx+C'\int_{X_{\mathsf{reg,ell}}}\rho^{d-s}\gamma^{-s}\,dx \tag{9.2.2}$$

with C' depending on s in the cases described in Examples 9.2.1–9.2.5 below. We assume that $\mathsf{D}<\infty$ is fixed.

Example 9.2.1. Assume that the regularity conditions

$$|D^\beta a_\alpha|\le c\varrho\rho^{-|\alpha|}\gamma^{-|\beta|}\qquad \forall\alpha:|\alpha|\le m, \tag*{(9.2.3)$_1$}$$

$$|D^\beta b_{j\alpha}|\le c\varrho_j\rho^{-|\alpha|}\gamma^{-|\beta|}\qquad \forall j=1,\dots,\mu\qquad \forall\alpha:|\alpha|\le m_j, \tag*{(9.2.3)$_2$}$$

$$|D^\beta J|\le c\varrho_0\gamma^{-|\beta|},\qquad \forall\beta:|\beta|\le K \tag*{(9.2.3)$_3$}$$

and

$$|D^\beta\phi|\le c\gamma^{1-|\beta|}\qquad \forall\beta:1\le|\beta|\le K \tag*{(9.2.3)$_4$}$$

are fulfilled; recall that a_α and $b_{j\alpha}$ are coefficients of A and B_j and ϕ appears in (9.1.15)–(9.1.16). Moreover assume that

$$|J^{-1}|\le c\varrho_0^{-1} \tag{9.2.4}$$

and operator $\varrho^{-1}A$ is microhyperbolic at energy level 0 (uniformly, after rescaling $x \mapsto x\gamma^{-1}$, $\xi \mapsto \xi\rho^{-1}$).

Further let us assume that operator $\varrho^{-1}A$ and boundary value problem $\{\varrho^{-1}A, \varrho_1^{-1}B_1, \dots, \varrho_\mu^{-1}B_\mu\}$ are uniformly elliptic in the classical sense. Recall that in virtue of Theorem 7.5.15 we do not need to assume microhyperbolicity of the boundary value problem.

Example 9.2.2. Consider in dimension $d \geq 2$ Schrödinger operator (possibly with magnetic field)

$$A = \sum_{j,k} (D_j - V_j)g^{jk}(D_k - V_k) + V \tag{9.2.5}$$

with real-valued $g^{jk} = g^{kj}$, V_j, V such that

$$c^{-1}|\eta|^2 \leq \sum_{j,k} g^{jk}\eta_j\eta_k \leq c|\eta|^2 \qquad \forall \eta \in \mathbb{R}^d \tag{9.2.6}$$

and

$$|D^\beta g^{jk}| \leq c\gamma^{-|\beta|} \qquad \forall j,k, \tag*{(9.2.7)$_1$}$$

$$|D^\beta F_{jk}| \leq c\rho\gamma^{-|\beta|-1} \qquad \forall j,k, \tag*{(9.2.7)$_2$}$$

$$|D^\beta V| \leq c\rho^2\gamma^{-|\beta|}, \qquad \forall \beta : |\beta| \leq K \tag*{(9.2.7)$_3$}$$

where $F_{jk} = \partial_k V_j - \partial_j V_k$ is the tensor of magnetic field intensity.

Recall that due to results of Section 8.2 we do not need microhyperbolicity.

Furthermore, due to Sections 4.6, 7.5 and 8.2 we can relax smoothness conditions in Examples 9.2.1 and 9.2.2:

Example 9.2.3. Assume ξ-microhyperbolicity in example 9.2.1. Then one can replace condition $(9.2.3)_1$ by

$$|D^\beta a_\alpha| \leq c\varrho\beta!\rho^{-|\alpha|}\gamma^{-|\beta|}(1 + \varepsilon^{1-|\beta|}) \qquad \text{with } \varepsilon = C(\rho\gamma|\log \rho\gamma|)^{-1}$$
$$\forall \alpha : |\alpha| \leq m, \forall \beta : |\beta| \leq \epsilon\rho\gamma; \tag*{(9.2.8)$_1$}$$

we leave to the reader to rewrite the smoothness conditions to ϕ, $b_{j,\alpha}$ (starting from (7.5.42)) according to Section 7.5 [8].

[8] One can relax smoothness conditions to J as well but this is not important since it is an auxiliary weight we introduce as needed and can smoothen it.

Example 9.2.4. Assume that either $d \ge 3$ or that $d = 2$ and

$$|V| + |\nabla V|\gamma \ge \epsilon\rho^2 \tag{9.2.9}$$

in Example 9.2.2. Then we can relax conditions $(9.2.7)_1$–$(9.2.7)_3$ in the similar way.

Example 9.2.5. Furthermore, again due to Sections refsect-4-6, 7.5 and 8.2 we can relax smoothness conditions even further: instead of $(9.2.8)_1$ we can assume $(9.2.3)_1$ with $K = 1$ and

$$|\nabla a_\alpha(x) - \nabla a_\alpha(y)| \le C|\log \mathsf{dist}(x,y)|^{-1}. \tag{9.2.10}$$

Then we can find operators A^+ and A^- bracketing A[9], both satisfying $(9.2.8)_1$ and such that

$$\mathcal{N}^{\mathsf{W}}(A^-) - CR_1 \le \mathcal{N}^{\mathsf{W}}(A) \le \mathcal{N}^{\mathsf{W}}(A^+) \le CR_1. \tag{9.2.11}$$

Other such examples following immediately from Parts II and III we leave to the reader.

9.2.2 Regular Zone: Sharper Asymptotics

In Parts II and III we also derived sharper asymptotics. Their main parts contained next terms some of which were expressed as integrals of functions supported in

$$\Sigma_\tau = \mathsf{supp}\bigl(\partial_\tau \boldsymbol{n}(x,\xi,\tau)\bigr) \tag{9.2.12}$$

and others as integrals over

$$\iota\bigl(\mathsf{supp}\bigl(\boldsymbol{n}(x,\xi,\tau)\bigr) \cap T^*X|_Y\bigr); \tag{9.2.13}$$

recall that $\iota : T^*X|_Y \to T^*Y$. We recall the exact expressions of them in Part V as needed.

Meanwhile the remainder estimate contains two terms:

$$\frac{C}{T_j}\partial_\tau \iint_{\Lambda_j} \boldsymbol{n}(x,\xi,\tau)\, dxd\xi \tag{9.2.14}$$

[9] I.e. such that $A^+ \ge A \ge A^-$ and therefore $\boldsymbol{A}^+ \ge \boldsymbol{A} \ge \boldsymbol{A}^-$ as we use the same matrix J satisfying $(9.2.3)_3$ for all three operators.

and

$$C'' \int_{X_{\mathsf{reg}}} \rho^{d-1-\delta}\gamma^{-1-\delta}\, dx \tag{9.2.15}$$

where (generalized) Hamiltonian trajectories (or billiards) issued from Λ_j in an appropriate time direction should satisfy certain assumptions of Part II or III. Again we leave exact form of those assumptions to more concrete Part V but we need to mention that while the geometric shape of these trajectories on the energy level 0 does not depend on J, the temp of movement depends on ϱ_0. Really, consider just the case of scalar symbols $a(x,\xi)$ and $J(x)$. Then trajectory is given by equation

$$\begin{cases} \dfrac{dx}{dt} = & \varrho_0(x)^{-1}\kappa\partial_\xi a(x,\xi) \\ \dfrac{d\xi}{dt} = & -\varrho_0(x)^{-1}\kappa\partial_x a(x,\xi) \end{cases} \tag{9.2.16}$$

where

$$\kappa = \varrho^{-1}\varrho_0\rho\gamma\bigr|_{x=x(0)} \tag{9.2.17}$$

is a normalization factor, so that initial ball $B(x,\gamma)$ is passed with time $T_0 \asymp 1$. Here we renormalized time with our usual scaling. So, these conditions and estimates are affected by the choice of ϱ_0.

9.2.3 Singular zZone

All results in the singular zone are due to Appendix 9.A.1. We mention only few of them.

Example 9.2.6. Let us assume that

$$(Au,u) \ge \epsilon\|u\|_r^2 - \|\rho^r u\|^2 \qquad \forall u \in \mathfrak{D}(A_B) : \mathsf{supp}(u) \subset \bar{X}_{\mathsf{sing}} \tag{9.2.18}$$

where $m = 2r$ is an order of operator. Therefore due to the Rozenblioum estimate (9.A.11) if on $\partial X_{\mathsf{sing}}$ $\mathfrak{D}(A_B)$ is defined by Dirichlet boundary conditions and if $d > m$ then

$$R_0 = C\int \bigl(\rho^d + \varrho_0^{d/m}\bigr)\, dx. \tag{9.2.19}$$

Meanwhile R_2 is less than R_0, so R_0 estimates the total contribution of X_{sing} to the remainder. Let us recall that the standard remainder estimate in the regular zone is given by (9.2.2) but the first term may be sharpened; then the total remainder estimate is given by the sum of (9.2.2) and (9.2.19):

$$C\int_{X_{\mathsf{reg}}\setminus X_{\mathsf{reg,ell}}} \rho^{d-1}\gamma^{-1}\,dx + C'\int_{X_{\mathsf{reg,ell}}} \rho^{d-s}\gamma^{-s}\,dx + C\int_{X_{\mathsf{sing}}} \bigl(\rho^d + \varrho_0^{d/m}\bigr)\,dx. \tag{9.2.20}$$

Since the answer should not depend on J we can select $\varrho_0 \le \rho^m$ thus arriving to

$$C\int_{X_{\mathsf{reg}}\setminus X_{\mathsf{reg,ell}}} \rho^{d-1}\gamma^{-1}\,dx + C'\int_{X_{\mathsf{reg,ell}}} \rho^{d-s}\gamma^{-s}\,dx + C\int_{X_{\mathsf{sing}}} \rho^d\,dx. \tag{9.2.21}$$

Note however that selection of J affects sharper asymptotics.

Example 9.2.7. Let all assumptions of the previous example are fulfilled with the exception that on part Y_{N} of $(\partial X)\cap \bar{X}_{\mathsf{sing}}$ $\mathfrak{D}(A_B)$ is defined by some other boundary condition, which does not violate (9.2.18).

Assume that the cone condition (9.A.23)–(9.A.26) is satisfied with the cone of some fixed angle ϕ and height h_j (depending on $X_{\mathsf{sing},j}$). Then in virtue of (9.A.32) remainder estimate is still valid with an extra term

$$C\sum_j h_j^{-d}\,\mathsf{mes}(X_j). \tag{9.2.22}$$

Example 9.2.8. Consider *quasiregular zone* X_{qreg} in which all assumptions of the regular zone are fulfilled with the exception of those to the boundary which may be rather irregular there (see Figure 9.3).

Let us include the *inner zone* $\mathcal{X} = \{x \in X_{\mathsf{qreg}},\ \mathsf{dist}(x,\partial X) \ge \rho^{-1}(x)\}$ into X_{reg} and the *boundary strip* $\mathcal{Y} = \{x \in X_{\mathsf{qreg}},\ \mathsf{dist}(x,\partial X) \le 2\rho^{-1}(x)\}$ into X_{sing}.

Assume that

(9.2.23) After rescaling $x \mapsto (x-y)\gamma(y)^{-1}$ in (rescaled) $\mathcal{B}(y,\gamma(y),X)$ condition (7.5.28) is fulfilled uniformly with respect to y.

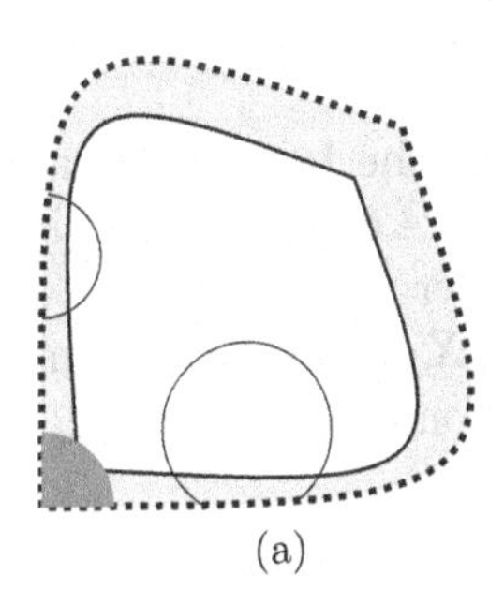

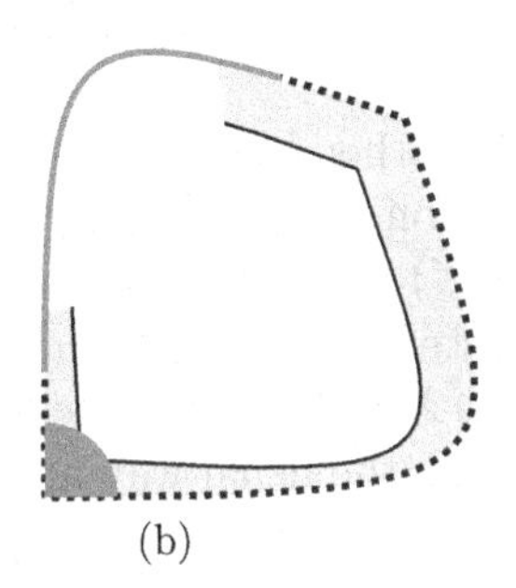

(a) (b)

Figure 9.3: On both pictures dotted line shows the part of $\partial X \cap \bar{X}_{\mathsf{qreg}}$ while solid gray line shows the smooth part of the boundary; dark shade indicates $X_{\mathsf{sing}} \setminus X_{\mathsf{qreg}}$ and light shade indicates $\mathcal{Y}$. Finally, on (a) two balls $\mathcal{B}(y, \gamma(y), X)$ are drawn.

Then due to Theorem 7.5.10 and Proposition 7.5.6(iii) contribution of $\mathcal{B}(y, \gamma(y), X) \cap \mathcal{X}$ to the remainder does not exceed $C\rho^{d-1}\gamma^{d-1}|\log\log(\rho\gamma)|$ [10]. Therefore the total contribution of $\mathcal{X}$ to the remainder does not exceed

$$C\int_{X_{\mathsf{qreg}}} \rho^{d-1}\gamma^{-1}|\log\log(\rho\gamma)|\, dx. \tag{9.2.24}$$

Assume that the Dirichlet boundary conditions are fulfilled there. Then as $d > m$ we can apply Rozenblioum estimate (9.2.12) and conclude that the contribution of X_{sing} to the remainder does not exceed contribution of $X_{\mathsf{sing}} \setminus X_{\mathsf{qreg}}$ plus

$$C\int_{\{x\in X_{\mathsf{qreg}},\ \mathsf{dist}(x,\partial X)\le 2\rho^{-1}(x)\}} \rho^{d}\, dx \tag{9.2.25}$$

(provided $\varrho_0 \le \rho^d$ there). Due to (9.2.23) contribution of $\mathcal{B}(y, \gamma(y), X)$ to this expression does not exceed $C\rho^{d-1}\gamma^{d-1}$ and therefore (9.2.25) does not

[10]. Really, as $\gamma(y) = 1$, $\rho(y) = h^{-1}$ we are exactly in the framework of this section. General case is reduced to this by rescaling. Obviously γ at this moment is the radius of the ball rather then scaling function inside it.

exceed

$$C\int_{X_{\mathsf{qreg}}} \rho^{d-1}\gamma^{-1}\,dx. \tag{9.2.26}$$

On the other hand, if $d\le m$ we can apply the boxing described in Subsubsection *9.A.1.2 Bracketing* and since ρ has a constant magnitude in $\mathcal{B}(y,\gamma(y),X)$ we again recover the same remainder estimate.

So, instead of the assumption $X = X_{\mathsf{reg}}\cap X_{\mathsf{sing}}$ we can assume that $X = X_{\mathsf{reg}}\cap X_{\mathsf{qreg}}\cap X_{\mathsf{sing}}$ and calculate contribution of X_{reg} and X_{sing} to the remainder in the usual way and of X_{qreg} by (9.2.24).

Example 9.2.9. (i) If after rescaling $x\mapsto x\gamma^{-1}$ we find ourselves in the framework of condition (7.5.39) uniform with respect to y, we can skip factor $|\log\log(\rho\gamma)|$ in (9.2.24).

(ii) Further, assume that in X_{qreg} cone condition (9.A.23)–(9.A.26) is satisfied with the cone of fixed angle ϕ and height $h=\epsilon\gamma(x)$. Then condition (9.2.23) is fulfilled automatically. In this case we do not need to assume that we have Dirichlet boundary condition but only that (9.2.18) holds.

(iii) Therefore under certain assumptions contribution of X_{qreg} to the remainder does not exceed (9.2.24).

Example 9.2.10. On the other hand, assume that the boundary is less regular; namely we assume only that

$$\begin{aligned}\mathsf{mes}\bigl(\{x\in X_{\mathsf{qreg}}\cap\mathcal{B}(y,\gamma(y),X),\,\mathsf{dist}(x,\partial X)\le t\}\bigr)\le ct^{\sigma}\gamma(y)^{d-\sigma}\\ \forall t:\rho(y)^{-1}\le t\le\gamma(y)\end{aligned} \tag{9.2.27}$$

is fulfilled uniformly with respect to $y\in X_{\mathsf{qreg}}$ with $\sigma\in(0,1)$ [11].

In this case we can relax smoothness conditions to coefficients as well and due to Theorem 7.5.22 the contribution of $\mathcal{X}$ to the remainder does not exceed

$$C\int_{X_{\mathsf{qreg}}} \rho^{d-\sigma}\gamma^{-\sigma}\,dx. \tag{9.2.28}$$

[11] As $\sigma=1$ this condition coincides with (9.2.23).

On the other hand, expression (9.2.25) in this case does not exceed (9.2.28) as well and so the contribution of X_{qreg} to the remainder is estimated by (9.2.28).

This allows us to treat fractal boundaries in Chapter 11.

Remark 9.2.11. (i) So far we considered operators which in the regular zone X_{reg} are either regular or "rough". However bracketing operator with the irregular coefficients A between two rough operators $A^- \le A \le A^+$ we conclude that $\mathsf{N}^-(A^+) \le \mathsf{N}(A)\mathsf{N}(A^-)$ and estimating difference $\mathsf{N}^-(A^-) - \mathsf{N}^-(A^+)$ we get remainder estimate for $\mathsf{N}^-(A)$ as well. Actually since $\mathsf{N}^-(A^-)$ and $\mathsf{N}^-(A^+)$ have the same remainder estimates we need to estimate $\mathcal{N}(A^-) - \mathcal{N}(A^+)$.

(ii) The similar approach works for Riesz means as well; we can also apply results of Section 4.7.

9.2.4 Riesz Means

Example 9.2.12. In the framework of Example 9.2.6 remainder estimate (9.2.21) is replaced by

$$C\int_{X_{\mathsf{reg}}\setminus X_{\mathsf{reg,ell}}} \varrho^{\omega}\rho^{d-1-\omega}\gamma^{-1-\omega}\,dx + C'\int_{X_{\mathsf{reg,ell}}} \varrho^{\omega}\rho^{d-s}\gamma^{-s}\,dx + C\int_{X_{\mathsf{sing}}} \varrho^{\omega}\rho^{d}\,dx. \tag{9.2.29}$$

Note however that selection of J affects sharper asymptotics.

Example 9.2.13. In the framework of Example 9.2.7 an extra term (9.2.22) is replaced by

$$C\sum_j h_j^{-d}\int_{X_j} \varrho^{\omega}\,dx. \tag{9.2.30}$$

Example 9.2.14. In the framework of Example 9.2.8 the contributions of X_{qreg} to both the remainder and the principal part are estimated by

$$C\int_{X_{\mathsf{qreg}}} \varrho^{\omega}\rho^{d-1}\gamma^{-1}\,dx. \tag{9.2.31}$$

Remark 9.2.15. In Section 12.6 we discuss Riesz means in domains with corners, angles and critical points and discover that those singularities can produce non-Weyl correction terms in the main part of asymptotics.

9.3 Estimates for Quadratic Forms

Let us prove another version of the basic theorems of this chapter. These new theorems deal with the quadratic forms instead of operators and N^- is now the maximal dimension of the negative subspace lying in $\mathscr{C}_0^K(\mathbb{R}^d,\mathbb{H})\cap\mathfrak{D}(Q)$. The advantage is that conditions to J become more transparent and we are able to select it properly in X_{sing}.

So, we have two quadratic forms, namely $Q(u)=(Au,u)$ and $Q_0(u)=(Ju,u)$, but we are interested in the negative subspaces of $Q(u)$, and the answer cannot depend on $Q_0(u)$ [12].

Let us introduce space $\mathcal{H}=\mathcal{H}_{Q_0}$ which is the closure of $\mathscr{C}_0^K(\mathbb{R}^d,\mathbb{H})\cap\mathfrak{D}(Q)$ in the norm $\|u\|_0=Q_0(u)^{\frac{1}{2}}$. If the form $Q(u)$ generates a self-adjoint operator $\boldsymbol{A}$ in this space then

$$\mathsf{N}^-(\boldsymbol{A})=\mathsf{N}^-(Q) \tag{9.3.1}$$

and under appropriate conditions we can apply Theorems 9.1.7 and 9.1.13.

Let us discuss how the answer depends on J. Let us observe first that if we do not consider trajectories then neither conditions nor estimates depend on the choice of J outside of X_{sing}. Surely, if we were to assume that $\boldsymbol{A}$ has a scalar principal symbol (or something like this) then we should take J satisfying this condition because then we have special theorems; so, let us assume that if A has a scalar principal symbol then J is a scalar matrix.

Note that existence of self-adjoint operator requires that

(9.3.2) $\mathfrak{D}(Q)$ is a dense subset of $\mathcal{H}$.

We try to take J in X_{sing} as small as possible (in order to decrease R_0 in estimate (9.1.23)) but we need to fulfill $(9.2.3)_3$ and (9.1.24)–(9.1.25). Note that if we assume that

$$\varrho=\varrho_0\rho\gamma \qquad \text{in } X_{\mathsf{reg}}\setminus X_{\mathsf{reg,ell}} \tag{9.3.3}$$

[12] Nevertheless the estimates we will derive can depend on J.

then in the equations of trajectories (9.2.16) $\kappa = 1$ and therefore these equations and conditions will be more natural. Despite these simplifications the constant R_0 has not yet been calculated and whether the quadratic form $Q(u)$ generates a self-adjoint operator in $\mathcal{H}$ is not yet clear.

However, in the lower bounds of $\mathsf{N}^-(Q)$ the latter condition is not necessary. Really, these estimates are non-trivial only if $\mathsf{N}^-(Q) < \infty$ but in this case the quadratic form Q is semibounded from below on $\mathcal{H}$ and due (9.3.3) this form really generates an operator on $\mathcal{H}$ which is semibounded from below. Therefore the lower and the upper estimates are provided and in the upper estimates the constant R_0 is deduced from the quadratic form $Q_t(u) = Q(u) - tQ_0(u)$, i.e., conditions (9.1.23) and (9.1.60) should be replaced by

$$\mathsf{N}^-(Q - tQ_0) \le R_0(t+1)^p \qquad \forall t \ge 0 \tag{9.3.4}$$

with Q here restricted to the domain X_{sing}.

Let us consider upper estimates. Namely, let us assume that Q is given by the formula

$$Q(u) = \int_X \sum_{|\alpha|,|\beta|\le r} \langle a_{\alpha\beta} D^\beta u, D^\alpha u\rangle\, dx + \int_{\partial X} \sum_{|\alpha|,|\beta|\le r-1} \langle b_{\alpha\beta} D^\beta u, D^\alpha u\rangle\, dx' \tag{9.3.5}$$

with coefficients satisfying

$$a_{\alpha\beta} = a^\dagger_{\beta\alpha}, \quad b_{\alpha\beta} = b^\dagger_{\beta\alpha} \qquad \forall \alpha, \beta, \tag{9.3.6}$$

and

$$\sum_{\alpha,\beta:|\alpha|=|\beta|=r} \langle a_{\alpha\beta}\xi^{\alpha+\beta} v, v\rangle \ge \epsilon_0 \varrho \rho^{-2r} |\xi|^{2r} \|v\|^2 \qquad \forall \xi \in \mathbb{R}^d \quad \forall v \in \mathbb{H}. \tag{9.3.7}$$

Further, let us also assume that

(9.3.8) Function $u \in \mathscr{C}_0^K(\mathcal{X}, \mathbb{H})$ belongs to $\mathfrak{D}(Q)$ if and only if $B_j u|_{\partial X} = 0$ $\forall j = 1, \dots, \mu'$ where B_j are differential operators of orders $m_j \le r-1$ (where $m = 2r$) and that Q is originally defined on these functions and is expanded to $\mathcal{H}$ as a closure.

Thus the boundary conditions for operator $\boldsymbol{A}$ are $B_j u|_{\partial X} = 0$ with $j = 1, \dots, \mu$ of two types: original conditions (imposed to the domain of the quadratic form) with $j = 1, \dots, \mu'$ and $m_j \le r - 1$ and additional conditions with $j = \mu' + 1, \dots, \mu$ and $m_j \ge r$ which follow from the variational problem.

Furthermore, let us assume that

(9.3.9) $J, J^{-1} \in \mathscr{L}^{\infty}_{\mathsf{loc}}(\mathcal{X})$ where $\mathcal{X} \supset \bar{X}_{\mathsf{reg}}$ and $\mathcal{X}$ is dense in X.

Then

(9.3.10) The subspaces in which we test whether Q is negative definite are the same for the chosen J and for $J = I$.

All these conditions should be assumed from the beginning which we will do in the statements of theorems. To proceed further, let us first note that under the assumptions of our theorems the contribution of every ball $\mathcal{B}(x, \gamma(x), X)$ with $x \in X_{\mathsf{reg}}$ to $\mathcal{N} + \mathsf{R}$ is either larger than a constant ϵ_1 or is $O(h^s)$ where h means $h_{\mathsf{eff}} = (\rho\gamma)^{-1}$.

We will call these balls *balls of the first type* and *balls of the second type* respectively. Therefore

(9.3.11) If the right-hand expression is finite then the number of balls of the first type is finite and $\rho, \gamma, \varrho, \varrho_0, J$ on the union of these balls are disjoint from 0 and ∞.

Thus, in order to obtain a theorem similar to Theorems 9.1.7 and 9.1.13 but with $\mathsf{N}^-(Q)$ instead of $\mathsf{N}^-(\boldsymbol{A})$ we need only to prove that Q generates a self-adjoint operator in $\mathcal{H}$. Surely it is sufficient to prove that $\mathsf{N}^-(Q_T) < \infty$ for some $T > 0$ with $Q_T(u) = Q(u) + T(Ju, u)$.

Moreover, note that estimate $N^-(Q) \le M$ is due to the uniform estimate $N^-(Q_t) \le M$ for $Q_t(u) = Q(u) + t(J_1 u, u)$ and for all $t > 0$ and for some function J_1 satisfying the same assumptions as J [13].

Further, if Q_t generates a self-adjoint operator A_t then one can apply the Theorems 9.1.7 and 9.1.13 to this operator.

Let us note that in balls of the first type for small enough $t > 0$ the same conditions are fulfilled[14] which were assumed to be fulfilled for $t = 0$. Estimates in balls of the second type and in the singular zone are obvious.

---

[13] With another weight function ϱ_0' instead of ϱ_0 and with a non-essential constant c' instead of c.

[14] With constants $2c$ and $\frac{1}{2}\epsilon_0$ instead of c and ϵ.

Therefore we get the estimate $N^{-}(Q_t) \le M + \varepsilon(t)$ for $t > 0$, $\varepsilon = 0(1)$ as $t \to +0$, which yields the estimates in question.

Thus, it is sufficient to assume that

(9.3.12) Form $Q_{t,T}(u) = Q(u) + t(J_1 u, u) + T(Ju, u)$ is non-negative-definite for every $t > 0$ and sufficiently large $T = T_t$.

But this assumption holds provided

(9.3.13) The quadratic form $Q_T(u) = Q(u) + (J_0 u, u)$ is non-negative definite for some function J_0 satisfying the same assumptions as J[13].

Really, under this condition one can pick $J_1 = J_0^2 J^{-1}$ and $T = t^{-1}$. Therefore,

(9.3.14) Under condition (9.3.13) one can pass from operators to quadratic forms.

There are many situations in which condition (9.3.13) is obviously fulfilled. Let us list some of them. Assume that

(9.3.15) $\{\psi_n^2\}_{n\ge 0}$ is a partition of unity such that ψ_n satisfy (9.1.19) in X_{reg} uniformly on n, $\operatorname{supp}(\psi_0) \subset X_{\mathsf{sing}} \setminus Y_{\mathsf{reg},\epsilon}$, $\operatorname{supp}(\psi_n) \subset X_{\mathsf{reg}} \setminus Y_{\mathsf{sing},\epsilon}$ where $Y_{\mathsf{reg},\epsilon}$ is $\epsilon\gamma$-neighborhood of $X \setminus X_{\mathsf{sing}}$ and $Y_{\mathsf{sing},\epsilon}$ is $\epsilon\gamma$-neighborhood of $X \setminus X_{\mathsf{reg}}$, $\operatorname{diam}\operatorname{supp}(\psi_n) \le \epsilon_2\gamma$ for $n \ge 1$ and the multiplicity of this partition does not exceed $C_0 = C_0(d)$.

Then

$$(9.3.16)\quad Q(u) = \sum_{n\ge 0}\Bigl(\int_X \psi_n^2 \sum \langle a_{\alpha\beta} D^\beta u, D^\alpha u\rangle\, dx + \int_{\partial X} \psi_n^2 \sum \langle b_{\alpha\beta} D^\beta u, D^\alpha u\rangle\, dx\Bigr)$$

which in turn equals

$$\begin{aligned}\sum_{n\ge 0}\Bigl(Q(\psi_n u) - 2\operatorname{Re}\int_X \sum \langle a_{\alpha\beta}[D^\beta, \psi_n]u, D^\alpha \psi_n u\rangle\, dx &+ \\ \operatorname{Re}\int_X \sum \langle a_{\alpha\beta}[D^\beta, \psi_n]u, [D^\alpha, \psi_n]u\rangle dx &- \\ 2\operatorname{Re}\int_{\partial X} \sum \langle b_{\alpha\beta}[D^\beta, \psi_n]u, D^\alpha \psi_n u\rangle\, dx' &+ \\ \operatorname{Re}\int_{\partial X} \sum \langle b_{\alpha\beta}[D^\beta, \psi_n], [D^\alpha \psi_n]u\rangle\, dx'\Bigr)&\end{aligned}$$

where we sum on α, β inside. Therefore (9.3.16) yields that

$$Q(u) \geq \sum_{n\geq 0} Q(\psi_n u) - \sum_{|\alpha|\leq r-1} C \int_X \chi\varrho(\rho\gamma)^{-2}\rho^{-2|\alpha|}|D^\alpha u|^2\,dx - \sum_{|\alpha|\leq r-1} C \int_{\partial X} \chi\varrho\rho^{-2|\alpha|-1}|D^\alpha u|^2\,dx' \geq \sum_{n\geq 0} \tilde{Q}(\psi_n u) \tag{9.3.17}$$

where

$$\tilde{Q}(v) = Q(v) - \sum_{|\alpha|\leq r-1} C \int_X \chi\varrho(\rho\gamma)^{-2}\rho^{-2|\alpha|}|D^\alpha v|^2\,dx - \sum_{|\alpha|\leq r-1} C \int_{\partial X} \chi\varrho\rho^{-2|\alpha|-1}|D^\alpha v|^2\,dx', \tag{9.3.18}$$

χ is the characteristic function of $X \setminus (2\epsilon\gamma\text{-neighborhood of } (X_{\mathsf{reg}} \setminus X_{\mathsf{sing}}))$ and we assume from the beginning that

$$|D^\sigma a_{\alpha\beta}| \leq c\varrho\rho^{-|\alpha|-|\beta|}\gamma^{1-|\sigma|}, \tag{9.3.19}$$

$$|D^\sigma b_{\alpha\beta}| \leq c\varrho\rho^{-|\alpha|-|\beta|}\gamma^{1-|\sigma|} \qquad \forall \sigma : |\sigma| \leq 1 \quad \forall \alpha, \beta. \tag{9.3.20}$$

One can easily see that under conditions (9.3.8)

$$\tilde{Q}(u) + 2(J_0 u, u) \geq \frac{1}{2} Q(u) + (J_0 u, u) \tag{9.3.21}$$

for appropriate function J_0.

Let us consider $\tilde{Q}(\psi_n u)$ with $n \geq 1$ first. In the case of balls of the first type we consider quadratic forms without boundary conditions and one can see easily that for a large enough T the corresponding quadratic form is positive definite.

Let us consider balls of the second type. The following new difficulty arises: $\psi_n u$ does not necessarily satisfy the boundary conditions defining the domain of the quadratic form Q.

However, this difficulty does not arise provided one of the following conditions is fulfilled:

$$B(\bar{x}, \gamma(\bar{x})) \subset X \qquad \forall \bar{x} \in X_{\mathsf{reg}}; \tag{9.3.22$_1$}$$

$$m_j \leq \frac{m}{2} - 1 \implies m_j = 0; \tag{9.3.22$_2$}$$

in particular, this condition is fulfilled if $m=2$;

$$B_j = D_1^{j-1} \qquad \forall j=1,\dots,\mu'. \tag{9.3.22$_3$}$$

Moreover, this difficulty does not arise in some other cases. Finally, one can avoid this difficulty provided

(9.3.22)$_4$ $\forall j$ with $m_j \le \frac{m}{2}-1$ operator B_j is a $\mathsf{D}_j\times \mathsf{D}$ matrix operator with $\mathsf{D}_j\le \mathsf{D}$ and the coefficient of $D_1^{m_j}$ has rank D_j at any point of ∂X (where in the local coordinates $X=\{x_1>0\}$). Further, (9.3.20) holds for all $\sigma:|\sigma|\le K$ (in contrast to original (9.3.20) with $K=1$).

Indeed, one can prove easily that under this condition there exists an operator $\mathcal{L}$ such that

$$\operatorname{supp}(v)\subset \mathcal{B}(\bar{x},(1-2\epsilon)\gamma(\bar{x}),X) \implies \operatorname{supp}(\mathcal{L}v)\subset \mathcal{B}(\bar{x},(1-\epsilon)\gamma(\bar{x}),X), \tag{9.3.23}$$

$$\sum_{|\alpha|\le q}\rho^{-|\alpha|}\|D^\alpha \mathcal{L}v\| \le C_q \sum_{|\alpha|\le q-1}\rho^{-|\alpha|-1}\gamma^{-1}\|D^\alpha v\| \tag{9.3.24}$$

where K in (9.3.20) also depends on q and $q\ge p$ is arbitrary,

$$B_j(I+\mathcal{L})v|_{\partial X}=0 \qquad \forall j: m_j\le \frac{m}{2}-1 \tag{9.3.25}$$

and

$$(I+\mathcal{L})v=0,\ \operatorname{supp}(v)\subset \mathcal{B}(\bar{x},(1-\epsilon)\gamma(\bar{x}),X) \implies v=0. \tag{9.3.26}$$

An easy proof is left to the reader.

Further, a similar operator $\mathcal{L}$ is also defined for $n=0$ with

$$\operatorname{supp}(v)\subset \mathcal{Z}_2 \implies \operatorname{supp}(\mathcal{L}v)\subset \mathcal{Z}_1, \qquad v=\mathcal{L}v \ \text{ in } \mathcal{Z}_4, \tag{9.3.27}$$

$$\sum_{|\alpha|\le q}\|\varrho^{\frac12}\rho^{-|\alpha|}D^\alpha \mathcal{L}v\| \le C_q \sum_{|\alpha|\le q-1}\|\varrho^{\frac12}\rho^{-|\alpha|-1}\gamma^{-1}D^\alpha v\| \tag{9.3.28}$$

$$B_j(I+\mathcal{L})v|_{\partial X}=0 \quad \forall j: m_j\le \frac{m}{2}-1 \quad \text{provided } B_jv|_{\partial X\cap\mathcal{Z}_3}=0 \quad \forall j: m_j\le \frac{m}{2}-1, \tag{9.3.29}$$

$$(I+\mathcal{L})v = 0,\ \mathsf{supp}(v) \subset \mathcal{Z}_1 \implies v = 0 \tag{9.3.30}$$

where $\mathcal{Z}_k$ is $\bar{X} \setminus (k\epsilon\gamma$-neighborhood of $X \setminus X'_{\mathsf{reg},0})$ and $\epsilon > 0$ is a small enough constant. This construction is also left to the reader. Recall that ψ_0 differs from 1 only in the regular zone.

Finally, we obtain estimate

$$Q(u) \ge \sum_{n\ge 0} \tilde{Q}((I+\mathcal{L})\psi_n u). \tag{9.3.31}$$

Therefore in all the cases $(9.3.22)_{1-4}$ the following estimate

$$\mathsf{N}^-(Q) \le \sum_{n\ge 1} \mathsf{N}^-(\tilde{Q}; \mathcal{B}(x^n(1-\epsilon)\gamma^n, X) + \mathsf{N}^-(\tilde{Q}, \bar{X}_{\mathsf{sing}}) \tag{9.3.32}$$

holds where Z denotes the geometrical domain on which the quadratic form Q is defined (i.e., all the test functions are supported) in the definition of $\mathsf{N}^-(\tilde{Q}, Z)$, $x^n \in \mathsf{supp}(\psi_n)$ and $\gamma^n = \gamma(x^n)$ as $n \ge 1$.

Let us recall that under the above conditions estimate

$$\mathsf{N}^-(\tilde{Q}, \mathcal{B}(x^n, (1-\epsilon)\gamma^n, X) \le M(\mathcal{N}_n^- + R_n)$$

holds where $\mathcal{N}_n^-$ and R_n are the contributions of the ball $\mathcal{B}(x^n, \gamma^n, X)$ to the principal part and the remainder estimate of the corresponding asymptotics. To prove this one should consider balls of the first and second type separately. Recall that there is only a finite number of balls of the first type.

Therefore the total contributions of these elements is finite provided the right-hand expression in the upper estimate is finite. Furthermore, condition (9.3.4) yields that the contribution of $n = 0$ is finite as well. Therefore,

(9.3.33) In the framework of our assumptions for appropriate function J_0 the number $N^-(Q_1)$ is finite and therefore $Q_T(u) = Q(u) + T(J_0 u, u)$ is positive definite for some T.

Thus we arrive to

Theorem 9.3.1. *In the framework of our assumptions Theorems 9.1.7 and 9.1.13 hold for quadratic forms as well.*

9.A Appendices

9.A.1 Variational Estimates

Rozenblioum estimates

Rozenblioum estimates are far the most important variational estimates we use. Let us consider the following two quadratic forms defined on scalar functions:

$$Q_1(u) = \sum_{|\alpha|=m} (a(x)D^\alpha u, D^\alpha u), \tag{9.A.1}$$

and

$$Q_0(u) = \sum_{|\alpha_1|\le m, |\alpha_2|\le m, |\alpha_1|+|\alpha_2|=2r} (V_{\alpha_1,\alpha_2}(x)D^{\alpha_1}u, D^{\alpha_2}u) \tag{9.A.2}$$

where

$$m > r \ge 0, \quad m \in \mathbb{Z}, \quad r \in \frac{1}{2}\mathbb{Z}, \tag{9.A.3}$$

$$a \ge 0, \quad V_{\alpha_1,\alpha_2} = V^\dagger_{\alpha_2,\alpha_1}. \tag{9.A.4}$$

Let us assume that

$$a^{-1} \in \mathscr{L}^p, \quad V_{\alpha_1,\alpha_2} \in \mathscr{L}^q, \tag{9.A.5}$$

$$1 \le p \le \infty, \quad 1 \le q \le \infty, \quad \frac{1}{p} + \frac{1}{q} = \frac{1}{\theta} = \frac{2}{d}(m-r), \tag{9.A.6}$$

$$m - r \le d, \ p \ge 1; \quad \theta > 1 \implies q > 1. \tag{9.A.7}$$

Moreover, let us assume that

(9.A.8) Q_0 contains only terms with

$$2(m - |\alpha_i|) < d(1 + \frac{1}{p}) \qquad \text{for} \quad p > 1,$$

$$m - |\alpha_i| \le d \qquad \text{for} \quad p = 1.$$

Let us introduce main objects of the study:

(9.A.9) Let $\mathsf{N}_0^-(Q, X)$ be a maximal dimension of Q-negative subspace of $\mathscr{C}_0^\infty(X)$

and

(9.A.10) Let $\mathsf{N}^-(Q, X)$ be a maximal dimension of Q-negative subspace of $\mathscr{C}^\infty(\bar{X})$ having compact support where integration is taken over X in definition of Q.

Then the following is estimate due to G. Rozenblioum [5]:

Theorem 9.A.1. *In assumptions* (9.A.1)–(9.A.8) *estimate*

$$\mathsf{N}^-(Q_1 - Q_0, \mathbb{R}^d) \le C\|a^{-1}\|^{\theta}_{\mathscr{L}^p} \cdot \|V\|^{\theta}_{\mathscr{L}^q} \tag{9.A.11}$$

holds where

$$\|V\|_* := \sum_{\alpha_1,\alpha_2} \|V_{\alpha_1,\alpha_2}\|_*$$

and $C = C(d, m, r, p, q)$.

Example 9.A.2. The most well known case is $m = 1, r = 0$ when for $d \ge 3$ one can take $p = \infty, q = \frac{1}{2}d$. This is the famous Cwikel-Lieb-Rozenblioum estimate (CLR) (see M. Reed & B. Simon [1] and for $d = 2$ one can take arbitrary $q > 1$ and $p = q/(q-1)$.

Corollary 9.A.3. *Estimate* (9.A.11) *holds for* $\mathsf{N}^-_0(Q_1 - Q_0, X)$ *as* $X \subset \mathbb{R}^d$.

Other estimates of this type are discussed in survey M. S. Birman & M. Z. Solomyak [5].

Bracketing

Let γ be a scaling function on X, and X be covered by balls $\mathcal{B}(x_n, \gamma(x_n), X)$ of multiplicity not exceeding c. Let $\sum_n \psi_n^2 = 1$ where ψ_n be real-valued functions supported in $\mathcal{B}(x_n, \gamma(x_n), X)$ and satisfy

$$|D^\beta \psi_n| \le c\gamma_n^{-|\beta|} \tag{9.A.12}$$

Note that

$$Q_1(u) \ge \sum_n \Bigl(\frac{1}{2} Q_1(\psi_n u) - Q_1'(\psi_n u)\Bigr)$$

with

$$Q_1'(u) = C(a\gamma^{-2m}u, u)$$

and as $r = 0$

$$Q_0(u) = \sum_n Q_0(\psi_n u).$$

Then

$$\mathsf{N}_0^-(Q_1 - Q_0, X) \le \sum_n \mathsf{N}_0^-\bigl(Q_1 - 2Q' - 2Q_0, \mathcal{B}(x_n, \gamma(x_n), X)\bigr). \tag{9.A.13}$$

Similar inequalities hold for $\mathsf{N}^(Q_1 - Q_0, X)$

$$\mathsf{N}^-(Q_1 - Q_0, X) \le \sum_n \mathsf{N}^-\bigl(Q_1 - 2Q' - 2Q_0, X; \mathcal{B}(x_n, \gamma(x_n), X)\bigr) \tag{9.A.14}$$

where $\mathsf{N}^-(Q, X; Y_n)$ denotes Q-negative subspace of $\mathscr{C}^\infty(Y_n)$ supported in Y_n where integration is taken over X in definition of Q.

Neumann Type Problems

Let us now consider these quadratic forms on the space $\mathcal{V} \subset \mathscr{C}^m(\bar{X})$. Let us assume that $X = \bigcup_{j\in\mathcal{J}} X_j$ and $X_j \subset \mathcal{X}_j$ where X_j and $\mathcal{X}_j$ are open domains such that for every j there exists a linear operator $\mathcal{L}_j : \mathcal{V} \to \mathscr{C}_0^m(\mathcal{X}_j)$ satisfying

$$\mathcal{L}_j u = u \qquad \text{in} \quad X_j \quad \forall u \in \mathcal{V}, \tag{9.A.15}$$

with

$$Q_{1j}(\mathcal{L}_j u) \le Q_{1j}'(u) + Q_{2j}'(u) \qquad \forall u \in \mathcal{V} \qquad \forall j \tag{9.A.16}$$

where Q_{1j} and Q_{1j}' are forms of (9.A.1)-type with a replaced by $a_j \ge 0$, $a_j' \ge 0$ respectively and Q_{2j}' are forms of (9.A.2)-type.

Then summing (9.A.16) over j we obtain from (9.A.15) that if the quadratic form $Q - Q_0$ is negative definite on the subspace $\mathcal{W} \subset \mathcal{V}$ then on this subspace

$$\sum_j Q_{1j}(\mathcal{L}_j u) \le Q_1(u) + \sum_j Q_{2j}'(u) \le \\ Q_0(u) + \sum_j Q_{2j}'(u) \le \sum_j Q_0'(\mathcal{L}_j u) \tag{9.A.17}$$

provided

$$\sum_j a'_j = a, \tag{9.A.18}$$

and

(9.A.19) Forms Q_0, Q'_{2j} contain only terms with $\alpha_1 = \alpha_2$, $|\alpha_1| = r$ and $V_{\alpha_1,\alpha_1} \geq 0$ (and the same is true for the coefficients of Q'_{2j}) and

$$Q'_0 = Q_0 + \sum_k Q'_{2k}. \tag{9.A.20}$$

Therefore $\mathsf{N}^-(Q_1 - Q_0, \mathcal{V})$ does not exceed the maximal dimension of the subspace $\mathcal{W} \subset \bigoplus_j \mathscr{C}_0^m(\mathcal{X}_j)$ with the following property:

(9.A.21) For every $w = (w_j)_{j\in\mathcal{J}} \in \mathcal{W}$ there exists $u \in V$ with $w_j = \mathcal{L}_j u$ for all j on which the quadratic form $\sum_j (Q_{1j} - Q'_{0j})(w_j)$ is negative definite.

Dropping the condition $\mathcal{L}_j u = w_j$ on w_j we obtain that

$$\mathsf{N}^-(Q_1 - Q_0, \mathcal{V}) \leq \sum_j \mathsf{N}^-(Q_{1j} - Q'_{0j}, \mathcal{W}_j) \tag{9.A.22}$$

with $\mathcal{W}_j = \mathscr{C}_0^m(\mathcal{X}_j)$.

Now one can apply Rozenblioum estimate (9.A.11) in order to estimate every term in the right-hand expression. One should take $\mathscr{L}^p$- and $\mathscr{L}^q$-norms on $\mathcal{X}_j$ in these estimates.

In particular, let $\mathcal{V}$ be the space of all compactly supported $\mathscr{C}^{m+1}$ functions satisfying the Dirichlet condition of order $(m+1)$ on $Y_{\mathsf{D}} \subset \partial X$. Let us assume that

$$Y_{\mathsf{N}} = \partial X \setminus Y_{\mathsf{D}}, \tag{9.A.23}$$

$$X = \bigcup_{j\geq 0} X_j, \tag{9.A.24}$$

$$Z_j = Y_{\mathsf{N}} \cap \bar{X}_j, \qquad Y_{\mathsf{N}} \cap \bar{X}_0 = \emptyset. \tag{9.A.25}$$

Let Γ_j be open cones with vertex 0 in $\mathbb{R}^d$ with angle $\phi > 0$ and altitudes h_j such that the following *cone condition* is fulfilled:

(9.A.26) For every point $\bar{x} \in X : \mathsf{dist}(x, X_j) \le \epsilon_0 h_j$ the connected component of $(\bar{x} + \Gamma_j) \setminus Y_D$ containing $\bar{x}$ does not intersect Y_N.

Recall that $\mathsf{dist}(x, y)$ is a path distance in X.

Then taking the Sobolev extension operators $\mathcal{L}_j$ (see f.e. S. L. Sobolev [1]) associated with cones $\epsilon\Gamma_j$ with a small enough constant $\epsilon > 0$ and cutting of on the distances exceeding $\epsilon_1 h_j$ from X we see that for $j \ge 1$

$$\sum_{|\alpha|=m} \|D^\alpha \mathcal{L}_j u\| \le C \sum_{|\alpha|=m} \|D^\alpha u\|_{X'_j} + C h_j^{-m} \|u\|_{X'_j} \tag{9.A.27}$$

where X'_j is a connected component of $(X_j + 2\epsilon\Gamma_j) \setminus Y_D$, containing $\bar{x}$.

Assuming that

(9.A.28) X'_j form no more than a c-multiple covering of $X \setminus X_0$

one can take $Q_1 \asymp \|.\|_m^2$, $Q'_{1j} \asymp \|.\|_m^2$, $Q_{1j} \asymp \|.\|_m^2$ in the domains X, X'_j and $\mathbb{R}^d$ respectively and $Q'_{2j} = C h_j^{-2m} \|.\|_{X'_j}^2$.

Moreover, assume that

(9.A.29) $X''_j = \{x \in X, \mathsf{dist}(x, Z_j) < h_j\}$ form no more than a c-multiple covering of $X \setminus \mathcal{X}_0$.

Then as $\mathcal{L}_0$ one can take the operator of multiplication by a function ψ such that $\psi = 1$ in X_0, $\psi = 0$ in the ϵh_j-neighborhood of Z_j (for every $j \ge 1$) satisfying the inequalities $|D^\alpha \psi| \le c\gamma^{-|\alpha|}$ $\forall \alpha : |\alpha| \le m$ with $\gamma = \mathsf{dist}(x, Z)$.

In this case one can take $Q_1(u) \asymp \sum_{|\alpha|=m} (aD^\alpha u, D^\alpha u)$, Q'_{10} and Q_{10} given by the same formula in X, X_0 and $\mathbb{R}^d$ respectively and $Q'_{20} = C \sum_j h_j^{-2m} \|.\|_{X''_j}^2$ provided

$$\epsilon_0 \le a \le c \quad \text{in} \quad Y'_j \quad \forall j \ge 1. \tag{9.A.30}$$

So, we obtain that $Q'_{0j}(v) = Q_0(v) + (W_j v, v)$ with $W_j = h_j^{-2m} \chi_{X''_j}$ for $j \ge 1$ and $W_0 = \sum_{j \ge 1} h_j^{-2m} \chi_{X''_j}$ where χ_M is the characteristic function of the set M. In fact, without any loss of the generality one can assume that X_j are contained in the $\frac{2}{3} h_j$-neighborhoods of Z_j.

Let us apply the following inequality

$$\mathsf{N}^-(Q_{(1)} + Q_{(2)}) \le \mathsf{N}^-(Q_{(1)}) + \mathsf{N}^-(Q_{(2)}) \tag{9.A.31}$$

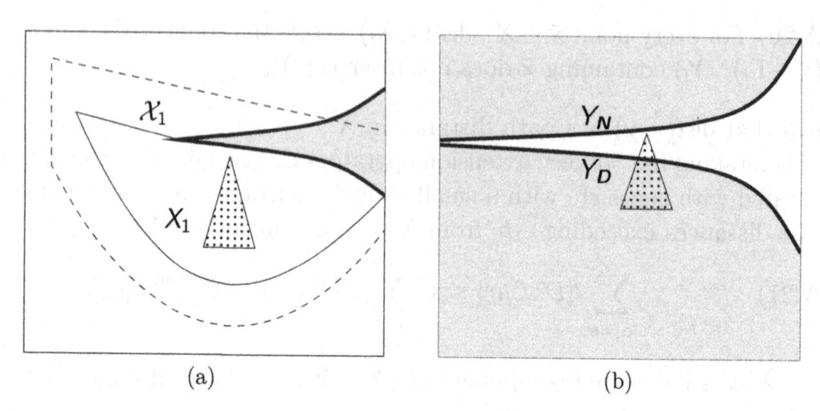

Figure 9.4: Domain X is white, cone is covered by dots. On (a) this cone serves X_1; flipping vertically we get X_2 and corresponding cone. On (b) pretty large cone serves thin domain because of Y_D making the lower part of the cone irrelevant.

(on the same functional spaces) with $Q_{(1)} = \frac{1}{2}Q_{1j} - Q_0$ and $Q_{(2)} = \frac{1}{2}Q_{1j} - Q'_{2j}$.
First of all, let us apply Rozenblioum estimate (9.A.11) to $\mathsf{N}^-(Q_{(1)})$.

(i) Assume that $d > 2m$. Then we can apply the same estimate (9.A.11) to $\mathsf{N}^-(Q_{(2)})$ with $p = \infty$, $\theta = q = d/2m$. After summation over j we end up with estimate

(9.A.32) $$\mathsf{N}^-(Q_1 - Q_0) \le C\|a^{-1}\|^{\theta}_{\mathscr{L}^p}\|V\|^{\theta}_{\mathscr{L}^q} + C\sum_j h_j^{-d}\operatorname{mes}(X''_j).$$

(ii) For $d \le 2m$ let us assume that

(9.A.33) $j = 0, \dots, J < \infty$ in the covering in question.

In this case let us again apply (9.A.31) and take a covering of X''_j by h_j-cubes of multiplicity less than C_0. Furthermore, let us apply estimate $\mathsf{N}^-_(Q_1 - 2Q_0) \le C_1$ in every cube. Then we again obtain estimate (9.A.32) with a constant C depending on J.

So the following statement has been proven:

Proposition 9.A.4. *(i) Under the above conditions estimate* (9.A.32) *holds for* $d > 2m$.

(ii) For $d \le 2m$ *this estimate holds under additional assumption* (9.A.33) *with* C *depending on* J.

Some Useful Estimates

In the end of this subsection let us assume that $a \ge \epsilon_0$ in X_{sing} and let us estimate $\mathsf{N}^-(Q_1 - Jt + J_3)$ in the domain X_{sing} with $J = J_1 + J_2$ where $J_1 = \chi_{X_{\mathsf{reg}} \cap X_{\mathsf{sing}}} \rho^{-2s}\gamma^{-2s-2m}$ $(s \in \mathbb{R})$, $J_2 \ge 0$, $J_3 \ge 0$, with thea Dirichlet condition on $\partial X_{\mathsf{sing}} \cap X$. Here ρ and γ satisfy all assumptions of Section 9.1.

Let us apply estimate (9.A.31). Then in order to estimate $\mathsf{N}^-(Q_1 - 2J_1 t)$ one can use a covering of $X_{\mathsf{reg}} \cap X_{\mathsf{sing}}$ by balls $\mathcal{B}(x, \gamma(x), X)$ with multiplicity less than C_0.

Then in every ball without boundary conditions (because ∂X is regular in the intersection with this ball)

$$\mathsf{N}^-(Q_1 - 2J_1 t + J_3) \le C(t\rho^{-2s}\gamma^{-2s})^{\frac{d}{2m}} + C_0 \tag{9.A.34}$$

in the general case and the last term is absent provided

$$x \in X_{\mathsf{reg},+} = \{x \in X_{\mathsf{reg}},\, J_3 \ge \epsilon\rho^{2m} \text{ in } \ \mathcal{B}(x, \bar{\gamma}(x))\}. \tag{9.A.35}$$

Therefore their sum does not exceed

$$Ct^{\frac{d}{2m}} \int_{X_{\mathsf{reg}} \cap X_{\mathsf{sing}}} \rho^{-\frac{ds}{m}}\gamma^{-\frac{ds}{m}-d}\, dx + C\int_{X_{\mathsf{reg},-} \cap X_{\mathsf{sing}}} \gamma^{-d}\, dx \tag{9.A.36}$$

where $X_{\mathsf{reg},-} = X_{\mathsf{reg}} \setminus X_{\mathsf{reg},+}$.

Moreover, the contribution of $X_{\mathsf{sing}} \setminus X_{\mathsf{reg}}$ is equal to 0 (because the form $Q_1 - 2J_1 t + J_3$ is non-negative definite on functions supported there). So in this case we obtain the estimate

$$\begin{gathered}\mathsf{N}^-(Q_1 - Jt + J_3) \le \\ \mathsf{N}^-(Q_1 - 2J_2 t + J_3) + Ct^{\frac{d}{2m}} \int_{X_{\mathsf{reg}} \cap X_{\mathsf{sing}}} \rho^{\frac{ds}{m}}\gamma^{\frac{ds}{m}-d}\, dx + C\int_{X_{\mathsf{reg},-} - \cap X_{\mathsf{sing}}} \gamma^{-d}\, dx.\end{gathered} \tag{9.A.37}$$

9.A.2 Riesz Means Estimates

Assume that $r = 0$, $Q_0 = (Wu, u)$ with $W \ge 0$ and $\bar{Q} = \|u\|^2$. Let us consider

$$\mathsf{N}^-_\omega := \int_{-\infty}^{0} t_-^\omega \, d_t \mathsf{N}^-(Q_1 - Q_0 - t\bar{Q}) = \omega \int_{-\infty}^{0} t_-^{\omega-1} \mathsf{N}^-(Q_1 - Q_0 - t\bar{Q})\, dt. \tag{9.A.38}$$

Then, according to Rozenblioum estimate (9.A.11)

$$\mathsf{N}^-_\omega \le C\|a^{-1}\|^\theta_{\mathscr{L}^p} \int_0^\infty \|(V-t)_+\|^\theta_{\mathscr{L}^q} \cdot t^{\omega-1}\, dt. \tag{9.A.39}$$

In the case $p = \infty$, $q = \theta$ [15] we immediately obtain estimate

$$\mathsf{N}^-_\omega \le C\|a^{-1}\|^{\frac{d}{2m}}_{\mathscr{L}^\infty} \int W^{\frac{d}{2m}+\omega}\, dx. \tag{9.A.40}$$

On the other hand, let $p < \infty$; then $q > \theta$. Applying the Hölder inequality to the right-hand expression of estimate (9.A.11) we get estimate

$$\mathsf{N}^-_\omega \le C\|a^{-1}\|^\theta_{\mathscr{L}^p} \cdot \Bigl(\int \phi(t)\, dt\Bigr)^{1-\frac{\theta}{q}} \Bigl(\iint t^{\frac{q}{\theta}(\omega-1)} (W-t)_+^q \phi(t)^{\frac{1-q}{\theta}}\, dxdt\Bigr)^{\frac{\theta}{q}}$$

with arbitrary function $\phi(t) \ge 0$; here we integrate with respect to $t \in (0, \infty]$. Picking up $\phi(t) = t^{-1+\delta}(t+\tau)^{-\delta-\delta'}$ with $\delta > 0, \delta' > 0$ and parameter $\tau \ge 0$ we then obtain that

$$\mathsf{N}^-_\omega \le C\|a^{-1}\|^\theta_{\mathscr{L}^p} \Bigl[\Bigl(\int W^{l_1} dx\Bigr)^{\frac{\theta}{q}} \tau^{-\delta'(1-\frac{\theta}{q})} + \Bigl(\int W^{l_2} dx\Bigr)^{\frac{\theta}{q}} \tau^{\delta(1-\frac{\theta}{q})}\Bigr]$$

with

$$l_1 = q(1+\frac{\omega}{\theta}) + \delta'(\frac{q}{\theta}-1),\quad l_2 = q(1+\frac{\omega}{\theta}) - \delta(\frac{q}{\theta}-1).$$

Minimizing with respect to τ and denoting $\rho = (\delta+\delta')(\frac{q}{\theta}-1)$, $\nu = \delta(\delta+\delta')^{-1}$ we obtain the following

Proposition 9.A.5. *(i) As $d > 2m$ estimate* (9.A.40) *holds.*

[15] This is possible if and only if $d > 2m$.

(ii) As $d \le 2m$ *estimate*

$$\mathsf{N}^-_\omega \le C\|a^{-1}\|^{\theta}_{\mathscr{L}^p}\Bigl(\int W^{l_1}\,dx\Bigr)^{\frac{\nu\theta}{q}}\Bigl(\int W^{l_2}\,dx\Bigr)^{\frac{(1-\nu)\theta}{q}} \tag{9.A.41}$$

holds with arbitrary $\nu \in (0,1)$, $\rho > 0$

$$l_1 = q(1+\frac{\omega}{\theta}) + (1-\nu)\rho,\quad l_2 = q(1+\frac{\omega}{\theta}) - \nu\rho.$$

(iii) In particular, as $d = 2m$ *estimate*

$$\mathsf{N}^-_\omega \le C_q\|a^{-1}\|_{\mathscr{L}^\infty}\bigl(\mathsf{mes}(X)\bigr)^{1-\frac{1}{q}}\Bigl(\int W^{(\omega+1)q}\,dx\Bigr)^{\frac{1}{q}} \tag{9.A.42}$$

holds with arbitrarily chosen $q > 1$.

We leave to the reader the following

Problem 9.A.6. Investigate averaging with function φ satisfying (4.2.73)–(4.2.75).

The following problem is really hard (see Comments).

Problem 9.A.7. Investigate if estimate (9.A.40) holds for some $q > 0$ even if $d \le 2m$.

Comments

The principal idea of this chapter was first used in the papers V. Ivrii [13] and V. Ivrii & S. Fedorova [1] where a second order operator was treated via the wave equation, and the finite spatial velocity of the propagation of supports was used instead of the finite spatial velocity of propagation of singularities. The next paper V. Ivrii [14] and all consecutive papers were using semiclassical asymptotics in the regular zone and variational estimates in the singular zone and were based on the slightly more complicated (and essentially more universal and flexible) proof presented here.

In all my papers and in Part V just the Rozenblioum spectral estimate (see G. Rozenblioum [2]) was used. Sometimes even the earlier Birman–Solomyak estimate (see M. S. Birman & M. Z. Solomyak [3–6]) is sufficient.

But we do not want to "close the door" to using other estimates and hence a more abstract form of the basic theorems is presented.

Note that in the special case of the Schrödinger operator the Rozenblioum estimate is the famous Lieb–Cwikel–Rozenblioum estimate, at least five different proofs of which are known (J. Conlon [1], M. Cwikel [1], E. Lieb [1], P. Li & S. T. Yau [1], G. Rozenblioum [2].

Essential in these estimates are Dirichlet boundary conditions. It is well known that Neumann Laplacian could differ drastically from Dirichlet Laplacian in the domains which are not "bounded domains with smooth boundaries". This difference will be exploited in Section 12.7. But in more general setting not just asymptotics and estimates but the whole character of the spectrum could be different. What we have in Appendix 9.A.1 is the only patch. We can think about more systematic use of the methods and results of Yu. Netrusov & Yu. Safarov [1, 2].

Estimates to Riesz means were investigated in many papers, see f.e. E. H. Lieb & W. Thirring [1], L. Erdös [1], A. Laptev & T. Weidl [1], Y. Netrusov & T. Weidl [1].

Surely, in applications in which an operator depends on parameters, the auxiliary functions $\rho, \varrho, \ldots$ (and maybe γ) also depend on the parameters, and thus regular and singular zones, etc., also depend on the parameters but constants do not.

Chapter 10

Estimates of the Spectrum in the Interval

In this chapter we obtain upper and lower estimates for

$$\mathsf{N}(A;\tau_1,\tau_2) := \operatorname{Tr} E(A;\tau_1,\tau_2) \tag{10.0.1}$$

with $-\infty < \tau_1 < \tau_2 < \infty$.

Here $E(A;\tau_1,\tau_2)$ is the spectral projector corresponding to the *open interval* (τ_1,τ_2) rather than to more standard semi-open interval $[\tau_1,\tau_2)$ as before[1] and $\mathsf{N}(A;\tau_1,\tau_2)$ is number of eigenvalues in (τ_1,τ_2) counting their multiplicities, unless $(\tau_1,\tau_2)\cap \operatorname{Spec}_{\mathsf{ess}}(A)\neq\emptyset$ in which case $\mathsf{N}(A;\tau_1,\tau_2)=\infty$.

These results are useful in the case when $\mathsf{N}^-(A-\tau_1)=\infty$ and also $\mathsf{N}^-(-A+\tau_2)=\infty$. Our analysis is very similar to the analysis of Chapter 9 but here an important role is played by Birman-Schwinger principle. It is rather strange that this extremely powerful and simple tool has been used much less than it should be in the papers devoted to spectral asymptotics.

Further, in this chapter we do not consider Riesz means-like mollifications with respect to spectral parameters[2].

[1] This adjustment allows us to cover the case $\tau_1\in \operatorname{Spec}_{\mathsf{ess}}(A)$.

[2] Riesz means themselves are usually infinite.

V. Ivrii, *Microlocal Analysis, Sharp Spectral Asymptotics and Applications II*, https://doi.org/10.1007/978-3-030-30541-3_10

10.1 Statement of the Main Theorems

10.1.1 Preliminary Remarks

First of all let us just mention a series of results which follows trivially from the results of the previous chapter applied to operators $A-(\tau_1+0)$ and $A-\tau_2$ with corresponding functions J_1 and J_2, and from the formula

$$\mathsf{N}(A;\tau_1,\tau_2)=\mathsf{N}^-(A-\tau_2)-\mathsf{N}^-(A-(\tau_1+0)). \tag{10.1.1}$$

These results do not work, however, in the numerous cases when there is an indefiniteness of the form "$\infty-\infty$" in the right-hand expression.

Similarly, replacing A, τ_1 and τ_2 by $-A$, $-\tau_2$ and $-\tau_1$ respectively we could apply formula

$$\mathsf{N}(A;\tau_1,\tau_2)=\mathsf{N}^-(-A+\tau_1)-\mathsf{N}^-(-A+(\tau_2-0)) \tag{10.1.2}$$

but again these results do not work, however, in the numerous cases when there is an indefiniteness of the form "$\infty-\infty$" in the right-hand expression.

Formula

$$\mathsf{N}(A;\tau_1,\tau_2)=\mathsf{N}^-\bigl((2A-(\tau_2+\tau_1))^2-(\tau_2-\tau_1)^2\bigr) \tag{10.1.3}$$

may be more useful than both (10.1.1) and (10.1.2) but it may ignore that levels τ_1 and τ_2 are not "of equal rights".

Further, let us mention some results which follow trivially from Parts II and III: if operators $A-\tau_1$ and $A-\tau_2$ satisfy assumptions of the previous chapter with $J_1=J_2=I$ (this is essential only in the restrictions to trajectories) and with the auxiliary functions $\gamma_{(1)}$, $\rho_{(1)}$, $\varrho_{(1)1}$, ... and $\gamma_{(2)}$, $\rho_{(2)}$, $\varrho_{(2)1}$, ... respectively and if in both cases $X_{\mathsf{sing}(1)}=X_{\mathsf{sing}(2)}=\emptyset$ then under additional restriction $\gamma_{(1)}=\gamma_{(2)}=\gamma$, $\rho_{(1)}=\rho_{(2)}=\rho$, ... we can apply a γ-admissible partition of unity for

$$\mathsf{N}(A;\tau_1,\tau_2)=\int \operatorname{tr}\bigl(e(x,x,\tau_1,\tau_2)\bigr)\,dx \tag{10.1.4}$$

and obtain estimate

$$|\mathsf{N}(A;\tau_1,\tau_2)-\mathcal{N}(\tau_1,\tau_2)|\le C\bigl(R_{(1)}+R_{(2)}\bigr) \tag{10.1.5}$$

where $R_{(i)}$ are remainder estimates for operators $(A-\tau_i)$ while $\mathcal{N}$ is a main part of asymptotics due to Part II ; writing $\mathcal{N}(\tau_1,\tau_2)=\mathcal{N}_{(2)}-\mathcal{N}_{(1)}$ we may get indefiniteness of the form "$\infty - \infty$". Recall that remainder estimates were corresponding to the energy levels rather than to the error $|\mathsf{N} - \mathcal{N}|$ itself.

Note that again $e(.,.,\tau_1,\tau_2)$ is a Schwartz kernel of the modified spectral projector $E(A;\tau_1,\tau_2)$ but as one can notice easily the results of Parts II and III are equally valid for intervals (τ_1,τ_2), $(\tau_1,\tau_2]$, $[\tau_1,\tau_2)$ and $[\tau_1,\tau_2]$.

Furthermore, restriction $\gamma_{(1)} = \gamma_{(2)} = \gamma$, $\rho_{(1)} = \rho_{(2)} = \rho$, ... in this analysis is not necessary.

Indeed, let us take a γ-admissible partition of unity with $\gamma = \max(\gamma_{(1)},\gamma_{(2)})$. Then for every element of this partition we have an estimate like

$$|\int \frac{1}{L}\phi(\frac{\tau-\tau_i}{L})\Bigl(\int_X \operatorname{tr}(e(x,x,\tau,\tau_i))\psi(x)\,dx - \tilde{\mathcal{N}}_\psi(\tau,\tau_i)\Bigr)\,d\tau| \le CR_{(i),\psi} \tag{10.1.6}$$

where $\tilde{\mathcal{N}}_\psi$ may also contain extra terms in comparison with $\mathcal{N}_\psi$ as L is large enough.

Estimate (10.1.6) is definitely correct as $\gamma_{(i)} \ge \epsilon_0\gamma$ on the given element; if this condition is violated there one can make an additional $\gamma_{(i)}$-admissible subpartition on this element, apply on every element of this subpartition results of Parts II and III and then sum.

Note that as $L \ge |\tau_2 - \tau_1|$ we can replace there τ_i by $\bar{\tau} = \frac{1}{2}(\tau_1+\tau_2)$. Then adding these estimates with the subtraction of the estimated quantities we arrive to

$$\begin{aligned}|\int_X &\operatorname{tr}(e(x,x,\tau_1,\tau_2))\psi(x)\,dx-\\ &\int \frac{1}{L}\phi(\frac{\tau-\bar{\tau}}{L})\Bigl(\tilde{\mathcal{N}}_\psi(\tau_1,\tau) - \tilde{\mathcal{N}}_\psi(\tau_2,\tau)\Bigr)\,d\tau| \le CR_{(1),\psi} + CR_{(2),\psi}\end{aligned} \tag{10.1.7}$$

where we assumed that $\int \phi(\tau)\,d\tau = 1$ and used equality $E(\tau_1,\tau) - E(\tau_2,\tau) = E(\tau_1,\tau_2)$.

Note that

$$\tilde{\mathcal{N}}_\psi(\tau_1,\tau) - \tilde{\mathcal{N}}_\psi(\tau_2,\tau) = \tilde{\mathcal{N}}_\psi(\tau_1,\tau_2) = \mathcal{N}_\psi(\tau_1,\tau_2). \tag{10.1.8}$$

Really, the first equality is due to construction of $\tilde{\mathcal{N}}$. The second equality is true since if $|\tau_2 - \tau_1| \ge C_0 \max(\varrho_{(1)} + \varrho_{(2)})$ then $\tilde{\mathcal{N}}_\psi(\tau_1',\tau_2') = 0$ as

$$\tau_1 + C_0\varrho_{(1)} \le \tau_1' \le \tau_2' \le \tau_2 - C_0\varrho_{(2)}$$

for sufficiently large C_0.

Now (10.1.5) follows from (10.1.7), (10.1.8) after summation over partition.

The results of this type are not very exciting because in the real situations condition $X_{\mathsf{sing}(1)} = X_{\mathsf{sing}(2)} = \emptyset$ is not necessarily fulfilled and that in the conditions on Hamiltonian trajectories restriction $J_1 = J_2 = I$ is not necessarily natural.

In order to circumvent the first restriction we can apply the arguments of Section 9.1 (assuming that $X_{\mathsf{reg}(1)} = X_{\mathsf{reg}(2)} = X_{\mathsf{reg}}$ and $X_{\mathsf{sing}(1)} = X_{\mathsf{sing}(2)} = X_{\mathsf{sing}}$ in the upper estimates; one picks up $J_1 = J_2 = I$). Therefore the following statement holds:

Theorem 10.1.1. *Let the operators $(A - \tau_i)$ satisfy assumptions of Theorem 9.1.7 in $X_{\mathsf{reg}(i)}$ with $J_i = \lambda I$ (where λ is a constant parameter) and with auxiliary functions $\gamma_{(i)}, \rho_{(i)}, \dots$ instead of $\gamma, \rho, \dots$; conditions depending on j should be fulfilled for both $i = 1$ and $i = 2$.*

Let $\tau_1 < \tau_2$ and $X_{\mathsf{reg}(1)} = X_{\mathsf{reg}(2)} = X_{\mathsf{reg}}$. Then

(i) Estimate

$$\mathsf{N}(A; \tau_1, \tau_2) \geq \mathcal{N}(A; \tau_1, \tau_2) - C\bigl(R_{*(1)} + R_{*(2)}\bigr) \tag{10.1.9}$$

holds where $\mathcal{N}(A; \tau_1, \tau_2)$ is prescribed by Parts II and III and $R_{*(1)}$ and $R_{*(2)}$ are taken from the lower estimates for $\mathsf{N}^-(A - \tau_1)$ and $\mathsf{N}^-(A - \tau_2)$ respectively, obtained in the Theorem 9.1.7 in which the conditions were fulfilled.

(ii) Moreover, let $X_{\mathsf{sing}(1)} = X_{\mathsf{sing}(2)} = X_{\mathsf{sing}}$ and let assumptions (9.1.24)–(9.1.25) be fulfilled in $X_{\mathsf{reg}} \cap X_{\mathsf{sing}}$ with $\varrho_{(i)0} = \lambda$; further, let condition (9.1.22) be fulfilled with inequality (9.1.23) replaced by inequality

$$\mathsf{N}(A'; \tau_1 - t\lambda, \tau_2 + t\lambda) \leq R_0(1 + t)^l \qquad \forall t \geq 0 \tag{10.1.10}$$

with some (arbitrarily large) exponent l [3]*. Then estimate*

$$\mathsf{N}(A; \tau_1, \tau_2) \leq \mathcal{N}(A; \tau_1, \tau_2) + C\bigl(R^*_{(1)} + R^*_{(2)}\bigr) \tag{10.1.11}$$

holds with $R^*_{(1)}$ and $R^*_{(2)}$ taken from the upper estimates for $\mathsf{N}^-(A - \tau_1)$ and $\mathsf{N}^-(A - \tau_2)$ respectively, and these estimates are not less than CR_0.

[3] We denote the auxiliary operator coinciding with A in the $\epsilon\gamma$-vicinity of $X_{\mathsf{reg}} \setminus X_{\mathsf{sing}}$ by A' instead of $\boldsymbol{A}'$ because $J = I$.

Remark 10.1.2. This theorem does not follow from Theorem 9.1.7 because $\mathcal{N}(A;\tau_1,\tau_2)$, $R_{(1)}$ and $R_{(2)}$ can be finite and $\mathcal{N}^-(A-\tau_i)$ can be infinite at the same time. Usually this occurs in the case when $\tau_1<\tau_2$ belong to the same gap in the semiclassical approximation to the spectrum for $|x|\ge C_0$; examples provided by the Dirac operator be given in Subsection 11.1.3.

Similar examples for the Schrödinger and Dirac operators with constant magnetic field in dimension 2 and higher even dimensions will be provided in Parts VI–X.

10.1.2 Birman-Schwinger Principle

Main Formula

In order to obtain less trivial results let us apply the following identity which is trivial due to *Birman-Schwinger principle:*

$$\mathsf{N}(A;\tau_1,\tau_2)=\mathsf{N}(A-tJ;\tau_1,\tau_2)+\\ \mathsf{N}\bigl(J^{-1}(A-\tau_1);0,t\bigr)-\mathsf{N}\bigl(J^{-1}(A-\tau_2);0,t\bigr) \tag{10.1.12}$$

provided $\tau_1<\tau_2$, $t>0$ and J is a positive definite operator such that

$$J^{-1}(A-\tau_i) \text{ are self-adjoint operators in the space } \mathcal{H}=\mathcal{H}_J. \tag{10.1.13}$$

As we will prove in Section 10.2 it is sufficient to assume that J is a symmetric, positive[4], bounded and A-compact[5] operator. We will justify formula (10.1.12) in Section 10.2. Then this formula (10.1.12) is also valid for $t<0$ where we define

$$\mathsf{N}(B;0,t):=-\mathsf{N}(B;0,-t) \qquad \text{as } t<0. \tag{10.1.15}$$

Then formula (10.1.12) yields more general formula

$$\mathsf{N}(A;\tau_1,\tau_2)=\mathsf{N}(A-t_1J_1-\ldots-t_nJ_n;\tau_1,\tau_2)-\\ \sum_{1\le k\le n}\mathsf{N}\bigl(J_k^{-1}(A-t_1J_1-\ldots-t_{k-1}J_{k-1}-\tau);0,t_k\bigr)\Bigr|_{\tau=\tau_1}^{\tau=\tau_2} \tag{10.1.16}$$

---

[4] I.e. $(Jv,v)>0\quad \forall u\ne 0$.

[5] I.e. $J(A+i)^{-1}$ is a compact operator. This condition is stronger than the standard definition

$$\|Jv\|\le\epsilon\|Av\|+C_\epsilon\|v\| \qquad \forall v\in\mathfrak{D}(A)\qquad\forall\epsilon>0 \tag{10.1.14}$$

which will be also useful.

where we assume that

(10.1.17) All operators J_k ($k = 1, \dots, n$) are symmetric, positive definite, bounded and A-compact.

Artificial energy levels in (10.1.16) may spoil our estimate; to overcome this we will mollify with respect to them arriving to the following final formula which will be used:

$$\begin{aligned}(10.1.18)\quad & \mathsf{N}(A;\tau_1,\tau_2) = \\ & \int_{\mathbb{R}^n} \mathsf{N}(A - t_1J_1 - \dots - t_nJ_n; \tau_1,\tau_2)\phi_1(t_1)\cdots\phi_n(t_n)\,dt_1\cdots dt_n - \\ & \sum_{1\le k\le n}\int_{\mathbb{R}^k} \mathsf{N}(J_k^{-1}(A - t_1J_1 - \dots - t_{k-1}J_{k-1} - \tau); 0, t_k)\Bigr|_{\tau=\tau_1}^{\tau=\tau_2} \times \\ & \qquad\qquad \phi_1(t_1)\cdots\phi_k(t_k)\,dt_1\cdots dt_k\end{aligned}$$

provided

$$\phi_k \in \mathscr{C}_0^K(\mathbb{R}), \quad \int_{\mathbb{R}} \phi_k(t)dt = 1 \qquad \forall k. \tag{10.1.19}$$

Partition and Scaling

Let $\{\psi_j^2\}_{j\ge 0}$ be a partition of unity of the same type as in Chapter 9; then the right-hand expression of (10.1.18) is a sum of terms of the type

$$\int_{\mathbb{R}^n} \operatorname{tr}(\Gamma\psi^2 e_{A-t_1J_1-\dots-t_nJ_n})(\tau_1,\tau_2)\phi_1(t_1)\cdots\phi_n(t_n)\,dt_1\cdots dt_n \tag{10.1.20$_{n+1}$}$$

and

$$\begin{aligned}(10.1.21)_k\quad \mp\int_{\mathbb{R}^k} & \operatorname{tr}(\Gamma\psi^2 e_{J_k^{-1}(A-t_1J_1-\dots-t_{k-1}J_{k-1}-\tau)})(0,t_k)\times \\ & \qquad \phi_1(t_1)\cdots\phi_k(t_k)\,dt_1\dots dt_k\end{aligned}$$

with $k = 1, \dots, n$.

Remark 10.1.3. (i) In these expressions for the sake of simplicity we skip the subscript j of ψ and also the subscript $i = 1, 2$ of τ in terms with $k \le n$; in these terms $\tau = \tau_{1,2}$ correspond to signs $\mp$.

(ii) Obviously we can take different partitions of unity for different $k = 1, \dots, n+1$ and for different $i = 1,2$ if $k \le n$.

Let us assume that J_k are Hermitian positive definite matrices (bounded and A-compact as operators) and that in $\mathcal{B}(y, \gamma(y), X)$

$$|D^\alpha J_k| \le \varrho'_k \gamma^{-|\alpha|}, \qquad |J_k^{-1}| \le \varrho'^{-1}_k \qquad \forall \alpha : |\alpha| \le K \tag{10.1.22}$$

with appropriate ϱ'_k and $\gamma = \gamma_{(i)}$, $i = 1,2$. Later ϱ'_k will be scaling functions. These conditions must be fulfilled in addition to all "standard" conditions which will be formulated later and invoking $\gamma = \gamma_{(i)}$, $\rho = \rho_{(i)}$, $\varrho = \varrho_{(i)}$ and possibly $\varrho_l = \varrho_{l(i)}$, associated with the boundary conditions.

General Operators. Consider first term $(10.1.20)_{n+1}$. Then $h_{\mathsf{eff}} = (\rho\gamma)^{-1}$ is an effective semiclassical parameter. Let us note first of all that after rescaling we get the mollification parameter $L_{\mathsf{eff}} = \varrho'_k \varrho^{-1}$ for t_k and our results of Section 6.5 on spectral asymptotics with the mollification[6] yield that we have a complete[7] asymptotics of term in question provided $L_{\mathsf{eff}} \ge h_{\mathsf{eff}}^{\kappa-\delta}$ i.e.

$$\max_{r\le k} \varrho'_r \ge \varrho h^{\kappa-\delta}, \qquad h = (\rho\gamma)^{-1}, \qquad \text{with } \kappa = \frac{1}{2} \tag*{$(10.1.23)_k$}$$

where *here* $k = n$ and arbitrarily small exponent $\delta > 0$. Further, in the scalar case we can replace exponent $(\frac{1}{2} - \delta)$ by $(1 - \delta)$ i.e. take $\kappa = 1$ in condition $(10.1.23)_k$.

Moreover, for $k \ge 2$ we have a complete asymptotics of $(10.1.21)_k$ provided condition $(10.1.23)_{k-1}$ is fulfilled; here do not need microhyperbolicity assumption with respect to (x, ξ) because these problems are automatically microhyperbolic with respect to $t_1, \dots, t_k$ and condition $(10.1.23)_k$ means that there is sufficient mollification with respect to t_r with appropriate $r \le k-1$ is necessary.

Finally, consider term $(10.1.21)_1$. Then under condition $(10.1.23)_1$ we need microhyperbolicity condition[8] as $t_1 = 0$ and we do not have complete asymptotics. So let us now assume that

(10.1.24) For $\tau = \tau_i$ $(i = 1,2)$ microhyperbolicity condition[8] is fulfilled for the original problem $(A - \tau_i, B)$.

---

[6] Which obviously hold near boundary as well.

[7] I.e., modulo $O(h^s)$ with arbitrary s.

[8] Or ξ-microhyperbolicity for "rough" operators in the sense of Sections 4.6 and 7.5.

Then modulo $O(h^{1-d})$ we have asymptotics of the term in question.

On the other hand, if condition $(10.1.23)_1$ is violated then this microhyperbolicity condition is fulfilled for the problem $(A - t_1 J_1 - \tau_i, B)$ $(i = 1, 2)$ and we have asymptotics of the term $(10.1.21)_2$ modulo $O(h^{1-d})$ also. Repeating these arguments we conclude that under the microhyperbolicity condition we have asymptotics modulo $O(h^{1-d})$ of all the terms $(10.1.21)_k$ with $k = 1, \dots, n$ (and we have no need to check conditions $(10.1.23)_k$.

Furthermore, the same is true for the modified term $(10.1.20)_{n+1}$ now defined by

$$(10.1.25)_{n+1} \quad \frac{1}{L}\int_{\mathbb{R}^{n+1}} \operatorname{tr}\bigl(\Gamma\psi^2 e_{A-t_1J_1-\ldots-t_nJ_n}\bigr)(\tau',\tau_i)\times \\ \phi_0\bigl(\frac{1}{L}(\tau'-\tau_i)\bigr)\phi_1(t_1)\cdots\phi_n(t_n)\,d\tau'dt_1\cdots dt_n$$

with $L \ge h^{\frac{1}{2}-\delta}$ (or $L \ge h^{1-\delta}$ in the scalar case).

Moreover, the same assertion is true if the microhyperbolicity condition for $(A-\tau_i, B)$ is replaced by the ellipticity condition for the operator $(A-\tau_i)$; this condition does not include the assumption that the boundary value problem is elliptic (or microhyperbolic).

Finally, if we assume that the boundary value problem $(A - \tau_i, B)$ is also elliptic then we can obtain all the terms $(10.1.21)_k$ $(k = 1, \dots, n)$ and $(10.1.25)_{n+1}$ modulo $O(h^s)$.

Dirac Operator. Let us now treat the Dirac operator. Let us recall that according to Subsections 5.3.2 (smooth case) and 5.4.3 ("rough case") Dirac operator

$$(10.1.26) \qquad A = \sum_{j,l} \frac{1}{2}\sigma_l(\omega^{jl}P_j + P_j\omega^{jl}) + \sigma_0 m + V\cdot I, \qquad P_j = D_j - V_j$$

the usual smoothness (or "roughness") conditions are fulfilled but with

$$(10.1.27) \qquad \varrho = \min\bigl(\frac{\rho^2}{m},\rho\bigr)$$

$V \pm m$ instead of V and weakened condition to V_j:

$$(10.1.28) \qquad |F_{jl}| \le c\rho\gamma^{-1}, \qquad F_{jl} = F_{jk} = \partial_l V_j - \partial_j V_l$$

instead of $|V_j| \le c\rho$. Corresponding conditions should be fulfilled for their derivatives.

Also Dirac operator is accessible to the scaling technique and one does not need any microhyperbolicity condition as $d \ge 3$, or even $d = 2$ but under some smoothness or weakened nondegeneracy conditions, and similarly for $d = 1$.

Remark 10.1.4. We need matrix functions J_k such that operators

$$J_k^{-\frac{1}{2}}\bigl(A - t_1 J_1 - \ldots - t_{k-1}J_{k-1} - \tau_i\bigr)J_k^{-\frac{1}{2}} \tag{10.1.29}$$

are generalized Dirac operators (5.3.71) i.e. that matrices

$$W_k = -t_1 J_1 - \ldots - t_{k-1}J_{k-1} \tag{10.1.30}$$

satisfy after rescaling $(5.3.77)_5$ in the smooth case or similar condition in the "rough" case.

Scalar-Like Operators. Finally, let us treat the arbitrary matrix operator and let us assume that (after normalization, i.e., multiplication by an appropriate number if necessary) this operator satisfies the following condition:

(10.1.31) Interval $[\tau_1 - \epsilon_0, \tau_2 + \epsilon_0]$ contains no eigenvalues of the principal symbol of A different from the eigenvalue $\lambda(x, \xi)$ of constant multiplicity.

Then this condition is also fulfilled for operator $J_k^{-1}A$ provided one of the following three conditions is fulfilled:

$(10.1.32)_1$ λ is a simple eigenvalue of the principal symbol of A;

$(10.1.32)_2$ J_k is scalar;

$(10.1.32)_3$ The restriction of J_k to $\mathsf{Ker}(a^0 - \lambda)$ is scalar (a^0 is the principal symbol of A).

Remark 10.1.5. (i) Condition $(10.1.32)_3$ is more general than $(10.1.32)_{1,2}$ but it is difficult to fulfill $(10.1.32)_3$ outside of the framework of $(10.1.32)_{1,2}$ because J_k does not depend on ξ and $(a^0 - \lambda)$ depends on ξ.

(ii) Application of the matrix J_k which does not satisfy one of the conditions $(10.1.32)_{1-3}$ is possible only if before there was mollification with parameter $L_{\mathsf{eff}} \ge h_{\mathsf{eff}}^{\frac{1}{2}-\delta}$.

Thus, let us assume that one of conditions $(10.1.32)_{1-3}$ is fulfilled. Assume that second derivatives of coefficients are bounded (after rescaling); then it is true for λ as well. Let us make a γ-admissible partition of unity with

$$\gamma = \epsilon_0(|\lambda - \tau_i| + |\nabla\lambda|^2)^{\frac{1}{2}} + h^{\frac{1}{2}}. \tag{10.1.33}$$

Then for $\gamma \ge 2h^{\frac{1}{2}}$ microhyperbolicity condition is fulfilled for the original operator after the rescaling-multiplication procedure; therefore in this case our remainder estimate is justified. For $\gamma \le h^{\frac{1}{2}-\delta'}$ the condition $L' \ge h'^{\frac{1}{2}-\delta'}$ is fulfilled where $\delta' = \delta'(\delta) > 0$ is small enough, h' and L' are effective values of h and L after the dilatation-multiplication procedure and we assume that $L \ge h^{1-\delta}$. So in this we can replace condition $L \ge h^{\frac{1}{2}-\delta}$ by $L \ge h^{1-\delta}$ and thus we can take $\kappa = 1$ in $(10.1.23)_k$.

On the other hand, for rough operators one needs ξ-microhyperbolicity and thus

$$\gamma = \epsilon_0(|\lambda - \tau_i| + |\nabla_\xi\lambda|^2)^{\frac{1}{2}} + h^{\frac{1}{2}}. \tag{10.1.34}$$

which leads us to $\kappa = \frac{2}{3}$ in $(10.1.23)_k$. In a more regular case we need some $\kappa \in (\frac{2}{3}, 1)$ depending on regularity.

Finally, we obtain that in the general case the microhyperbolicity condition can be replaced by some non-degeneracy condition of Sections 5.2, 5.4 and

$(10.1.35)_p$ For every $k = 1, \dots, p$ the matrix J_k satisfies one of conditions $(10.1.32)_{1-3}$ and there exists $k \le p$ such that $(10.1.23)_k$ holds with $\kappa \in [\frac{2}{3}, 1]$ defined by regularity

with some $p \le n+1$.

Moreover, let conditions (9.1.31) and $(9.1.32)_1$ be fulfilled. Then $(10.1.35)_p$ is reduced to the condition

$(10.1.36)_p$ There exists $k \le p$ such that $\varrho'_k \ge \varrho h^{\kappa-\delta} = \varrho(\rho\gamma)^{\delta-\kappa}$.

So far contribution of the ball $\mathcal{B}(y, \gamma(y), X)$ to the remainder is $O(h_{\mathsf{eff}}^{1-d})$. Furthermore, under appropriate conditions to the trajectories contribution of the ball $\mathcal{B}(y, \gamma(y), X)$ to the remainder estimate is $O(h_{\mathsf{eff}}^{1-d+\delta})$ with some small exponent $\delta > 0$.

Recall that while geometry of trajectories do not depend on the choice of J_k but the temp of movement depends on it. Not to invoke multiple matrices we should assume that condition $(10.1.23)_1$ is fulfilled.

Remark 10.1.6. In the lacunary zone, which appears in the case of periodic trajectories, restrictions of the type $\varrho'_k \le \eta$ and $\varrho'_k \ge \eta h^{1-\delta}$ are also possible; here $\eta \in (h', h)$ is a width of the spectral gap.

Calculations in the Regular zZone

Let us now treat the terms of the derived asymptotics. For the sake of the notational simplicity we do not write the mollification on $t_1, \dots, t_k$. Then expression $(10.1.21)_k$ with $k = 1, \dots, n$ generates

$$(10.1.37)_k \qquad \int_0^{t_k} \operatorname{Tr}\Bigl(\operatorname{Res}_{\mathbb{R}}\bigl(z - J_k^{-1}(A - t_1 J_1 - \dots - t_{k-1}J_{k-1} - \tau)\bigr)^{-1}\psi\Bigr)\, dz$$

with $\tau = \tau_1, \tau_2$ and expression $(10.1.20)_{n+1}$ generates

$$(10.1.38)_{n+1} \qquad \int_{\tau_1}^{\tau_2} \operatorname{Tr}\bigl(\operatorname{Res}_{\mathbb{R}}(z - A + t_1 J_1 + \dots + t_n J_n)^{-1}\psi\bigr)\, dz$$

where $\operatorname{Res}_{\mathbb{R}}$ is the difference of the limits from the upper and lower complex half-planes with respect to z, we skip numerical factors depending on d only and G^{-1} means the formal resolvent of G (in the context of the formal calculus of pseudodifferential operators) derived according to the rules of Parts II and III.

We need to treat the case $n = 1$ only: the general case can be reduced to this case if we replace A by $A - t_1 J_1 - \dots - t_{k-1}J_{k-1}$ and sum with respect to k. Therefore we can skip the subscript "1" of t and J.

Then term $(10.1.37)_1$ is replaced by

$$(10.1.39)_1 \qquad \int_0^{t} \operatorname{Tr}\bigl(\operatorname{Res}_{\mathbb{R}}(Jz - A + \tau)^{-1} J\psi\bigr)\, dz$$

(it is easy to see that the equality $(z - J^{-1}A)^{-1} = (Jz - A)^{-1}J$ holds for formal resolvents also) and term $(10.1.38)_2$ is replaced by

$$\int_{\tau_1}^{\tau_2} \operatorname{Tr}\bigl(\operatorname{Res}_{\mathbb{R}}(z - A + tJ)^{-1}\psi\bigr)\, dz. \tag{10.1.40$_2$}$$

Denote by $\Theta(t)$ and $\Theta'_i(t)$ be expressions $(10.1.40)_2$ and $(10.1.39)_1$ at $\tau = \tau_i$ respectively. We claim that

$$\begin{aligned}\Theta(0) - \Theta(t) - \Theta'_2(t) + \Theta'_1(t) &= \\ -\int_0^t \int_{\tau_1}^{\tau_2} &\operatorname{Tr}\bigl(\operatorname{Res}_{\mathbb{R}}(z - A + tJ)^{-2}J(z - A + tJ)^{-1}[A, \psi]\bigr)\, dz\end{aligned} \tag{10.1.41}$$

Indeed, the derivative with respect to t of the left-hand expression is equal to

$$-\int_{\tau_1}^{\tau_2} \operatorname{Tr}\bigl(\operatorname{Res}_{\mathbb{R}}(z - A + tJ)^{-2}J(z - A + tJ)^{-1}[A, \psi]\bigr)\, dz$$

because of the chain of equalities

$$\begin{aligned}&\partial_t \operatorname{Tr}\bigl(\operatorname{Res}_{\mathbb{R}}(z - A + tJ)^{-1}\psi\bigr) = \\ &\qquad - \operatorname{Tr}\bigl(\operatorname{Res}_{\mathbb{R}}(z - A + tJ)^{-1}J(z - A + tJ)^{-1}\psi\bigr) = \\ &- \operatorname{Tr}\bigl(\operatorname{Res}_{\mathbb{R}}(z - A + tJ)^{-2}J\psi\bigr) - \operatorname{Tr}\bigl(\operatorname{Res}_{\mathbb{R}}(z - A + tJ)^{-2}J(z - A + tJ)^{-1}[A, \psi]\bigr)\end{aligned}$$

which follow from the properties of the trace and from the fact that J and ψ commute. The first term in the right-hand expression is equal to

$$\partial_z \operatorname{Tr}\bigl(\operatorname{Res}_{\mathbb{R}}(z - A + tJ)^{-1}J\psi\bigr)$$

and after integration with respect to z it reduces with $\partial_t(-\Theta'_2(t) + \Theta'_1(t))$. So, derivatives with respect to t of both sides of (10.1.41) and for $t = 0$ both expressions vanish.

In the framework of condition $(10.1.35)_1$ these calculations are justified by microhyperbolicity for small ϱ'_1 and by microhyperbolicity and mollifications otherwise; justifications in the frameworks of other conditions are similarly.

Finally, after adding all terms due to $(10.1.21)_k$ $(k = 1, \dots, n)$ and $(10.1.20)_{n+1}$ we obtain that the right-hand expression of (10.1.18) contains main parts of asymptotics of $(\Gamma\psi e_A)(\tau_1, \tau_2)$ described in Parts II and III plus a remainder estimate also described there plus extra terms of the form

$$\sum_{1\le r\le s} h^{r-d} \int (\mathcal{L}_r\psi)\, dx.$$

Here $\mathcal{L}_r$ are differential operators such that $\mathcal{L}_r 1 = 0$. Let us note that these operators $\mathcal{L}_r$ depend uniquely on A, $J_1, \dots, J_n$, $\phi_1, \dots, \phi_n$ and hence after summation with respect to partition of unity all these extra terms disappear.

All these assertions have been proven provided $\varrho'_k \le \varrho$ for all $k = 1, \dots, n$. On the other hand, as it is true for $k < k_0$ only and $\varrho'_k \ge \varrho$, one can replace in all the terms with $k \ge k_0 + 1$ h by $h' = h(\varrho/\varrho'_k)^{\frac{1}{m}}$, etc.

Therefore after summation with respect to partition of unity we obtain that only terms supported in $X_{\mathsf{reg}} \cap X_{\mathsf{sing}}$ are not completely reduced, and the remaining terms do not exceed

$$Ch^{1-d}\Bigl(1 + \frac{(\varrho'_1 + \dots + \varrho'_n)}{\varrho}\Bigr)^{\frac{(d-1)}{m}}. \tag{10.1.42}$$

Finally, let us note that if in some ball $\mathcal{B}(y, \gamma(y), X)$ condition

(10.1.43) τ_i lies in the spectral gap of the problem $(A - t'_1 J_1 - \dots - t'_k J_k, B)$ for all $t'_k \in [0, t_k]$ (or $t'_k \in [t_k, 0]$ for $t_k < 0$) for all $k = 1, \dots, n$ and for all $t_k \in \mathsf{supp}(\phi_k)$

is fulfilled then all the operators $\mathcal{L}_r$ vanish in this ball.

Analysis in the Singular Zone

Let us now treat the zone X_{sing}. First of all we must assume that in every ball in $X_{\mathsf{reg}} \cap X_{\mathsf{sing}} \cap \{\rho\gamma \ge c\}$ [9)] either an ellipticity condition or condition

$$\varrho'_1 \ge \varrho h \tag{10.1.44}$$

are fulfilled.

Furthermore, we need to introduce condition similar to (9.1.22). Namely, let us assume that

(10.1.45) There exist a self-adjoint operator A' and symmetric operators J'_k coinciding with A and J_k respectively in X_{sing} and such that for all k the operators J'_k are A'-compact and

$$\begin{gathered}\mathsf{N}\bigl({J'_k}^{-1}(A' - \tau_i - t_1 J'_1 - \dots - t_{k-1} J'_{k-1}); -t, t\bigr) \le R_0(|t| + 1)^l \\ \forall k \quad \forall t_j \in \mathsf{supp}(\phi_j) \quad (j = 1, \dots, k-1) \quad \forall t \in \mathbb{R}.\end{gathered} \tag{10.1.46}$$

---

9) In the general case; modifications will be given below for the Schrödinger and Dirac operators.

Moreover, we assume that

(10.1.46)* condition (10.1.46) is fulfilled for some sequences $\tau_{i,n} \to \tau_i$ with $\tau_{i,n} \in [\tau_1, \tau_2] \setminus \mathsf{Spec}_{\mathsf{ess}}(A)$.

Finally, let us assume that

$$\int_{X_{\mathsf{sing}}} \mathsf{tr}\bigl(e_{A-t_1J_1-\ldots-t_nJ_n}(x,x,\tau_1,\tau_2)\bigr)\,dx \le R_0 \qquad \forall t_j \in \mathsf{supp}(\phi_j) \quad (j=1,\ldots,n). \tag{10.1.47}$$

Instead of this condition we will usually use the condition

$$\mathsf{N}(A - t_1J_1 - \ldots - t_nJ_n; \tau_1, \tau_2) \le R_0 \qquad \forall t_j \in \mathsf{supp}(\phi_j) \quad (j=1,\ldots,n). \tag{10.1.48}$$

This condition may seem to be too restrictive because it is associated with X instead of X_{sing}, but in fact this is not so.

10.1.3 Main Theorems

Now we can formulate and prove our main theorems.

General oOperators

We start from the general operators.

Theorem 10.1.7. *Let X, X_{reg}, X_{sing}, $\gamma_{(i)}$, $\rho_{(i)}$, $\varrho_{(i)j}$ and ϱ'_k satisfy conditions* (9.1.5)–(9.1.8) *for both $i=1,2$* [10].

Let operators A and B be of the forms (9.1.9) *and* (9.1.10) *respectively. Let us assume that in X_{reg}*

$(10.1.49)_1$ *Operators $\rho^m_{(i)}\varrho^{-1}_{(i)1}A$ are classically uniformly elliptic for $i=1,2$* [11];

$(10.1.49)_2$ *Boundary value problems*

$$\bigl(\rho^m_{(i)}\varrho^{-1}_{(i)1}A,\ \rho^{m_j}_{(i)}\varrho^{-1}_{(i)j}B_j,\ j=1,\ldots,\mu\bigr)$$

are classically uniformly elliptic for $i=1,2$.

[10] Here X_{reg}, X_{sing} and ϱ'_k do not depend on i.

[11] In which case $\rho^m_{(1)}\varrho^{-1}_{(1)1} \asymp \rho^m_{(2)}\varrho^{-1}_{(2)1}$ and a similar conclusion holds for the next assumption $(10.1.49)_2$.

Further, let us assume that $\tau_1 < \tau_2$ *and*

(10.1.50) *For both* $i = 1, 2$, *for every ball* $\mathcal{B}(x, \gamma^{(i)}(x), X)$ *in*

$$X_{\mathsf{reg}} \cap X_{\mathsf{sing}} \cap \{\rho_{(i)}\gamma_{(i)} \ge c\}$$

either the boundary value problem $\bigl(\varrho_{(i)1}^{-1}(A - \tau_i),\ \varrho_{(i)j}^{-1}B_j,\ j = 1, \dots, \mu\bigr)$ *is elliptic (as an* $h_{(i)}$*-boundary value problem with* $h_{(i)} = (\rho_{(i)}\gamma_{(i)})^{-1}$*) or condition* $(10.1.36)_1$ *is fulfilled*[12].

Furthermore, let us assume that J_k *are bounded and* A*-compact operators. Finally, let us assume that there exist operator* $A'_{(i)}$ *coinciding with* A *in* $X_{(i)\mathsf{sing}}$, *self-adjoint and such that for all* k *the operators* J_k *are* A'*-compact and conditions* (10.1.46)*, (10.1.48) *are fulfilled.*

Then

(i) The following estimate holds:

$$\mathsf{N}(A; \tau_1, \tau_2) \ge \mathcal{N}(A; \tau_1, \tau_2) - C\sum_{i=1,2} \bigl(R_{(i)1} + R_{(i)2} + R_{(i)3}\bigr) - CR_0 \tag{10.1.51}$$

where $\mathcal{N}$ *is derived as a semiclassical approximation in construction in* X_{reg} *for the sum of* $\mathsf{N}\bigl(J_k^{-1}(A - t_1J_1 - \dots - t_{k-1}J_{k-1}, 0, t_k\bigr)$ *with mollification, according to rules of the Part II,* $R_{(i)1}$ *is the corresponding remainder estimate,* $R_{(i)2}$ *is given by* (9.1.30) *i.e. one can define*

$$R_{(i)2} = \int_{X_{\mathsf{reg}} \cap X_{\mathsf{sing}}} \rho_{(i)}^{d-s}\gamma_{(i)}^{-s}\,dx, \tag{10.1.52}$$

$$R_{(i)3} = \int_{X_{\mathsf{reg}} \cap X_{\mathsf{sing}}} \bigl(\varrho'_1 + \dots + \varrho'_n\bigr)^{\frac{(d-1)}{m}}\varrho^{-\frac{(d-1)}{m}}\rho_{(i)}^{d-1}\gamma_{(i)}^{-1}\,dx, \tag{10.1.53}$$

(ii) On the other hand, if also condition (10.1.47) *is fulfilled then*

$$\mathsf{N}(A; \tau_1, \tau_2) \le \mathcal{N}(A; \tau_1, \tau_2) + C\sum_{i=1,2} \bigl(R_{(i)1} + R_{(i)2} + R_{(i)3}\bigr) + CR_0. \tag{10.1.54}$$

The proof which, after (10.1.12) is established, repeats almost exactly the proofs of Theorems 9.1.7 and 9.1.13 will be presented in Subsection 10.2.3.

[12] I.e., $\varrho'_1 \ge \varrho_{(i)1}h_{(i)}^{1-\delta}$ with $\delta > 0$.

Dirac Operator

In the case of the Dirac operator one can get effective expressions for $R_{j(i)}$ and deduce the following statements:

Theorem 10.1.8. *Let X, X_{reg}, X_{sing}, γ and ρ satisfy conditions* (9.1.5)–(9.1.8). *Let operator A be of the form* (10.1.26), $d \ge 2$, $\tau_1 < \tau_2$. *Let in X_{reg} let the following conditions be fulfilled:*

$$|D^\alpha \omega^{jk}| \le c\gamma^{-|\alpha|}, \tag{10.1.55}$$

$$|D^\alpha F_{jk}| \le c\rho\gamma^{-|\alpha|-1}, \tag{10.1.56}$$

$$|D^\alpha V| \le c\min(\rho, \frac{1}{m}\rho^2)\gamma^{-|\alpha|} \qquad (\alpha \ne 0) \qquad \forall \alpha : |\alpha| \le K, \tag{10.1.57}$$

and

$$(V - \tau_2 - m)_+ \le c\min(\rho, \frac{1}{m}\rho^2), \tag*{(10.1.58)$_2$}$$

$$(V - \tau_1 + m)_- \le c\min(\rho, \frac{1}{m}\rho^2), \tag*{(10.1.58)$_1$}$$

$$|\det(\omega^{jk})| \ge \epsilon_0, \tag{10.1.59}$$

$$B(x, \gamma(x)) \subset X. \tag{10.1.60}$$

Then

(i) Estimate

$$\mathsf{N}(A, \tau_1, \tau_2) \ge \mathsf{N}^{\mathsf{W}}(A; \tau_1, \tau_2) - C(R_1 + R_2) \tag{10.1.61}$$

holds with

$$R_1 = \int_{X_{\mathsf{reg}} \setminus X_{\mathsf{reg,ell}}} \rho^{d-1}\gamma^{-1}\, dx, \tag{10.1.62}$$

$$R_2 = \int_{X_{\mathsf{reg,ell}}} \rho^{d-s}\gamma^{-s}\, dx, \tag{10.1.63}$$

with arbitrarily large exponent s where $X_{\mathsf{reg,ell}}$ consists of points X_{reg} satisfying both of the following conditions

$$V - \tau_2 - m \le -\epsilon_0 \min\bigl(\rho, \frac{1}{m}\rho^2\bigr), \tag*{(10.1.64)$_2$}$$

$$-V + \tau_1 - m \le -\epsilon_0 \min\bigl(\rho, \frac{1}{m}\rho^2\bigr). \tag*{(10.1.64)$_1$}$$

(ii) Estimate

$$\mathsf{N}(A;\tau_1,\tau_2)\le \mathsf{N}^{\mathsf{W}}(A;\tau_1,\tau_2)+C\bigl(R_1+R_2+R_3+R_4+R_5\bigr) \tag{10.1.65}$$

holds with R_1, R_2 *defined above,*

$$R_3=\int_{X_{\mathsf{reg}}\cap X_{\mathsf{sing}}}\rho^d\,dx+\left\{\begin{aligned} &C\int_{X_{\mathsf{sing}}}\tilde\rho^d\,dx && \text{for } d\ge 3,\\ &C_p\Bigl(\int_{X_{\mathsf{sing}}}\tilde\rho^{2p}\,dx\Bigr)^{\frac1p}(\mathsf{mes}(X_{\mathsf{sing}}))^{1-\frac1p}\\ &&& \text{for } d=2,\ p>1,\end{aligned}\right. \tag{10.1.66}$$

$$R_4=\Bigl(\int(\rho+m)^{\delta}\chi_Y\gamma^{-d}\,dx\Bigr)^{\frac12}\Bigl(\int(\rho+m)^{-\delta}\chi_Y\gamma^{-d}\,dx\Bigr)^{\frac12} \tag{10.1.67}$$

with arbitrarily small $\delta>0$ *and we assume that*

$$\bar X_{\mathsf{sing}}\cap\partial X=\emptyset, \tag{10.1.68}$$

and in X_{sing} *conditions* $(10.1.58)_{1,2}$ *with* ρ *replaced by* $\tilde\rho$ *and* (10.1.59) *are fulfilled and also there*

$$|V_j|\le c\tilde\rho, \tag{10.1.69}$$

$$|D_j\omega^{kl}|\le c\tilde\rho. \tag{10.1.70}$$

Finally,

$$R_5=\min\Bigl(\frac{Mm^{\frac d2}}{|\tau_2-\tau_1|},\, M^{\frac12}m^{\frac d2}(\mathsf{mes}(X_{\mathsf{reg}}\cap X_{\mathsf{sing}}))^{\frac12}\Bigr), \tag{10.1.71}$$

$$M=\int_{X_{\mathsf{reg}}\cap X_{\mathsf{sing}}}\gamma^{-d}\,dx \tag{10.1.72}$$

and only those connected components of X_{sing} *where condition* $\rho\le cm$ *is violated are taken into account.*

The proof of this theorem will be given in Subsection 10.2.4.

Remark 10.1.9. (i) For $d \geq 3$ estimates (10.1.61), (10.1.65) hold with

$$R_1 = \sum_{\pm} \int_{\Sigma_\pm \cap T^*X_{\mathsf{reg}}} dxd\xi : dH_\pm + R'_1, \tag{10.1.73}$$

$$R'_1 = \int_{X_{\mathsf{reg}} \setminus X_{\mathsf{reg,ell}}} \rho^{d-1-\delta}\gamma^{-1-\delta}\, dx, \tag{10.1.74}$$

$$H_\pm = V \pm \bigl(m^2 + g^{jk}(\xi_j - V_j)(\xi_k - V_k)\bigr)^{\frac{1}{2}}, \tag{10.1.75}$$

$\Sigma_\pm = \{H_\pm = 0\}$. Moreover, thus modified estimate remains true for $d = 2$ provided either $V_j = 0$ or microhyperbolicity condition is fulfilled. Finally, under appropriate modifications (see Section 5.3.2) this estimate remains true for $d = 2$ in the general case.

(ii) One can improve R_1 in the standard way under assumptions for trajectories. While ϱ'_1 defining temp along trajectories could be selected in different ways, it affects contribution of $X_{\mathsf{reg}} \cap X_{\mathsf{sing}}$.

(iii) All these results hold for "rough" Dirac operator (see Subsection 5.4.3) as long as contribution of X_{reg} to the remainder estimate does not change.

(iv) Moreover, in the case when $X_{\mathsf{reg}} \cap X_{\mathsf{sing}}$ contains few connected components one can calculate R_4 and R_5 for every component and then sum the results. As $d = 2$ the same statement holds for R_3.

(v) There will be some improvement for R_4 (see the second term in (10.2.73)).

(vi) Dirac and Schrödinger operators with the strong magnetic field will be covered later.

(vii) The method of this chapter and the previous chapters provide estimates for $\mathsf{N}(A; \tau_1, \tau_2)$ averaged with respect to τ_1 or (and) τ_2 with improved contribution of X_{reg} to the remainder estimate.

(viii) For principally massless Dirac operator one can just take $J = I$.

Consider principally massless operator

$$A = \sum_{j,l} \frac{1}{2}\sigma_l(\omega^{jl} P_j + P_j \omega^{jl}) + V \cdot I, \qquad P_j = D_j - V_j \tag{10.1.76}$$

which is not necessarily of the form (10.1.26) with $m=0$ since matrix σ_0 may not exist.

It is not necessarily of the type (10.1.26) since σ_0 does not necessarily exist. However we can always extend σ_j by $\begin{pmatrix}\sigma_j & 0\\ 0 & \sigma_j\end{pmatrix}$ and $\sigma_0=\begin{pmatrix}0 & i\\ -i & 0\end{pmatrix}$ fits the bill. Therefore we arrive to

Corollary 10.1.10. *Let all assumptions of Theorem 10.1.8 be fulfilled with $m=0$. Then estimates* (10.1.61) *and* (10.1.65) *hold with $R_5=0$.*

10.2 Functional-Analytic Arguments and Proofs of Main Theorems

10.2.1 Self-Adjointness of the Dirac Operator

Let us begin with the self-adjointness of the (generalized) Dirac operator. First of all let us prove

Theorem 10.2.1. *Let us assume that A is an operator of the form* (10.1.26) *or* (10.1.76) *with real-valued coefficients ω^{jl}, V_j and V. Let us assume that* (10.1.59) *holds and*

(10.2.1) $\omega^{jl}\in\mathscr{C}(X)$, $D_j\omega^{jl}, V_j, V\in\mathscr{L}^p_{\mathsf{loc}}(X)$ *with $p=d$ for $d\ge 3$ and with some $p>2$ for $d=2$.*

Then the operator A with the domain

$$\mathfrak{D}(A)=\bigl\{u\in\mathscr{L}^2(X,\mathbb{H})\cap\mathscr{H}^1_{\mathsf{loc}}(X,\mathbb{H}) : Au\in\mathscr{L}^2(X,\mathbb{H})\bigr\}\tag{10.2.2}$$

is self-adjoint in $\mathscr{L}^2(X,\mathbb{H})$.

Proof. (a) Let us first observe that if $u\in\mathscr{H}^1_{\mathsf{loc}}$ then Sobolev embedding theorem yields that $u\in\mathscr{L}^q_{\mathsf{loc}}$ and

$$\|u\|_{\mathscr{L}^q(Y)}\le C_Y\|u\|_{\mathscr{H}^1(Y)}\tag{10.2.3}$$

on every compact Y where here and below $q=d/(d-2)$ for $d\ge 3$ and $q=p/(p-2)$ with arbitrary $p>2$ for $d=2$. Therefore (10.2.1) yields that $V_ju, Vu, \omega' u\in\mathscr{L}^2_{\mathsf{loc}}(x,\mathbb{H})$ where $\omega'=i\sum_j D_j\omega^{jl}$.

(b) Next let us prove that the operator A with the indicated domain is closed. Let us first assume that $\max_j \bar{\gamma}_j < \infty$ [13].

Let us rewrite V, V_j in the form $V = V' + V''$ with $\|V''\|_{\mathscr{L}^p} \le \varepsilon$, $\|V'\|_{\mathscr{L}^\infty} \le C_\varepsilon$, etc., and let us rewrite ω^{jl} in the form $\omega^{jl} = \omega^{jl\prime} + \omega^{jl\prime\prime}$ with $\|\omega^{jl\prime\prime}\|_{\mathscr{L}^\infty} \le \varepsilon$, $\|\omega^{jl\prime}\|_{\mathscr{C}^2} \le C_\varepsilon$ with arbitrarily small $\varepsilon > 0$. Then

$$\|Au - A_1 u\| \le C_0\varepsilon \|u\|_{\mathscr{H}^1} + C'_\varepsilon \|u\| \qquad \forall u \in \mathfrak{D}(A) \tag{10.2.4}$$

where $A_1 = \sum_{j,l} \omega^{jl\prime}\sigma_l hD_j$ and C_0 does not depend on ε. Moreover, properties of $\omega^{jl\prime}$ and σ_l yield that

$$\Bigl| \|A_1 u\|^2 - \int \sum_{j,k} g^{jk\prime}(D_k u)^\dagger (D_j u)\, dx \Bigr| \le C'_\varepsilon \varepsilon' \|u\|^2_{\mathscr{H}^1} + C''_{\varepsilon\varepsilon'}\|u\|^2 \qquad \forall u \in \mathfrak{D}(A)$$

where $g^{jk\prime} = \sum_l \omega^{jl\prime}\omega^{kl\prime}$ and therefore $|g^{jk} - g^{jk\prime}| \le C_0\epsilon$; here $\varepsilon' > 0$ is arbitrarily small and C'_ε does not depend on ε'. This inequality and (10.2.4) yield inequality

$$\Bigl| \|Au\|^2 - \int \sum_{j,k} g^{jk}(D_k u)^\dagger (D_j u)\, dx \Bigr| \le C_0\varepsilon \|u\|^2_{\mathscr{H}^1} + C'_\varepsilon \|u\|^2 \qquad \forall u \in \mathfrak{D}(A)$$

and then (10.1.59) yields estimate

$$\|u\|_{\mathscr{H}^1} \le C\bigl(\|Au\| + \|u\|\bigr) \qquad \forall u \in \mathfrak{D}(A). \tag{10.2.5}$$

This estimate immediately yields that the operator A is closed provided $\max_j \bar{\gamma}_j < \infty$.

Now let us consider the case $\max_j \bar{\gamma}_j = \infty$. Let $\psi \in \mathscr{C}^1_0(X)$ be a scalar function. Then $A\psi u = \psi A u + \psi' u$ with a matrix-valued function $\psi' \in \mathscr{C}_0(X)$. Then the above arguments yield the estimate

$$\|\psi u\|_{\mathscr{H}^1} \le C\bigl(\|Au\| + \|u\|\bigr) \qquad \forall u \in \mathfrak{D}(A) \tag{10.2.6}$$

where C depends on ψ. But then A is again obviously closed.

(c) Let us prove that the operator A is symmetric. This is obvious for $\max_j \bar{\gamma}_j < \infty$. For $\max_j \bar{\gamma}_j = \infty$ equality

$$(Au, v) = (u, Av) \qquad u, v \in fD(A) \tag{10.2.7}$$

[13] Recall that $X = \mathbb{R}^d/(\bar{\gamma}_1\mathbb{Z} \times \ldots \times \bar{\gamma}_d\mathbb{Z})$.

holds for compactly supported v. Let us replace v by $\psi(x't^{-1})v$ where $\psi \in \mathscr{C}_0^1(X')$ equals 1 in a neighborhood of 0 and $X' \ni x' = (x_j)_{j\in J}$ with $J = \{j, \bar{\gamma}_j = \infty\}$. For t tending to $+\infty$ we obtain the same equality (10.2.7) for arbitrary $v \in \mathfrak{D}(A)$.

(d) To prove that A is self-adjoint let us consider the problem

$$D_t u = Au, \qquad u|_{t=0} = f. \tag{10.2.8}$$

We are going to prove that for every $f \in \mathfrak{D}(A)$ there exists a solution $u \in \mathscr{C}(\mathbb{R}, \mathscr{L}^2(X, \mathbb{H})) \cap \mathscr{L}^\infty(\mathbb{R}, \mathfrak{D}(A))$ of this problem.

Let us assume first that $\max_j \bar{\gamma}_j < \infty$. Then problem (10.2.8) is obviously well-posed provided $\omega^{jl}, V_j, V \in \mathscr{C}^2$. In this case $\mathfrak{D}(A) = \mathscr{H}^1(X, \mathbb{H})$.

Let $\omega_\delta^{jl}, V_{j\delta}, V_\delta$ be sequences of mollifications of ω^{jl}, V_j, V respectively with mollifying function $\psi_\delta(x) = \delta^{-d}\psi(x\delta^{-1})$ where $\psi \in \mathscr{C}_0^\infty(\mathbb{R}^d)$ is a real-valued non-negative function with $\int \psi\, dx = 1$. Then for $\delta \to +0$, ω_δ^{jl} tends to ω^{jl} in $\mathscr{C}$ and ω_δ^l, $V_{j\delta}$, V_δ tend to ω^l, V_j, V respectively in $\mathscr{L}^p$. Moreover, $\|V_\delta''\|_{\mathscr{L}^p} \le C_0\varepsilon$ and $\|V_\delta'\|_{\mathscr{L}^\infty} \le C_\varepsilon$, etc., where V' and V'' were introduced in part (b).

Let u_δ be a sequence of solutions of problem (10.2.8) with the operator A_δ instead of A. Then the arguments of part (b) and the equalities $\|u_\delta\| = \|f\|$, $\|A_\delta u_\delta\| = \|A_\delta f\|$ $\forall t$ yield that for all $\delta > 0$ small enough,

$$\|D_t u_\delta\| + \|u_\delta\|_{\mathscr{H}^1} \le C(\|Af\| + \|f\|) \qquad \forall t \in \mathbb{R}$$

where the constant C does not depend on δ. Then the standard compactness arguments yield that one can choose a subsequence u_δ tending to a function u in $\mathscr{C}(\mathbb{R}, \mathscr{L}^2(X, \mathbb{H}))$ which lies in the required solution space and, moreover, $A_\delta u_\delta \to Au$ in $\mathscr{D}'(\mathbb{R} \times X, \mathbb{H})$ and therefore u is a solution of (10.2.8). The symmetry of A yields that this solution is unique and

$$\|u\| = \|f\|, \ \|Au\| = \|Af\| \qquad \forall t \in \mathbb{R}. \tag{10.2.9}$$

Now let $\max_j \bar{\gamma}_j = \infty$. Let us first take f supported in the domain $\{x\colon |x'| \le R\}$. Let us replace $\bar{\gamma}_j = \infty$ by $\bar{\gamma}_j = CR^2$ and let us solve problem (10.2.8). Then $|\omega^{jl}| \le c$ yields that the propagation speed for operators A_δ does not exceed C_0 and therefore the same is true for the operator A. Then u is supported in $\{|x'| \le \frac{1}{2}CR^2\}$ for $|t| \le R$ and sufficiently large R;

moreover, $u(.,t)$ does not depend on $R \geq |t|$. Letting R tend to $+\infty$ we conclude that problem (10.2.8) has a solution in the required space for every compactly supported $f \in \mathfrak{D}(A)$ and the propagation speed does not exceed C_0 and estimate (10.2.9) holds. But then the compactness condition is not necessary.

Now Theorem 10.2.1 is due to Proposition 10.2.2 below. □

Proposition 10.2.2. *Let A be a symmetric closed operator with the dense domain in the Hilbert space $\mathcal{H}$ and let $U(t)$ be a group of unitary operators in $\mathcal{H}$ such that for $f \in \mathfrak{D}(A)$*

(a) $U(t)f \in \mathfrak{D}(A) \quad \forall t \in \mathbb{R}$ and $\|AU(t)f\| \leq M_f e^{ct}$ with some constant $M = M_f$;

(b) $U(t)f \to f \quad$ as $t \to 0$;

(c) $D_t U(t)f = AU(t)f$ in the weak sense.

Then A is a self-adjoint generator of $U(t)$.

Proof. Let us set

$$R_\pm = \mp i \int_0^\infty U(\pm t)e^{-\lambda t}dt$$

with $\lambda > c$. Obviously, the operator norm of $R_\pm$ in $\mathcal{H}$ does not exceed λ^{-1} and for $f \in \mathfrak{D}(A)$ $R_\pm f \in \mathfrak{D}(A)$ and $(A \pm i\lambda)R_\pm f = f$. Then $R_\pm f \in \mathfrak{D}(A)$ $\forall f \in \mathcal{H}$ and both defect indices vanish; here we used that A is closed and $\mathfrak{D}(A)$ is dense. Therefore A is self-adjoint. But then a generator of $U(t)$ coincides with A. □

Remark 10.2.3. (i) Obviously, symmetric operator A is defined under conditions weaker than in Theorem 10.2.1. For example, one can take its domain $\mathfrak{D}(A) = \mathscr{C}_0^1(X)$ provided $\omega^{jl}, \omega^l, V_j, V \in \mathscr{L}^2_{\mathrm{loc}}(X)$.

Moreover, let Y be a measure zero closed subset of X. Assume that $\omega^{jl}, \omega^l, V_j, V \in \mathscr{D}'(X) \cap \mathscr{L}^2_{\mathrm{loc}}(X \setminus Y)$. Then one can take $\mathfrak{D}(A) = \mathscr{C}_0^1(X \setminus Y)$. Obviously, A is a symmetric operator with domain dense in $\mathscr{L}^2(X, \mathbb{H})$.

(ii) Further, if $V_j = 0$ and <u>either</u> d is odd <u>or</u> d is even and $m = 0$ then operator (10.1.66) has a self-adjoint extension. In fact, there exists a unitary matrix β such that $\beta^\dagger \bar{\sigma}_j \beta = \sigma_j$ with $j = 0, \dots, d$ as d is odd and with

$j = 1, \dots, d$ as d is even where the bar means a complex conjugation[14]. But then $\beta^\dagger \bar{A} \beta = A$ and therefore both defect indices of A coincide.

(iii) Similarly, if $V = 0$, $m = 0$, then $\beta^\dagger A \beta = -A$ with $\beta = \sigma_0$ and therefore both defect indices of A again coincide[14].

On the other hand, if $V = 0$ and $m > 0$ then

$$\|Au\|^2 = \|(A - m\sigma_0)u\|^2 + m^2\|u\|^2$$

and inequality

$$\|Au\| \geq m\|u\| \qquad \forall u \in \mathfrak{D}(A) \tag{10.2.10}$$

holds. Therefore both defect indices of A coincide in this case also; moreover, there exists a self-adjoint extension of A satisfying (10.2.10).

Thus, if $V = 0$ then there exists a self-adjoint extension of A satisfying (10.2.10)[13].

(iv) For $d = 2$ let us consider an operator A of the form

$$A_1 = \sigma_1(D_r + \frac{i}{2}r^{-1}) + \sigma_2 r^{-1}(D_\phi + \alpha) + \sigma_0 m \tag{10.2.11}$$

in the polar coordinates with $\mathfrak{D}(A) = \{u \in \mathscr{H}^1(\mathbb{R}^2, \mathbb{H}), r^{-1}u \in \mathscr{L}^2(\mathbb{R}^2, \mathbb{H})\}$. One can see easily that A is self-adjoint for $\alpha \notin \mathbb{Z}$ and that

$$q^\dagger(\sigma_1 \cos\phi - \sigma_2 \sin\phi)q = \sigma_1, \quad q^\dagger(\sigma_1 \sin\phi + \sigma_2 \cos\phi)q = \sigma_2, \quad q^\dagger \sigma_0 q = \sigma_0$$

for the unitary matrix-valued function $q = \operatorname{diag}(e^{i\zeta\phi}, 1)$ and σ_0, σ_1, σ_2 defined by (5.A.28). Moreover,

$$A_2 = q^\dagger A_1 q = \sigma_1(D_1 - \alpha x_2|x|^{-2}) + \sigma_2(D_2 + \alpha x_1|x|^{-2}) + \sigma_0 m. \tag{10.2.12}$$

Then for $\alpha \notin \mathbb{Z}$ the operator A_2 (with the same domain as A_1) is self-adjoint. Moreover, magnetic intensity $F = 0$ for $x \neq 0$ but for $\alpha \notin \mathbb{Z}$ one can remove the vector potential by means the of a gradient transform only in a simply connected subdomain in $\mathbb{R}^2 \setminus 0$.

Furthermore, one can easily see that $A_2(\alpha')$ and $A_2(\alpha)$ are unitarily equivalent if and only if $\alpha' - \alpha \in \mathbb{Z}$.

(v) Other examples of self-adjoint Dirac operators with the magnetic field will be given in Parts VI–X.

---

[14] These arguments however do not work as d is odd but operator A is principally massless operator (10.1.76).

10.2.2 Abstract Theory

Let us now consider the abstract theory. We will use the following definition:

Definition 10.2.4. Let A be a self-adjoint operator and J a symmetric operator in the separable Hilbert space $\mathcal{H}$ such that $\mathfrak{D}(J) \supset \mathfrak{D}(A)$. Then

(i) Operator J is A-*bounded* if

$$\|Ju\| \le C\|Au\| + C'\|u\| \quad \forall u \in \mathfrak{D}(A). \tag{10.2.13}$$

The smallest C_0 such that (10.2.13) holds with $C = C + \epsilon$ and $C' = C'_\epsilon$ for arbitrarily small $\epsilon > 0$ is called A-*bound* of J.

(ii) Operator J is A-*compact* if $J(A+i)^{-1}$ is a compact operator.

Proposition 10.2.5. *Let A be a self-adjoint operator and J_k be symmetric operators in the separable Hilbert space $\mathcal{H}$. Assume that operator J_1 is A-compact and A-bound of J_2 is less than 1. Then*

(i) A-bound of J_1 equals 0 and therefore operators $A + J_k$ are self-adjoint.

(ii) Operator $J_1(A + J_2 + i)^{-1}$ is compact.

(iii) $\operatorname{Spec}_{\mathsf{ess}}(A + J_1) = \operatorname{Spec}_{\mathsf{ess}}(A)$.

Proof. Recall that operator J_1 is A-compact if $J_1(A+i)^{-1}$ is compact; then for every $\epsilon > 0$ there exists a finite-dimensional operator $K = K_\epsilon$ such that $\|J_1(A+i)^{-1} - K\| \le \epsilon$. This operator is of the form

$$Ku = \sum_{1 \le j \le N} f_j \langle u, g_j \rangle$$

and without any loss of the generality one can assume that $g_j \in \mathfrak{D}(A)$ (otherwise they can be replaced by nearby elements of $\mathfrak{D}(A)$). But then $K(A+i)$ is a bounded operator and therefore

$$\|J_1 u\| \le \|K(A+i)u\| + \epsilon\|(A+i)u\| \le \epsilon\|Au\| + C'_\epsilon\|u\| \qquad \forall u \in \mathfrak{D}(A)$$

and therefore A-bound of J_1 equals 0. Statement (i) is proven.

Statement (ii) is due to equality

$$J_1(A+i)^{-1} - J_1(A + J_2 + i)^{-1} = J_1(A+i)^{-1} \cdot J_2(A + J_2 + i)^{-1}$$

with compact operator $J_1(A+i)^{-1}$ and bounded operator $J_2(A+J_2+i)^{-1}$ (because the A-bound of J_2 is less than 1).

Let us prove that if $\tau \in \mathsf{Spec}_{\mathsf{ess}}(A)$ then $\tau \in \mathsf{Spec}_{\mathsf{ess}}(A+J_1)$; this assertion combined with Statements (i) and (ii) with $J_2 = J_1$ yield the inverse implication and thus Statement (iii). If $\tau \in \mathsf{Spec}_{\mathsf{ess}}(A)$ then there exists a sequence $u_n \in \mathfrak{D}(A)$ such that $\|u_n\| = 1$, $u_n \overset{w}{\to} 0$ and $\|(A-\tau)u_n\| \to 0$. Let $v_n = (A+i)u_n$. Then $1 \le \|v_n\| \le M$ obviously and

$$\langle v_n, f\rangle = \langle u_n, (A-i)f\rangle \to 0 \qquad \text{as } f \in \mathfrak{D}(A).$$

Then $v_n \overset{w}{\to} 0$ and then $J_1 u_n = J_1(A+i)^{-1} v_n \overset{s}{\to} 0$ because operator $J_1(A+i)^{-1}$ is compact. Hence $\|(A+J_1-\tau)u_n\| \to 0$ and therefore $\tau \in \mathsf{Spec}_{\mathsf{ess}}(A+J_1)$.□

Proposition 10.2.6. *Let A be a self-adjoint operator and let J be a symmetric operator in the separable Hilbert space $\mathcal{H}$. Assume that J is A-bounded. Then*

(i) Let $\mathsf{dist}(\tau_0, \mathsf{Spec}(A)) \ge \epsilon > 0$. Then $\mathsf{dist}(\tau_0, \mathsf{Spec}(A+tJ)) \ge \frac{1}{2}\epsilon$ for $|t| \le \delta = \delta(\epsilon) > 0$.

(ii) If $\mathsf{dist}(\tau_0, \mathsf{Spec}(A) \setminus \tau_0) \ge \epsilon > 0$ then

$$\mathsf{Spec}(A+tJ) \cap [\tau_0 - \frac{1}{2}\epsilon, \tau_0 + \frac{1}{2}\epsilon] \subset [\tau_0 - C|t|, \tau_0 + C|t|]$$

for $|t| \le \delta = \delta(\epsilon) > 0$.

(iii) Moreover, in the framework of (ii) the spectral projector of the operator $A+tJ$ corresponding to the interval $[\tau_0 - \frac{1}{2}\epsilon, \tau_0 + \frac{1}{2}\epsilon]$ depends analytically on t.

(iv) In particular, if τ_0 is an eigenvalue of A with multiplicity $n < \infty$ then for $|t| \le \delta$

$$\mathsf{Spec}(A+tJ) \cap [\tau_0 - \frac{1}{2}\epsilon, \tau_0 + \frac{1}{2}\epsilon] = \{\tau_1(t), \dots, \tau_n(t)\}$$

where $\tau_k(t)$ are Lipschitz functions and the number of eigenvalues of $A+tJ$ in the interval $[\tau_0 - \frac{1}{2}\epsilon, \tau_0 + \frac{1}{2}\epsilon]$ (counting multiplicities) is equal to n.

Proof. Both Statements (i) and (ii) immediately follow from the inequality

$$\|(A-\tau)u\| \ge \mathsf{dist}(\tau, \mathsf{Spec}(A)) \cdot \|u\| \qquad \forall u \in \mathfrak{D}(A)$$

for self-adjoint operators.

Rewriting the spectral projector in question in the form

$$\Pi(t) = \frac{1}{2\pi i}\int_\Gamma (z - A - tJ)^{-1}dz$$

where $\Gamma = \{z \in \mathbb{C}, |z - \tau_0| = \frac{3}{4}\epsilon\}$ is a counter-clockwise oriented contour we obtain Statement (iii) and the assertion concerning the number of eigenvalues in Statement (iv). Moreover, we conclude that

$$\mathsf{Spec}_{\mathsf{ess}}(A + tJ) \cap [\tau_0 - \frac{1}{2}\epsilon, \tau_0 + \frac{1}{2}\epsilon] = \emptyset.$$

Finally, the Lipschitz property of τ_k easily follows from the arguments of the proof of Statements (i) and (ii). □

Proposition 10.2.7. *In the framework of Proposition 10.2.6(iv) assume that*

$$\langle Jv, v\rangle \ge \epsilon\|v\|^2 \qquad \forall u \in \mathsf{Ker}(A - \tau_0). \tag{10.2.14}$$

Then

$$\tau_k(t') \ge \tau_k(t) + \delta_1(t' - t) \qquad \forall k = 1, \dots, n \quad \forall t, t' \in [-\delta, \delta] : t' > t \tag{10.2.15}$$

with $\delta_1 = \delta_1(\epsilon) > 0$.

Proof. Obviously it is sufficient to prove (10.2.15) under additional condition $t' < t + \delta_2(t)$ with $\delta_2(t) > 0$.

Consider $t = \bar{t}$. Let $\bar{\tau}$ be an eigenvalue of $A + \bar{t}J$ of multiplicity r and let $\bar{\Pi}$ be the corresponding spectral projector.

Let τ' be an eigenvalue of $A + t'J$ close to $\bar{\tau}$ and let $|t' - \bar{t}|$ be small enough; then Proposition 10.2.6(iii) yields that

$$\|(I - \bar{\Pi})u\| \le M|t' - \bar{t}| \cdot \|u\| \qquad \text{with} \quad M = M(\bar{t}). \tag{10.2.16}$$

We know that $|\tau' - \bar{\tau}| \le C|t' - \bar{t}|$. This inequality combined with (10.2.16), equalities

$$(A + t'J - \tau')v = 0, \quad (A + \bar{t}J - \bar{\tau})\bar{\Pi}v = 0 \quad \text{for} \quad v \in \mathsf{Ker}(A + t'J - \tau')$$

and the A-boundedness of the operator J yield that

$$\|A(I-\bar{\Pi})v\| \le M|t'-\bar{t}|\cdot\|v\| \quad \text{for} \quad v\in \mathsf{Ker}(A+t'J-\tau'). \tag{10.2.17}$$

Then the same inequality (10.2.17) holds for J instead of A (we increase M if necessary). These inequalities and the equality $(A+\bar{t}J-\bar{\tau})\bar{\Pi}=0$ yield that

$$\begin{aligned}\bigl|(t'-\bar{t})\langle Jv,v\rangle+(\tau'-\bar{\tau})\|v\|^2\bigr| &= \\ &|\langle(A+\bar{t}J-\bar{\tau})v,v\rangle| \le M|t'-t|^2\|v\|^2.\end{aligned} \tag{10.2.18}$$

On the other hand, inequality (10.2.14) remains true (with $\frac{1}{2}\epsilon$ instead of ϵ) for $u\in\mathsf{Ker}(A+t'J-\tau')$ also with $|t'|\le\delta$, $|\tau'-\bar{\tau}|\le\delta$. This inequality, (10.2.18) and Proposition 10.2.6(iii) yield that

$$\mathsf{Spec}(A+t'J)\cap[\tau_0-\delta_3,\tau_0+\delta_1(t'-\bar{t})]=\emptyset \qquad \text{for } t'\in[\bar{t},\bar{t}+\delta_2(\bar{t})]$$

where $\delta_3=\delta_3(\bar{t})>0$ while $\delta_1>0$ does not depend on $\bar{t}$. So (10.2.15) is proven for $t'<\bar{t}+\delta_2(\bar{t})$. □

Now we are ready to prove the main statement:

Theorem 10.2.8 (Birman-Schwinger principle). *Let A be self-adjoint operator and let J be a symmetric operator in the separable Hilbert space $\mathcal{H}$. Assume that J is A-compact operator.*

Let us assume that $\bar{\tau}_1,\bar{\tau}_2\in\mathbb{R}$, $\bar{\tau}_1<\bar{\tau}_2$, $\bar{t}>0$ and

$$(A+tJ-\bar{\tau}_i)u=0 \implies \varsigma_i\langle Ju,u\rangle\ge\epsilon\|u\|^2 \qquad \forall i=1,2 \quad \forall t\in[0,\bar{t}] \tag{10.2.19}$$

with $\varsigma_i=\pm1$, $\epsilon>0$. Then

(i) The following equality holds:

$$\begin{aligned}\mathsf{N}(A;[\bar{\tau}_1,\bar{\tau}_2]) &= \mathsf{N}(A+\bar{t}J;[\bar{\tau}_1,\bar{\tau}_2])+ \\ &\varsigma_2\tilde{\mathsf{N}}((A-\bar{\tau}_2),-J;[0,\bar{t}])-\varsigma_1\tilde{\mathsf{N}}((A-\bar{\tau}_1),-J;[0,\bar{t}])\end{aligned} \tag{10.2.20}$$

where $\tilde{\mathsf{N}}(B,-J;\mathcal{J})$ is the number of eigenvalues $t\in\mathcal{J}$ of the problem

$$(B+tJ)u=0 \qquad u\in\mathfrak{D}(B)\setminus 0 \tag{10.2.21}$$

counting their multiplicities[15]).

On the other hand, $\mathsf{N}(B, \mathcal{J})$ *is the cardinality of the spectrum of* B *in* $\mathcal{J}$ *and* $\mathsf{N}(B, \mathcal{J}) = \infty$ *if* $\operatorname{Spec}_{\mathsf{ess}}(B) \cap \mathcal{J} \neq \emptyset$.

(ii) Further, equality (10.2.20) *remains true if one replaces the interval* $\mathcal{J}$ *in the left-hand and right-hand expressions: one* <u>*removes*</u> $\bar{\tau}_1$ *and (or)* $\bar{\tau}_2$ *from* $\mathcal{J}$ *and* <u>*replaces*</u> $[0, \bar{t}]$ *by* $(0, \bar{t})$ *in the third and (or) second terms in the right-hand expression; however, in order to avoid complications with the domain of* u, *one should assume that* $\bar{\tau}_i \notin \operatorname{Spec}_{\mathsf{ess}}(A)$.

(iii) Furthermore, $\tilde{\mathsf{N}}(A - \bar{\tau}_i, -J; [0, \bar{t}]) < \infty$ *provided* $\bar{\tau}_i \notin \operatorname{Spec}_{\mathsf{ess}}(A)$.

Remark 10.2.9. If $\operatorname{Spec}_{\mathsf{ess}}(A) \cap [\bar{\tau}_1, \bar{\tau}_2] \neq \emptyset$ then $\operatorname{Spec}_{\mathsf{ess}}(A + tJ) \cap [\bar{\tau}_1, \bar{\tau}_2] \neq \emptyset$ due to Proposition 10.2.5(iii) and both sides of (10.2.20) are infinite: indeed, it will follow from Statement (iii) of the Theorem that $\tilde{\mathsf{N}}((A - \bar{\tau}_i), -J; [0, \bar{t}]) < \infty$ for $i = 1, 2$ and therefore there is no indefiniteness in the right-hand expression).

Proof of Theorem 10.2.8. In virtue of Remark 10.2.9 in the proof of Statements (i) and (i) we can assume without any loss of the generality that

$$\operatorname{Spec}_{\mathsf{ess}}(A) \cap [\bar{\tau}_1, \bar{\tau}_2] = \emptyset. \tag{10.2.22}$$

Then

$$\operatorname{Spec}(A + tJ) \cap [\bar{\tau}_1, \bar{\tau}_2] = \{\tau_k(t), t \in [t'_k, t''_k], k = 0, \dots, N\}$$

where $N \leq \infty$ and for each $\tau \in [\bar{\tau}_1, \bar{\tau}_2]$ $\#\{k\colon \tau_k(t) = \tau\}$ is equal to the multiplicity of τ as an eigenvalue of $(A + tJ)$.

Furthermore, due to Propositions 10.2.5 and 10.2.5 $0 \leq t'_k < t''_k \leq \bar{t}$ and $\tau_k(t)$ are Lipschitz functions on the intervals $(\tilde{t}'_k, \tilde{t}''_k) \supset [t'_k, t''_k]$ such that for $t \in (\tilde{t}'_k, \tilde{t}''_k)$ $\tau_k(t) \in [\bar{\tau}_1, \bar{\tau}_2]$ if and only if $t \in [t'_k, t''_k]$. Finally, $\varkappa_i \tau_k(t)$ are monotone increasing function of t in neighborhoods of $t_{ik} \in [t'_k, t''_k]$ with $\tau_k(t_{ik}) = \bar{\tau}_i$.

Then the various terms in (10.2.20) are the numbers of graphs $\{\tau = \tau_k(t)\}$ intersecting respectively the upper, lower, right and left sides of the rectangle $\{0 \leq t \leq \bar{t},\ \bar{\tau}_1 \leq \tau \leq \bar{\tau}_1\}$ on Figure 10.1, taken with the correct signs and

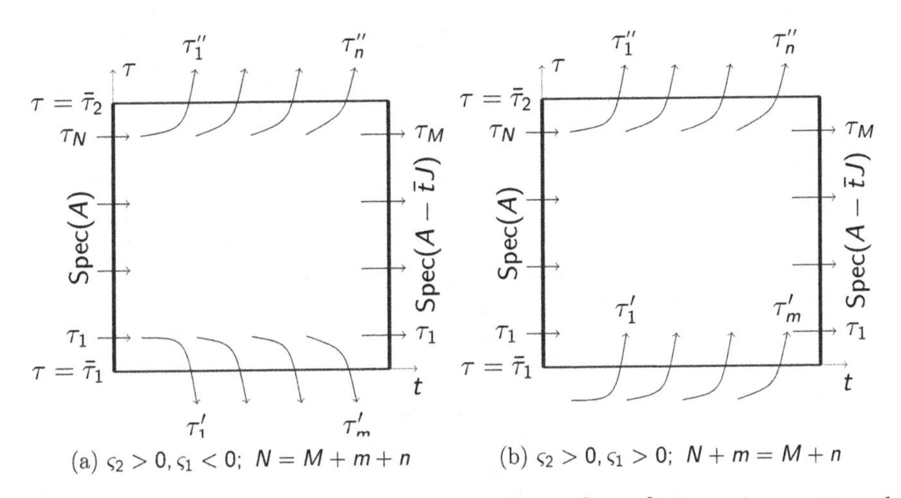

(a) $\varsigma_2 > 0, \varsigma_1 < 0;\ N = M + m + n$ (b) $\varsigma_2 > 0, \varsigma_1 > 0;\ N + m = M + n$

Figure 10.1: : As $\varsigma_2 \gtrless 0$ eigenvalues exit/enter $[\bar\tau_1, \bar\tau_2]$ through $\tau = \bar\tau_2$ and as $\varsigma_1 \lessgtr 0$ eigenvalues exit/enter $[\bar\tau_1, \bar\tau_2]$ through $\tau = \bar\tau_1$ as t grows. As $\varsigma_2 < 0$ pictures should be modified in the obvious way.

equality (10.2.20) and its modifications in Statement (ii) are due to obvious geometric arguments.

Statements (i) and (ii) are proven; let us prove Statements (iii). If $\tilde{\mathsf{N}}(A - \bar\tau, -J; [0, \bar t]) = \infty$ then there would exist sequences $t_n \in [0, \bar t]$ and $u_n \in \mathfrak{D}(A)$ such that $(A - \bar\tau + t_n J)u_n = 0$ and $\langle J u_n, u_m\rangle = \delta_{nm}$. Then $\|u_n\| \le C$ by (10.2.19) and then the equality $(A + t_n J - \bar\tau)u_n = 0$ and the fact that the A-bound of J is 0 imply that $\|Au_n\| \le C_1$. Then the compactness of the operator $J(A + i)^{-1}$ yields that the sequence Ju_n is strongly convergent (we take a subsequence if necessary) but this contradicts the equality $\langle Ju_n, u_m\rangle = \delta_{nm}$. □

Remark 10.2.10. (i) As A is semi-bounded from below we are in the framework of Chapter 9 and Birman-Schwinger principle is illustrated on Figure 10.2(a); we can take $\bar\tau_1 = -\infty$ resulting in

$$\mathsf{N}\big(A; [\bar\tau_1, \bar\tau_2]\big) = \mathsf{N}\big(A + \bar t J; [\bar\tau_1, \bar\tau_2]\big) + \varsigma_2 \tilde{\mathsf{N}}\big((A - \bar\tau_2), -J; [0, \bar t]\big). \tag{10.2.20$'$}$$

[15] I.e., the dimensions of the subspaces of $\mathcal{H}$ on which (10.2.21) holds with given t; in this definition neither continuous nor fancier types of the spectrum are discussed, just the bare dimension of the null space.

(ii) And then we can take $\bar{t} = \infty$ important subcase is of $\bar{t} = \infty$ as illustrated on Figure 10.2(b):

$$\mathsf{N}(A; [\bar{\tau}_1, \bar{\tau}_2]) = \varsigma_2 \tilde{\mathsf{N}}((A - \bar{\tau}_2), -J; [0, \bar{t}]). \tag{10.2.20$''$}$$

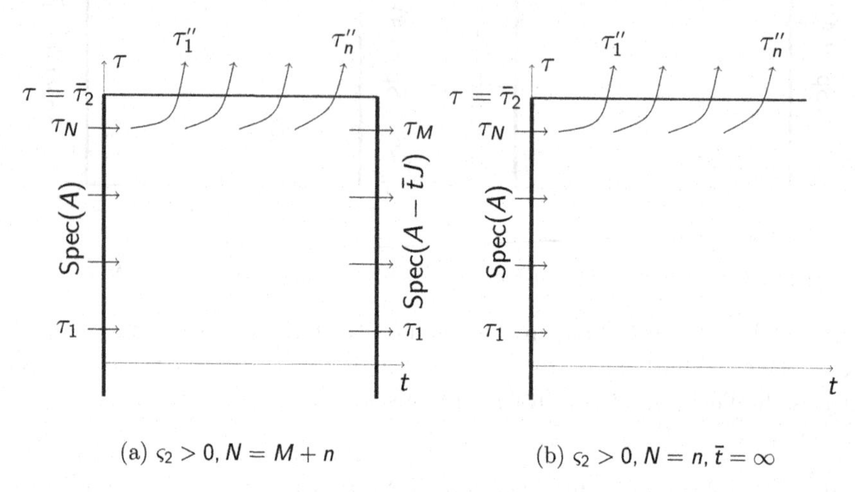

(a) $\varsigma_2 > 0, N = M + n$ (b) $\varsigma_2 > 0, N = n, \bar{t} = \infty$

Figure 10.2: : Case of A bounded from below.

Remark 10.2.11. Let A be a self-adjoint operator and J be a symmetric operator in the separable Hilbert space $\mathcal{H}$. Let $\mathfrak{D}(J) \supset \mathfrak{D}(A)$ and $J(A+i)^{-1}$ be a compact operator. Moreover, let us assume that $J \geq 0$, $\operatorname{Ker} J = 0$ and

$$\|(A - \bar{\tau}_i)u\| + \|Ju\| \geq \epsilon_1 \|u\| \qquad \forall u \in \mathfrak{D}(A) \tag{10.2.23}$$

with $\epsilon_1 > 0$. Then (10.2.19) holds.

In fact, otherwise there exist sequences $t_n \in [0, \bar{t}]$ and $u_n \in \mathfrak{D}(A)$ such that $Au_n = -t_n J u_n$ (we assume for the sake of simplicity that $\bar{\tau}_i = 0$), $\|u_n\| = 1$, $\langle Ju_n, u_n\rangle \to 0$. Then estimate $\|Ju\| \leq \delta\|Au\| + C_\delta\|u\|$ (with arbitrarily chosen $\delta > 0$) yields that $\|Au_n\| \leq C$ and since $\|u\|_n = 1$ we obtain that the sequence Ju_n is strongly convergent (we take a subsequence if necessary). Then the sequence Au_n is also strongly convergent. Then (10.2.23) yields that the sequence u_n is also strongly convergent (because a Cauchy sequence is a converging sequence and conversely).

Let $u = \lim u_n$. Then $u \in \mathfrak{D}(A)$, $\|u\| = 1$ and $\langle Ju, u\rangle = \lim\langle Ju_n, u_n\rangle = 0$; this contradicts the conditions $J \geq 0$ and $\operatorname{Ker} J = 0$.

Proposition 10.2.12. *Let assumptions of Remark 10.2.11 be fulfilled. Let us replace assumption* (10.2.23) *by assumption*

$$\|(A-\bar{\tau}_i)u\| + \|J^{\frac{1}{2}}u\| \ge \epsilon_1\|u\| \qquad \forall u \in \mathfrak{D}(A); \tag{10.2.24}$$

one can easily see that this condition is stronger. Then for small enough $\delta>0$ operators $((A-\bar{\tau}_i)\pm i\delta J)^{-1}$, $(A-\bar{\tau}_i)((A-\bar{\tau}_i)\pm i\delta J)^{-1}$ and $J((A-\bar{\tau}_i)\pm i\delta J)^{-1}$ are bounded (their norms depend on δ) and therefore operator $J((A-\bar{\tau}_i)\pm i\delta J)^{-1}$ is compact.

Moreover, one can take arbitrarily large $\delta>0$ provided J is a bounded operator.

Proof. Without any loss of the generality one can assume that $\bar{\tau}_i=0$. If $(A+i\delta J)u=f$ then

$$\|J^{\frac{1}{2}}u\|^2 \le \delta^{-1}\|u\|\cdot\|f\|, \quad \|Au\| \le \delta\|Ju\| + \|f\| \le C\delta(\|Au\|+\|u\|)+\|f\|$$

and hence $\|u\|+\|Au\| \le C_\delta\|f\|$. This inequality yields the boundedness of the indicated operators. Moreover, if J is a bounded operator then $\|Ju\| \le C\|J^{\frac{1}{2}}u\|$ and therefore $\|Au\| \le C\delta\|J^{\frac{1}{2}}u\| + \|f\|$; this inequality combined with the inequalities $\|J^{\frac{1}{2}}u\|^2 \le \delta^{-1}\|u\|\cdot\|f\|$ and (10.2.24) yield that $\|u\|+\|Au\| \le C_\delta\|f\|$ for arbitrary $\delta>0$. □

Proposition 10.2.13. *Let assumptions of Remark 10.2.11 be fulfilled and let J be a bounded operator (one can easily check that conditions* (10.2.23) *and* (10.2.24) *are equivalent in this case). Then*

(i) Operator $B=J^{-\frac{1}{2}}(A-\bar{\tau}_i)J^{-\frac{1}{2}}$ with domain

$$\mathfrak{D}(B)=\bigl\{u\in\mathfrak{D}(J^{-\frac{1}{2}}):\ J^{-\frac{1}{2}}u\in\mathfrak{D}(A)\ \text{ and }\ (A-\bar{\tau}_i)J^{-\frac{1}{2}}u\in\mathfrak{D}(J^{-\frac{1}{2}})\bigr\}$$

is self-adjoint.

(ii) Operators $(B\pm i)^{-1}$ are compact.

(iii) Eigenvalues of the operator $-B$ and those of the problem

$$(A-\bar{\tau}_i)u + tJu = 0 \qquad u\in\mathfrak{D}(A) \tag{10.2.25}$$

coincide (counting multiplicities) and

$$\operatorname{Ker}(B+tJ) = J^{\frac{1}{2}}\operatorname{Ker}((A-\bar{\tau}_i)+tJ). \tag{10.2.26}$$

Proof. The obvious proof is left to the reader. □

10.2.3 Proof of Theorem 10.1.7

Thus *Theorem 10.1.7 has been proven under additional condition* (10.2.23) *which should be checked for operator A replaced by*

$$A'_k := A - t_1J_1 - \dots - t_{k-1}J_{k-1},$$

$J = J_k$ *and* $\bar{\tau}_i = \tau_i$ *for every* $k = 1,\dots,n$*:*

$$\|(A-\tau_i - t_1J_1 - \dots - t_{k-1}J_{k-1})u\| + \|J_k u\| \ge \epsilon_1\|u\| \qquad u\in\mathfrak{D}(A)\quad \forall k=1,\dots,n. \tag{10.2.27}$$

Really, by the construction operators $J_k \ge 0$ are bounded[16] operators with $\operatorname{Ker} J_k = 0$ and therefore in virtue of Proposition 10.2.11

$$\tilde{\mathsf{N}}((A-\tau_2), -J; [0,\bar{t}]) = \mathsf{N}(-J^{-\frac{1}{2}}(A-\tau_2)J^{-\frac{1}{2}}; [0,\bar{t}]) = \mathsf{N}(-J^{-1}(A-\tau_2); [0,\bar{t}]) \tag{10.2.28}$$

and formula (10.2.20) becomes

$$\mathsf{N}(A;[\tau_1,\tau_2]) = \mathsf{N}(A+\bar{t}J;[\tau_1,\tau_2]) + \varsigma_2\mathsf{N}(-J^{-\frac{1}{2}}(A-\tau_2)J^{-\frac{1}{2}};[0,\bar{t}]) - \varsigma_1\mathsf{N}(J^{-\frac{1}{2}}(A-\tau_1)J^{-\frac{1}{2}};[0,\bar{t}]). \tag{10.2.29}$$

Due to (10.2.27) we can replace here A by A'_k and J_1 by J_k and $\bar{t}$ by $-t_k$.

Therefore we justified (10.1.12) and thus (10.1.18) and then the arguments of Subsection 10.1.2 prove the statement of Theorem 10.1.7.

One can easily check that

(10.2.30) Condition (10.2.23) is fulfilled provided

$$\tau_1, \tau_2 \notin \operatorname{Spec}_{\mathsf{ess}}(A). \tag{10.2.31}$$

---

[16] One can see easily that otherwise $R_{(i)1} = \infty$ and then the statement of the theorem is empty.

Moreover, Proposition 10.2.5(iii) yields that in this case condition (10.2.27) is fulfilled as well. Therefore *Theorem 10.1.7 has been proven under additional condition* (10.2.31).

Finally, let us consider the general case. Let us replace condition (10.2.31) by condition

(10.2.32) There exist sequences $\tau_{1,n} \to \tau_1 + 0$ and $\tau_{2,n} \to \tau_2 - 0$ with $\tau_{1,n}, \tau_{2,n} \notin \mathsf{Spec}_{\mathsf{ess}}(A)$;

this condition is fulfilled in virtue of (10.2.24). Under this condition we can apply Theorem 10.1.7 to intervals $(\tau_{1,n}, \tau_{2,n})$.

Let us note that $\mathsf{Spec}\, a^0(x,\xi) = \{\lambda_1(x,\xi), \dots, \lambda_D(x,\xi)\}$ with Lipschitz functions λ_j such that $|\lambda_j(x,\xi)| \to \infty$ as $|\xi| \to \infty$ and therefore

$$x \in X_{(1)\mathsf{reg,ell}} \cap X_{(2)\mathsf{reg,ell}} \implies \mathsf{Spec}\, a^0(x,\xi) \cap [\tau_1, \tau_2] = \emptyset. \tag{10.2.33}$$

On the other hand, it is easy to check (e.g., see the next Subsection 10.2.4) that if $R_{(1)1} + R_{(2)1} < \infty$ (otherwise the statement of Theorem 10.1.7 is empty) then $X_{(1)\mathsf{reg}} \setminus X_{(1)\mathsf{reg,ell}}$ and $X_{(2)\mathsf{reg}} \setminus X_{(2)\mathsf{reg,ell}}$ are compact sets. Therefore all conditions in $X_{(i)\mathsf{reg}} \setminus X_{(i)\mathsf{reg,ell}}$ (i.e., the microhyperbolicity condition, etc.) remain true (with $\gamma' = \frac{1}{2}\gamma$ and the same $\rho, \dots$) for $\tau_{i,n}$ instead of τ_i provided n is large enough.

Moreover, the ellipticity condition which defines $X_{(i)\mathsf{reg,ell}}$ remains true (with $\gamma' = \frac{1}{2}\gamma$ and $\rho' \geq \rho$ and appropriately changed other parameters) for τ_i' instead of τ_i provided $|\tau_i' - \tau_i|$ is small enough.

Therefore every term of the remainder estimate in the zone $X_{(i)\mathsf{reg}}$ for the interval $(\tau_{1,n}, \tau_{2,n})$ tends to the corresponding term of the remainder estimate for the interval (τ_1, τ_2) and $\mathcal{N}(\tau_{1,n}, \tau_{2,n})$ tends to $\mathcal{N}(\tau_1, \tau_2)$ (even if it is infinite).

Moreover, we can take the same R_0 for the interval $(\tau_{1,n}, \tau_{2,n})$ as for the original interval (τ_1, τ_2) (see conditions (10.1.44)–(10.1.46)). Therefore *Theorem 10.1.7 has been proven in the full generality.*

10.2.4 Proof of Theorem 10.1.8

Functional-Analytic Arguments

Let us return to the Dirac operator. In order to prove Theorem 10.1.8 one should derive some estimates of (10.1.46)- and (10.1.48)-type and construct the admissible matrix functions J_k.

Let us first consider the Dirac operator in the context of Theorem 10.1.7. Let us note first of all that if $R_1 < \infty$ (otherwise all the statements of this theorem are void) then $X_{\mathsf{reg}} \setminus X_{\mathsf{reg,ell}}$ are compact sets and on these sets ρ and γ are bounded and disjoint from 0. Really, otherwise there exists a family of non-intersecting balls $B(x^n, \gamma(x^n))$ with $n = 1, 2, \dots$, lying in this set, the contribution of each of these balls to R_1 is larger than $\epsilon > 0$.

On the other hand, the definition of $X_{\mathsf{reg,ell}}$ immediately yields that $\rho \le c'm$ on this set. So we can assume that

$$\rho \le M,\ \gamma \ge \epsilon \qquad \text{at } X_{\mathsf{reg,ell}} \tag{10.2.34}$$

and

$$\varepsilon \le \rho \le M,\ \epsilon \le \gamma \le M \qquad \text{at } X_{\mathsf{reg}} \tag{10.2.35}$$

where ε, M should not enter in the estimates.

Now we need to prove the A-compactness property.

Proposition 10.2.14. *Let all the conditions of Theorem 10.1.8 associated with the zone X_{reg} be fulfilled. Let us assume that the operator A is self-adjoint. Moreover, let $\rho \le M$ at X_{reg} and $J \in \mathscr{L}^\infty(X)$ be supported in X_{reg}. Let*

$$\forall \delta > 0 \qquad \mathsf{mes}(\{x, |J(x)| > \delta\}) < \infty. \tag{10.2.36}$$

Then operator $J(A+i)^{-1}$ is compact in $\mathscr{L}^2(X, \mathbb{H})$.

Proof. The results of Subsection 5.3.2 and the boundedness of ρ yield the estimate

$$|e(x, x, 0, \tau)| \le M(|\tau|^d + 1) \qquad \forall x \in X_{\mathsf{reg}} \quad \forall \tau \in \mathbb{R} \tag{10.2.37}$$

where the constant M is increased if necessary and e assumed without losing generality that $\tau_1 = -\tau_2$.

Therefore $Q := J^{\frac{1}{2}}\bigl(E(\tau) - E(0)\bigr)J^{\frac{1}{2}}$ is a trace operator with the trace norm not exceeding $M(|\tau|^d + 1)$ provided $J \in \mathscr{L}^1 \cap \mathscr{L}^\infty$ where $E(\tau)$ is the spectral projector of A; recall that J is supposed to be supported in X_{reg} here.

This operator is of the form $Q = SS^*$ and therefore $S = J^{\frac{1}{2}}\bigl(E(\tau) - E(0)\bigr)$ is a Hilbert-Schmidt operator with the Hilbert-Schmidt norm not exceeding $M(|\tau|^d + 1)$. Then equalities

$$(A+i)^{-p} = \int_{-\infty}^{\infty} (\tau + i)^{-p} d_\tau E(\tau) = \int_{-\infty}^{\infty} p(\tau + i)^{-p-1}\bigl(E(\tau) - E(0)\bigr)\, d\tau$$

yield that

(10.2.38) For $p > d$ operator $J(A+i)^{-p}$ is also a Hilbert-Schmidt operator.

Let us note that for $\varepsilon > 0$

$$J(A+i)^{-1} = J(A+i)^{-1}(I+i\varepsilon A)^{-d} + JB_\varepsilon \tag{10.2.39}$$

with

$$B_\varepsilon = (A+i)^{-1}(I+i\varepsilon A)^{-d}\bigl((I+i\varepsilon A)^d - I\bigr); \tag{10.2.40}$$

then due to (10.2.38) the first term in the right-hand expression of (10.2.39) is a Hilbert-Schmidt operator and B_ε is a product of $\varepsilon A(A+i)^{-1}$ and an operator with the uniformly bounded operator norm. Hence the operator norm of B_ε does not exceed $C\varepsilon$. Therefore $J(A+i)^{-1}$ is approximated in the operator norm by compact operators and therefore is also compact. *Thus, for $J \in \mathscr{L}^1 \cap \mathscr{L}^\infty$ the proposition is proven.*

In the general case, for every $\varepsilon > 0$ there exists a representation $J = J' + J''$ with $J' \in \mathscr{L}^1 \cap \mathscr{L}^\infty$ and $\|J''\|_{\mathscr{L}^\infty} \le \varepsilon$. Then $J'(A+i)^{-1}$ is a compact operator and $\|J''(A+i)^{-1}\| \le \varepsilon$ and so $J(A+i)^{-1}$ is approximated by compact operators and so is also compact. □

Proposition 10.2.15. *Let all the conditions of Theorem 10.1.8 associated with the zone X_{sing} be fulfilled. Let $R_3 < \infty$ and $\gamma \ge \epsilon$ on $X_{\mathsf{reg}} \cap X_{\mathsf{sing}}$. Let $J \in \mathscr{L}^p(X)$ with $p = d$ for $d \ge 3$ and with $p > 2$ for $d = 2$*[17]*.*

Suppose $\mathsf{supp}(J) \subset X_{\mathsf{sing}}$ and $\mathsf{dist}(\mathsf{supp}(J), \partial X_{\mathsf{sing}} \setminus X_{\mathsf{reg}}) \ge \epsilon$. Then operator $J(A+i)^{-1}$ is compact.

Proof. Conditions of this proposition yield that there exists a function ϕ supported in the $\frac{2}{3}\epsilon$-neighborhood of $\mathsf{supp}(J)$ and equal 1 in the $\frac{1}{3}\epsilon$-neighborhood of $\mathsf{supp}(J)$ such that $0 \le \psi \le 1$ and $|D^\alpha \phi| \le C \quad \forall \alpha : |\alpha| \le K$.

Let us consider equation $(A+i)v = f \in \mathscr{L}^2(X, \mathbb{H})$ with v supported in X_{sing}. Let us apply the arguments used in the proof of Theorem 10.2.1, and also condition $R_3 < \infty$. One can prove easily that if $v \in \mathscr{H}^1$ then

$$\|Dv\| \le M(\|f\| + \|v\|). \tag{10.2.41}$$

Let us consider $v \in \mathscr{H}^1_{\mathsf{loc}} \cap \mathscr{L}^2(X, \mathbb{H})$ and take a function ψ with $|\psi| + |\nabla\psi| \le C$. Then $(A+i)\psi v = \psi f + [A, \psi]v$ and (10.2.41) yields

[17] Let us recall that for $d = 2$ the condition $R_3 < \infty$ also means that $\mathsf{mes}(X_{\mathsf{sing}}) < \infty$.

estimate $\|D\psi v\| \le M(\|f\| + \|v\|)$. Since this estimate holds for every ψ described above we obtain that $v \in \mathscr{H}^1$ and estimate (10.2.41) holds.

Similarly, if $u \in \mathfrak{D}(A)$ and $(A+i)u = f$ then

$$\|\phi u\|_{\mathscr{H}^1} \le M(\|f\| + \|u\|) \le (M+1)\|f\|$$

and therefore the operator $\phi(A+i)^{-1} : \mathscr{L}^2(X,\mathbb{H}) \to \mathscr{H}^1(X,\mathbb{H})$ is bounded. On the other hand, it is well known that if $J \in \mathscr{L}^p$ with p indicated above (and $\mathsf{mes}(X_{\mathsf{sing}}) < \infty$ for $d = 2$) then the operator $J : \mathscr{H}^1(X_{\mathsf{sing}},\mathbb{H}) \to \mathscr{L}^2(X,\mathbb{H})$ is compact. □

The final arguments of the proof of Proposition 10.2.14 obviously yield

Corollary 10.2.16. *The assertion of Proposition 10.2.15 remains true if condition $J \in \mathscr{L}^p$ is replaced by condition*

(10.2.42) $V = V' + V''$, $V \in \mathscr{L}^p$, $V'' \in \mathscr{L}^\infty$ *and satisfies*(10.2.36).

Estimates for a Number of Crossing Eigenvalues

Let us start from the construction of matrices J_k. One can see easily that there exists a function ζ' on X_{reg} such that $\zeta'(x) > 0$, $c^{-1} \le \zeta'(x)/\zeta'(y) \le c$ for $y \in B\big(x, \epsilon_1\gamma(x)\big)$, $x \in X'_{\mathsf{reg}}$ where X'_{reg} is a domain intermediate between X_{reg} and $X_{\mathsf{reg}} \setminus X_{\mathsf{sing}}$, $\zeta' \asymp \min(\rho, \frac{1}{m}\rho^2)$. Let us set $\zeta = \zeta'$ in X_{reg} and $\zeta = \eta\zeta''$ in $X_{\mathsf{sing}} \setminus X_{\mathsf{reg}}$ where $\zeta'' \in \mathscr{S}(X)$ is a fixed function, $\zeta''(x) > 0$ and $\mathbb{R} \ni \eta \to +0$ in what follows. Let us note that $\zeta \in \mathscr{L}^\infty$ provided $R_1 < \infty$.

Let us choose

$$J_1 = \begin{pmatrix} \zeta I & 0 \\ 0 & \epsilon_2\zeta I \end{pmatrix}, \qquad J_2 = \begin{pmatrix} \epsilon_2\zeta I & 0 \\ 0 & \zeta I \end{pmatrix}$$

and let us take ϕ_1 supported in $[c_0, 2c_0]$ and ϕ_2 supported in $[-2c_0, -c_0]$, where $\epsilon_2 > 0$ is small enough and c_0 is large enough constants, such that

$$m - V - \tau_2 + t_1 j + \epsilon_2 t_2 \zeta \ge \frac{\epsilon}{2}\min(\rho, \frac{1}{m}\rho^2), \tag*{$(10.2.43)_2$}$$

$$-m + V - \tau_1 + t_1\epsilon_2 j + t_2\zeta \le -\frac{\epsilon}{2}\min(\rho, \frac{1}{m}\rho^2) \tag*{$(10.2.43)_1$}$$

$$\text{in } X_{\mathsf{reg}} \qquad \forall t_k \in \mathsf{supp}(\phi_k).$$

Let us apply the arguments of Section 10.1. Then in order to estimate $\mathsf{N}(A, \tau_1, \tau_2)$ from below we need to estimate R_0 in condition (10.1.46); the compactness of the operator $J(A+i)^{-1}$ obviously follows from $R_1 < \infty$ and Proposition 10.2.14.

Further, in order to estimate R_0 we need to estimate $\mathsf{N}(\tilde{A}-\tau_i, -\tilde{J}_1, (-t,t))$ and $\mathsf{N}(\tilde{A}+t\tilde{J}_1-\tau_i, -J_2, (-t,t))$ where Proposition 10.2.13(iii) yields that we can replace $\tilde{\mathsf{N}}$ by N. Here $\tilde{A}$ and $\tilde{J}_k$ mean corresponding operators extended from X_{sing} [18]. Let us consider the first number; one can analyze the second number in the same way. Let us assume for the sake of simplicity that $\tau_i = 0$ (otherwise we replace V by $V - \tau_i$).

We know that $\mathsf{N}(\tilde{A}, -\tilde{J}_1; (-t,t))$ for $t \geq 0$ is the maximal dimension of the subspace $\mathcal{L}' \subset \mathscr{C}^1_0(X, \mathbb{H})$ on which inequality

$$\|\tilde{J}_1^{-\frac{1}{2}}\tilde{A}u\|^2 \leq t^2\|\tilde{J}_1^{\frac{1}{2}}u\|^2 \tag{10.2.44}$$

holds. Therefore we should estimate the maximal dimension of the subspace $\mathcal{L} \subset \mathscr{C}^1_0(X, \mathbb{H})$ on which inequality

$$\|\tilde{\zeta}^{-\frac{1}{2}}\tilde{A}u\|^2 \leq C_0t^2\|\tilde{\zeta}^{\frac{1}{2}}u\|^2 \tag{10.2.45}$$

holds where $\tilde{\zeta}$ is an admissible continuation of ζ from X''_{sing} (which is a domain intermediate between X_{sing} and $X_{\mathsf{sing}} \setminus X'_{\mathsf{reg}}$) to $\mathbb{R}^d$ and none of the constants below depend on η.

Let $\tilde{A}$ be an operator of (5.3.71) type:

$$\tilde{A} = \tilde{A}_0 + \tilde{V}, \qquad \tilde{A}_0 = \sum \frac{1}{2}\sigma_l(\omega^{kl}\tilde{P}_j + \tilde{P}_k\omega^{kl}) + \sigma_0(m+W) \tag{10.2.46}$$

where the coefficients ω^{kl} are the same as in A,

$$\tilde{P}_k = D_k - \tilde{V}_k, \quad |\tilde{V}_k| \leq c\tilde{\rho}, \quad |\tilde{V}| \leq m + c\min(\rho, \frac{1}{m}\rho^2) \tag{10.2.47}$$

and $\tilde{V}_k$, $\tilde{V}$ are supported in X_{sing},

$$W \geq 0, \qquad |D^\alpha W| \leq c\rho\gamma^{-|\alpha|} \quad \forall \alpha : |\alpha| \leq K, \tag{10.2.48}$$

$W \asymp \rho$ for $\rho \geq cm$ outside of X_{sing} and W vanishes in X'_{sing} and also W vanishes for $\rho \leq \epsilon_1 m$.

---

[18] Or rather from its connected component, and later we sum the results obtained for these components.

(10.2.49) *For notational simplicity in what follows we skip tilde writing* A, P, W *instead of* $\tilde{A}$, $\tilde{P}$, $\tilde{V}$.

Let us consider a partition of unity $1=\sum_{\nu\ge 0}\psi_\nu^2$ in X_{sing} where ψ_ν are supported in the $\epsilon\gamma$-vicinity of X_{reg} and γ-admissible for $\nu\ge 1$, and where ψ_0 is supported in X_{sing}. Then the left-hand expression in (10.2.45) is equal to

$$2\|\varsigma^{-\frac{1}{2}}Au\|^2 \ge 2\sum_{\nu\ge 0}\lambda_\nu\|\psi_\nu Au\|^2 \ge \sum_{\nu\ge 0}\lambda_\nu\|A\psi_\nu u\|^2 - 2\sum_{\nu\ge 0}\lambda_\nu^2\|[A,\psi_\nu]u\|^2 \tag{10.2.50}$$

provided

$$\varsigma\lambda_\nu^2\le 1 \quad \text{on} \quad \mathsf{supp}(\psi_\nu). \tag{10.2.51}$$

Note that

$$[A,\psi_\nu]=\sum_{k,l}\sigma_l\omega^{kl}[P_j,\psi_\nu]$$

and therefore

$$[A,\psi_\nu]^*[A,\psi_\nu]\asymp\sum_l\|[P_l,\psi_\nu]\|^2\le C|\nabla\psi_\nu|^2$$

and

$$\sum_{\nu\ge 0}\lambda_\nu^2\|[A,\psi_\nu]u\|^2\le C\sum_{\nu\ge 0}\lambda_\nu^2\|(\nabla\psi_\nu)u\|^2 = C\sum_{\mu,\nu\ge 0}\lambda_\mu^2\|(\nabla\psi_\mu)\psi_\nu u\|^2. \tag{10.2.52}$$

Using the finite multiplicity of the covering we can estimate the sum of terms with $\mu\ge 1,\nu\ge 1$ and $\nu=\mu=0$ by

$$C\sum_{\nu\ge 0}\lambda_\nu^2\|\chi_Y\gamma^{-1}\psi_\nu u\|^2 \tag{10.2.53}$$

provided

$$\mathsf{supp}(\psi_\nu)\cap\mathsf{supp}(\psi_\mu)\ne\emptyset \implies \lambda_\nu\le c\lambda_\mu \qquad \forall\mu\ge 1,\nu\ge 1 \tag{10.2.54}$$

where Y is $\epsilon\gamma$-vicinity of X_{reg} intersected with X_{sing} and χ_Y is its characteristic function. However the sum in the right-hand expression of (10.2.52) contains also terms with $\nu = 0$ and $\mu \ge 1$ and inversely and we want to "untangle" these indices. We assume that

$$|\nabla\psi_\nu| \le \bigl(\alpha + \Phi(\alpha)\psi_\nu\bigr)\gamma^{-1} \qquad \text{for } \alpha \in (0,1), \tag{10.2.55}$$

with slowly growing to $+\infty$ function $\Phi(\alpha) \ge 1$ as $\alpha \to +0$. We discuss the possible choice of Φ and its implication later.

Then the sums of the terms with $\nu = 0, \mu \ge 1$ and with $\nu \ge 1, \mu = 0$ in the right-hand expression of (10.2.52) do not exceed

$$\begin{multline} C\sum_{\nu\ge 1}\bigl(\lambda_\nu\Phi(\alpha_\nu) + \lambda_0\beta_\nu\bigr)^2 \|\chi_Y\gamma^{-1}\psi_\nu u\|^2 + \\ C\sum_{\nu\ge 1}\bigl(\lambda_0\Phi(\beta_\nu) + \lambda_\nu\alpha_\nu\bigr)^2 \|\chi_Y\gamma^{-1}\bar{\psi}_\nu\psi_0 u\|^2 \end{multline} \tag{10.2.56}$$

where $\bar{\psi}_\nu$ is a characteristic function of $\mathsf{supp}(\psi_\nu)$.

This expression is greater than (10.2.53). Combining with (10.2.50) we conclude that

$$\begin{multline} \sum_{\nu\ge 0}\lambda_\nu^2 \|A\psi_\nu u\|^2 \le 2\|\zeta^{-\frac12}Au\|^2 + \\ C\sum_{\nu\ge 1}\bigl(\lambda_\nu\Phi(\alpha_\nu) + \lambda_0\beta_\nu\bigr)^2 \|\chi_Y\gamma^{-1}\psi_\nu u\|^2 \ + \\ C\sum_{\nu\ge 1}\bigl(\lambda_0\Phi(\beta_\nu) + \lambda_\nu\alpha_\nu\bigr)^2 \|\chi_Y\gamma^{-1}\bar{\psi}_\nu\psi_0 u\|^2. \end{multline} \tag{10.2.57}$$

This conclusion allows us to "localize" variational problem. Note that

$$\|A_0\psi_\nu u\|^2 = \|A\psi_\nu u\|^2 + \|V\psi_\nu u\|^2 - 2\operatorname{Re}(\psi_\nu Au, V\psi_\nu u), \tag{10.2.58}$$

because $[A, \psi_\nu]$ is a skew-Hermitian matrix valued function and V, ψ_ν are real-valued and scalar. Then

$$\begin{multline} \|A_0\psi_\nu u\|^2 \le 2\|A\psi_\nu u\|^2 + \|V\psi_\nu u\|^2 + \\ \lambda_\nu^{-2}\|\zeta^{-\frac12}\psi_\nu Au\|^2 + \lambda_\nu^2\|\zeta^{\frac12}V\psi_\nu u\|^2. \end{multline} \tag{10.2.59}$$

Multiplying by λ_ν^2, summing on ν and applying (10.2.57) we conclude that

$$\begin{multline}\sum_{\nu\ge 0}\lambda_\nu^2\|A_0\psi_\nu u\|^2\le 2\|\varsigma^{-\frac{1}{2}}Au\|^2+\sum_{\nu\ge 0}\lambda_\nu^2\|V\psi_\nu u\|^2+\\ C\sum_{\nu\ge 0}\lambda_\nu^4\|\varsigma^{\frac{1}{2}}V\psi_\nu u\|^2+C\sum_{\nu\ge 1}\bigl(\lambda_\nu\Phi(\alpha_\nu)+\lambda_0\beta_\nu\bigr)^2\|\chi_Y\gamma^{-1}\psi_\nu u\|^2+\\ C\sum_{\nu\ge 1}\bigl(\lambda_0\Phi(\beta_\nu)+\lambda_\nu\alpha_\nu\bigr)^2\|\chi_Y\gamma^{-1}\bar{\psi}_\nu\psi_0 u\|^2.\end{multline}\tag{10.2.60}$$

and we can replace in the right-hand expression $\|\varsigma^{-\frac{1}{2}}Au\|^2$ by $t^2\sum_{\nu\ge 0}\|\varsigma^{\frac{1}{2}}\psi_\nu u\|^2$ provided $\|\varsigma^{-\frac{1}{2}}Au\|\le t\|\varsigma^{\frac{1}{2}}u\|$.

Therefore

(10.2.61) Maximal dimension $\mathsf{N}(t)$ of subspace $\mathcal{L}\subset\mathscr{C}_0^1(X_{\mathsf{sing}})$ on which $\|\varsigma^{-\frac{1}{2}}Au\|\le t\|u\|$ does not exceed the sum of maximal dimensions $\mathsf{N}_\nu(t)$ of subspaces $\mathcal{L}_\nu\subset\mathscr{C}_0^1(\mathsf{supp}(\psi_\nu))$ on which

$$\begin{multline}\|A_0u_\nu\|^2\le 2t^2\lambda_\nu^{-2}\|\varsigma^{\frac{1}{2}}u_\nu\|^2+\|V\psi_\nu u\|^2+C\lambda_\nu^2\|\varsigma^{\frac{1}{2}}Vu_\nu\|^2+\\ C\bigl(\Phi(\alpha_\nu)+\lambda_\nu^{-1}\lambda_0\beta_\nu\bigr)^2\|\chi_Y\gamma^{-1}u_\nu\|^2\qquad\text{as }\nu\ge 1\end{multline}\tag{10.2.62}$$

and

$$\begin{multline}\|A_0u_0\|^2\le 2t^2\lambda_0^{-2}\|\varsigma^{\frac{1}{2}}u_0\|^2+\|Vu_0\|^2+\\ C\lambda_0^2\|\varsigma^{\frac{1}{2}}Vu_0\|^2+C\sum_{\nu\ge 1}\bigl(\Phi(\beta_\nu)+\lambda_\nu\lambda_0^{-1}\alpha_\nu\bigr)^2\|\chi_Y\gamma^{-1}\bar{\psi}_\nu u_0\|^2.\end{multline}\tag{10.2.63}$$

One can see easily that

$$A_0^2=\sum_{l,k}P_lg^{lk}P_k+(m+W)^2+\sum_l(\theta^lP_l+P_l\theta^l)+\theta\tag{10.2.64}$$

where θ^l,θ are Hermitian matrices such that $|\theta^l|\le C\rho$, $|\theta|\le C\tilde{\rho}^2+C\rho\gamma^{-1}\chi_Y$. Therefore

$$\begin{multline}\|A_0v\|^2\ge\\ \epsilon_1\sum\|P_jv\|^2+\|(m+W)v\|^2-C\|\tilde{\rho}v\|^2-C\|\rho^{\frac{1}{2}}\gamma^{-\frac{1}{2}}\chi_Yv\|^2\end{multline}\tag{10.2.65}$$

with $\tilde{\rho}$ supported in X_{sing} and due to (10.2.47)

(10.2.66) Estimate (10.2.65) remains true with P_j replaced by D_j in the right-hand expression.

Then on $\mathcal{L}_\nu$ with $\nu \ge 1$

$$\epsilon_1\|\nabla v\|^2 \le 2t^2\lambda_\nu^{-2}\|\zeta^{\frac{1}{2}}v\|^2 + C\bigl(\Phi(\alpha_\nu)+\lambda_\nu^{-1}\lambda_0\beta_\nu\bigr)^2\|\chi_Y\gamma^{-1}v\|^2 + C\lambda_\nu^2\|\zeta^{\frac{1}{2}}Vv\|^2 + C\|\tilde{\rho}v\|^2 + C\|\chi_Y\rho^{\frac{1}{2}}\gamma^{-\frac{1}{2}}v\|^2 \tag{10.2.67}$$

where originally we had $\|(m+W)u_\nu\|^2$ and $\|Vu\|^2$ in the left- and right-hand expressions respectively and used that $V^2 \le m^2 + c\tilde{\rho}^2$, $W \ge 0$.

Let us apply Rozenblioum variational estimate (9.A.11) . As $d \ge 3$ we conclude that for $\nu \ge 1$

$$\mathsf{N}_\nu(t) \le C\int \Bigl(t^d\lambda_\nu^{-d}\zeta^{\frac{d}{2}} + \lambda_\nu^d\zeta^{\frac{d}{2}}|V|^d + \tilde{\rho}^d + \rho^{\frac{d}{2}}\gamma^{-\frac{d}{2}} + \bigl(\Phi(\alpha_\nu)+\lambda_\nu^{-1}\lambda_0\beta_\nu\bigr)^d\gamma^{-d}\Bigr)\bar{\psi}_\nu\,dx. \tag{10.2.68}$$

Similarly starting from (10.2.63) instead of (10.2.62) we conclude that

$$\mathsf{N}_0(t) \le C\int \Bigl(t^d\lambda_0^{-d}\zeta^{\frac{d}{2}} + \lambda_0^d\zeta^{\frac{d}{2}}|V|^d + \tilde{\rho}^d + \rho^{\frac{d}{2}}\gamma^{-\frac{d}{2}}\chi_Y\Bigr)\bar{\psi}_0\,dx + C\sum_{\nu\ge 1}\int \bigl(\Phi(\beta_\nu)+\lambda_0^{-1}\lambda_\nu\alpha_\nu\bigr)^d\gamma^{-d}\bar{\psi}_\nu\bar{\psi}_0\,dx. \tag{10.2.69}$$

Then

$$\mathsf{N}(t) \le \sum_{\nu\ge 0}\mathsf{N}_\nu \le C(t^d+1)R \tag{10.2.70}$$

with

$$R = \int \Bigl(\lambda_0^{-d}\zeta^{\frac{d}{2}} + \lambda_0^d\zeta^{\frac{d}{2}}|V|^d + \tilde{\rho}^d + \rho^{\frac{d}{2}}\gamma^{-\frac{d}{2}}\chi_Y\Bigr)\bar{\psi}_0\,dx + \sum_{\nu\ge 1}\int \Bigl(\lambda_\nu^{-d}\zeta^{\frac{d}{2}} + \lambda_\nu^d\zeta^{\frac{d}{2}}|V|^d + \tilde{\rho}^d + \rho^{\frac{d}{2}}\gamma^{-\frac{d}{2}}\chi_Y + \bigl(\Phi(\alpha_\nu)+\lambda_\nu^{-1}\lambda_0\beta_\nu + \Phi(\beta_\nu)+\lambda_0^{-1}\lambda_\nu\alpha_\nu\bigr)^d\gamma^{-d}\Bigr)\bar{\psi}_\nu\,dx. \tag{10.2.71}$$

Our purpose now is to make (10.2.71) as small as possible in the framework of our construction. First of all recall that we can make ζ on $\bar{\psi}_0$ arbitrarily small function without affecting estimates; so we can skip first two terms in the first line. Since Φ is slowly growing at 0 almost optimal choice is $\alpha_\nu = \min(\lambda_0\lambda_\nu^{-1}, 1)$ and $\beta_\nu = \min(\lambda_0^{-1}\lambda_\nu, 1)$ and (10.2.71) becomes

$$R = \int \bigl(\tilde{\rho}^d + \rho^{\frac{d}{2}}\gamma^{-\frac{d}{2}}\chi_Y\bigr)\, dx + \\ \sum_{\nu\ge 1}\int \Bigl(\lambda_\nu^{-d}\zeta^{\frac{d}{2}} + \lambda_\nu^d \zeta^{\frac{d}{2}}|V|^d + \bar{\Phi}(\lambda_0^{-1}\lambda_\nu)^d\gamma^{-d}\Bigr)\bar{\psi}_\nu\, dx$$

with

$$\bar{\Phi}(t) = \Phi\bigl(\min(t, t^{-1})\bigr). \tag{10.2.72}$$

Estimating $|V| \le c(m+\rho)$, $\zeta \le c\rho^2(m+\rho)^{-1}$ and selecting $\lambda_\nu \asymp (m+\rho)^{-\frac{1}{2}}$ on $\mathsf{supp}(\psi_\nu)$ to optimize two first terms in the second line here we arrive to

$$R = \int \bigl(\tilde{\rho}^d + \rho^d\chi_Y\bigr)\, dx + \int \bar{\Phi}\bigl(\lambda_0(\rho+m)^{\frac{1}{2}}\bigr)^d\gamma^{-d}\chi_Y\, dx. \tag{10.2.73}$$

Estimates for a Number of Crossing Eigenvalues (End)

Let us discuss the possible choice of Φ. Note that if ϕ_ν^2 is some partition then partition ψ_ν^2 with $\psi_\nu = \varphi(\phi_\nu)\bigl(\sum\varphi(\phi_\nu)^2\bigr)^{-\frac{1}{2}}$ satisfies (10.2.55) provided

$$\varphi(t) = 0 \text{ for } t<0, \quad \varphi>0 \text{ for } t>0, \quad |\frac{d}{dt}\varphi(t)| \le c \tag{10.2.74}$$

and

$$|\frac{d}{dt}\varphi(t)| \le \alpha + \Phi(\alpha)\varphi(t) \qquad \forall \alpha, t\in (0,1). \tag{10.2.75}$$

Remark 10.2.17. One can prove easily that function φ satisfying (10.2.74)–(10.2.75) exists provided

$$\int_0^1 \frac{d\beta}{\min_{\alpha>0}(\alpha+\Phi(\alpha)\beta)} < \infty. \tag{10.2.76}$$

Therefore one can take $\Phi(\alpha) = C_\delta \alpha^{-2\delta}$ with arbitrarily small exponent $\delta > 0$ in which case the second term in (10.2.73) becomes

$$\lambda_0^{2\delta} \int (\rho + m)^\delta \chi_Y \gamma^{-d}\, dx + \lambda_0^{-2\delta} \int (\rho + m)^{-\delta} \chi_Y \gamma^{-d}\, dx.$$

Optimizing by λ_0 we arrive to term R_4 defined by (10.1.67).

However, one can pick $\Phi(\alpha)$ $\Phi(\alpha) = \log^{1+\delta} \alpha^{-1}$ or even

$$\Phi(\alpha) = \log \alpha^{-1} \cdot \log_{(2)} \alpha^{-1} \cdots \log_{(n)} \alpha^{-1} \cdot \log_{(n+1)}^{1+\delta} \alpha^{-1}$$

with arbitrarily small exponent $\delta > 0$, arbitrary n and $\log_{(n)}$ denoting the n-tuple logarithm (as $\alpha \ll 1$) which implies

Remark 10.2.18. If

$$c^{-1}\lambda \le \rho + m \le c\lambda \qquad \text{in } X_{\mathsf{reg}} \cap X_{\mathsf{sing}} \tag{10.2.77}$$

with some parameter λ then the second term optimizes to $C \int \chi_Y \gamma^{-d}\, dx$ which is less than the first term.

Further, if X_{sing} consists of several connected components $X_{\mathsf{sing},l}$ then one can use different parameters $\lambda = \lambda_l$ in $X_{\mathsf{reg}} \cap X_{\mathsf{sing},l}$.

Remark 10.2.19. For $d = 2$ we should slightly modify Rozenblioum estimate; however this modification affects only term with $\nu = 0$ and only contribution of $X_{\mathsf{sing}} \setminus X_{\mathsf{reg}}$. However, applying the same arguments as in Appendix *9.A.1.2 Bracketing* we see that one needs only to replace the first term in (10.2.73) by (10.1.66).

Further, if $d = 2$ and X_{sing} consists of several connected components one can instead calculate R_3 for each of them and then add results.

Remark 10.2.20. Let us recall that deriving (10.2.67) we used estimate $(m + W)^2 - V^2 \le c\rho^2$ in X_{reg}. However in $X_{\mathsf{reg,ell}}$ one can replace estimate by $(m + W)^2 - V^2 \le c - \epsilon\rho^2$ and, moreover, from the very beginning one can take $\zeta = \rho\gamma^{-1}m^{-1}$ here and therefore one can get slightly better remainder estimate than (10.2.73): from integral of ρ^d is excluded $X_{\mathsf{reg,ell}} \cap \{\rho\gamma \ge C_\epsilon\}$ and from integral of $\bar{\Phi}\bigl(\lambda_0(\rho + m)^{\frac{1}{2}}\bigr)^d \gamma^{-d}$ is excluded $X_{\mathsf{reg,ell}} \cap \{\rho\gamma \ge C_\epsilon\}$ where $\bar{\Phi}\bigl(\lambda_0(\rho + m)^{\frac{1}{2}}\bigr) \le \epsilon\rho\gamma$.

Estimates for a Number of Remaining Eigenvalues

Thus, we derived asymptotics for terms counting eigenvalues crossing τ_1 and τ_2 i.e. the last line in the right-hand expression of (10.1.18). The main part of the result is given by Weyl formula (cut outside of X_{reg}) and the remainder is estimated.

In order to prove the upper estimate (10.1.65) we need also to derive an upper estimate for $\mathsf{N}(B;\tau_1,\tau_2)$ which is the second line in (10.1.18) where here and in what follows

$$B = A - t_1 J_1 - \ldots - t_n J_n. \tag{10.2.78}$$

So B is a generalized Dirac operator (5.3.71) with $J = I$ satisfying all the conditions of Subsection 5.3.2 with the following modifications: in conditions on $D^\alpha V$, $D^\alpha W$ with $|\alpha|\geq 1$ there is an additional factor $\bar{t}$ in the right-hand expression, conditions on V (without derivatives) remain true and

$$|V| \leq \bigl(m - \bar{t}\min(\rho, \frac{1}{m}\rho^2)\bigr)_+ + \min(\rho, \frac{1}{m}\rho^2) \qquad \text{in } X'_{\mathsf{reg}} \tag{10.2.79}$$

$$\bar{t}\min(\rho, \frac{1}{m}\rho^2) > m \implies W \geq \bar{t}\min(\rho, \frac{1}{m}\rho^2) \qquad \text{in } X'_{\mathsf{reg}}, \tag{10.2.80}$$

$$c\bar{t}\min(\rho, \frac{1}{m}\rho^2) \geq W \geq 0 \qquad \text{in } X \tag{10.2.81}$$

where $\bar{t}$ is arbitrarily large and c does not depend on $\bar{t}$, X'_{reg} is the $\epsilon\gamma$-neighborhood of $X_{\mathsf{reg}}\setminus X_{\mathsf{sing}}$; however we can still denote it by X_{reg} without breaking estimates. All these conditions are provided by an appropriate choice of $J_1,\ldots,J_n$ and $\phi_1,\ldots,\phi_n$. Also in X_{sing} one should replace ρ by $\tilde{\rho}$.

Note that conditions (10.2.79)–(10.2.79) mean that ellipticity condition in X_{reg} is fulfilled. More precisely:

$$\tau_1 - \frac{k}{m}\rho^2 + m \geq V \geq \tau_2 - m + \frac{k}{m}\rho^2 \qquad \text{as } m \geq k_1\rho, \tag{10.2.82}$$

$$W \geq k\rho \qquad \text{as } m \leq k_1\rho, \tag{10.2.83}$$

and thus in both cases

$$\tau_1 - k\min(\rho, \frac{1}{m}\rho^2) + m + W \geq V \geq \\ \tau_2 - m - W + k\min(\rho, \frac{1}{m}\rho^2). \tag{10.2.84}$$

Let us divide connected components of X_{sing} into two types. Components with $\tilde{\rho}\le cm$ at every point are components of the first type.

Remark 10.2.21. Note that we could choose $J_1,\dots,J_n$ such that conditions (10.2.79)–(10.2.81) are also fulfilled on components of the first type but with ρ replaced by $\tilde{\rho}$ and all estimates of the previous subsubsection remain true.

One can see easily that $\mathsf{N}(B;(\tau_1,\tau_2))$ equals the maximal dimension of the subspace $\mathcal{L}\subset\mathscr{C}^1_0(X,\mathbb{H})$ on which

$$\|(B-\bar{\tau})u\|^2<\tau^2\|u\|^2\qquad (u\ne 0) \tag{10.2.85}$$

where $\bar{\tau}=\frac{1}{2}(\tau_2+\tau_1)$, $\tau=\frac{1}{2}(\tau_2-\tau_1)$.

Let $\{\psi_\nu^2\}_{k=1,2}$ be a partition of unity such that $|\nabla\psi_\nu|\le c\gamma^{-1}$ and ψ_ν be supported in X_ν where X_1 is the union of X_{reg} and all components of X_{sing} the first type and X_2 is the union of all components of X_{sing} of the second type.

Then one can see easily that

$$\sum_\nu\|(B-\bar{\tau})\psi_\nu u\|^2=\|(B-\bar{\tau})u\|^2+\sum_\nu\|[B,\psi_\nu]u\|^2, \tag{10.2.86}$$

$$\|[B,\psi_\nu]u\|\le C_0\|\gamma^{-1}\chi_Y u\| \tag{10.2.87}$$

where χ_Y is the characteristic function of $Y=X_{\mathsf{reg}}\cap X_{\mathsf{sing}}$.

Therefore

(10.2.88) $\dim\mathcal{L}$ does not exceed $\max\dim\mathcal{L}_1+\max\dim\mathcal{L}_2$ where $\mathcal{L}_\nu$ is a subspace of $\mathscr{C}^1_0(X_\nu,\mathbb{H})$ on which inequality

$$\|(B-\bar{\tau})u\|^2<\tau^2\|u\|^2+C_0\|\gamma^{-1}\chi_Y u\|^2\quad (u\ne 0) \tag{10.2.89}$$

is fulfilled.

Obviously there exists an extension B' of B from X_1 to X such that conditions (10.2.79)–(10.2.81) are fulfilled on X. Then $\dim\mathcal{L}_1$ does not exceed the cardinality of the spectrum of the problem

$$(B'-\bar{\tau})v+\lambda Jv=0\qquad v\in\mathfrak{D}(B') \tag{10.2.90}$$

in the interval $(-1,1)$ where $J=(\tau^2+c\gamma^{-2}\chi_Y)^{\frac{1}{2}}$ and $J\in\mathscr{L}^\infty$ provided the right-hand expression of the estimate in question is finite.

Dividing v into $\frac{1}{2}$D-dimensional components v_1 and v_2 we can rewrite operator $B' - \bar{\tau} + \tau J$ in the form

$$B' - \bar{\tau} + \tau J = \begin{pmatrix} \zeta_1 & T \\ T^* & -\zeta_2 \end{pmatrix}$$

where conditions (10.2.79)–(10.2.81) fulfilled in X yield that

$$\zeta_j > \epsilon_0(1 - |\tau|), \ |\tau| \le 1 \qquad \forall j = 1, 2$$

provided $\bar{t}$ is large enough. Then $(-1, 1)$ does not intersect the spectrum of the problem (10.2.90) and therefore $\dim \mathcal{L}_1 = 0$.

Let us estimate $\dim \mathcal{L}_2$. Let us note that the conditions on V can be rewritten in the form

$$\tau - m - c\min(\tilde{\rho}, \frac{1}{m}\tilde{\rho}^2) \le V - \bar{\tau} \le m - \tau + c\min(\tilde{\rho}, \frac{1}{m}\tilde{\rho}^2). \tag{10.2.91}$$

In particular

$$\tau \ge m \implies |V - \tau| + |\tau - m| \le c\min(\tilde{\rho}, \frac{1}{m}\tilde{\rho}^2). \tag{10.2.92}$$

Let us consider operator $B_0 = B - V \cdot I - \sigma_0 W$. Standard arguments and assumptions (10.1.69), (10.1.70) easily yield that

$$\|B_0 u\|^2 \ge \epsilon_1 \|Du\|^2 + m^2\|u\|^2 - C\|\tilde{\rho}u\|^2 \tag{10.2.93}$$

On the other hand, on $\mathcal{L}_2$ estimates

$$\begin{aligned}\|B_0 u\|^2 \le (1+\varepsilon^{-2})\|(B-\bar{\tau})u\|^2 + (1+\varepsilon^2)\|(V + \sigma_0 W - \bar{\tau})u\|^2 \le \\ \bigl((1+\varepsilon^{-2})\tau^2 + (1+\varepsilon^2)(m-\tau)^2\bigr)\|u\|^2 + \\ C(1+\varepsilon^{-2})\|\chi\gamma^{-1}u\|^2 + C(1+\varepsilon^2)\|\tilde{\rho}u\|^2\end{aligned}$$

are fulfilled where we have applied (10.2.91) and the restrictions on W. These inequalities, (10.2.93) and (10.1.69) yield that

$$\|Du\|^2 \le C(\mathcal{K}u, u) \qquad \text{on} \quad \mathcal{L}_2 \tag{10.2.94}$$

where

$$\mathcal{K} = 9\bigl((m-\tau)\varepsilon - \varepsilon^{-1}\tau\bigr)^2 + \tilde{\rho}^2(1+\varepsilon^2) + \chi\gamma^{-2}(1+\varepsilon^{-2}). \tag{10.2.95}$$

If $\tau \geq 2m$ then (10.2.91) yield that $m + \tau \leq c\tilde{\rho}$ in X_2 and therefore for $\varepsilon = 1$ one can skip the first and the third terms in $\mathcal{K}$; then Rozenblioum estimate yields that $\dim \mathcal{L}_2$ does not exceed CR_3 defined by (10.1.66); as $d = 2$ a standard trick with an extra partition is needed to get $\int \rho^d\, dx$ rather than the same expression as with $\tilde{\rho}$.

Let us consider the case $\tau \leq 2m$. One can assume without any loss of the generality that $\tau_2 - \tau_1 \leq \frac{1}{2}m$; otherwise one can divide the interval (τ_1, τ_2) into fewer than 10 intervals (τ_1', τ_2') with $\frac{1}{4}m \leq |\tau_1' - \tau_2'| \leq \frac{1}{2}m$. Moreover, if $\tau_2 - \tau_1 \geq \frac{1}{4}m$ then the value of $\varepsilon > 0$ for which the first term in $\mathcal{K}$ vanishes is disjoint from 0 and ∞ and in this case estimate $\dim \mathcal{L}_2 \leq CR_3$ remains true.

Let us now consider the case $\tau \leq \frac{1}{4}m$. Then Rozenblioum estimate yields

$$\dim \mathcal{L}_2 \leq C\Bigl(\bigl((m-\tau)\varepsilon - \varepsilon^{-1}\tau\bigr)^2 + a^2\varepsilon^2 + b^2\varepsilon^{-2}\Bigr)^{\frac{d}{2}}(\mathrm{mes}(X_2)) + CR_3 \tag{10.2.96}$$

with

$$a = \begin{cases} \Bigl(\dfrac{1}{\mathrm{mes}(X_2)}\displaystyle\int_{X_2} \rho^d\, dx\Bigr)^{\frac{1}{d}} & (d \geq 3), \\ \Bigl(\dfrac{1}{\mathrm{mes}(X_2)}\displaystyle\int_{X_2} \rho^{2p}\, dx\Bigr)^{\frac{1}{2p}} & (d = 2, p > 1), \end{cases} \tag{10.2.97}$$

$$b = \Bigl(\frac{1}{\mathrm{mes}(X_2)}\int_{X_2} \chi\gamma^{-d}\, dx\Bigr)^{\frac{1}{d}} \qquad (d \geq 2) \tag{10.2.98}$$

(for $d = 2$ the procedure of covering of $X_1 \cap X_2$ by balls $B(x_n, \epsilon\gamma(x_n))$ is applied; see above).

Optimizing with respect to ε we obtain from (10.2.96) the final estimate

$$\dim \mathcal{L}_2 \leq CR_3 + CR_5 \tag{10.2.99}$$

with R_5 defined by (10.1.68).

For $\tau \geq \epsilon_0 m$ this estimate coincides with (10.2.96).

Theorem 10.1.8 has been proven.

Remark 10.2.22. The contribution of X_2 to the remainder in Theorem 10.1.8 is estimated reasonably as $\tilde{\rho} \geq cm$ everywhere on X_2. One can get better estimates using Birman-Schwinger principle while reducing $\tilde{\rho}$ and estimating number of "escaping" eigenvalues. Further development in this direction is left to the reader.

10.2.5 From Rough to Irregular Operators

Remark 10.2.23. (i) While we formulated our results for smooth operators, the same results hold for rough operators (in the assumptions of the corresponding theorems).

(ii) Approximation of operators with irregular coefficients by rough operators is more subtle in the case when we calculate $\mathsf{N}(A,\tau_1,\tau_2)$ because in contrast to the case $\mathsf{N}^-(A-\tau)$ simple bracketing does not work here.

However if we know that $A'-\eta<A<A'+\eta$[19] we conclude that

$$\mathsf{N}(A',\tau_1+\eta,\tau_2-\eta)\le \mathsf{N}(A,\tau_1,\tau_2)\le \mathsf{N}(A',\tau_1-\eta,\tau_2+\eta) \tag{10.2.100}$$

and therefore

$$\begin{aligned} -\mathsf{N}(A',\tau_1,\tau_1+\eta)-\mathsf{N}(A',\tau_2-\eta,\tau_2)\le{}&\\ \mathsf{N}(A,\tau_1,\tau_2)-\mathsf{N}(A',\tau_1,\tau_2)\le{}&\\ &\mathsf{N}(A',\tau_1-\eta,\tau_1)+\mathsf{N}(A',\tau_2,\tau_2+\eta)\end{aligned} \tag{10.2.101}$$

and this conclusion could be applied either directly or through Birman-Schwinger principle.

Comments

Due to its nature and to the fact that M. Sh. Birman for many years has the leader of the school using variational approach to eigenvalue asymptotics Birman-Schwinger principle has been used mainly in such papers. One should mention many papers of M. Sh. Birman and M. Z. Solomyak and their students: G. Rozenblioum, Yu. Safarov, A. Sobolev and many others.

Extremely surprising is that this powerful method had not find its way to the papers studying sharp asymptotics much sooner; at least the author is not aware of any such papers prior to his own. Furthermore, it took some time and advice of M. Z. Solomyak to realize the tremendous power of Birman-Schwinger principle which allows to analyze often very easily many asymptotics which either could not be analyzed directly or at least require some difficult and ingenious arguments. With Birman-Schwinger principle

[19] Here A is an original operator with the irregular coefficients, A' is its rough approximation and $\eta>0$ is a small parameter.

we just reduce asymptotics with respect to some parameters (f.e. spectral parameter) to asymptotics with respect to others, most notably semiclassical parameter.

In the framework of this chapter this method was first used by V.Ivrii [17]. We will use Birman-Schwinger principle systematically either directly or through Chapters 9 and 10.

Part V

Asymptotics of Spectra

Chapter 11

Weyl Asymptotics of Spectra

This chapter is devoted to applications of the results of Parts Parts II and III and IV to the eigenvalue asymptotics. Now we consider operator A depending on parameters; then all the functions ρ, $\tilde{\rho}$, ϱ, and γ with different indices and zones X_{reg} and X_{sing} depend on these parameters as well. Further, the main part of the asymptotics N^{W} (given by Weyl formula but may be with corrections) and the remainder R depend on this parameter as well; both of them tend to infinity but under appropriate assumptions $R \ll \mathsf{N}^{\mathsf{W}}$. Further, as parameters tend to their limits X_{sing} shrinks.

In Section 11.1 we treat semiclassical asymptotics. In this case we consider operator $A_h = A(x, hD, h)$ and select $\rho_h = h^{-1}\rho$, $\gamma_h = \gamma$ and $h_{\mathsf{eff}} = (\rho_h\gamma_h)^{-1} = h\rho^{-1}\gamma^{-1}$; then $X_{\mathsf{reg},h} = \{\rho_h\gamma_h \ge 1\} = \{x : \rho\gamma \ge h\}$ and $X_{\mathsf{sing}} = \{\rho\gamma \le 2h\}$.

In Sections 11.2–11.6 we consider asymptotics with respect to the spectral parameter τ; in this case $A_\tau = A - \tau$ and usually $\rho_\tau = |\tau|^{1/m} + \rho$, $\gamma_\tau = \gamma$ as we assume that A satisfies some conditions with functions ρ, γ and $\varrho = \rho^m$. Then $h_{\mathsf{eff}} = (|\tau|^{1/m} + \rho)^{-1}\gamma^{-1}$ and $X_{\mathsf{reg},\tau} = \{(|\tau|^{1/m} + \rho)\gamma \ge 1\}$ and $X_{\mathsf{sing},\tau} = \{(|\tau|^{1/m} + \rho)\gamma \le 2\}$.

More precisely, in Sections 11.2–11.5 we consider asymptotics of the eigenvalues tending to infinity, then $\tau \to +\infty$ but there are different scenarios:

- in Section 11.2 domain X is "essentially" bounded (i.e., exits to infinity are thin enough) and the coefficients are weakly singular;

- in Section 11.4 the coefficients are strongly singular and the coercivity condition is fulfilled but the domain is still "essentially" bounded;

V. Ivrii, *Microlocal Analysis, Sharp Spectral Asymptotics and Applications II*, https://doi.org/10.1007/978-3-030-30541-3_11

- in Section 11.5 the domain is "very unbounded" and the spectrum is discrete only because the lower order terms are coercive. In particular, we treat the Schrödinger operator with potential V tending to $+\infty$ at infinity and with the spectral parameter τ tending to $+\infty$, usually like $|x|^{2\mu}$ with $\mu > 0$ and the growths of both $\mathsf{N}^{\mathsf{W}}_{\tau}$ and R_{τ} depend on μ.

Further, in Section 11.6 we treat asymptotics of the eigenvalues tending to the boundary of the essential spectrum. In particular, we treat the Schrödinger and Dirac operators with potentials tending to 0 at infinity, usually like $|x|^{2\mu}$ with $\mu \in (-1, 0)$ and the spectral parameter τ tending to -0 and $M - 0$ or $-M + 0$ respectively (where M is the mass in the case of the Dirac operator). Then the rates of the growth of both $\mathsf{N}^{\mathsf{W}}_{\tau}$ and R_{τ} depend on μ.

Section 11.7 is devoted to the multiparameter asymptotics; for example, there is the semiclassical parameter $h \to +0$ and the spectral parameter τ tending either to $+\infty$ or $-\infty$ (for the Schrödinger and Dirac operators in bounded domains or for the Schrödinger operator in unbounded domains), or to -0, $M - 0$ or $-M + 0$ (for the Schrödinger and Dirac operators in unbounded domains), or to $\inf V + 0$, $\inf V + M + 0$, $\sup V - M - 0$ (for the Schrödinger and Dirac operators) where for the Schrödinger operator $\inf V$ is not necessarily finite. Moreover, the case of a vector semiclassical parameter $h = (h_1, \dots, h_d)$ is treated as well.

For semi-bounded operators, Riesz means are also treated.

Finally, in Section 11.8 we consider asymptotics of the spectrum of the negative order pseudodifferential operators tending to 0. While these operators are not exactly covered by the general theory and require special analysis, some special cases (like $\Delta^{-m} V$) are and the general results are the same.

For simplicity we assume the sufficient smoothness but the reader can generalize easily these results applying results for rough and irregular operators.

11.1 Semiclassical Eigenvalue Asymptotics

In this section the operator is fixed and only the semiclassical parameter $h \to +0$ varies.

11.1.1 General Analysis

Standard Remainder Estimates

Let us consider first operators which are semi-bounded from below and apply the results of Chapter 9 derived on the basis of the results of Parts II and III. This means that $\tau_1 = -\infty$ and without any loss of the generality one can assume that $\tau_2 = 0$.

Let us consider operator

$$A_h = \sum_{|\alpha|,|\beta|\le m} (hD)^\alpha a_{\alpha,\beta}(hD)^\beta \tag{11.1.1}$$

with coefficients satisfying

$$a_{\alpha,\beta} = a^\dagger_{\beta,\alpha}, \tag{11.1.2}$$

$$\|D^\sigma a_{\alpha,\beta}\| \le c\varrho\rho^{-|\alpha|-|\beta|}\gamma^{-|\sigma|} \qquad \forall \alpha,\beta : |\alpha|\le m, |\beta|\le m\ \forall\sigma : |\sigma|\le K \tag{11.1.3}$$

and

$$\sum_{|\alpha|=|\beta|=m} \langle a_{\alpha,\beta}v_\alpha, v_\beta\rangle \ge \epsilon_0\varrho\rho^{-2m}\sum_{|\alpha|=m}\|v_\alpha\|^2 \qquad \forall v=(v_\alpha)_{|\alpha|=m}\quad \forall x:\rho\gamma\le h_0 \tag{11.1.4}$$

where scaling function γ satisfies $(9.1.6)_1$ on X and

$$\operatorname{mes}(\mathcal{X}_{\mathsf{sing}}) = 0 \qquad \text{where}\quad \mathcal{X}_{\mathsf{sing}} = \{x, \gamma(x)=0\}, \tag{11.1.5}$$

functions ρ, ϱ satisfying $(9.1.6)_2$ and $(9.1.6)_3$ are defined in $\mathcal{X}_{\mathsf{reg}} = X\setminus \mathcal{X}_{\mathsf{sing}}$, and $h_0 = \mathsf{const} > 0$.

The sets $\mathcal{X}_{\mathsf{reg}}$ and $\mathcal{X}_{\mathsf{sing}}$ are called *regular* and *singular sets* respectively. These sets do not coincide with but are obviously related to the *regular* $X_{\mathsf{reg},h} = \{x, \rho\gamma\ge h\}$ and *singular zones* $X_{\mathsf{sing},h} = \{x,\rho\gamma\le 2h\}$ and respectively: $X_{\mathsf{reg},h}\subset \mathcal{X}_{\mathsf{reg}}$ and expands to it, $X_{\mathsf{sing},h}\supset\mathcal{X}_{\mathsf{sing}}$ and shrinks to it as $h\to +0$.

In particular, for the Schrödinger operator

$$A = \sum_{1\le j,k\le d} (hD_j - V_j)g^{jk}(hD_k - V_k) + V \tag{11.1.6}$$

with real-valued functions $g^{jk}=g^{kj}$, V_j and V we always take $\varrho=\rho^2$ and conditions (11.1.3)–(11.1.4) become

$$|D^\alpha g^{jk}|\le c\gamma^{-|\alpha|},\quad |D^\alpha V_j|\le c\rho\gamma^{-|\alpha|},\quad |D^\alpha V|\le c\rho^2\gamma^{-|\alpha|} \tag{11.1.7}$$

and

$$\epsilon_0|\xi|^2\le\sum_{j,k} g^{jk}\xi_j\xi_k\le c|\xi|^2\qquad \forall\xi\in\mathbb{R}^d. \tag{11.1.8}$$

Moreover, we assume that conditions (9.1.15) (the boundary regularity condition) are fulfilled in $\mathcal{X}_{\mathsf{reg}}$. For the sake of simplicity we assume that the Dirichlet boundary conditions are given in which case the ellipticity condition for the boundary value problem is fulfilled automatically. However later we discuss Neumann-like problems as well.

Then one can apply Theorem 9.1.7 with $X_{\mathsf{reg}}=X_{\mathsf{reg},h}=\{x\in X,\rho\gamma>h\}$ and $X_{\mathsf{sing}}=X_{\mathsf{sing},h}=\{x\in X,\rho\gamma\le 2h\}$ provided $h\le h_0$. Here conditions (11.1.2)–(11.1.4) yield the following estimate on X_{sing}:

$$(A_hu,u)\ge \epsilon_0\int\varrho\Bigl(h^{2m}\rho^{-2m}\sum_{|\alpha|=m}|D^\alpha u|^2-C_0\sum_{|\alpha|\le m-1}h^{2|\alpha|}\rho^{-2|\alpha|}|D^\alpha u|^2\Bigr)\,dx \tag{11.1.9}$$

for all compactly supported functions $u\in\mathscr{C}^{2m}(\bar X_{\mathsf{sing}},\mathbb{H})$ satisfying the Dirichlet boundary condition. Therefore Rozenblioum estimate (9.A.11) yields that the contribution of the zone X_{sing} to the upper estimate of N_h^- in Theorem 9.1.7 does not exceed CR_0h^{-d} with

$$R_0= C\Bigl(\int_{X_{\mathsf{sing}}}\varrho^{-p}\rho^{2mp}\,dx\Bigr)^{\frac{d}{2mp}}\Bigl(\int_{X_{\mathsf{sing}}}\varrho^{q}\,dx\Bigr)^{\frac{d}{2mq}}\Bigl(\operatorname{mes}(X_{\mathsf{sing}})\Bigr)^{1-\frac{d}{2m}(\frac{1}{p}+\frac{1}{q})} \tag{11.1.10}$$

where p,q are arbitrarily chosen numbers such that

$$\frac{1}{p}+\frac{1}{q}\le\frac{2m}{d},\qquad p\ge 1, q\ge 1 \tag{11.1.11}$$

and

$$\text{either}\quad \frac{1}{p}+\frac{1}{q}<\frac{2m}{d},$$
$$\text{or}\quad \frac{1}{p}+\frac{1}{q}=\frac{2m}{d}, d\geq m \text{ and } d>m \implies q>1 \text{ and } p>1 \implies 2m<\frac{d}{p}(p+1). \tag{11.1.12}$$

Recall that X_{sing} does not contribute to the lower estimate of N_h^-.

At the same time the contribution of the zone X_{reg} to both lower and upper estimates does not exceed $Ch^{1-d}R_1 + Ch^{s-d}R_2$ where

$$R_1 = \int_{X_{\mathsf{reg},h}\setminus X_{\mathsf{reg,ell},h}} \rho^{d-1}\gamma^{-1}\,dx, \tag{11.1.13}$$

$$R_2 = \int_{X_{\mathsf{reg},h}} \rho^{d-s}\gamma^{-s}\,dx \tag{11.1.14}$$

with arbitrarily chosen s and we assume that in every ball $\mathcal{B}(y,\gamma(y),X)$ with $y\in X_{\mathsf{reg}}$ one of the conditions of Part Parts II and III providing remainder estimate $O(h_{\mathsf{eff}}^{1-d})$ [1)] is fulfilled; here $h_{\mathsf{eff}} = h(\rho\gamma)^{-1}$.

Further, the principal term of the upper and lower estimates is $\mathsf{N}_h^{\mathsf{W}} = h^{-d}\mathsf{N}^{\mathsf{W}}$ with the *cut Weyl expression*

$$\mathsf{N}^{\mathsf{W}} = (2\pi)^{-d}\int_{T^*X_{\mathsf{reg},h}} \boldsymbol{n}(x,\xi,-\infty,0)\,dxd\xi \tag{11.1.15}$$

where $\boldsymbol{n}(x,\xi,.,.)$ is the eigenvalue counting function for the symbol

$$a(x,\xi)=\sum_{\alpha,\beta} a_{\alpha,\beta}\xi^{\alpha+\beta}; \tag{11.1.16}$$

recall that both X_{reg} and X_{sing} and thus N^{W} depend on h.

So the following statement is proven:

Theorem 11.1.1. *Let A be a self-adjoint operator (with the Dirichlet boundary conditions) of the form* (11.1.1) *in $\mathscr{L}^2(X,\mathbb{H})$ and let conditions* (11.1.2)–(11.1.5) *be fulfilled. Moreover, let us assume that, in every ball $\mathscr{B}(y,\gamma(y),X)$*

---

1) To be discussed in Remarks 11.1.3 and 11.1.4.

with $y \in \mathcal{X}_{\mathsf{reg}}$, *one of the conditions of Parts II and III providing remainder estimate* $O(h_{\mathsf{eff}}^{1-d})$ [1] *is fulfilled. Then the estimates*

$$h^{-d}\mathsf{N}^{\mathsf{W}} - Ch^{1-d}R_1 \le \mathsf{N}^-(A_h) \le h^{-d}\mathsf{N}^{\mathsf{W}} + Ch^{1-d}R_1 + Ch^{s-d}R_2 + CR_0h^{-d} \tag{11.1.17}$$

hold with N^{W}, R_1, R_2 *and* R'' *given by* (11.1.15), (11.1.13), (11.1.14) *and* (11.1.10) *respectively.*

Corollary 11.1.2. *(i) For* N^{W} *in formula* (11.1.15) *let us replace the domain* $X_{\mathsf{reg},h}$ *by the domain* $\mathcal{X}_{\mathsf{reg}}$ *(or* X*); then the error does not exceed* $C\int_{X_{\mathsf{sing},h}} \rho^d\,dx \le CR_0$ *due to Hölder inequality. Therefore*

$$|\mathsf{N}^-(A_h) - \mathsf{N}^{\mathsf{W}}h^{-d}| \le Ch^{1-d}R_1 + Ch^{s-d}R_2 + Ch^{-d}R_0 \tag{11.1.18}$$

where N^{W} *is now given by (uncut) Weyl expression and no longer depends on* h.

(ii) This estimate yields that if

$$\rho^{d-s}\gamma^{-s} \in \mathscr{L}^1(\{X,\ \rho\gamma \ge h_0\}), \tag{11.1.19}$$

$$\rho^{d-1}\gamma^{-1} \in \mathscr{L}^1(X \setminus X_{\mathsf{reg,ell},h_0}) \tag{11.1.20}$$

and

$$R_0 = O(h) \qquad \text{as } h \to +0, \tag{11.1.21}$$

then

$$\mathsf{N}^-(A_h) = h^{-d}\mathsf{N}^{\mathsf{W}} + O(h^{1-d}). \tag{11.1.22}$$

These asymptotics will be improved and generalized below, where other asymptotics due to the results of Chapter 9 will be derived.

(iii) Let us assume that

$$\varrho \asymp \rho^k \quad \text{in } \mathcal{X}_{\mathsf{reg}} \tag{11.1.23}$$

and that one of the following conditions are fulfilled:

$$d > 2m, \qquad k \in [0, 2m]; \tag*{(11.1.24)$_1$}$$

$$m < d \le 2m, \qquad k \in [2m-d, d); \tag*{(11.1.24)$_2$}$$

$$d = m, \qquad k = d. \tag*{(11.1.24)$_3$}$$

Then one can choose p, q *satisfying condition* (11.1.11) *with equality and condition* (11.1.12) *and such that*

$$R_0 \asymp \int_{X_{\mathsf{sing},h}} \rho^d\, dx \le 2h \int_{X_{\mathsf{sing},h}} \rho^{d-1}\gamma^{-1}\, dx.$$

This inequality and (11.1.20) *yield that* $R_0 = o(h)$ *and therefore condition* (11.1.21) *is fulfilled. In particular, for* $d > 2m$ *one can take* $k = 2m$.

Remark 11.1.3. Let us list some cases when according to Parts II and III contribution of $\mathcal{B}(y, \gamma(y), X)$ to the remainder is $O(h^{1-d}\rho^{d-1}\gamma^{d-1})$; these conditions are uniform with respect to y (after rescaling):

(11.1.25) Assumptions (11.1.3), (11.1.4) and (9.1.15) are fulfilled with $K = K(d)$ and operator A is microhyperbolic on the energy level 0;

(11.1.26) A is Schrödinger operator, $d \ge 2$, assumptions (11.1.7), (11.1.8) and (9.1.15) are fulfilled with $K = K(d)$;

(11.1.27) A is "rough" operator described in Section 4.6 as $B(y, \gamma(y)) \subset X_{\mathsf{reg}}$ and in Section 7.5 otherwise.

Through bracketing $A^- \le A \le A^+$ between two rough operators such that contributions to N^{W} of $\mathcal{B}(y, \gamma(y), X)$ for A^- and A^+ differ by $O(h_{\mathsf{eff}}^{1-d})$ we cover also

(11.1.28) Assumptions (11.1.3), (11.1.4) are fulfilled with $K = 1$, $\nabla a_{\alpha,\beta}$ are continuous with continuity modulus $C\varrho\rho^{-|\alpha|-|\beta|}\gamma^{-1}|\log(|x-y|\gamma^{-1})|^{-1}$ and either $B(y, \gamma(y)) \subset X_{\mathsf{reg}}$ or extra conditions of Theorem 7.5.15 to the boundary are fulfilled, operator A is ξ-microhyperbolic on the energy level 0;

(11.1.29) A is Schrödinger operator, assumptions (11.1.7), (11.1.8) are fulfilled with $K = 1$, ∇g^{kl}, ∇V_k and ∇V are continuous with continuity moduli $C\rho^p\gamma^{-1}|\log(|x-y|\gamma^{-1})|^{-1}$, $p = 0, 1, 2$ respectively and either $B(y, \gamma(y)) \subset X_{\mathsf{reg}}$ or extra conditions of Theorem 7.5.15 to the boundary are fulfilled, $d \ge 3$ or $|V| + \gamma|\nabla V| \ge \epsilon\rho^2$;

(11.1.30) A is Schrödinger operator, $d=2$, assumptions (11.1.7), (11.1.8) are fulfilled with $K=2$, $\nabla^2 g^{kl}$, $\nabla^2 V_k$ and $\nabla^2 V$ are continuous with continuity moduli $C\rho^p\gamma^{-2}|\log(|x-y|\gamma^{-1})|^{-2}$, $p=0,1,2$ respectively and either $B(y,\gamma(y))\subset X_{\mathsf{reg}}$ or extra conditions of Theorem 7.5.15 to the boundary are fulfilled.

Remark 11.1.4. Actually we can weaken assumptions (11.1.28)–(11.1.30) because we can bracket symbol $a(x,\xi)$ between $a(x,\xi)\pm b_\varepsilon(x)$ where

$$b_\varepsilon(x)=C_0\max_{\alpha,\beta}\operatorname{osc}(\nabla a_{\alpha,\beta},x,\varepsilon)*\phi_\varepsilon \tag{11.1.31}$$

where

$$\operatorname{osc}(a,x,\varepsilon)=\max_{y:|x-y|\le\varepsilon}a-\min_{y:|x-y|\le\varepsilon}a \tag{11.1.32}$$

ϕ_ε is a standard mollification function. An easy proof is left to the reader.

Then for ξ-microhyperbolic operator contributions of $\mathcal{B}(y,\gamma(y),X)$ to N^{W} for A^- and A^+ differ by no more than

$$Ch_{\mathsf{eff}}^{-d}\max_{\alpha,\beta}\int\operatorname{osc}(\nabla a_{\alpha,\beta},x,\varepsilon)\,dx \tag{11.1.33}$$

with an integral taken over $\mathcal{B}(y,\gamma(y),X)$ rescaled and in the end of the day results in an additional error

$$R_3=Ch_{\mathsf{eff}}^{1-d}\max_{\alpha,\beta}\int_{X_{\mathsf{reg}}\setminus X_{\mathsf{reg,ell}}}\rho^d\operatorname{osc}\bigl(\nabla a_{\alpha,\beta},x,\rho^{-1}\log(\rho\gamma)\bigr)\,dx. \tag{11.1.34}$$

So, to estimate (11.1.33) by $O(h_{\mathsf{eff}}^{1-d})$ we actually need to impose estimate $O(|\log\varepsilon|^{-1})$ to $\mathscr{L}^1$-norm rather than $\mathscr{L}^\infty$-norm of $\operatorname{osc}(\nabla a_{\alpha,\beta},x,\varepsilon)$.

We leave to the reader the following easy problem:

Problem 11.1.5. Apply rescaling to modified expressions (11.1.33) and (11.1.34) for Schrödinger operator ($d\ge3$) without condition $|V|\ge\epsilon\rho^2$.

Remark 11.1.6. Some of the conditions of Theorem 11.1.1 and Corollary 11.1.2 are too restrictive but we can weaken them.

(i) First of all, one can assume that $X=\mathcal{X}_{\mathsf{reg}}\cup\mathcal{X}_{\mathsf{sing}}\cup\mathcal{Y}$ where the domains $(\mathcal{X}_{\mathsf{reg}}\cup\mathcal{X}_{\mathsf{sing}})$ and $\mathcal{Y}$ are properly overlapping (with respect to the scaling function γ), and that in $(\mathcal{X}_{\mathsf{reg}}\cup\mathcal{X}_{\mathsf{sing}})$ all the previous conditions are fulfilled and we assume that in $\mathcal{Y}$ the inequality

$$\sum_{\alpha,\beta}\langle a_{\alpha,\beta}v_\alpha, v_\beta\rangle \ge \epsilon_0\sum_{|\alpha|=m}\varrho_1\|v_\alpha\|^2 - c\sum_{|\alpha|\le m-1}\varrho_1\rho_1^{2|\alpha|}\|v_\alpha\|^2 \tag{11.1.35}$$

holds with appropriate measurable functions ϱ_1 and ρ_1. Then one can take $X_{\mathsf{reg}}=\{x\in\mathcal{X}_{\mathsf{reg}}, \rho\gamma\ge h\}$ and $X_{\mathsf{sing}}=\{x\in\mathcal{X}_{\mathsf{reg}}, \rho\gamma\le 2h\}\cup X_{\mathsf{sing}}\cup\mathcal{Y}$ in which case R_0 is given by the formula

$$\begin{aligned}R_0 = C\Bigl(\int_{\mathcal{X}_{\mathsf{reg}}\cap\{\rho\gamma\le 2h\}}\rho^{-p}\rho^{2mp}\,dx + \int_{\mathcal{Y}}\varrho_1^{-p}dx\Bigr)^{\frac{d}{2mp}}\times\\ \Bigl(\int_{\mathcal{X}_{\mathsf{reg}}\cap\{\rho\gamma\le 2h\}}\varrho^{q}\,dx + \int_{\mathcal{Y}}\varrho_1^{q}\rho_1^{2mq}\,dx\Bigr)^{\frac{d}{2mq}}\times\\ \bigl(\mathrm{mes}(\mathcal{X}_{\mathsf{reg}}\cap\{\rho\gamma\le 2h\}\cup\mathcal{Y})\bigr)^{1-\frac{d}{2m}(\frac{1}{p}+\frac{1}{q})}.\end{aligned}\tag{11.1.36}$$

(ii) Further, one can assume that the Dirichlet boundary condition is given only on $\partial X\cap(\bar{\mathcal{X}}_{\mathsf{reg}}\cap\{\rho\gamma\le h_0\}\cup\bar{\mathcal{Y}})$; on other part of ∂X it could be different (but still appropriate boundary condition).

Furthermore, assume that even in $\partial X_{\mathsf{sing},h}$ Dirichlet boundary condition is assumed only on Y_{D} while on $Y_{\mathsf{N}}=\partial X_{\mathsf{sing},h}\setminus Y_{\mathsf{D}}$ another boundary condition is given, not necessary Neumann one but satisfying

$$\begin{aligned}(Au,u)_{X_{\mathsf{sing},h}}\ge\\ \epsilon\sum_{\alpha:|\alpha|=m}h^{2m}\|D^\alpha u\|^2 - C\sum_{\alpha:|\alpha|\le m-1}h^{2|\alpha|}\|\rho_1^{m-|\alpha|}D^\alpha u\|^2\\ \forall u\in\mathfrak{D}(A):\ u=0 \qquad\text{on}\quad X\setminus X_{\mathsf{sing},h}\end{aligned}\tag{11.1.37}$$

Finally, let us assume that conditions (9.A.23)–(9.A.26) are fulfilled for X replaced by $X_{\mathsf{sing},h}$; recall that the inner boundary $\partial X_{\mathsf{sing},h}\cap\overset{\circ}{X}$ is considered as part of Y_{D}.

Then the previous estimate remain true with $\varrho_1=1$ and with ρ_1 replaced by

$$\rho_2=\rho_1+\sum_j h_j^{-1}\chi_j \tag{11.1.38}$$

in the estimates from above where χ_j is a characteristic function of X_j. In some examples specific arguments will improve even this estimate.

(iii) All these results remain true if we replace hD_j by $hD_j - V_j$ in the operator A where V_j are real-valued functions supported in $X_{\mathsf{reg},h}$ such that

$$|D^\alpha (D_j V_k - D_k V_j)| \le c\rho\gamma^{-|\alpha|-1} \qquad \forall \alpha : |\alpha| \le K. \tag{11.1.39}$$

In fact, after an appropriate gradient transform in every ball $\mathcal{B}(x, \epsilon\gamma(x), X)$ lying in $X_{\mathsf{reg},h}$ condition

$$|D^\alpha V_j| \le c\rho\gamma^{-|\alpha|} \qquad \forall \alpha : |\alpha| \le K \tag{11.1.40}$$

is fulfilled.

(iv) If the operator $(A - \tau)$ satisfies all the conditions of Theorem 11.1.1 uniformly with respect to $\tau \in [\tau_1, 0]$ with $\tau_1 < 0$ then all the estimates and asymptotics of Theorem 11.1.1 and Corollary 11.1.2 remain true for this operator uniformly with respect to $\tau \in [\tau_1, 0]$. In particular, for the Schrödinger operator with $d \ge 2$ the condition of this statement is fulfilled automatically for $\tau_1 = -\infty$.

Remark 11.1.7. Let us consider the Riesz means

$$\mathsf{N}^-_{(\vartheta)}(A) = \int \tau_-^{\vartheta-1} \mathsf{N}^-(A - \tau)\, d\tau \tag{11.1.41}$$

of order $\vartheta > 0$ (i.e., we calculate them for the spectral parameter $\tau = 0$). Let us include in $\mathsf{N}^{\mathsf{W}}_{(\vartheta)}$ (depending on h for $A = A_h$) a sufficient number of terms. So, in principle $\mathsf{N}^{\mathsf{W}}_{(0)}$ does not necessarily coincide with N^{W} (in more sharp asymptotics it may contain one extra term).

(i) Then the remainder estimates in the local semiclassical spectral asymptotics zone $X_{\mathsf{reg},h}$ yield that terms $h^{1-d}R_1$ and $h^s R_2$ should be replaced by $h^{1+\vartheta-d}R_{1(\vartheta)}$ and $h^s R_{2(\vartheta)}$ with

$$R_{1(\vartheta)} = \int_{X_{\mathsf{reg},h} \setminus X_{\mathsf{reg,ell},h}} \varrho^\vartheta \rho^{d-\vartheta-1} \gamma^{-\vartheta-1}\, dx, \tag{11.1.42}$$

$$R_{2(\vartheta)} = \int_{X_{\mathsf{reg}}} \varrho^\vartheta \rho^{d-s} \gamma^{-s}\, dx. \tag{11.1.43}$$

To do this we need $O(h_{\mathsf{eff}}^{1+\vartheta-d})$ remainder estimate which requires a stronger regularity: $K = 1 + \lfloor \vartheta \rfloor$ and the K-th derivatives should be continuous with the continuity modulus $O(|x-y|^{\vartheta - \lfloor \vartheta \rfloor} |\log |x-y||^{-1})$ and the boundary should be a bit smoother.

(ii) Further, in virtue of SubSection 9.2.4 and Appendix 9.A.2 term $h^{-d}R_0$ with R_0 defined by (11.1.10) should be replaced by $h^{-d}R_{0(\vartheta)}$ with $R_{0(\vartheta)}$ defined by the same formula but with the second factor in the right-hand expression should be given in the form

$$\int_0^\infty \tau^{\vartheta-1}\Bigl(\int_{X_{\mathsf{sing},h}} (\varrho_1-\tau)_+^q\,dx\Bigr)^{\frac{d}{2mq}}\,d\tau. \tag{11.1.44}$$

In particular, if one can take $q=\frac{d}{2m}$ and $p=\infty$ then

$$R_{0(\vartheta)} = C\int_{X_{\mathsf{sing},h}} \varrho^{\frac{d}{2m}+\vartheta}\,dx \times \sup_{X_{\mathsf{sing},h}} (\varrho^{-\frac{d}{2m}}\rho^d). \tag{11.1.45}$$

Furthermore, in virtue of Appendix 9.A.2 we can use this formula even if $d=2m$ but $\vartheta>0$ which is the case for the Schrödinger operator with $d=2$.

(iii) In particular, replacing conditions (11.1.19)–(11.1.21) by conditions

$$\varrho_1^\vartheta\rho^{d-s}\gamma^{-s}\in\mathscr{L}^1(\{X,\ \rho\gamma\ge h_0\}), \tag{11.1.46}$$

$$\varrho_1^\vartheta\rho^{d-1-\vartheta}\gamma^{-1-\vartheta}\in\mathscr{L}^1(X\setminus X_{\mathsf{reg,ell},h_0}) \tag{11.1.47}$$

and

$$R_{0(\vartheta)} = O(h^{1+\vartheta})\qquad\text{as }\ h\to+0, \tag{11.1.48}$$

we obtain t asymptotics

$$\mathsf{N}^-_{(\vartheta)}(A_h) = h^{-d}\mathsf{N}^{\mathsf{W}}_{(\vartheta)} + O(h^{1+\vartheta-d}). \tag{11.1.49}$$

In particular, in the framework of (11.1.23)–(11.1.24) (upgraded to $d\ge 2m, k\in[0,2m]$ as $\vartheta>0$) conditions (11.1.48) and (11.1.50) below follow from (11.1.47).

Non-standard Remainder Estimates

So far we assumed that the remainder estimate in the local spectral asymptotics is $O(h_{\mathsf{eff}}^{1-d+\vartheta})$ as $\vartheta\ge 0$; $\vartheta=0$ refers to the most interesting case of the asymptotics without averaging.

Sharp and Sharper Asymptotics However the remainder estimate in the local spectral asymptotics could be better than this: it could be $o(h_{\mathsf{eff}}^{1+\vartheta-d})$ or even $O(h_{\mathsf{eff}}^{\kappa+\vartheta-d})$ with $\kappa > 1$. In the former case conditions (11.1.46)–(11.1.47) imply that the contribution of $X_{\mathsf{reg},h}$ to the remainder is $o(h^{1-d})$. Adding assumption

$$R_{0(\vartheta)} = o(h^{1+\vartheta}) \qquad \text{as } h \to +0, \tag{11.1.50}$$

we arrive to asymptotics

$$\mathsf{N}^-_{(\vartheta)}(A) = h^{-d}\mathsf{N}^{\mathsf{W}}_{(\vartheta)} + o(h^{1+\vartheta-d}). \tag{11.1.51}$$

In the latter case conditions (11.1.47)–(11.1.48) should be upgraded to

$$\varrho_1^{\vartheta}\rho^{d-\kappa-\vartheta}\gamma^{-\kappa-\vartheta} \in \mathscr{L}^1(X \setminus X_{\mathsf{reg,ell},h_0}) \tag{11.1.52}$$

and

$$R_{0(\vartheta)} = O(h^{\kappa+\vartheta}) \qquad \text{as } h \to +0 \tag{11.1.53}$$

respectively. Note that in the framework of (11.1.23)–(11.1.24) (upgraded to $d \geq 2m$, $k \in [0, 2m]$ as $\vartheta > 0$) condition (11.1.53) follows from (11.1.52).

Then we arrive to

$$\mathsf{N}^-_{(\vartheta)}(A) = h^{-d}\mathsf{N}^{\mathsf{W}}_{(\vartheta)} + O(h^{\kappa+\vartheta-d}). \tag{11.1.54}$$

Remark 11.1.8. (i) Let us recall that sharp local asymptotics are due to corresponding non-periodicity condition:

$$\mu_0\big((\Pi_+ \cup \Lambda_+) \cap (\Pi_- \cup \Lambda_-)\big) = 0 \tag{11.1.55}$$

where μ_0 is a measure introduced on Σ_0 and

$$\Pi_\pm = \{z : \exists t : \pm t > 0 : \Psi_t(z) \ni z\} \tag{11.1.56}$$

is the set of points "periodic" with respect to Hamiltonian flow Ψ_t with reflections, possibly branching and $\Lambda_\pm$ is the set of dead-end trajectories.

(ii) Recall, that we need also a bit more smoothness: for example, as $\vartheta = 0$ the first derivatives of the coefficients should be continuous with the continuity modulus $o(|\log|x-y||^{-1})$ rather than $O(|\log|x-y||^{-1})$ (after rescaling) and condition to the smoothness of the boundary should be also a bit stronger.

As $\vartheta > 0$ smoothness conditions grow stronger: $K = 1 + \lfloor \vartheta \rfloor$ and the K-th derivatives should be continuous with the continuity modulus $o(|x-y|^{\vartheta - \lfloor \vartheta \rfloor} |\log|x-y||^{-1})$.

We refer to these assumptions as *improved regularity conditions*.

(iii) Also a bit less degeneration is allowed: as we consider Schrödinger operator with $d=2$ and $\vartheta = 0$ we need to assume that

$$\operatorname{mes}(\{x, V(x)=0, F=|D_1V_2 - D_2V_1| \ne 0\}) = 0. \tag{11.1.57}$$

We refer to these assumptions as *improved non-degeneracy condition*

Remark 11.1.9. Recall that for sharper remainder estimates these improved conditions are required in the stronger form but since non-periodicity condition is so difficult to check we will consider only some special cases.

Less Sharp Asymptotics However, either due to irregularity (as in Theorem 4.6.21 and similar) or to degeneration or both the local asymptotics may have remainder estimate not as good as $O(h_{\mathsf{eff}}^{1-d+\vartheta})$; for example it could be $O(h_{\mathsf{eff}}^{\kappa-d+\vartheta})$ with $\kappa < 1$. Then again under assumptions (11.1.46), (11.1.52), (11.1.53) we arrive to asymptotics (11.1.54), this time with $\kappa < 1$.

While estimates from above relied upon taking $T \asymp 1$ in $X_{\mathsf{reg}} \cap X_{\mathsf{sing}}$ which is not necessary the case now, it does not matter as $\rho\gamma \asymp h$ in $X_{\mathsf{reg}} \cap X_{\mathsf{sing}}$.

11.1.2 Examples

Let us consider a series of examples.

Example 11.1.10. Let $d \ge 2$ and let X be a compact domain (or manifold) with the boundary. Let assumptions of Theorem 11.1.1 be fulfilled with $\rho \asymp 1$ and $\gamma \asymp 1$. Then asymptotics (11.1.22) holds.

Further, under appropriate smoothness condition (see Remark 11.1.8(i)) asymptotics (11.1.49) holds.

Furthermore, under non-periodicity condition (11.1.55)–(11.1.56), improved smoothness and non-degeneracy conditions of Remark 11.1.8 asymptotics (11.1.54) holds. Here $\Lambda_\pm = \emptyset$.

Finally, under corresponding stronger non-periodicity conditions sharper asymptotics hold.

The conditions of our theorems are local and all our analysis is local (until the end of the subsection). So we assume that outside of the singularities indicated below everything is well (in the sense of Example 11.1.10).

Example 11.1.11 (Conical vertex). Let $d \geq 2$ and in a neighborhood of 0 let the domain X coincide with the cone $X^0 = \{(r,\theta) \in \mathbb{R}^+ \times U\}$ where (r,θ) are spherical coordinates, U is a domain on the sphere $\mathbb{S}^{d-1}$ with the boundary ∂U.

Moreover, let the conditions of Theorem 11.1.1 be fulfilled with $\varrho = \rho^{2m}$, $\rho = r^\mu$, $\gamma = \epsilon r$ where here and below $\epsilon > 0$ is a small constant.

(i) Then ρ^d and (or) $\rho^{d-1}\gamma^{-1}$ belong to $\mathscr{L}^1(X)$ if and only if $\mu > -1$. Let us assume that

$$\mu > \max(-1, -\frac{d}{2m}). \tag{11.1.58}$$

Then asymptotics (11.1.22) holds.

(ii) Further, under non-periodicity condition and improved regularity and non-degeneracy condition asymptotics (11.1.54) holds. Here $\Lambda_\pm$ are sets of points such that $\Psi_t(z)$ "falls" to 0 in the finite time. For example, if problem is spherically symmetric then $\Lambda_\pm$ consists of points with $\xi \parallel x$ (i.e. with angular momentum equal 0).

(iii) On the other hand, $\varrho^\vartheta \rho^{d-1-\vartheta}\gamma^{-1-\vartheta} \in \mathscr{L}^1(X)$ if and only if

$$(\mu + 1 - 2\mu m)\vartheta < (d-1)(\mu+1) \tag{11.1.59}$$

and if this condition is fulfilled and if also $\mu > -\frac{d}{2m}$ then asymptotics (11.1.49) holds for given ϑ.

Moreover, under non-periodicity condition (11.1.55)–(11.1.56), improved smoothness and non-degeneracy conditions of Remark 11.1.8 asymptotics (11.1.54) holds.

(iv) Furthermore, in the case

$$(\mu + 1 - 2\mu m)\vartheta = (d-1)(\mu+1) \tag{11.1.60}$$

we obtain asymptotics

$$\mathsf{N}^-_{(\vartheta)}(A) = h^{-d}\mathsf{N}^{\mathsf{W}}_{(\vartheta)} + O(h^{1-d+\vartheta}\log h); \tag{11.1.61}$$

here we use Theorem 11.1.1 instead of Corollary 11.1.2 [2].

(v) Finally, in the case

$$(\mu + 1 - 2\mu m)\vartheta > (d-1)(\mu+1), \quad 2\mu m\vartheta + d(\mu+1) > 0 \tag{11.1.62}$$

we obtain asymptotics

$$\mathsf{N}^-_{(\vartheta)}(A) = h^{-d}\mathsf{N}^{\mathsf{W}}_{(\vartheta)} + O(h^p) \tag{11.1.63}$$

with

$$p = \frac{2\mu}{\mu+1}m\vartheta \in (-d,\ 1+\vartheta-d) \tag{11.1.64}$$

where only terms with the degree of h less than $p+d$ are included in $\mathsf{N}^{\mathsf{W}}_{(\vartheta)}$.

Remark 11.1.12. (i) We assumed that $\mu > -d/2m$ only to guarantee that

$$a_{\alpha\beta} \in \mathscr{L}^1(X \cap \{|x| \le \epsilon\}) \tag{11.1.65}$$

and one can replace it by the latter, while also taking care that Rozenblioum estimate works; we leave details to the reader.

(ii) On the other hand, as $\mu \le -1$ it may happen that the corresponding quadratic form is not semi-bounded from below and thus operator A is not defined; even if we define it, its essential spectrum may occupy the whole $\mathbb{R}$.

Remark 11.1.13. (i) One can easily see that in the domain $X_{\mathsf{sing},h} = \{x \in X, r \le \epsilon h^{\frac{1}{\mu+1}}\}$ inequality

$$(A_h u, u) \ge \epsilon_0 \sum_{|\alpha|=m} h^{2m}\|D^\alpha u\|^2 \qquad \forall u \in \mathscr{C}^m(\bar{X}) : \operatorname{supp} u \subset B(0, \epsilon h^{\frac{1}{\mu+1}}) \tag{11.1.66}$$

---

[2] Here and below $\mathsf{N}^{\mathsf{W}}_{(\vartheta)}$ is given by the Weyl formula but in general we only integrate over the domain $X_{\mathsf{reg},h} = \{x \in X, |x| \ge \epsilon h^{1/(\mu+1)}\}$ so we have cut Weyl expression. Moreover, one can skip all the terms with $n-d > p$ in the local semiclassical spectral asymptotics and in the terms with $n-d < p$ integrate over X; $p = 1-d+\vartheta$ under condition (11.1.60).

holds. This inequality follows from the inequalities

$$h^k \sum_{|\alpha|=k} \|D^\alpha u\| \ge \epsilon_0 \varepsilon^{-\delta} \|r^{\mu k} u\| \qquad \forall k=1,\dots,m \tag{11.1.67}$$

for the same u as in (11.1.66). Here $\delta>0$ is a small enough exponent and ϵ_0 does not depend on ε.

In fact, for $h=\epsilon=1$ estimate (11.1.67) follows easily from condition (11.1.58), and the estimate in the general case follows from the estimate in this special case by means of rescaling.

Therefore one can take $\rho=0$, $\varrho=r^{2\mu m}$ in $X_{\mathsf{sing},h}$ and this simplifies the estimate of R_0. Further, we obtain that for

$$m\vartheta + d(\mu+1)=0 \tag{11.1.68}$$

asymptotics (11.1.63) holds with $p=-d$ and with $\mathsf{N}^{\mathsf{W}}_{(\vartheta)}$ given only by the integration of the $n=0$ term over the domain $X_{\mathsf{reg},h}$. Here $\mathsf{N}^{\mathsf{W}}_{(\vartheta)}=O(\log h)$; moreover, if

$$\inf_{v:\|v\|=1,\xi\in\mathbb{R}^d} \langle a(x,\xi)v,v\rangle \le -\epsilon_0\rho^{2m} \qquad \forall x\in\Gamma\cap\{|x|\le\epsilon\} \tag{11.1.69}$$

for some non-empty cone $\Gamma\subset\mathbb{R}^+\times U$ with vertex in 0 then $\mathsf{N}^{\mathsf{W}}_{(\vartheta)}\asymp\log h$.

(ii) Further, let us note that under condition (11.1.60) the main contribution to the remainder estimate in (11.1.63) is delivered by the zone

$$X_{h,\varepsilon} := \{X,\, ch^{\frac{1}{\mu+1}} \le |x| \le \varepsilon_h\} \tag{11.1.70}$$

with $\varepsilon_h=o(1)$ for $h\to+0$ (the contribution of the other zones is "o" with respect to the contribution of this zone). In order to improve the remainder estimate let us assume that the operator A stabilizes at 0 to the quasihomogeneous operator A^0, i.e., that

$$D^\alpha(a_\beta - a^0_\beta) = o(|x|^{\mu|\beta|-|\alpha|}) \qquad \forall\beta:|\beta|\le 2m \quad \forall\alpha:|\alpha|\le K \tag{11.1.71}$$

for $|x|\to 0$ where a^0_β are positively homogeneous of degrees $\mu|\beta|$. Further, let us assume that the symbol

$$a^0(x,\xi)=\sum_\beta a^0_\beta(x)\xi^\beta \tag{11.1.72}$$

satisfies non-periodicity condition (11.1.55)–(11.1.56) to Hamiltonian flow. Here and below Σ^0_τ, μ^0_τ, $\Pi^0_\pm$, $\Lambda^0_\pm$ are constructed for a^0.

Then asymptotics

$$\mathsf{N}^-_{(\vartheta)}(A) = h^{-d}\mathsf{N}^{\mathsf{W}}_{(\vartheta)} + o(h^{1-d+\vartheta}\log h) \tag{11.1.73}$$

holds. Here in the term with $n = -d+1+\vartheta$ we integrate over the domain $X_{\mathsf{reg},h}$ and this term is $O(h^{1+\vartheta}\log h)$; in all the other terms we can integrate over X.

Example 11.1.14. We can treat the case when the singularity is more complicated. Namely, for the Schrödinger operator we can consider positively homogeneous g^{jk}, V_j, V and also ∂U which are not necessarily as regular as required by Theorem 11.1.1 on $\mathbb{S}^{d-1}$.

Let us assume that on $\mathbb{S}^{d-1}$ all the conditions are fulfilled with $\hat{\rho}$ and $\hat{\gamma}$ instead of ρ and γ. Then we can take $\rho = \hat{\rho}(\theta)r^\mu$ and $\gamma = \epsilon\hat{\gamma}(\theta)r$ with a small enough constant $\epsilon > 0$.

Let us note that $\rho^{2\vartheta+d} \in \mathscr{L}^1$ if and only if $\hat{\rho}^{2\vartheta+d} \in \mathscr{L}^1(U)$ and $(2\vartheta + d)\mu + d > 0$; moreover, let us note that $\rho^{\vartheta+d-1}\gamma^{-\vartheta-1} \in \mathscr{L}^1$ if and only if $\hat{\rho}^{\vartheta d-1}\hat{\gamma}^{-\vartheta-1} \in \mathscr{L}^1(U)$ and $(\mu-1)\vartheta + (\mu+1)(d-1) > 0$.

Therefore a more complicated structure is possible: one can treat a hierarchy of singularities: "vertices of singularities," "edges of singularities," "faces of singularities," etc.

We leave to the reader

Problem 11.1.15. Analyze the case in which

$$\gamma = \epsilon\min(|x_{(1)}|,\dots,|x_{(l)}|),\quad \rho = \prod_{S\subset\{1,\dots,l\}}\zeta_S^{\mu_S},\quad \zeta_S = \sum_{j\in S}|x_{(j)}| \tag{11.1.74}$$

where $x = (x_{(1)},\dots,x_{(l)}) \in \mathbb{R}^{d_1}\times\ldots\times\mathbb{R}^{d_l}$ is an appropriate partition.

Example 11.1.16 (Conical edge). Let us consider a more general case. Let X be a bounded domain locally coinciding with $\mathbb{R}^{d'}\times\mathbb{R}^+\times U$ where spherical coordinates (r,θ) are introduced in $\mathbb{R}^{d''}$ $(d''=d-d')$, and where $U\subset\mathbb{S}^{d''-1}$, $0\le d'\le d-1$. Let conditions of Theorem 11.1.1 be fulfilled with $\varrho=\rho^{2m}$, $\rho = r^\mu$, $\gamma=\epsilon r$ and let

$$\mu > \max(-1, -\frac{d''}{2m}). \tag{11.1.75}$$

Then

(i) For

$$(\mu + 1 - 2\mu m)\vartheta < (\mu + 1)(d - 1) - d' \tag{11.1.76}$$

asymptotics (11.1.49) holds. Further, under non-periodicity condition and improved regularity and non-degeneracy condition asymptotics (11.1.54) holds. Here $\Lambda_\pm$ are sets of points such that $\Psi_t(z)$ "falls" to $\mathbb{R}^{d'} \times 0$ in the finite time.

(ii) On the other hand, for

$$(\mu + 1 - 2\mu m)\vartheta = (\mu + 1)(d - 1) - d' \tag{11.1.77}$$

asymptotics (11.1.61) holds and for

$$(\mu + 1 - 2\mu m)\vartheta > (\mu + 1)(d - 1) - d', \quad 2\mu m\vartheta + d\mu + d'' \geq 0 \tag{11.1.78}$$

asymptotics (11.1.63) holds with

$$p = \frac{2\mu}{\mu + 1}(m\vartheta - d') \tag{11.1.79}$$

where for

$$2\mu m\vartheta + d\mu + d'' = 0 \tag{11.1.80}$$

only one term in $\mathsf{N}^{\mathsf{W}}_{(\vartheta)}$ remains and integration is taken over $X_{\mathsf{reg},h}$. In this case $\mathsf{N}^{\mathsf{W}}_{(\vartheta)} = O(\log h)$ and $\mathsf{N}^{\mathsf{W}}_{(\vartheta)} \asymp \log h$ under condition (11.1.69) fulfilled in $V \times \Gamma \cap \{x, |x''| \leq \epsilon\}$ where V is a neighborhood of 0 in $\mathbb{R}^{d'}$ and Γ is a non-empty cone in $\mathbb{R}^+ \times U$ with vertex at 0.

Let us note that in this example (in contrast to Example 11.1.11) the variants are possible even for $\vartheta = 0$.

(iii) Let $\vartheta = 0$. Then asymptotics (11.1.22) and (11.1.61) are replaced by asymptotics (11.1.63) with $p = -d + 1 - \delta$ with arbitrarily small exponent $\delta > 0$ while asymptotics (11.1.63) with $p < -d + 1$ remains true.

Consider conical singularity with the critical exponents:

Example 11.1.17. Let $0 \in \bar{X}$ and all the conditions be fulfilled with $\gamma = \epsilon r$, $\rho = r^{\mu}|\log r|^{-\nu}$, $r = |x|$ where we assume that

$$\mu > -\frac{2m}{d},\ \mu \geq -1,\ \vartheta \geq 0,\ \nu > 0,\qquad 2m\mu\vartheta + (d - \vartheta - 1)(\mu + 1) = 0. \tag{11.1.81}$$

Then repeating the arguments of the previous examples with

$$X_{\mathsf{sing},h} = \{x \in X, |x| \leq \epsilon h^{\frac{1}{\mu+1}} |\log h|^{\frac{\nu}{\mu+1}}\} \qquad \text{for } \mu > -1 \tag{11.1.82}$$

and

$$X_{\mathsf{sing},h} = \{x \in X, \log|x| \leq -Ch^{-\frac{1}{\nu}}\} \qquad \text{for } \mu = -1 \tag{11.1.83}$$

and with corresponding $X_{\mathsf{reg},h}$ we conclude that

(i) Under condition

$$(2m\vartheta + d - \vartheta - 1)\nu > 1 \tag{11.1.84}$$

asymptotics (11.1.49) holds; moreover, under standard non-periodicity condition and improved regularity and non-degeneracy conditions asymptotics (11.1.51) holds.

(ii) On the other hand, under condition

$$(2m\vartheta + d - \vartheta - 1)\nu = 1 \tag{11.1.85}$$

asymptotics (11.1.61) holds.

(iii) Finally, if

$$(2m\vartheta + d - \vartheta - 1)\nu < 1 \tag{11.1.86}$$

and if either

$$\mu > -1 \tag*{(11.1.87)$_1$}$$

or

$$\mu = -1 (\implies \vartheta = 0),\ \nu \geq \frac{1}{d} \tag*{(11.1.87)$_2$}$$

then asymptotics (11.1.63) holds with

$$p = 2m\vartheta - \vartheta - \frac{1}{\nu}. \tag{11.1.88}$$

For $\mu = -1$, $\vartheta = 0$, $\nu = \frac{1}{d}$ we see that $\mathsf{N}^{\mathsf{W}} = O(\log h)$ and $\mathsf{N}^{\mathsf{W}} \asymp \log h$ provided condition (11.1.69) is fulfilled in some non-empty cone with vertex at 0 (for $|x| \le \epsilon$).

(iv) Moreover, in the first critical case (11.1.85) one can improve asymptotics (11.1.61). We do not do a careful analysis of the Hamiltonian trajectories here but refer to the arguments of the proofs of Example 11.5.21 (see below). Let us assume that the coefficients a_β of A stabilize to functions $a^0_\beta \bigl|\log |x|\bigr|^{-\nu|\beta|}$ in the sense that

$$D^\sigma\bigl(a_\beta \bigl|\log |x|\bigr|^{\nu|\beta|} - a^0_\beta\bigr) = o(|x|^{\mu|\beta| - |\sigma|}) \qquad \forall \beta : |\beta| \le 2m\ \forall \sigma : |\sigma| \le 1 \tag{11.1.89}$$

where a^0_β are positively homogeneous of degree $\mu|\beta|$. Moreover, let us assume that the quasihomogeneous symbol (11.1.72) satisfies the standard non-periodicity condition.

Then one can easily prove that asymptotics (11.1.73) holds with all the explanations made in Remark 11.1.13.

(v) For $\nu > 0$ satisfying the critical condition (11.1.85) one can treat the case in which $\rho = r^\mu |\log r|^{-\nu} \bigl|\log|\log r|\bigr|^{-\nu'}$ and analyze all the possible cases. In this case the conditions on ν are replaced by similar conditions on ν'. One can continue this process of complication.

(vi) On the other hand, one can treat the case in which $\rho = r^\mu |\log r|^{-\nu}$ even if equality in (11.1.81) is violated, i.e., μ is not the critical value. In this case for $2m\mu\vartheta + (d - \vartheta - 1)(\mu + 1) > 0$ asymptotics (11.1.49) (and even (11.1.51) under the standard non-periodicity condition to the Hamiltonian flow) holds and for $2m\mu\vartheta + (d - \vartheta - 1)(\mu + 1) < 0$ asymptotics (11.1.63) holds with the remainder estimate $O(h^p |\log h|^q)$ instead of $O(h^p)$. The calculation of q is left to the reader.

(vii) In all the above statements the condition $\nu > 0$ is not necessary if $\mu > -1$.

(viii) Statement (ii) of Example 11.1.16 remains true (with obvious modifications).

Different combinations of Statements (v) and (vi) are possible.

Example 11.1.18 (Quasiconical singularity). Let $0 \in \bar{X}$ and let us consider operator with coefficients $a_{\alpha\beta}$ which are *L-quasihomogeneous of degrees* $(2m - |\alpha| - |\beta|)\mu$ respectively, smooth away from 0. In a neighborhood of zero let X be an *L-quasiconical* domain and let ∂X be smooth away from 0. Here $L = (L_1, \dots, L_d)$ is an exponent vector with $L_j > 0\ \forall j$ and $\max_j L_j = 1$.

Let us assume that either a Dirichlet or some other appropriate boundary condition is given and in the second case let us assume that the domain X satisfies the standard cone condition.

In this case $\gamma(x) = \epsilon[x]_L$, $\rho(x) = \gamma^\mu$, $\varrho(x) = \gamma^{2m\mu}$ with *L-quasihomogeneous norm* $[x]_L = \sum_j |x_j|^{\frac{1}{L_j}}$. Then repeating the arguments of Example 11.1.11 and assuming that

$$\mu > \max(-1, -\frac{L^*}{2m}), \qquad L^* = L_1 + \dots + L_d \tag{11.1.90}$$

we conclude that

(i) For

$$(\mu + 1 - 2\mu m)\vartheta < (d-1)\mu + L^* \tag{11.1.91}$$

asymptotics (11.1.49) holds.

Moreover, under non-periodicity condition and improved regularity and non-degeneracy conditions asymptotics (11.1.51) holds.

On the other hand, for

$$(\mu + 1 - 2\mu m)\vartheta = (d-1)\mu + L^* \tag{11.1.92}$$

asymptotics (11.1.61) holds and for

$$(\mu + 1 - 2\mu m)\vartheta > (d-1)\mu + L^*, \quad 2m\mu\vartheta + d\mu + L^* > 0 \tag{11.1.93}$$

asymptotics (11.1.63) holds with $p = (2m\mu\vartheta + L^* - d)(\mu + 1)^{-1}$.

However the author have failed to solve

Problem 11.1.19. (i) Apply more advanced arguments of Remark 11.1.13.

(ii) Investigate the second critical case $2m\mu\vartheta + d\mu + L^* = 0$.

We leave to the reader to

Problem 11.1.20. Investigate the case $\rho = r^{\mu}$, $\gamma = \epsilon r^{1+\nu}$ with $\nu > 0$ arising in the analysis of the Schrödinger operator with potential $V = \rho^{2\mu} \sin r^{-\nu}$ and other examples of this type.

Let us treat exits to infinity. We assume that

(11.1.94) Either $X_{\mathsf{sing},h} = \emptyset$ or a Dirichlet boundary condition is given on $\partial X \cap \{|x| \geq c\}$.[3)]

Example 11.1.21 (Exit to infinity). (a) In a neighborhood of infinity let X coincide with $\{x, x_1 > 0, x' = F(x_1)\Omega\}$ where $\Omega \subset \mathbb{R}^{d-1}$ is a bounded domain satisfying assumptions of Theorem 11.1.1 with $\gamma(x') \asymp 1$, $x' = (x_2, \dots, x_d)$, $t\Omega = \{tx', x' \in \Omega\}$ and

$$F(x_1) \asymp x_1^{\beta}, \qquad \beta \leq 1, \tag{11.1.95}$$

$$|D^{\alpha} F| \leq cF^{1-|\alpha|} \qquad \forall \alpha : |\alpha| \leq 2. \tag{11.1.96}$$

For $\beta = 1$ we have a conical exit, for $\beta \in (0, 1)$ we have a slowly expanding quasiconical exit, for $\beta = 0$ we have a cylindrical exit and for $\beta < 0$ we have a constricting exit and these exits could be corrugated along the x_1 axis.

(b) Similarly, let us consider the case in which X coincides with $\{x, x' = F(|x''|)\Omega\}$ in a neighborhood of infinity where $x = (x'', x') \in \mathbb{R}^{d''} \times \mathbb{R}^{d'}$, $0 < d' < d = d' + d''$ and the function F satisfies conditions (11.1.95)–(11.1.96).

Then geometry yields that (at least for $|x''| \geq c$)

(11.1.97) X satisfies conditions of Theorem 11.1.1 with $\gamma \asymp F(|x''|)$.

Actually all we need is (11.1.96) with $K = 1$ and (11.1.97). Moreover, let us assume that the other conditions are fulfilled with $\gamma = \epsilon F(|x''|)$ and $\rho = |x''|^{\mu}$. Then

(i) If $\mu + \beta \geq 0$ one can take $X_{\mathsf{reg},h} = X$ and $X_{\mathsf{sing},h} = \emptyset$. Then under the condition

$$2m\mu\vartheta + d\mu + d'\beta + d'' < 0 \tag{11.1.98}$$

---

3) In the next sections we discuss how to weaken this assumption (albeit it is impossible to get rid of it).

(which provides that $\mathsf{N}^{\mathsf{W}}_{(\vartheta)}$ converges) asymptotics (11.1.49) holds. Note that $\mu+\beta\ge 0$ and (11.1.98) imply that $\mu<0$, $\vartheta>0$.

Moreover, under the standard non-periodicity condition and improved regularity and non-degeneracy conditions asymptotics (11.1.51) holds.

(ii) Let $\mu+\beta<0$. Then one can take $X_{\mathsf{reg},h}=X\cap\{r\le Ch^{1/(\mu+\beta)}\}$ and $X_{\mathsf{sing},h}=X\cap\{r\ge\frac{1}{2}Ch^{1/(\mu+\beta)}\}$ where $r=|x''|$. Let us assume that a Dirichlet condition is given on $\partial X\cap\{|x''|\ge C\}$ [3]. Then inequality

$$h^m\sum_{\alpha:|\alpha|=m}\|D'^\alpha u\|_{(x'')}\ge\epsilon_0F(r)^{-m}\|u\|_{(x'')} \tag{11.1.99}$$

(where $\|.\|_{(x'')}$ means the $\mathscr{L}^2$-norm over sections of X at points x'') yields that under condition

$$(2m\mu-\beta-\mu)\vartheta+(d-1)\mu-\beta+d'\beta+d''<0 \tag{11.1.100}$$

asymptotics (11.1.49) holds.

Moreover, under standard non-periodicity condition and improved regularity and non-degeneracy conditions asymptotics (11.1.51) holds.

(iii) On the other hand, for

$$(2m\mu-\beta-\mu)\vartheta+(d-1)\mu-\beta+d'\beta+d''=0 \tag{11.1.101}$$

asymptotics (11.1.61) holds and for

$$\begin{aligned}&(2m\mu-\beta-\mu)\vartheta+(d-1)\mu-\beta+d'\beta+d''>0,\\ &\qquad\qquad 2m\mu\vartheta+d\mu+d'\beta+d''\le 0\end{aligned} \tag{11.1.102}$$

asymptotics (11.1.63) holds with $p=(2m\mu\vartheta+d''-d''\beta)(\mu+\beta)^{-1}$; here in the case of equality in the last condition the integral in $\mathsf{N}^{\mathsf{W}}_{(\vartheta)}$ is calculated over the domain $X_{\mathsf{reg},h}$ and $\mathsf{N}^{\mathsf{W}}_{(\vartheta)}=O(\log h)$.

Moreover, let condition (11.1.69) be fulfilled for $|x''|\ge c, x''\in\Gamma$ and $x'\in F(|x''|)\Omega'$ where $\Gamma\subset\mathbb{R}^{d''}$ is a non-empty cone and $\Omega'\subset\Omega$ is a non-empty domain. Then $\mathsf{N}^{\mathsf{W}}_{(\vartheta)}\asymp\log h$.

We leave to the reader to analyze the case of more general dependence of F on x''.

Example 11.1.22 (Conical exit to infinity). (i) In a neighborhood of infinity let the domain X coincide with the cone $X^0 = \{(r, \theta) \in \mathbb{R}^+ \times U\}$ and let the conditions be fulfilled with $\gamma = \epsilon r$, $\rho = r^\mu$. Then if

$$\mu \geq -1, \qquad 2m\mu\vartheta + (\mu + 1)d < 0 \tag{11.1.103}$$

then asymptotics (11.1.49) holds.

Moreover, under standard non-periodicity condition and improved regularity and non-degeneracy conditions asymptotics (11.1.51) holds.

(ii) Let us treat the case

$$\mu < \min(-1, -\frac{2m}{d}); \tag{11.1.104}$$

then estimates (11.1.66) and (11.1.67) hold for functions u supported in $\mathbb{R}^d \setminus B(0, \epsilon^{-1}h^{1/(\mu+1)})$ and therefore under condition (11.1.59)$^\#$ asymptotics (11.1.49) holds.

Moreover, under standard non-periodicity condition and improved regularity and non-degeneracy conditions asymptotics (11.1.51) holds.

Here the superscript "#" means that the condition is replaced by its opposite (i.e., ">" by "<" and v.v., and "=" remains), $|x| \to 0$ is replaced by $|x| \to \infty$ and $|x| \leq \epsilon$ is replaced by $|x| \geq C$.

(iii) Further, under condition (11.1.60)$^\#$ we obtain asymptotics (11.1.61) and under condition (11.1.62)$^\#$ we obtain asymptotics (11.1.63) with p given by the same formula as before.

Finally, under condition (11.1.68)$^\#$ asymptotics (11.1.63) remains true with the same clarifications as before.

Moreover, under conditions (11.1.60)$^\#$, (11.1.71)$^\#$ and the standard non-periodicity condition and improved regularity and non-degeneracy conditions we obtain asymptotics (11.1.73).

Example 11.1.23 (Conical exit to infinity. II). Let $\gamma = \epsilon r$, $\rho = r^\mu \log^{-\nu} r$ where

$$\mu \leq -1. \qquad \mu < -\frac{2m}{d} \tag{11.1.105}$$

satisfies (11.1.59).

Then all the results of Example 11.1.17 remain true and the inequalities on ν do not change and in the stabilization conditions and in (11.1.69) $|x| \to 0$ is replaced by $|x| \to \infty$ and $|x| \le \epsilon$ is replaced by $|x| \ge C$. Again we need to change all inequalities to their opposites.

Example 11.1.24 (Quasiconical exit to infinity). In a neighborhood of infinity let X be an L-quasiconical domain and $\partial X \in \mathscr{C}^2$. Then in a neighborhood of infinity one can take $\gamma = \epsilon [x]^{\delta}_{\delta L}$ where we assume that $L = (L_1, \dots, L_d)$, $L_j \ge 1\ \ \forall j$ and $\delta \le \min_j L_j^{-1}$. Let us assume that $\rho = \gamma^{\mu}$ and $\varrho = \gamma^{2\mu m}$.

In this case all the statements of Example 11.1.18 remain true with the inequalities on μ replaced by their opposites.

We leave to the reader the following rather easy problem:

Problem 11.1.25. Investigate the case in which $X = \mathbb{R}^{d'} \times \mathbb{T}^{d''}$ in a neighborhood of infinity and $\gamma = 1$.

11.1.3 Non-Semi-Bounded Operators

Let us now consider operators which are non-semi-bounded. In this case we assume that $\vartheta = 0$ unless otherwise indicated. Since the order of operator is not necessarily even anymore, denote it by m rather than by $2m$.

Theorem 11.1.1 and Corollary 11.1.2, as well as Example 11.1.10, remain true in this case with obvious modifications. Moreover, one can treat the Dirac operator in $X = \mathbb{R}^d$ assuming that for $|x| \ge c$

$$V + M - \tau_2 \ge \epsilon_0 \rho^2, \qquad M - V + \tau_1 \ge \epsilon_0 \rho^2 \tag{11.1.106}$$

and

$$\rho\gamma \ge \epsilon_0, \tag{11.1.107}$$

$$\int \rho^{d-s}\gamma^{-s}\,dx < \infty \tag{11.1.108}$$

because these conditions ensure that $\{|x| \ge c\} \subset X_{\mathsf{reg,ell}}$. Recall that M is the mass.

We leave to the reader

Problem 11.1.26. (i) Include a neighborhood of infinity in $X_{\mathsf{sing},h}$ and skip condition (11.1.108);

(ii) Treat the case $X=\mathbb{R}^{d'}\times\mathbb{T}^{d''}$ with $d'>0, d''>0$.

Furthermore, one can treat the *generalized Dirac* operator

$$A=\begin{pmatrix} V_1 & \mathcal{L}\\ \mathcal{L}^* & V_2\end{pmatrix} \tag{11.1.109}$$

where V_k are matrix-valued functions, $\mathcal{L}$ is a uniformly elliptic operator of order m and for $|x|\ge c$

$$V_2-\tau_2\ge \epsilon_0\rho^{2m},\qquad V_1-\tau_1\le -\epsilon_0(\rho+1)^{2m},\qquad \rho\gamma\ge\epsilon_0. \tag{11.1.110}$$

Moreover, we assume that the coefficients of $\mathcal{L}$ satisfy conditions

$$\|D^\alpha b_\beta\|\le c\rho^{m-|\beta|}\gamma^{-|\alpha|}\qquad \forall\beta:|\beta|\le m \tag{11.1.111}$$

and

$$\|D^\alpha V_2\|\le c\rho^{2m}\gamma^{-|\alpha|},\qquad \|D^\alpha V_1\|\le c(\rho+1)^{2m}\gamma^{-|\alpha|}\\ \forall\alpha:1\le|\alpha|\le K \tag{11.1.112}$$

We can assume instead that the corresponding roughness conditions are fulfilled.

Then

(11.1.113) All examples with singularities at 0 [4] remain true for $X=\mathbb{R}^d$ if in the inequalities for exponents we replace $2m$ by m; however, we do not modify the inequalities (11.1.58) and (11.1.75) to reflect that m rather than $2m$ is an order of the operator.

Further,

(11.1.114) All examples with singularities at infinity[5] remain true without modification in the inequalities on exponents provided in a neighborhood of infinity

$$V_1-\tau_1\le-\epsilon_0,\qquad V_2-\tau_2\le-\epsilon_0\rho^{2m}. \tag{11.1.115}$$

Recall that in these examples $\rho\le c$.

[4] I.e. Examples 11.1.10, 11.1.11 and 11.1.14–11.1.18.

[5] I.e. Examples 11.1.21–11.1.24.

Finally, let us note that under conditions (in X)

$$V_1 - \tau_1 \le -\epsilon_0, \qquad V_2 - \tau_1 \ge \epsilon_0, \qquad \rho \le c \tag{11.1.116}$$

one can treat

$$\mathsf{N}^-_{(\vartheta)}(A) = \int_{\tau_1}^{\tau_2} (\tau_2 - \tau)_+^{\vartheta-1}\, d\tau \tag{11.1.117}$$

with $\vartheta > 0$. Then Example 11.1.10 and all the examples with singularities at infinity[5] remain true.

We leave to the reader

Problem 11.1.27. (i) Justify these claims (11.1.113) and (11.1.114) basing on the results of Chapter 10.

(ii) Based on Remark 10.2.23 pass from operators with rough coefficients to operators with irregular coefficients. In particular, using the fact that eigenvalues are monotone with respect to V_1, V_2 consider the case when only V_1 and V_2 are irregular and either τ_1 or τ_2 belong to spectral gap.

11.2 Large Eigenvalues for Operators with Weakly Singular Potentials

This section is devoted to the General scheme and main theorems.

11.2.1 General Scheme

In this and the following two sections we consider the asymptotics of the eigenvalues of operator A, tending to $+\infty$. In fact, one can treat more general problem

$$(A - \lambda J)u = 0 \qquad u \in \mathfrak{D}(A) \tag{11.2.1}$$

where A and J are operators of orders $2m$ and $n < 2m$ respectively ($m \in \frac{1}{2}\mathbb{Z}$). Recall that in this problem one should assume that

(11.2.2) A is a self-adjoint operator, J is a symmetric operator and J is A-compact[6] and at least one of operators A and J is positive definite.

---

[6] This condition will be fulfilled automatically in the frameworks of the theorems below.

Surely, the case $J = I$ is the most natural and interesting from the point of view of miscellaneous, in particular physical applications; in the next sections that case will be treated but in this section we consider the more general case (11.2.1) as well.

In this section we assume that A is semi-bounded from below; at the end of the section $J = I$ and A is not necessarily semi-bounded. Note that Birman-Schwinger principle yields that

(11.2.3) If the operator J is positive definite then the the number of eigenvalues of the problem (11.2.1) less than τ is equal to $\mathsf{N}^-(\tau)$ which is the number of negative eigenvalues of the operator $(A - \tau J)$

and

(11.2.4) If the operator A is positive definite then the number of eigenvalues $\lambda \in (0, \tau)$ of the problem (11.2.1) (with $\tau > 0$) is equal to $\mathsf{N}^-(\tau) - \mathsf{N}^-(+0)$ where $\mathsf{N}^-(\tau)$ is again the number of negative eigenvalues of the operator $(A - \tau J)$ and $\mathsf{N}^-(+0) = \lim_{\tau\to+0} \mathsf{N}^-(\tau)$.

There is no contradiction since as both A and J are positive definite problem (11.2.1) has only positive eigenvalues.

Therefore in the both cases we should consider the asymptotics of $\mathsf{N}^-(\tau)$ as $\tau \to +\infty$.

Let us assume that A is given by (11.1.1) with $h = 1$, $m \in \mathbb{Z}$, that conditions (11.1.2) and (11.1.3) are fulfilled and

$$J = \sum_{|\alpha|,|\beta|\le m, |\alpha|+|\beta|\le n} D^\alpha j_{\alpha,\beta} D^\beta, \tag{11.2.5}$$

with the coefficients satisfying

$$|D^\sigma a_{\alpha,\beta}| \le c\varrho\rho^{-|\alpha|-|\beta|}\gamma^{-|\sigma|}, \tag{11.2.6}$$

$$|D^\sigma j_{\alpha,\beta}| \le c\varrho_0\rho^{-|\alpha|-|\beta|}\gamma^{-|\sigma|} \qquad \forall \sigma : |\sigma| \le K \tag{11.2.7}$$

or corresponding "rough" conditions.

Further, let us assume that the boundary ∂X is smooth enough after rescaling of $\mathcal{B}(y, \gamma(y), X)$ and that the coefficients of the boundary operator $B = (B_1, \dots, B_l)$ satisfy condition $(9.2.3)_2$. Furthermore, let us assume that the operator $\varrho^{-1}\rho^{2m}A$ is uniformly elliptic and that the boundary value problem $(\varrho^{-1}\rho^{2m}A, \varrho_{(1)}^{-1}B_1, \dots, \varrho_{(l)}^{-1}B_l)$ is uniformly elliptic as well.

One can easily see that in the zone $\{x \in X, \rho_\tau\gamma_\tau \ge 1\}$ operator $(A - \tau J)$ satisfies all the conditions of Section 9.1[7] with functions

$$\gamma_\tau = \gamma, \quad \rho_\tau = \rho\bigl(1 + \frac{\tau\varrho_0}{\varrho}\bigr)^{\frac{1}{2m-n}}, \quad \varrho_\tau = \varrho\bigl(1 + \frac{\tau\varrho_0}{\varrho}\bigr)^{\frac{2m}{2m-n}}, \quad \varrho_{\tau(j)} = \varrho_{(j)}\bigl(\frac{\tilde{\rho}}{\rho}\bigr)^{-m_j} \tag{11.2.8}$$

instead of γ, ρ, ϱ and $\varrho_{(j)}$ respectively[8].

Let us treat the microhyperbolicity or non-degeneracy condition in the zone

$$X'_{\mathsf{reg}} = \bigl\{x \in X, \tau\varrho_0 \ge C\varrho\bigl(1 + (\rho\gamma)^{n-2m}\bigr)\bigr\} \tag{11.2.9}$$

in which

$$\rho_\tau\gamma \ge C' \quad \text{and} \quad \rho_\tau \ge C'\rho \tag{11.2.10}$$

provided $C = C(C')$. We easily see that for large enough C' in this zone one can replace the complete symbols of the operators A, B and J by their senior symbols[9] in the verification of such condition and after rescaling with respect to x and ξ[10] we easily see that

(11.2.11) One needs to check this non-degeneracy condition with $\gamma\partial_x$ instead of ∂_x and with ∂_ξ preserved for the symbol $\bigl(\tilde{a}(x,\xi) - \tilde{j}(x,\xi), \tilde{b}(x,\xi)\bigr)$

where tilde means that one needs to multiply a, j and b_k by ϱ^{-1}, ϱ_0^{-1} and $\varrho_{(k)}^{-1}$ respectively at point x; these factors are "well-differentiable" (in the corresponding scale) with respect to x.

Let us first check the conditions which are not associated with the symbol $\tilde{b}$ of the boundary operator. Note that for $|\xi| \ge \epsilon$ with an arbitrarily small constant $\epsilon > 0$ the ellipticity of $\tilde{a}(x,\xi)$, the homogeneity of $\tilde{a}(x,\xi)$ and $\tilde{j}(x,\xi)$

---

[7] With the possible exception of the microhyperbolicity, or ξ-microhyperbolicity, or some other non-degeneracy condition.

[8] We use the notation $\varrho_{(k)}$ instead of the ϱ_k of Section 9.1 in order to avoid misunderstanding.

[9] I.e., the higher degree homogeneous symbols.

[10] I.e., rescaling with respect to x and multiplication of the operators by appropriate factors.

of degrees $2m$ and n with respect to ξ and the Euler identity yield that ξ-microhyperbolicity condition is fulfilled. Therefore *the remaining check should be done for* $\xi=0$ *only*.

For general matrix operators one needs to check ξ-microhyperbolicity or microhyperbolicity condition and since $\tilde{a}$ with its first order derivatives vanishes at $\xi=0$ this condition should be checked for $\tilde{\jmath}(x,\xi)$.

The three following cases are possible:

(i) $n=0$; then microhyperbolicity condition means exactly that

(11.2.12) $n=0$ and for every $x\in X_{\mathsf{reg}}$ there exists $\ell\in T_xX$ such that $|\ell|=1$ and

$$\gamma\langle \ell j(x)v,v\rangle \ge \epsilon_0\varrho_0\|v\|^2 - c\varrho_0^{-1}\|j(x)v\|^2 \qquad \forall v\in \mathbb{H} \tag{11.2.13}$$

where $j(x)=J(x)$ and we returned to j from $\tilde{\jmath}$. In particular, this condition is fulfilled automatically if

$$n=0 \text{ and } \quad \|j(x)v\|\ge \epsilon\varrho_0\|v\| \qquad \forall x\in X' \quad \forall v\in\mathbb{H}; \tag{11.2.14}$$

(ii) $n=1$; in this case $\tilde{\jmath}(x,0)=0$ and one should differentiate with respect to ξ only; then microhyperbolicity condition means exactly that

(11.2.15) $n=1$ and for every $x\in X_{\mathsf{reg}}$ there exists $\xi\in T^*_xX$ such that $|\xi|=1$ and

$$\langle j(x,\xi)v,v\rangle \ge \epsilon_0\varrho_0\rho^{-1}\|v\|^2 \qquad \forall v\in\mathbb{H}$$

where $j(x,\xi)$ is the senior symbol of $J(x,D_x)$.

(iii) $n\ge 2$; in this case $\tilde{\jmath}$ and its first order derivatives vanish at $\xi=0$ and the microhyperbolicity condition is surely violated.

Note that ξ-microhyperbolicity condition is fulfilled in the cases (11.2.14) and (11.2.15) only; in this case remainder estimate $O(h_{\mathsf{eff}}^{1-d})$ is ensured under regularity conditions marginally stronger than $K=1$ (see Sections 4.6 and 7.5); we refer to them as *minimal regularity assumptions*.

For scalar symbols a and j [11] the following additional possibilities arise:

(iv) $m = 1$, $d \geq 2$; then $(A - \lambda J)$ after all rescaling is the Schrödinger operator and no microhyperbolicity is needed; so we assume that

(11.2.16) $m = 1$, $n = 0, 1$, a and j are scalar symbols and $d \geq 2$; further, as <u>either</u> $d \geq 3$ <u>or</u> $d = 2$ or (11.2.12) holds we recover local remainder estimate $O(h_{\mathsf{eff}}^{1-d})$ under minimal regularity assumptions;

(v) $m = 1$, $d = 1$, $n = 0$; we recover local remainder estimate $O(h_{\mathsf{eff}}^{1-d})$ as $K = K(p)$ provided

(11.2.17) $m = 1$, $d = 1$, $n = 0$ and for every $x \in X_{\mathsf{reg}}$

$$\sum_{\alpha:|\alpha|\leq p} \gamma^{|\alpha|}|D^\alpha j| \geq \epsilon_0\varrho_0; \tag{11.2.18}$$

as $p = 1$ we recover local remainder estimate $O(\log(h_{\mathsf{eff}}))$ under minimal regularity assumptions[12];

(vi) $m = 1$, $d = 1$, $n = 1$; then

$$J = \frac{1}{2}(\langle D_x, j(x)\rangle + \langle j(x), D_x\rangle) + j'(x) \tag{11.2.19}$$

and different variants of non-degeneracy condition are possible:

(11.2.20) $m = 1$, $d = 1$, $n = 1$ and $|j(x)| \geq \epsilon_0\varrho_0\rho^{-1}$ for every $x \in X_{\mathsf{reg}}$ ensures local remainder estimate $O(\log(h_{\mathsf{eff}}))$ under under minimal regularity assumptions;

---

11) It is sufficient to assume that either the eigenvalues of $a(x,\xi) - j(x,\xi)$ close to 0 are simple in a neighborhood of $\{\xi = 0\}$ or these eigenvalues are of constant multiplicity there and $\rho\gamma \leq C$ in the latter case. For $n \geq 1$ that automatically yields that $(a - j)$ is a scalar symbol and the homogeneity of a and j yield that a and j are scalar symbols.

For $n = 0$ this conjecture is not necessary. In this case in the appropriate local coordinates $j = \begin{pmatrix} j'' & 0 \\ 0 & j'I \end{pmatrix}$, $a = \begin{pmatrix} * & * \\ * & a'I \end{pmatrix}$ where j'' is a non-degenerate matrix and j' and a' are scalar symbols and under the above condition all the following conditions should be checked for symbols a' and j'.

12) As $p \geq 2$ we need to assume that A and J are scalar operators.

while

(11.2.21) $m=1$, $d=1$, $n=1$ and (11.2.18) ensures local remainder estimate $O(1)$ as $K=K(p)$ [12].

(vii) If $m\geq 2$ then the first and the second order derivatives of a vanish at $\xi=0$ and we obtain the following cases ensuring local remainder estimate $O(h_{\mathsf{eff}}^{1-d})$ under smoothness assumptions:

(11.2.22) $m\geq 2$, $d\geq 2$, $n=0$ and for every $x\in X_{\mathsf{reg}}$ inequality

$$|j(x)|+\gamma|\nabla j(x)|\leq \epsilon_0\varrho_0 \tag{11.2.23}$$

yields that $\mathsf{Hess}\, j(x)$ has two eigenvalues with absolute values greater than $\epsilon_0\varrho_0\gamma^{-2}$;

(11.2.24) $m\geq 2$, $d\geq 2$, $n=0$ and for every $x\in X_{\mathsf{reg}}$ (11.2.18) holds with $p=1$ or $p=2$ [12];

(11.2.25) $m\geq 2$, $n=1$ and for every $x\in X_{\mathsf{reg}}$ inequality

$$|j(x)|+\gamma|(Dj)(x)|\geq \epsilon_0\varrho_0\rho^{-1} \tag{11.2.26}$$

holds where here J is defined by (11.2.19), $j=j(x)$ is a vector field and (Dj) is the matrix of its derivatives, $j'=j'(x)$.

(viii) Finally, as $m\geq 2$ we can consider $n=2$ and arrive to

(11.2.27) $m\geq 2$, $d\geq 2$, $n=2$ and

$$\sum_{\alpha:|\alpha|=2}|j_\alpha|\geq \epsilon_0\varrho_0\rho^{-2} \tag{11.2.28}$$

for every $x\in X_{\mathsf{reg}}$ where j_α are the coefficients of the senior symbol of J.

Remark 11.2.1. (i) For $\vartheta>0$ we recover a local remainder estimate $O(h_{\mathsf{eff}}^{1-d+\vartheta}\tau^{\vartheta})$ under corresponding regularity assumptions.

(ii) Let both A and J be scalar operators. Then in the smooth case we do not need any non-degeneracy assumptions to recover a local remainder estimate $O(h_{\mathsf{eff}}^{1-d+\vartheta}\tau^{\vartheta})$ as $\vartheta>0$ and $O(h_{\mathsf{eff}}^{1-d-\delta})$ with arbitrarily small $\delta>0$ as $\vartheta=0$; "smooth" means "with $K=K(d,\vartheta)$ or $K=K(d,\delta)$ respectively".

Remark 11.2.2. To avoid $\mathsf{N}^{\mathsf{W}}=0$ we need to assume that

(11.2.29) $j(x,\xi)$ (the principal symbol of J) is not a non-positive definite matrix (as long as we assume that $a(x,\xi)$ is a positive definite matrix).

We leave to the the following

Problem 11.2.3. Investigate the opposite case when according to Chapter 7 the local remainder estimate may be $O(h_{\mathsf{eff}}^{2-d})$ and $\mathsf{N}^{\mathsf{W}}\asymp h_{\mathsf{eff}}^{1-d}$.

Let us discuss conditions associated with the boundary value problem. Recall that this condition for 1-dimensional operator $f=\tilde{a}-\tilde{j}$ is of the form

$$\operatorname{Re}\langle \ell f v, v\rangle_{+} + \operatorname{Re}\langle \ell\tilde{b}v, \beta v\rangle_{0} \ge \epsilon_0\|v\|_{+}^{2} \qquad \forall v\in\mathscr{S}'(\mathbb{R}^{+},\mathbb{H}) : fv=0 \text{ and } bv|_{x_1=0}=0 \tag{11.2.30}$$

where we use the notation of Chapter 7 and make a rescaling; here $f=f(z,D_1)$, $b=b(z,D_1)$ and $\beta=\beta(z,D_1)$ is defined by equality

$$\langle fu, v\rangle_{+} + \langle \tilde{b}u, \beta v\rangle_{0} = \langle u, fv\rangle_{+} + \langle \beta u, \tilde{b}v\rangle_{0} \qquad \forall u,v\in\mathscr{S}(\mathbb{R}^{+},\mathbb{H}) \tag{11.2.31}$$

where $z=(x',\xi')$.

Let us note that

(11.2.32) Condition (11.2.30) survives substitution $D_1\mapsto\theta(z)D_1$ with $\theta\asymp 1$.

In fact, it is obvious for $\theta=\mathsf{const}$ and therefore we need check this statement only for θ equal to 1 at the point in question. In this case additional term

$$(\ell\theta)(z)\times\operatorname{Re}\bigl(\langle f'v, v\rangle_{+} + \langle \tilde{b}'v, \beta v\rangle_{0}\bigr) \tag{11.2.33}$$

in the left-hand expression of (11.2.30) appears where the prime means derivative with respect to θ. Let us introduce $v_\theta(x_1)=v(\theta^{-1}x_1)$; then

$f_\theta v_\theta = 0$ and $\tilde{b}_\theta v_\theta|_{x_1=0} = 0$. Therefore treating θ as an independent variable, differentiating and taking $\theta = 1$ we obtain that

$$f'v + fv' = 0, \qquad \tilde{b}'v + \tilde{b}v' = 0 \tag{11.2.34}$$

where the prime on v means derivative on θ also and therefore $v' = -x_1\partial_1 v$. Then (11.2.34), (11.2.31) and the equalities $fv = 0$, $\tilde{b}v|_{x_1=0} = 0$ obviously yield that the second factor in (11.2.30) vanishes; (11.2.32) has been proven.

Let us consider first $|\xi'| \ge \epsilon_0$ and let us introduce $\ell = \langle \xi', \partial_{x'}\rangle$, $\theta = |\xi'|$. Then 1-dimensional operators $\tilde{a}(x', \xi', \theta D_1)$ and $\tilde{\jmath}(x', \xi', \theta D_1)$ are positive homogeneous of degrees $2m$ and n on ξ' respectively. Moreover, operators $\tilde{b}_k(x', \xi', \theta D_1)$ are positive homogeneous of degrees m_k on ξ' and $b_k v = 0$. Therefore the Euler identity yields that the left-hand expression of (11.2.30) is equal to

$$2m\langle \tilde{a}v, v\rangle_+ - n\langle \tilde{\jmath}v, v\rangle_+.$$

Since $(\tilde{a} - \tilde{\jmath})v = 0$ we obtain that this expression is equal to

$$(2m - n)\langle \tilde{a}v, v\rangle_+ = (2m - n)\langle \tilde{\jmath}v, v\rangle_+.$$

Therefore we arrive to

Proposition 11.2.4. *As* $|x'| \ge \epsilon_0$ *condition* (11.2.30) *is fulfilled provided either*

$$\langle \tilde{a}v, v\rangle_+ \ge \epsilon_0 \|v\|_+^2 \qquad \forall v \in \mathscr{S}(\mathbb{R}^+, \mathbb{H}) : \ \tilde{b}v = 0 \ \textit{at} \ \ x_1 = 0 \tag{11.2.35}$$

or the same condition is fulfilled for $\tilde{\jmath}$*:*

$$\langle \tilde{\jmath}v, v\rangle_+ \ge \epsilon_0 \|v\|_+^2 \qquad \forall v \in \mathscr{S}(\mathbb{R}^+, \mathbb{H}) : \ \tilde{b}v = 0 \ \textit{at} \ \ x_1 = 0 \tag{11.2.36}$$

If A is a semi-bounded operator (under the boundary condition $Bv = 0$) then the ellipticity of the boundary value problem (A, B) yields that condition (11.2.35) is fulfilled. *In what follows when we consider non-semi-bounded operators* A *we assume that* $J = I$ and then condition (11.2.36) is fulfilled automatically.

Let us consider the case $\xi' = 0$. Let us prove that

(11.2.37) As $|\xi'| = 0$ the kernel of the problem

$$fv = 0, \qquad \tilde{b}v = 0 \ \text{ at } \ x_1 = 0, \qquad v \in \mathscr{S}(\mathbb{R}^+, \mathbb{H}) \tag{11.2.38}$$

is trivial.

Then condition (11.2.30) will be automatically fulfilled.

Let us note that if $\xi' = 0$ then $\tilde{a} = a_0 D_1^{2m}$, $\tilde{\jmath} = j_0 D_1^n$, $\tilde{b}_k = b_{0k} D_1^{m_k}$ where a_0, j_0, b_{0k} are matrices (depending on z). Let us replace D_1 by θD_1 and let us introduce $v_\theta(x_1) = v(\theta^{-1}x_1)$ with an arbitrary parameter θ. The above arguments yield the following equality

$$\langle f'v, v\rangle_+ + \langle \tilde{b}'v, \beta v\rangle_0 = 0$$

for any solution of the problem (11.2.38). Let us note that in our case $\tilde{b}'v = 0$ at $x_1 = 0$. Then equalities $fv = 0$, $f = \tilde{a} - \tilde{\jmath}$ and $f' = 2m\tilde{a} - n\tilde{\jmath}$ obviously yield that

$$\langle \tilde{a}v, v\rangle_+ = 0, \qquad \langle \tilde{\jmath}v, v\rangle_+ = 0.$$

The first equality is impossible for $v \ne 0$ if A is a semi-bounded elliptic operator and the second equality is impossible for $v \ne 0$ if $J = I$. So, (11.2.37) and thus (11.2.30) are established.

Thus we arrive to the following

Proposition 11.2.5. *Let one of the following assumptions be fulfilled:*

(a) Either A and J are matrix matrix operators satisfying one of the conditions (11.2.12)–(11.2.15) *or A and J are operators with scalar senior symbols satisfying condition* (11.2.20)*;*

(b) $B(x, \gamma(x)) \cap \partial X = \emptyset$ and A and J are operators with with scalar senior symbols satisfying one of the conditions (11.2.16), (11.2.17), (11.2.21)–(11.2.15)*;*

(c) $d \ge 2$ and A is the Schrödinger operator with a Dirichlet or Neumann boundary condition[13] *on $\partial X \cap B(x, \gamma(x))$.*

Then the contribution of $\mathcal{B}(x, \gamma(x), X) \subset X'_{\mathsf{reg}}$ to the principal part of the asymptotics for $\mathsf{N}^{\mathsf{W}}_{(\vartheta)}(\tau)$ is $O(h_{\mathsf{eff}}^{-d}\tau^{\vartheta})$[14] *and its contribution to the remainder is $O(h_{\mathsf{eff}}^{1-d+\vartheta}\tau^{\vartheta})$ with*

$$h_{\mathsf{eff}} = \Bigl(\frac{\tau\varrho_0}{\varrho}\Bigr)^{-\frac{1}{2m-n}}\rho^{-1}\gamma^{-1} \tag{11.2.39}$$

[13] Or a Robin boundary condition with appropriate restrictions on lower order coefficients.

[14] Recall that under assumption (11.2.29) it is $\asymp h_{\mathsf{eff}}^{-d}\tau^{\vartheta}$.

and therefore the total contribution of X'_{reg} defined by (11.2.9) to the remainder does not exceed

$$R_{1(\vartheta)}(\tau)=\tau^{\frac{d-1-\vartheta}{2m-n}+\vartheta}\int_{X'_{\mathsf{reg}}}\bigl(\frac{\varrho_0}{\varrho}\bigr)^{\frac{d-1-\vartheta}{2m-n}}\rho^{d-1-\vartheta}\gamma^{-1-\vartheta}\,dx. \tag{11.2.40}$$

11.2.2 Weak Singularity

Now consider the contribution of

$$X''_{\mathsf{reg}}=X_{\mathsf{reg}}\setminus X'_{\mathsf{reg}}=\{\rho\gamma\ge 1,\ \tau\varrho_0\le C_0\varrho\}; \tag{11.2.41}$$

we estimate the contribution of $\mathcal{B}(x,\gamma(x),X)\subset X''_{\mathsf{reg}}$ to both the main part and remainder estimate by $O(h_{\mathsf{eff}}^{-d}\varrho^{\vartheta}\varrho_0^{-\vartheta})$ with $h_{\mathsf{eff}}=\rho^{-1}\gamma^{-1}$ and the total contribution of X''_{reg} does not exceed

$$R_{2(\vartheta)}(\tau)=\int_{X''_{\mathsf{reg}}}\rho^{d}\varrho^{\vartheta}\varrho_0^{-\vartheta}\,dx \tag{11.2.42}$$

provided

$$\vartheta>0\implies J=I\quad\text{i.e.}\quad n=0,\ \varrho_0=1. \tag{11.2.43}$$

As A and J are scalar operators we can use Remark 11.2.1 and vastly improve the estimate of the contribution of X''_{reg} to the remainder.

As we use such crude estimate in X''_{reg} and hope to recover a reasonable result we will need to assume that singularity is weak.

Further, we need to estimate the contribution of X_{sing}. If we can apply Rozenblioum estimate we arrive to the following

Proposition 11.2.6. *(i) As $\vartheta=0$ the contribution of X_{sing} to asymptotics does not exceed*

$$CR_0+CR'_0 \tag{11.2.44}$$

where R_0 is given by (11.1.10) with exponents satisfying (11.1.11) and (11.1.12) and

$$R'_0=\tau^{\frac{d}{2m-n}}\Bigl(\int_{X_{\mathsf{sing}}}\varrho^{-p'}\rho^{2mp'}\,dx\Bigr)^{\frac{d}{(2m-n)p'}}\times\Bigl(\int_{X_{\mathsf{sing}}}\varrho_0^{q'}\rho^{-nq'}\,dx\Bigr)^{\frac{d}{(2m-n)q'}}\times\bigl(\mathsf{mes}(X_{\mathsf{sing}})\bigr)^{1-\frac{d}{2mq'}(\frac{1}{p'}+\frac{1}{q'})} \tag{11.2.45}$$

with exponents p' *and* q' *satisfying conditions* (11.1.11)$^{\#}$ *and* (11.1.12)$^{\#}$ *where the* $^{\#}$ *means that we replace* $2m$, p *and* q *by* $2m-n$, p' *and* q' *respectively.*

We assume that a Dirichlet boundary condition is given on $\partial X\cap \bar{X}_{\mathsf{sing}}$. *Otherwise we should assume that a cone condition is fulfilled (with constant angle and height of the cone),* $\varrho\asymp\rho^{2m}$ *and* $\mathsf{mes}(X_{\mathsf{sing}})<\infty$.

(ii) As $\vartheta>0$ *and* $J=I$ *the contribution of the zone* X_{sing} *to the asymptotics does not exceed*

$$CR_{0(\vartheta)}+CR'_{0(\vartheta)} \tag{11.2.46}$$

where

$$\begin{aligned}R_{0(\vartheta)}=\int_0^\infty t^{\vartheta-1}\Bigl(\int_{X_{\mathsf{sing}}}(\varrho-t)_+^q\,dx\Bigr)^{\frac{d}{2mq}}dt\times\\ \Bigl(\int_{X_sing}\varrho^{-p}\rho^{2mp}\,dx\Bigr)^{\frac{d}{2mp}}\times\bigl(\mathsf{mes}(X_{\mathsf{sing}})\bigr)^{1-\frac{d}{2m}(\frac{1}{p}+\frac{1}{q})},\end{aligned} \tag{11.2.47}$$

$$R'_{0(\vartheta)}=\tau^{\vartheta+\frac{d}{2m}}\Bigl(\int_{X_{\mathsf{sing}}}\varrho^{-p}\rho^{2mp}\,dx\Bigr)^{\frac{d}{2mp}}\times\bigl(\mathsf{mes}(X_{\mathsf{sing}})\bigr)^{1-\frac{d}{2mp}} \tag{11.2.48}$$

and exponents p *and* q *satisfy conditions* (11.1.11)–(11.1.12).

11.2.3 Main Theorems

Let us start from the standard asymptotics. Then we improve the remainder estimate under conditions on the Hamiltonian trajectories in X'_{reg}. Resulting from this analysis we obtain the following statements:

Theorem 11.2.7. *Let* $X=\mathcal{X}_{\mathsf{reg}}\cup\mathcal{X}_{\mathsf{sing}}$, $\mathsf{mes}(\mathcal{X}_{\mathsf{sing}})=0$. *Let us assume that in* $\mathcal{X}_{\mathsf{reg}}$ *all the conditions* (9.1.5)–(9.1.13), (9.1.15) *(excluding restriction* $\rho\gamma\ge 1$*) are fulfilled.*

Further, let us assume that $\varrho^{-1}\rho^{2m}a^0(x,\xi)$ *is uniformly classically elliptic and the corresponding boundary value problem for it is also uniformly classically elliptic at the boundary points.*

Furthermore, let us assume that in each ball $\mathcal{B}(x,\gamma(x),X)\subset\mathcal{X}_{\mathsf{reg}}$ *one of the assumptions (a), (b) or (c) of Proposition 11.2.5 is fulfilled. Then*

(i) The following lower estimate holds:

$$N(0,\tau) \ge N^W(0,\tau) - CR_{1(0)}(\tau) \tag{11.2.49}$$

where $N^W(0,\tau)$ *is defined by cut Weyl formula (with integration over* X'_{reg}*):*

$$N^W(0,\tau) = \varkappa_0(\tau)\tau^{\frac{d}{2m-n}}, \tag{11.2.50}$$

$$\varkappa_0(\tau) = (2\pi)^{-d}\int_{X'_{reg}(\tau)} n(x,\xi;0,1)\,dxd\xi, \tag{11.2.51}$$

$n(x,\xi;0,1)$ *is the eigenvalue counting function for the symbol* $j^{-1}(x,\xi)a(x,\xi)$ *on the interval* $(0,1)$, $R_{1(0)}(\tau)$ *is defined by* (11.2.40) *with* $\vartheta = 0$; $X'_{reg}(\tau)$ *is defined by* (11.2.10).

(ii) Further, let us assume that in

$$X_{sing}(\tau) = X \setminus \bigl(X'_{reg}(\tau) \cup X''_{reg}(\tau)\bigr) \tag{11.2.52}$$

we can apply Rozenblioum estimate. Then the upper estimate holds

$$N(0,\tau) \le N^W(0,\tau) + CR_{1(0)}(\tau) + CR_{2(0)}(\tau) + CR_0(\tau) + CR'_0(\tau) \tag{11.2.53}$$

where $R_{2(0)}(\tau)$, $R_0(\tau)$, $R'_0(\tau)$ *are defined by* (11.2.42), (11.1.10), (11.2.45) *respectively with* $X''_{reg}(\tau)$ *and* $X_{sing}(\tau)$ *defined by* (11.2.42) *and* (11.2.52).

Similarly, as $\vartheta > 0$ we arrive to

Theorem 11.2.8. *Let assumptions of Theorem 11.2.7 be fulfilled. Moreover, let us assume that* $J = I$ *and* A *is semi-bounded from below. Then*

(i) The following lower estimate holds:

$$N^-_{(\vartheta)}(\tau) \ge N^W_{(\vartheta)}(\tau) - CR_{1(\vartheta)}(\tau) \tag{11.2.54}$$

where $N^W_{(\vartheta)}(\tau)$ *is defined by cut Weyl formula (with integration over* X'_{reg}*),* $R_{1(\vartheta)}(\tau)$ *is defined by* (11.2.40).

item Further, let us assume that in $X_{sing}(\tau)$ *defined by* (11.2.52) *we can apply Rozenblioum estimate. Then the upper estimate holds*

$$N^-_{(\vartheta)}(\tau) \ge N^W_{(\vartheta)}(\tau) + CR_{1(\vartheta)}(\tau) + CR_{2(\vartheta)}(\tau) + CR_{0(\vartheta)}(\tau) + CR'_{0(\vartheta)}(\tau) \tag{11.2.55}$$

where $R_{2(\vartheta)}(\tau)$, $R_{0(\vartheta)}(\tau)$, $R'_{0(\vartheta)}(\tau)$ *are defined by* (11.2.42), (11.2.46), (11.2.45) *respectively with* $X''_{reg}(\tau)$ *defined by* (11.2.42).

Obviously, if X is a compact domain with the smooth boundary and $J(x)=I$ we conclude that

$$\mathsf{N}^-_{(\vartheta)}(\tau)=\mathsf{N}^{\mathsf{W}}_{(\vartheta)}(\tau)+O\bigl(\tau^{\frac{d-1-\vartheta}{2m}+\vartheta}\bigr)\qquad \text{as}\quad \tau\to+\infty \tag{11.2.56}$$

with

$$\mathsf{N}^{\mathsf{W}}_{(\vartheta)}(\tau)\asymp \tau^{\frac{d}{2m}+\vartheta}. \tag{11.2.57}$$

Our goal now is to consider the *weakly singular* case when this conclusion holds. Since X'_{reg} extends to $\mathcal{X}_{\mathsf{reg}}$ as $\tau\to+\infty$, we need to assume that

$$\bigl(\frac{\varrho_0}{\varrho}\rho^{2m-n}\bigr)^{\frac{d}{2m-n}}\in\mathscr{L}^1(X) \tag{11.2.58}$$

and

$$\bigl(\frac{\varrho_0}{\varrho}\rho^{2m-n}\bigr)^{\frac{d-1-\vartheta}{2m-n}}\gamma^{-1-\vartheta}\in\mathscr{L}^1(X). \tag{11.2.59}$$

Corollary 11.2.9. *Assume that conditions of Theorem 11.2.7 as* $\vartheta=0$ *or Theorem 11.2.8 as* $\vartheta>0$ *are fulfilled. Further, let conditions* (11.2.58)–(11.2.59) *be fulfilled. Then*

$$R_{2(\vartheta)}(\tau)=o\bigl(\tau^{\frac{d-1-\vartheta}{2m-n}+\vartheta}\bigr)\qquad \textit{as}\quad \tau\to+\infty \tag{11.2.60}$$

and assuming that

$$R_{0(\vartheta)}(\tau)+R'_{0(\vartheta)}(\tau)=O\bigl(\tau^{\frac{d-1-\vartheta}{2m-n}+\vartheta}\bigr)\qquad \textit{as}\quad \tau\to+\infty \tag{11.2.61}$$

we conclude that (11.2.56)–(11.2.57) *hold.*

Remark 11.2.10. (i) One can easily see that extending integration in $\mathsf{N}^{\mathsf{W}}(\tau)$ from X_{reg} to X causes an error not exceeding (11.2.44). So we can replace cut Weyl expression by the full Weyl expression.

(ii) Also skipping lower degree terms in N^{W} leads to an error not exceeding $CR_{1(0)}(\tau)$; this leads to (11.2.50)–(11.2.51).

(iii) Combining Statements (i) and (ii) we arrive to (11.2.51) with integration over X (then $\varkappa_0$ does not depend on τ).

(iv) Some of these conclusions remain true for $\vartheta > 0$, at least with some modifications: for example, (11.2.50)–(11.2.51) becomes

$$\mathsf{N}^{\mathsf{W}}_{(\vartheta)}(0,\tau) = \sum_{0\le n<1+\vartheta} \varkappa_n(\tau)\tau^{\frac{d-n}{2m-n}+\vartheta} \tag{11.2.62}$$

but extension of integration in $\varkappa_n(\tau)$ terms is justified only under assumption (11.2.59) with ϑ replaced by n (and some other assumptions may be required as well).

(v) Theorems 11.2.7 and 11.2.8 and Corollary 11.2.9 remain true for compact $\mathscr{C}^K$ manifolds.

Remark 11.2.11. In the case $\varrho = \rho^{2m}$ and $2m - n < d$ one can set $q' = \frac{d}{2m-n}$ and then (11.2.59) yields (11.2.61).

We can slightly improve the results of above theorems under assumptions to Hamiltonian trajectories:

Theorem 11.2.12. *Assume that conditions of Theorem 11.2.7 as $\vartheta = 0$ or Theorem 11.2.8 as $\vartheta > 0$ are fulfilled.*

Further, let us assume that non-periodicity condition (11.1.55)–(11.1.56) *and improved smoothness and non-degeneracy conditions of Remark 11.1.8 hold for symbol $a^0(x,\xi) - j^0(x,\xi)$ on the energy level 0. Then sharp versions of asymptotics* (11.2.54), (11.2.55) *hold where "sharp version" means that $R_{1(\vartheta)}(\tau)$ is replaced by*

$$R'_{1(\vartheta)}(\tau) = o(1)R_{1(\vartheta)}(\tau) + R_{1(\vartheta-\delta)}(\tau) \tag{11.2.63}$$

where $\delta > 0$ is sufficiently small exponent and formula (11.1.39) *defining $R_{1(\vartheta)}$ is extended to $\vartheta < 0$ if needed; also in $\mathsf{N}^{\mathsf{W}}_{(\vartheta)}$ one extra term should be included if needed.*

Sharper asymptotics under more restrictive conditions on the Hamiltonian flow will be treated later.

11.2.4 Examples

Consider some examples. Let us start with examples in which $J = I$, $\varrho = \rho = 1$. Then conditions (11.2.58), (11.2.59) mean respectively that

$$1 \in \mathscr{L}^1(X) \qquad \text{i.e. } \mathsf{mes}(X) < \infty, \tag{11.2.64}$$

$$\gamma^{-1-\vartheta} \in \mathscr{L}^1(X). \tag{11.2.65}$$

In the first examples the coefficients of A are regular and γ is completely defined by the geometry of the domain X.

Example 11.2.13 (Quasipolygonal and quasipolyhedral domains). Let X be a bounded domain with ∂X contained in the union of a finite number of $\mathscr{C}^K$ hypersurfaces[15] $\tilde{S}_l$ (with $l = 1, \dots, L$).

Let $S_l = \tilde{S}_l \cap \partial X$ and δS_l be the relative boundary of S_l in $\tilde{S}_l$. Introduce the *singular set* of ∂X

$$\mathcal{K} = \bigcup_{l \ne k} (S_l \cap S_k) \cup \bigcup_l \delta S_k \tag{11.2.66}$$

Obviously, we can take

$$\gamma(x) \asymp \min\Bigl(\min_{l\ne k} \gamma_{lk}(x), \mathsf{dist}(x, \mathcal{K})\Bigr), \tag{11.2.67}$$

with

$$\gamma_{lk}(x) = \max\bigl(\gamma_l(x), \gamma_k(x)\bigr) \tag{11.2.68}$$

where $\gamma_l(x) = \mathsf{dist}(x, S_l)$.

Then condition (11.2.65) needs to be verified for γ_{lk} for all $l \ne k$ and for $\gamma'(x) = \mathsf{dist}(x, \mathcal{K})$ separately.

First of all,

(11.2.69) If the hypersurfaces $\tilde{S}_l$ and $\tilde{S}_k$ are transversal at $S_k \cap S_l$ then condition (11.2.65) is fulfilled for $\gamma = \gamma_{kl}$ and $\vartheta < \vartheta^*(= 1)$.

Moreover,

---

[15] Actually we need only regularity required to produce local remainder estimate $O(h_{\text{eff}}^{1-d+\vartheta})$ or $o(h_{\text{eff}}^{1-d+\vartheta})$ in case we are interested in the sharp asymptotics.

(11.2.70) If $\mathcal{K}$ is contained in the union of a finite number of $\mathscr{C}^K$ manifolds of codimension 2 then condition (11.2.65) is fulfilled for $\gamma = \gamma'$ for $\vartheta < \vartheta^* = 1$ as well.

One can see easily that the above assumptions yield the cone condition. Therefore

(i) For appropriate boundary conditions[16)] asymptotics (11.2.56) holds for $\vartheta < \vartheta^*$.

Moreover, under standard non-periodicity condition (11.1.55)–(11.1.56) to the Hamiltonian flow of the symbol $a^0(x,\xi)$ on the energy level 1 asymptotics

$$\mathsf{N}^-_{(\vartheta)}(\tau) = \mathsf{N}^\mathsf{W}_{(\vartheta)}(\tau) + o\bigl(\tau^{\frac{d-1-\vartheta}{2m}+\vartheta}\bigr) \qquad \text{as } \tau \to +\infty \tag{11.2.56*}$$

holds.

(ii) Further, our analysis yields that for $\vartheta = \vartheta^*$ asymptotics

$$\mathsf{N}^-_{(\vartheta)}(\tau) = \mathsf{N}^\mathsf{W}_{(\vartheta)}(\tau) + O\bigl(\tau^{\frac{1}{2m}(d-\vartheta-1)+\vartheta}\log\tau\bigr) \tag{11.2.71}$$

holds as $\tau \to +\infty$ and for $\vartheta > \vartheta^*$ asymptotics

$$\mathsf{N}^-_{(\vartheta)}(\tau) = \mathsf{N}^\mathsf{W}_{(\vartheta)}(\tau) + O\bigl(\tau^{\frac{1}{2m}(d-\vartheta^*-1)+\vartheta}\bigr) \tag{11.2.72}$$

holds as $\tau \to +\infty$.

Later we will show that under appropriate assumptions sharper asymptotics than (11.2.71)–(11.2.72) holds and in the latter case some extra non-Weyl term(s) associated with the singularities at $\mathcal{K}$ come into play.

Example 11.2.14 (Spikes and conjugations). Let us drop transversally condition (11.2.69) thus allowing *spikes*. Consider first spikes directed inward as on Figure 11.1(b).

---

16) Such as Dirichlet, Neumann and Robin boundary conditions, or boundary conditions which are smooth, elliptic, regular (see condition (7.1.14)) on $S_l \setminus \mathcal{K}$ and such that A is self-adjoint and $(Au,u) \ge \epsilon_0\|u\|_m^2 - c\|u\|_0^2$.

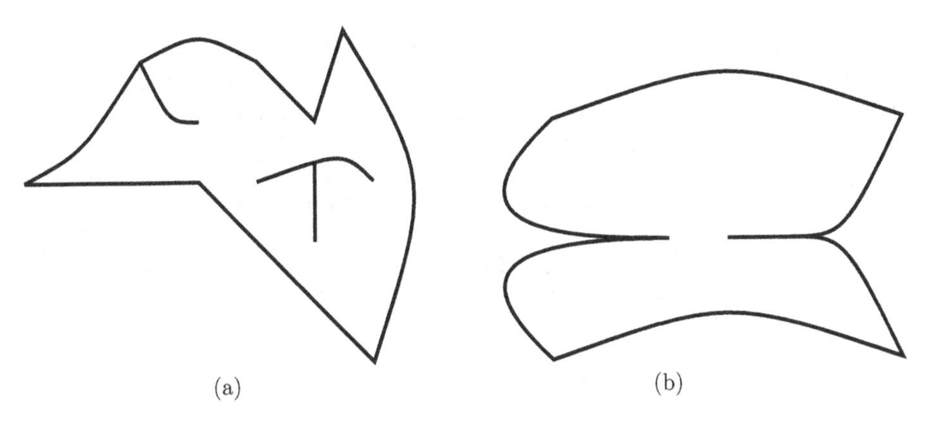

Figure 11.1: Conclusions of Example 11.2.13 holds despite on (b) transversally condition violated

(i) Assume first that locally only two surfaces S_k and S_l are involved and in appropriate coordinates

$$S_j = \{x_d = f_j(x'),\ x' \in X'\},\qquad (f_j - f_i)|_{\partial X'} = 0 \tag{11.2.73}$$

where $X' \subset \mathbb{R}^{d-1}$, $x' = (x_1, \dots, x_{d-1}$, $|\nabla f_j| \le \epsilon$. Then assuming that locally $\complement X$ is between surfaces

$$\complement X = \{f_k(x') \le x_d \le f_l(x'),\ x' \in X'\} \cup \{f_l(x') \le x_d \le f_k(x'),\ x' \in X'\} \tag{11.2.74}$$

we conclude that $\gamma(x) \asymp \gamma(x') = \mathsf{dist}(x, \mathcal{K})$ with

$$\mathcal{K} = \{x : x' = \mathcal{K}', x_d = f_j(x')\},\quad \mathcal{K}' = \partial X' \cup \{x', f_j(x') = f_k(x')\}. \tag{11.2.75}$$

Then condition (11.2.65) is fulfilled for $\vartheta < \vartheta^* = 1$ provided condition (11.2.70) is fulfilled. Moreover, cone condition is fulfilled as well and then all conclusions of Example 11.2.13 remain valid.

(ii) Further, the same conclusions obviously remain true as locally there is *conjugation* instead of spike

$$S_j = \{x_d = f_j(x'),\ x' \in X'_j\},\qquad (f_j - f_k)|_{\partial X'_j \cap \partial X'_k} = 0 \quad \text{for}\quad k \ne j \tag{11.2.76}$$

as X'_j form locally a partition of the vicinity of 0 in $\mathbb{R}^{d-1}$.

(iii) Furthermore, we can consider inward spike or conjugation and add extra surfaces to ∂X assuming that each new surface is either transversal to existing surfaces or further limits $\complement X$. For example, we can consider multidimensional inward spike:

$$\complement X = \{x_1 > 0, x' \in f(x_1)X'\} \tag{11.2.77}$$

with $f > 0, f_j' > 0$ as $x_1 > 0$, $f(0) = f'(0) = 0$ and X' bounded convex quasipolygon of Example 11.2.14.

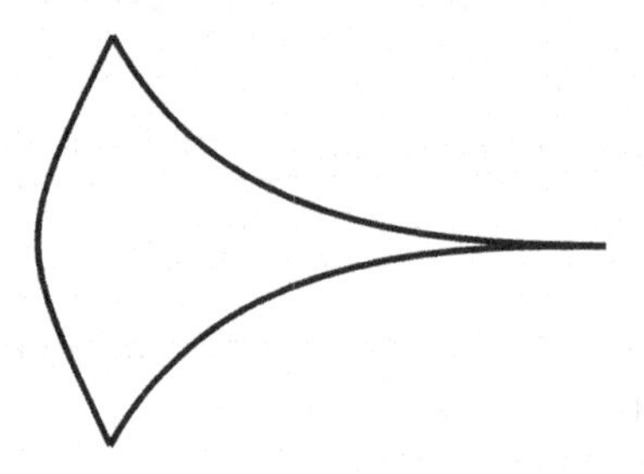

Figure 11.2: Outward spike

Example 11.2.15 (Spikes. II). Outward spike are completely different matter. Assume first that locally only two surfaces S_k and S_l are involved and in appropriate coordinates (11.2.73) and (11.2.74) are fulfilled with $\complement X$ replaced by X

$$X = \{f_k(x') < x_d < f_l(x'),\ x' \in X'\}\cup \\ \{f_l(x') < x_d < f_k(x'),\ x' \in X'\} \tag{11.2.78}$$

as on Figure 11.2. Obviously, without any loss of the generality one can assume that $f_k = f$, $f_l = -f$ and then $\gamma_{kl} \asymp |f|$. Therefore condition (11.2.65) becomes

$$|f|^{-\vartheta} \in \mathscr{L}^1(X') \tag{11.2.79}$$

which obviously is fulfilled for $\vartheta = 0$, violated for $\vartheta \geq 1$ and may be fulfilled or not for any particular $\vartheta \in (0, 1)$; f.e. as $f = f(x_1) \asymp x_1^r$ for $x_1 > 0$ $(r > 1)$ this condition holds for $\vartheta < \vartheta^* = r^{-1}$ while for $f = O(x_1^\infty)$ at 0 this condition holds only for $\vartheta = 0$.

So, if (11.2.79) is fulfilled then asymptotics (11.2.56) holds for appropriate boundary conditions. Moreover, under the standard non-periodicity condition (11.1.55)–(11.1.56) to the Hamiltonian flow of the symbol $a^0(x,\xi)$ on the energy level 1 asymptotics (11.2.56)* holds.

Problem 11.2.16. (i) Derive asymptotics similar to (11.1.69) or (11.1.70) for particular functions f as condition (11.2.79) is violated.

(ii) Consider multidimensional outward spikes like one similar to (11.2.77):

$$X = \{x_1 > 0, x' \in f(x_1)X'\} \tag{11.2.80}$$

with $f > 0, f'_j > 0$ as $x_1 > 0$, $f(0) = f'(0) = 0$ and X' bounded convex quasipolygon of Example 11.2.14.

(iii) Consider the mixture of inward-outward spikes like (11.2.80) but without any convexity assumptions.

Remark 11.2.17. (i) n the case of "outward spike" the cone condition now is not ensured, so "appropriate boundary conditions" in 11.2.16(i), (ii) includes cone condition. In particular cone condition is fulfilled if at least on one side of the spike we have a Dirichlet boundary condition. There is not much known about more general conditions. Still for 2-dimensional Laplace operator with Neumann boundary condition Yu. Netrusov and Yu. Safarov [1, 2] in pretty special case derived variational estimate which allows to allow Neumann boundary condition on the whole boundary.

(ii) Later in Section 12.7 we will derive spectral asymptotics for multidimensional Neumann Laplacian and those methods could be applied in some of the examples discussed here to estimate the contribution of zone $\{x : \gamma(x) \le \tau^{-1/2}\}$ to the asymptotics thus extending corresponding results (see Subsection 12.7.8). However there will be also plenty of examples when conclusions valid for Dirichlet Laplacian are wrong for Neumann Laplacian.

In the series of examples below $\gamma(x) \asymp \mathsf{dist}(x, \mathcal{K})$ where $\mathcal{K}$ is the union of a finite number of $\mathscr{C}^K$ manifolds of codimension $\delta \ge 2$. Namely, we consider domains with conical singularities of the boundary, domains the boundaries of which contain components of codimension $\delta \in [2, 4m-1]$ [17]. Then under appropriate assumptions critical exponent is $\vartheta^* = \delta - 1$.

[17] For operators of order $2m$ a boundary operator of degree less than $2m - \frac{1}{2}\delta$ can be considered on a δ-codimensional submanifold.

Example 11.2.18 (Scheme). Let X' be a domain in $\mathbb{R}^{d'}$. Assume that

(11.2.81) X' satisfies condition (9.1.15) with γ replaced by $\gamma' = \gamma'(x')$ such that $\gamma' \le c$ and $\gamma'^{-1-\vartheta} \in \mathscr{L}^1(X')$

Let $X = \{x = (x', x''), x'' \in \mathbb{R}^{d''}, x' \in f(x'')X'\}$ [18] where the function f satisfies condition

$$|D^\alpha f| \le cf^{1-|\alpha|} \qquad \forall \alpha : |\alpha| \le K. \tag{11.2.82}$$

Then for the domain X the function

$$\gamma(x) = \epsilon\gamma'(x'f^{-1}(x''))f(x'') \tag{11.2.83}$$

is admissible and in this case condition (11.2.65) is equivalent to

$$f^{d'-\vartheta-1} \in \mathscr{L}^1; \tag{11.2.84}$$

in particular, this condition as $\vartheta = 0$ is equivalent to the condition

(11.2.85) the $(d-1)$-dimensional measure of ∂X is finite.

Example 11.2.19 (Cusps). This scheme described in Example 11.2.18 includes cusps (exits to infinity): if $f \asymp |x''|^\nu$ with $\nu < 0$ for $|x''| \ge 1$ then the following cases are possible:

(a) Let $\vartheta < \vartheta^*$ with

$$\vartheta^* = d' - 1 - d''\nu^{-1} \iff (d' - 1 - \vartheta)\nu + d'' = 0. \tag{11.2.86}$$

Then asymptotics (11.2.56) holds.

Moreover, under standard non-periodicity condition (11.1.55)–(11.1.56) to the Hamiltonian flow of the symbol $a^0(x,\xi)$ on the energy level 1 asymptotics (11.2.56)* holds.

(b) For $\vartheta = \vartheta^*$ asymptotics with the remainder estimate $O\bigl(\tau^{\frac{1}{2m}(d-1-\vartheta)+\vartheta}\log\tau\bigr)$ holds.

---

[18] For $d'' = 1$ one can take only one semiaxis, i.e., replace $\mathbb{R}$ by $\mathbb{R}^\pm$.

(c) For $\vartheta > \vartheta^*$, $d'\nu + d'' < 0$, asymptotics with the remainder estimate $O\bigl(\tau^{\frac{1}{2m}(d-1-\vartheta^*)+\vartheta}\bigr)$ holds.

Moreover, if $\vartheta^* \in \mathbb{Z}^+$ then the last term of these asymptotics is of the form $\varkappa_{\vartheta,\vartheta^*+1}(\tau)\tau^{\frac{1}{2m}(d-1-\vartheta^*)+\vartheta}$ with $\varkappa_{\vartheta,\vartheta^*+1}(\tau) = O(\log\tau)$ while all other terms in this case and all terms in the other cases are of the form $\varkappa_{\vartheta,n}\tau^{\frac{1}{2m}(d-n)+\vartheta}$.

(d) If $d'\nu + d'' = 0$ then the previous statement remains true with only the term $\varkappa_{\vartheta,0}(\tau)\tau^{\frac{d}{2m}+\vartheta}$ in the principal part with $\varkappa_{\vartheta,0}(\tau) \asymp \log\tau$.

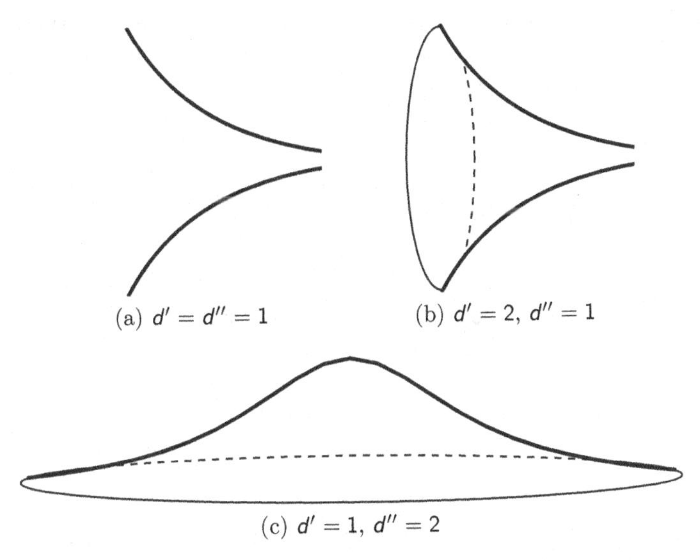

Figure 11.3: Cusps; in (c) X bounded by a drawn "infinitely wide hat" and horizontal plane

Remark 11.2.20. (i) As $d'' \geq 2$ we can consider more general cusp bases than $X'' = \mathbb{R}^{d''}$, most notably we can consider smooth consider a cone $X'' \subset \mathbb{R}^{d''}$ with a smooth boundary; rounding near $\partial X''$ here is obvious.

(ii) We treat Example 11.2.19 in Section 12.2 again and obtain non-Weyl asymptotics with a more precise remainder estimate in cases (b)–(d) and and also in the case $d'\nu + d'' > 0$. We will consider also a generalization mention above.

(iii) In Example 11.2.19 we need to assume that a Dirichlet boundary condition is given on a significant part of ∂X so the cone condition if fulfilled. For example, we can assume that the Dirichlet boundary condition is given as $x' \in Y_{\mathsf{D}} \subset \partial X'$ and the following infinite cone condition is fulfilled:

(11.2.87) there exist an infinite cone $\Gamma \subset \mathbb{R}^{d'}$ with a vertex at 0 such that for each $x' \in X'$ connected component of $(x' + \Gamma) \cap X'$ is disjoint from $\partial X' \setminus Y_{\mathsf{D}}$.

(iv) Without the assumption of (iii) (f.e. for Neumann Laplacian in domain with cusps we need to assume that the cusp is very thin: $|\log f| \asymp |x''|^{\kappa}$ with $\kappa > 0$. Otherwise massive essential spectrum appears. We treat the Neumann Laplacian in domain with "piggy-tail" cusps in Section 12.7 and derive possibly non-Weyl asymptotics.

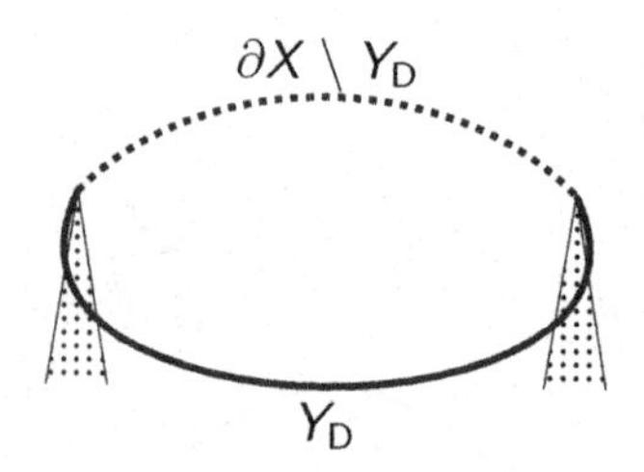

Figure 11.4: Infinite cone condition

Example 11.2.21 (Conical and sharp quasiconical singularities). (i) The scheme described in Example 11.2.18 covers conical singularities as well as sharp quasiconical singularities directed outward. Namely let

$$X = \{x = (x', x''),\ x' \in |x''|^{\nu} \cdot X',\ |x''| \le 1\} \tag{11.2.88}$$

with $\nu \ge 1$; for $d'' = 1$ one can take only half of this singularity (lying in $\{x, x_1 > 0\}$ where here $x'' = (x_1)$ for $d'' = 1$)). Then we have conical and quasiconical singularities as $\nu = 1$ and $\nu > 1$ respectively. Here we use X', γ' and γ described in Example 11.2.19.

The classification coincides with that in Example 11.2.19 but in all the inequalities on the exponents "$<$" should be replaced by "$>$" and v.v., and case (d) is impossible.

(ii) If the domain X described in (i) is contained in the ball $B(0,R)$ then $\gamma(x) = \max(\gamma_{\mathsf{old}}(x)), \operatorname{dist}(x, \partial X))$ works for the domain $\complement X$ where γ_{old} is defined as γ in (i). One can easily see that the classification given in (i) remains true. It allows us to cover sharp quasiconical singularities directed inward.

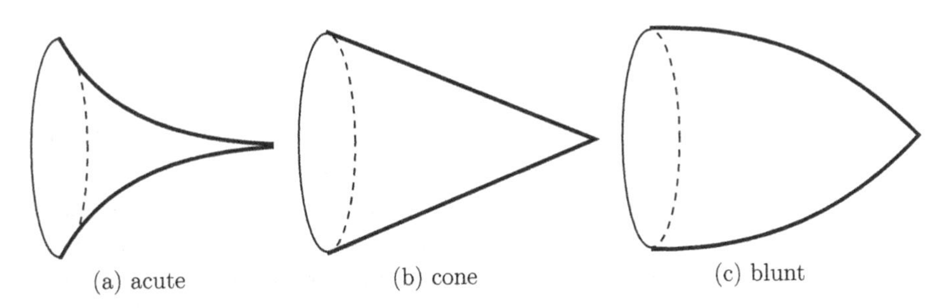

(a) acute (b) cone (c) blunt

Figure 11.5: Conical and quasiconical singularities, could be directed inward and outward

Example 11.2.22 (Blunt quasiconical singularity). (i) Let X' and γ' be as in Example 11.2.19 again, let $0 \in X'$ and let $X = \{x = (x', x''), x'' \in B(0,R), x' \in f(|x''|)X'\}$ where $f(0) = 0$ and f is a monotone decreasing function such that the inverse function f satisfies condition (11.2.82) in dimension 1. Then for $x'' \in B(0, \frac{1}{2}R)$ one can take $\gamma(x) = \epsilon |x| \gamma'(x' f^{-1}(|x''|))$. Then condition (11.2.65) fulfilled if and only if

$$f^{d'}(r) r^{-\vartheta - 1 + d'' - 1} \in \mathscr{L}^1([0,1]). \tag{11.2.89}$$

This example covers various blunt quasiconical singularities directed outward of the type $X = \{x = (x', x''), |x'| \le |x''|^{\nu}\}$ with $\nu \in (0, 1]$.

(ii) One can replace X by $B(0,R) \setminus X$ with the γ defined above. Again condition (11.2.89) is crucial. This example covers various blunt quasiconical singularities directed inward.

Remark 11.2.23. In Examples 11.2.21 and 11.2.22 cone condition poses restrictions, especially as $\nu > 1$ in their parts (i). And again under certain assumptions we will be able to allow Neumann Laplacian (see Section 12.7).

Example 11.2.24 (Clews). We can consider domain containing "disk with an infinite cut" $r = r(\theta)$ with $\theta_0 \le \theta < \infty$ with $r(\theta) = \theta^{-\alpha}$ as $\theta \to \infty$ (we use polar coordinates here). As $\alpha \le 1$ the length of the cut is infinite.

On this cut Dirichlet boundary condition should be imposed. We leave details to the reader.

Example 11.2.25 (Combs). (i) Let

$$X = \{x = (x_1, x_2, x''), 0 < x_1 < 1, \ -1 < x_2 < f(x_1), |x''| < 1\} \tag{11.2.90}$$

where $f(x_1)$ is the jump function given by $f(x_1) = h_i = \mathsf{const} \ge 0$ for $x_1 \in I_i$ and $f(x_1) = 0$ for $x_1 \in I_\infty$ where $I_i \cap I_j = \emptyset$ for $i \ne j$, I_i with $i < \infty$ are open intervals of lengths l_i, $I_\infty = [0, 1] \setminus \cup_{i<\infty} I_i$ and $\mathsf{mes}(I_\infty) = 0$.

Then the subdomain $X \cap \{x_2 > 0\}$ consists of separate "boxes" in each of which one should take γ according to scheme described in Example 11.2.18. Then for $\vartheta \in [0, 1)$ the contribution of the i-th box to $\int \gamma^{-1-\vartheta}\, dx$ is of the order $h_i l_i^{-\vartheta} + l_i^{1-\vartheta}$.

On the other hand, on $X \cap \{x_2 \le 0\}$ one can take $\gamma(x) = \frac{1}{2}(-x_2 + \delta(x_1))$ where $\delta(x_1) = \mathsf{dist}(x_1, I_\infty)$. Then the contribution of this subdomain is of the order

$$\int |\log \delta(x_1)|\, dx_1 \asymp \sum_i -l_i \log l_i \qquad \text{for } \vartheta = 0,$$

$$\int \delta(x_1)^{-\vartheta}\, dx_1 \asymp \sum_i l_i^{1-\vartheta} \qquad \text{for } \vartheta > 0.$$

Therefore condition (11.2.65) with $\vartheta = 0$ means that

$$\sum_i h_i < \infty, \quad \sum_i -l_i \log l_i < \infty \tag{11.2.91}$$

and condition (11.2.65) with $\vartheta > 0$ means that

$$\sum_i h_i l_i^{-\vartheta} < \infty, \qquad \sum_i l_i^{1-\vartheta} < \infty. \tag{11.2.92}$$

(ii) Recall that under the cone condition, a Dirichlet boundary value problem can be replaced by an arbitrary regular boundary value problem. Recall that this condition does not mean that the cone Γ lies completely in X; we assume that the intersection of Γ with X is connected and a Dirichlet condition is given on $\Gamma \cap \partial X$.

Example 11.2.26 (Trees). One can consider example when X is a *tree*: for example it consists of cylindrical pipes $P_{i,j}$ of the lengths $l_{i,j}$ and radii r_i ($d''=1$, $d'=d-1$) where $i=1,\dots,\infty$, $j=1,\dots,J(i)$ and each pipe $P_{i,*}$ has no more than c pipes $P_{i+1,j}$ on its end. Let us assume that $l_i\ge r_i$ and $r_{i+1}\le r_i\le cr_{i+1}$ for all i.

Assume that the cross-section of pipes and conjugations are regular enough so we can select $\epsilon_0 r_i$ as γ around $P_{i,*}$. Then the total volume is finite if and only if

$$\sum_i r_i^{d-1}L_i<\infty, \tag{11.2.93}$$

and condition (11.2.65) holds if and only if

$$\sum_i r_i^{d-2-\vartheta}L_i<\infty, \tag{11.2.94}$$

where $L_i=\sum_{1\le j\le J(i)} l_{i,j}$.

These pipes are not necessarily cylindrical: we can take any bounded domain with $\mathscr{C}^K$ boundary and make $P_{i,j}$ by scaling it l_i times along $x''=(x_1)$ and r_i times along x'. These trees could be bounded as $\sum_i l_i^*<\infty$ or extended to infinity as $\sum_i l_i^*=\infty$, $l_i^*=\max_{1\le j\le J(i)} l_{i,j}$.

Let us discuss mixed boundary value problems assuming the cone condition is fulfilled in the indicated form.

Let the geometry of the domain (and the singularities of the coefficients of the operator A) are compatible with the scaling function $\gamma(x)=\gamma_0(x)$. Then in order to take into account the singularities of the boundary value problem we should replace $\gamma_0(x)$ by $\min(\gamma_0(x),\gamma_1(x))$ where scaling function $\gamma_1(x)$ corresponds to the singularities of the boundary value problem.

If $\gamma_0(y)\le 2\mathsf{dist}(y,\partial X)$ then one can take $\gamma(x)\asymp\gamma_0(x)$ in $\bigl(y,\frac{1}{3}\gamma(y)\bigr)$. Otherwise one can straighten ∂X in $\mathcal{B}\bigl(y,\frac{2}{3}\gamma(y),X\bigr)$ by means of a map which is uniformly smooth in the γ_0-scale with Jacobian disjoint from 0. Therefore without any loss of generality one can assume that

$$X\cap\mathcal{B}\bigl(y,\gamma(y),X\bigr)=\{x_1>0\}\cap B\bigl(y,\gamma(y)\bigr). \tag{11.2.95}$$

Then $\gamma_1(x)\asymp x_1+\gamma_1(x')$ in $X\cap\mathcal{B}\bigl(y,\frac{1}{2}\gamma(y),X\bigr)$ where here $\gamma_1(x)$ is defined only on ∂X in order to take into account the singularities of the boundary

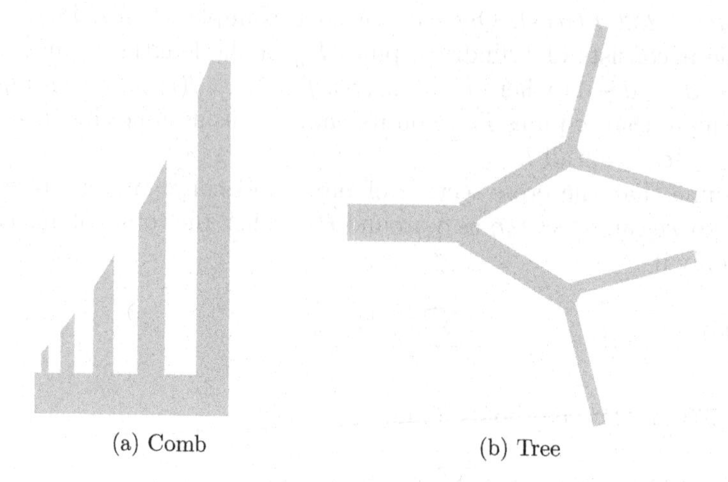

(a) Comb (b) Tree

Figure 11.6: Comb (Example 11.2.25) and tree (Example 11.2.26) with infinite numbers of teeth and branches

value problem. Assuming that $\gamma_0^{-1-\vartheta}\in \mathscr{L}^1(X)$ we conclude that

$$\gamma_1^{-1-\vartheta}\in \mathscr{L}^1(X) \iff \begin{cases} \log\gamma_1 \in \mathscr{L}^1(\partial\mathcal{X}_{\mathsf{reg}}) & \text{for } \vartheta=0,\\ \gamma_1^{-\vartheta}\in \mathscr{L}^1(\partial\mathcal{X}_{\mathsf{reg}}) & \text{for } \vartheta>0\end{cases} \tag{11.2.96}$$

where the standard surface measure is used on $\partial\mathcal{X}_{\mathsf{reg}}$. In particular, this condition is fulfilled for $\vartheta\in[0,\delta)$ as $\gamma_1(x)\ge \epsilon_0 \mathsf{dist}(x,\mathcal{K})$ where $\mathcal{K}$ is a subset of ∂X with the following property:

(11.2.97) For every $y\in\mathcal{X}_{\mathsf{reg}}$ $\mathcal{K}\cap B(y,\gamma_0(y),X)$ is contained in $\bigcup_{1\le l\le L}\mathcal{K}_l$ where $\mathcal{K}_l$ (depending on y also) are uniformly $\mathscr{C}^K$ submanifolds (in the γ_0-scale) of $\partial\mathcal{X}_{\mathsf{reg}}$ of codimension δ.

This analysis provides us immediately with a series of examples for the Laplacian when on part of the boundary the Dirichlet condition is given and on the remaining part the Neumann boundary condition is given and $\vartheta<1$.

Example 11.2.27. Let us consider a boundary value problem of variational origin with the functional

$$\begin{aligned}
(11.2.98)\quad \boldsymbol{\Phi}(u) = \boldsymbol{\Phi}_{\mathsf{inn}}(u) + \boldsymbol{\Phi}_{\mathsf{bound}}(u) = \\
\int_X \sum_{\alpha,\beta:|\alpha|\le m,|\beta|\le m} a_{\alpha,\beta}\langle a_{\alpha,\beta}D^\alpha u, D^\beta u\rangle\, dx + \\
\int_{\partial X} \sum_{\alpha,\beta:|\alpha|\le m-1,|\beta|\le m-1} a_{\alpha,\beta}\langle b_{\alpha,\beta}D^\alpha u, D^\beta u\rangle\, dS
\end{aligned}$$

where dS is a standard measure at ∂X. Let us assume that the boundary terms of this functional are regular in the sense that after the rescaling $x \mapsto (x - \bar{x})\gamma(\bar{x})^{-1}$ and multiplication by $\gamma(\bar{x})^{2m-d}$ we obtain a functional the coefficients of which (and all their derivatives of order $\le K$) are bounded in $B(0, 1)$. The same should be true for the boundary conditions which define $\mathfrak{D}(\boldsymbol{\Phi})$ but the l-th condition has its own weight $\varrho_{(l)}$ for multiplication and is given on the set $\Xi_l \subset \partial X$ $(l = 1, \dots, L)$.

Let us assume that the cone condition is fulfilled. Then $\mathcal{K}$ is the union of the relative (in ∂X) boundaries of Ξ_l. However, the possibly irregular structure of $\mathcal{K}$ prevents us from immediately applying the above arguments. Nevertheless, for $\vartheta = 0$ we avoid this difficulty by means of variational arguments. Namely, let us consider a covering of $\partial\Xi_l$ of finite multiplicity by balls $B(x^k, \epsilon\gamma(x^k))$ lying in the $2\epsilon\gamma$-neighborhood of $\partial\Xi_l$; let $\omega_{l,\epsilon}$ be the union of these balls. Let us replace Ξ_l by $\Xi'_l = \Xi_l \cap \omega_{l,\epsilon}$ or by $\Xi''_l = \Xi_l \setminus \omega_{l,\epsilon}$. We do the same with the set Ξ_0 where the Dirichlet boundary condition is given.

In the first case (when all the sets Ξ_l expand) the domain of $\boldsymbol{\Phi}$ shrinks and $\mathsf{N}^-(\tau)$ decreases, and in the second case (when all the sets Ξ_l shrink) the domain of $\boldsymbol{\Phi}$ expands and $\mathsf{N}^-(\tau)$ increases. Further, the cone condition remains true in both cases for small enough $\epsilon > 0$. Furthermore, in both cases the required regularity conditions are now fulfilled (because $\partial\Xi'_l$ and $\partial\Xi''_l$ intersected with ∂X_r lie in the union of spheres $S(x^k, \epsilon\gamma(x^k))$ and every ball $B(\bar{x}, \gamma(\bar{x}))$ intersects no more than $C = C_\epsilon$ of these spheres. Therefore

$$\mathsf{N}'^-(\tau) \le \mathsf{N}^-(\tau) \le \mathsf{N}''^-(\tau) \tag{11.2.99}$$

and $\mathsf{N}'^-(\tau) = \varkappa_0\tau^{\frac{d}{2m}} + O(\tau^{\frac{1}{2m}(d-1)})$ and the same is true for $\mathsf{N}''^-(\tau)$ with the same coefficient $\varkappa_0$ (because it does not depend on the boundary conditions).

Therefore *the same is true for the problem in question:*

$$\mathsf{N}^-(\tau) = \varkappa_0 \tau^{\frac{d}{2m}} + O\bigl(\tau^{\frac{1}{2m}(d-1)}\bigr) \qquad \text{as } \tau \to +\infty. \tag{11.2.100}$$

Moreover, under the standard non-periodicity condition to Hamiltonian trajectories[19] asymptotics

$$\mathsf{N}'(\tau) = \varkappa_0 \tau^{\frac{d}{2m}} + \varkappa_1' \tau^{\frac{1}{2m}(d-1)} + o\bigl(\tau^{\frac{1}{2m}(d-1)}\bigr)$$

holds and the same is true for $\mathsf{N}''(\tau)$. Here $\varkappa_1'$ is a monotone non-increasing and $\varkappa_1''$ is a monotone non-decreasing function of $\epsilon \in (0, \epsilon_0)$ and $\varkappa_1' \le \varkappa_1''$. Let us recall that the coefficient $\varkappa_1$ consists of two terms; one of them does not depend on boundary conditions and the other is an integral over ∂X_r of some function of the principal coefficients of the operator and the boundary operators. Therefore, assuming that

$$\mathsf{mes}_{\partial X}(\partial \Xi_l) = 0 \qquad \forall l = 0, \dots, L \tag{11.2.101}$$

we obtain that the limits of $\varkappa_1'$ and $\varkappa_1''$ as $\epsilon \to +\infty$ coincide and therefore the asymptotics

$$\mathsf{N}^-(\tau) = \varkappa_0 \tau^{\frac{d}{2m}} + \varkappa_1 \tau^{\frac{1}{2m}(d-1)} + o\bigl(\tau^{\frac{1}{2m}(d-1)}\bigr) \qquad \text{as } \tau \to +\infty \tag{11.2.102}$$

holds.

Let us treat the case in which the boundary is very irregular.

Example 11.2.28. Let us assume that the operator A satisfies our conditions with $\gamma = \mathsf{const} > 0$. Then one can take $\gamma(x) = \mathsf{dist}(x, \partial X)$ in which case the remainder estimate in the asymptotics of $\mathsf{N}^-_{(\vartheta)}(\tau)$ is "O" of

$$\begin{aligned} &\tau^{\frac{1}{2m}(d-1-\vartheta)+\vartheta} \int_{r \ge \tau^{\frac{1}{2m}}} r^{-1-\vartheta} d\mu(r) + \tau^{\frac{d}{2m}+\vartheta} \mu(\tau^{-\frac{1}{2m}}) \asymp \\ &\qquad \tau^{\frac{1}{2m}(d-1-\vartheta)+\vartheta} \int_{r \ge \tau^{\frac{1}{2m}}} r^{-2-\vartheta} \mu(r) dr + \tau^{\frac{d}{2m}+\vartheta} \mu(\tau^{-\frac{1}{2m}}) \end{aligned} \tag{11.2.103}$$

with $\mu(r) = \mathsf{mes}(\{x \in X, \mathsf{dist}(x, \partial X) \le r\})$; we assume that $\mathsf{mes}(X) < \infty$. In particular, if $\mu(r) = O(r^\delta)$ as $r \to +0$ then for $\vartheta < \delta - 1$[17] we obtain

[19] This condition is not associated with $\partial\Xi_l'$ or $\partial\Xi_l''$ because the $2d$-dimensional measures of points lying on trajectories issuing from these sets obviously vanish.

the remainder estimate $O\bigl(\tau^{\frac{1}{2m}(d-1-\vartheta)+\vartheta}\bigr)$; moreover, under standard non-periodicity condition to the Hamiltonian trajectories we obtain the remainder estimate $o\bigl(\tau^{\frac{1}{2m}(d-1-\vartheta)+\vartheta}\bigr)$.

On the other hand, we obtain the remainder estimate $O\bigl(\tau^{\frac{1}{2m}(d-1-\vartheta)+\vartheta}\log\tau\bigr)$ for $\vartheta=\delta-1$ and we obtain the remainder estimate $O\bigl(\tau^{\frac{1}{2m}(d-\delta)+\vartheta}\bigr)$ for $\vartheta>\delta-1$.

For $\vartheta=0,\delta=1$ and for $\vartheta=0,\delta\in(0,1)$ we obtain the results of the papers J. Brüning [1] and J. Fleckinger, M. Lapidus [1] respectively. Finally, in these two cases under the condition $\mu(r)=o(r^\delta)$ as $r\to+0$ we obtain "o" in the remainder estimates.

Remark 11.2.29. Condition $\mu(r)=O(r^\delta)$ as $r\to+0$ means exactly that *inner Minkowski dimension of ∂X does not exceed $d-\delta$*. So called *Weyl-Berry conjecture* (in sanitized form: originally M. Berry thought about Hausdorff dimension) claims the second term $\kappa_\delta\tau^{(d-\delta)/2}$ and the remainder estimate $o(\tau^{(d-\delta)/2})$ for Dirichlet (Euclidean) Laplacian.

However, as J. Brossard and R. Carmona [1] had shown it is almost never the case as capacity enters the game (recall that Weyl conjecture is almost always true for sure for Euclidean domains and may be even always true).

11.3 Large Eigenvalues for Operators with Weakly Singular Potentials (End)

11.3.1 Operators with Non-Periodic Flows

We start this subsection by discussing existence and some properties of the Hamiltonian billiard flow for second order operators.

Then we consider examples in which restrictive conditions on the Hamiltonian flow imply asymptotics with highly accurate remainder estimates (those we used to call *sharper*). $\vartheta=0$. Here we do not treat smooth closed manifolds without boundary; the reader can find all those results in A. Volovoy [1,2] and in the earlier papers P. Bérard [1], B. Randoll [1]. For the Laplace-Beltrami operator these statements yield the following

Example 11.3.1. (i) For the Laplace-Beltrami operator on a compact closed Riemannian $\mathscr{C}^K$ manifold of non-positive sectional curvature (for $d\ge 3$) or without conjugate points (for $d=2$) asymptotics of $\mathsf{N}^-(\tau)$ with remainder estimate $O\bigl(\tau^{\frac{1}{2}(d-1)}\log^{-1}\tau\bigr)$ holds.

(ii) For a certain class of surfaces of rotation and for a certain class of graded Lie algebras asymptotics of $\mathsf{N}^-(\tau)$ with the remainder estimate $O\bigl(\tau^{\frac{1}{2}(d-1-\delta)}\bigr)$ holds with $\delta>0$.

Remark 11.3.2. In the context of our book the results on Hamiltonian flows obtained in these papers yield asymptotics for $\mathsf{N}^-_{(\vartheta)}(\tau)$ with remainder estimates $O\bigl(\tau^{\frac{1}{2}(d+\vartheta-1)}\log^{-\vartheta-1}\tau\bigr)$ and $O\bigl(\tau^{\frac{1}{2}(d+\vartheta-1-\delta)}\bigr)$ respectively.

Existence of the Billiard Flow

Let us consider the Laplace(-Beltrami) operator in compact domains or on compact manifolds with singularities. The following statement is necessary:

Theorem 11.3.3. *Let X be a bounded domain in $\mathbb{R}^d$ which is a Riemannian manifold with boundary with metric $g(x,\xi)$ satisfying conditions*

$$g^{jk}=g^{kj},\qquad \epsilon_0|\xi|^2\le g(x,\xi)=\sum_{jk} g^{jk}\xi_j\xi_k\le c|\xi|^2, \tag{11.3.1}$$

$$|D^\alpha g^{jk}|\le c\gamma^{-|\alpha|}\qquad \forall\alpha:|\alpha|\le K\quad \forall j,k \tag{11.3.2}$$

and with scaling function $\gamma(x)$ such that $\gamma^{-1}\in\mathscr{L}^1(X)$.

Then there exist a set of measure 0 and also of the first Baire category $\Lambda\subset S^*\bar{X}$ and a continuous Hamiltonian flow Ψ_t on $S^*\bar{X}\setminus\Lambda$ with reflection at the boundary such that all its trajectories lie in $S^*\bar{X}_r$, $\bar{X}_{\mathsf{reg}}=\{x\in\bar{X},\gamma(x)>0\}$ and are transversal to ∂X; as usual $\bar{X}=X\cup\partial X$.

This flow conserves the natural measure $d\mu=dxd\xi:d\sqrt{g}$ on S^*X.

Proof. The proof coincides with the arguments of Chapter 6 in the book I. P. Cornfeld, S. V. Fomin and Y. G. Sinai, [1]. This proof follows the following scheme: first we establish that the flow where it is defined preserves measure μ.

Then we consider the set Λ_{t_1,t_2} of all points such that Ψ_t exists on time interval $[0,t_1]$ but breaks on $(t_1,t_2]$ (as $0<t_1<t_2$). Invariance of the measure μ implies that $\mu(\Lambda_{T,T+\varepsilon})=\mu(\Lambda_{0,\varepsilon})$ and therefore

$$\mu(\Lambda_{0,T})\le\mu(\Lambda_{0,\varepsilon})\lceil\frac{T}{\varepsilon}\rceil. \tag{11.3.3}$$

On the other hand, due to the bounded speed of propagation $\Lambda_{0,\varepsilon}$ is covered by $\{(x,\xi) : \gamma(x) \le C_1\varepsilon\}$ and by union of subsets

$$\{(x,\xi):\ \gamma(x) \asymp \gamma \mathsf{dist}(x,\partial X) \le C_0\varepsilon\}$$

which trajectories have an angle less than δ with the boundary; here $\delta = \delta(\varepsilon\gamma^{-1}) = o(1)$ as $\varepsilon \to +0$, with $\gamma \ge C_1\varepsilon$.

Actually those points belong to $C_0\delta\varepsilon$-vicinity of ∂X. Obviously μ-measure of the first set does not exceed $C\,\mathsf{mes}(\{x, \gamma(x) \le C_0\varepsilon\})$ and μ-measure of the second set does not exceed $C\delta^2\varepsilon\gamma^{-1}\,\mathsf{mes}(\{x, \gamma(x) \asymp \gamma\})$.

Note that in virtue of $\gamma^{-1} \in \mathscr{L}^1$ we estimate the measure of the first set

$$\mathsf{mes}(\{x, \gamma(x) \le C_1\varepsilon\}) = o(\varepsilon) \tag{11.3.4}$$

and the measure of the second set by

$$C\int_{x:\gamma(x)\ge C_1\varepsilon} \delta^2\bigl(\frac{\varepsilon}{\gamma(x)}\bigr)\gamma^{-1}(x)\,dx = o(\varepsilon);$$

so $\mu(\Lambda_{0,\varepsilon}) = o(\varepsilon)$ as $\varepsilon \to +0$.

Combining with (11.3.3) we conclude that $\mu(\Lambda_{0,T}) = o(1)$ and thus $\mu(\Lambda_{0,T}) = 0$. Negative time direction is covered by the same way. □

Remark 11.3.4. Let us note that if we are interested only in trajectories of length not exceeding T then in the statement of this lemma one can replace Λ by a closed nowhere dense subset Λ_T. However, $S^*\bar{X} \setminus \Lambda_T$ is not invariant with respect to the flow Ψ_t.

Properties of Hamiltonian Billiard Flows and Asymptotics

Recall that to recover sharper remainder estimate we need to estimate the "divergence" of the billiard flow and its non-periodicity. The first notion is covered by the following

Definition 11.3.5. Let X be a compact γ-admissible Riemannian manifold with boundary where γ is an admissible scaling function such that $\gamma^{-1} \in \mathscr{L}^1(X)$.

We say that

(i) Ψ_t is *a power geodesic flow with reflections*is given on X if for every $T \geq 0$ there exists a closed set $\Xi_T \subset \Sigma$ with $\mu(\Xi_T) \leq CT^{-1}$ such that through every point $z \in \Sigma \setminus \Xi_T$ there passes a geodesic trajectory $\Psi_t(z) = (x(t), \xi(t))$ with reflections from ∂X with $t \in [0, T]$ or $t \in [-T, 0]$, along which $\gamma(x(t)) \geq \varepsilon(T)$, the angle with ∂X is not less than $\varepsilon(T)$ at every point where the distance from the boundary is less than $\varepsilon(T)$, and also $|D\Psi_t| \leq \varepsilon(T)^{-1}$ where $\varepsilon(T) = T^{-L}$ with large enough C, L.

(ii) Similarly, we say that Ψ_t is *an exponential geodesic flow with reflections* if (i) holds with $\varepsilon(T) = e^{-CT}$ respectively.

We have a non-trivial *non-periodicity* condition. In order to take it into account we introduce

Definition 11.3.6. In the framework of Definition 11.3.5 we say that Ψ_t is a *non-periodic power (or exponential) geodesic flow* if in addition to the conditions of definition 11.3.5

$$\mu\bigl(\{z \in \Sigma \setminus \Xi_T, \exists t \in (0, T) : |t|^{-1}\mathsf{dist}(z, \Psi_t(z)) \leq \varepsilon(T)\}\bigr) = O(T^{-1}) \tag{11.3.5}$$

where we increase C and L in the definition of $\varepsilon(T)$ if necessary.

We conclude immediately that

Theorem 11.3.7. *(i) Let X be a domain with non-periodic power geodesic flow. Let γ_0 be scaling function describing singularities of domain and $\gamma \leq \gamma_0$ be scaling function describing singularities of operator and boundary conditions. Then as*

$$\gamma^{-1-\vartheta-\delta_1} \in \mathscr{L}^1(X) \tag{11.3.6}$$

with arbitrarily small exponent $\delta_1 > 0$, asymptotics

$$\mathsf{N}^-_{(\vartheta)} = \mathsf{N}^{\mathsf{W}}_{(\vartheta)} + O\bigl(\tau^{\frac{d-1-\vartheta}{2m}+\vartheta-\delta}\bigr) \tag{11.3.7}$$

holds with $\delta > 0$ as $\tau \to +\infty$.

(ii) Let X be a domain with non-periodic exponential geodesic flow. Let γ_0 be scaling function describing singularities of domain and $\gamma \leq \gamma_0$ be scaling

function describing singularities of operator and boundary conditions. Then as

$$\gamma^{-1-\vartheta}|\log\gamma|^{1+\vartheta}\in\mathscr{L}^1(X) \tag{11.3.8}$$

with arbitrarily small exponent $\delta_1>0$*, asymptotics*

$$\mathsf{N}^-_{(\vartheta)}=\mathsf{N}^{\mathsf{W}}_{(\vartheta)}+O\bigl(\tau^{\frac{d-1-\vartheta}{2m}+\vartheta}|\log\tau|^{-1-\vartheta}\bigr) \tag{11.3.9}$$

holds as $\tau\to+\infty$.

Examples

Polygons and Polyhedra Recall that according to Theorem 7.4.11 in a compact Euclidean polyhedral domain in $\mathbb{R}^d$ $|D\Psi_t(x,\xi)|\le C(|t|+1)$.

Theorem 11.3.8. *Let* X *be a polyhedral domain with cuts; assume that* X *is* γ*-admissible with* $\gamma^{-1-\sigma}\in\mathscr{L}^1(X)$ *with* $\sigma>0$*. Then there is a non-periodic power geodesic flow.*

Proof. Note first that

$$\operatorname{mes}\bigl(\{z=(x,\xi)\in\Sigma,\operatorname{dist}(x,\partial X)\le\varepsilon\}\bigr)\le C\varepsilon. \tag{11.3.10}$$

Let us cover set $\{z=(x,\xi)\in\Sigma,\operatorname{dist}(x,\partial X)\ge\varepsilon\}$ with no more than $C\varepsilon^{1-4d}$ balls of radius $O(\varepsilon^2)$ and let us divide the time interval $[0,T]$ into no more than $O(T\varepsilon^{-2})$ intervals of length $O(\varepsilon^2)$. Let $\varepsilon=O(T^{-L})$ with large enough exponent L.

Then if the distance between the Ψ_t-image of some ball and this ball is less than ε^2 for some t from the above subinterval then for all points (x,ξ) of this ball and for all t of this subinterval the geodesic flow is given by the formula

$$\Psi_t(x,\xi)=(y+\Pi x+t\Pi\xi,\Pi\xi)$$

with constant y and Π such that $\operatorname{dist}(\xi,\Pi\xi)\le\varepsilon$ for all ξ in question. Here $\Pi\in O(d)$ as in the proof of Theorem 7.4.11.

Taking into account that for every x the $\mathbb{S}^{d-1}$-measure of the set

$$\{\xi\in\mathbb{S}^{d-1},|y+\Pi x-x+t\Pi\xi|\le\varepsilon^N\quad\text{for some } t \text{ with } |t|\ge\varepsilon\}$$

does not exceed $O(\varepsilon^N)$ we obtain after summation over all the balls and subintervals that (11.3.5) holds. □

Example 11.3.9. Obviously in Euclidean polyhedral domains (without cuts) $\gamma_0^{-1-\vartheta} \in \mathscr{L}^1(X)$ as $\vartheta < 1$. So the only condition to be checked is $\gamma^{-1-\vartheta} \in \mathscr{L}^1(X)$ where $\gamma = \mathsf{dist}(x, \mathcal{K})$ and $\mathcal{K}$ is the set of the cut ends and transition points. This condition is equivalent to $\gamma^{-\vartheta} \in \mathscr{L}^1(\partial X)$ as $\vartheta > 0$ and $|\log \gamma| \in \mathscr{L}^1(\partial X)$ as $\vartheta = 0$.

Elliptical Domains According to Theorem 7.4.12 in Euclidean elliptical domains (or domains between two confocal ellipses) Ψ_t is power growth geodesic flow with singularities (see (7.4.32)– (7.4.34) for more details).

Theorem 11.3.10. *Let X be a domain in $\mathbb{R}^2$ bounded by one or two confocal ellipses. Then there is a non-periodic power geodesic flow.*

Proof. Obviously (11.3.5) is equivalent to

$$\mathsf{mes}_{\Sigma_2'}\bigl(\{z' \in \Sigma_2' : \exists k = 1, \dots, n : \mathsf{dist}(z', \Phi^k z') \le \varepsilon = n^{-M}\}\bigr) = O(n^{-\sigma}) \quad \text{as } n \to +\infty \tag{11.3.11}$$

for appropriate M and $\sigma > 0$. Recall that $\Sigma_2' = \Sigma|_{Y_2} = B^* Y_2$ where Y_2 is an outer ellipse and we use other notations of the proof of Theorem 7.4.12.

(a) Let us first consider the zone $\{(x, \xi), \beta > 0\}$. Let us consider the lines $\{\eta = \mathsf{const}\}$. These lines are analytically parameterized by a cyclic parameter $u \in \mathbb{R}/2\pi\mathbb{Z}$ (see the proof of theorem-7.4.12)[20].

Let $w \in \mathbb{R}/\mathbb{Z}$ be a new cyclic coordinate: $w = w_\eta(u) = \nu_\eta((0, u))m_\eta^{-1}$ for $u \in [0, 2\pi]$ and then by periodicity. Here $m_\eta = \nu_\eta([0, 2\pi])$. One can easily see that $m_\eta \asymp \log|\beta|$ as $\beta \to +0$ for $a_2 > b_2$ and $m_\eta \asymp \beta^{-1}$ as $\beta \to +0$ for $a_2 = b_2$.

For every fixed η the map Φ_η of this line is analytic and preserves the measure $dw_\eta = m_\eta^{-1} d\nu_\eta$ and the orientation[21] and therefore Φ_η is a shift $\Phi_\eta : w_\eta \to w_\eta + f(\eta)$.

[20] More precisely, this line consists of two connected components. Billiard map Φ transforms every component into itself and the x-projection is an analytic diffeomorphism of this component to the ellipse Y_2. Every component is parameterized by u. Permutation of these components is equivalent to replacing Φ by Φ^{-1}.

[21] Moreover, it is well known (see I.P. Cornfeld, S. V. Fomin and Y. G. Sinai, [1]) that for generic η, Φ_η is an ergodic map.

Then in the proof of (11.3.11) for $\Sigma'_2 \cap \{\beta > 0\}$ one needs to prove that

$$C n \operatorname{mes}\Bigl(\{\eta : \exists k, l = 1, \dots, n : |f(\eta) - \frac{k}{l}| \le \varepsilon = n^{-L}\}\Bigr) = O(n^{-\sigma}) \quad \text{as } n \to +\infty \tag{11.3.12}$$

for admissible L and $\sigma > 0$; here mes means Lebesgue measure.

Obviously, $f(\eta)$ is a continuous function. One can easily prove that

$$\epsilon \beta^{p} \le |\frac{df}{d\eta}| \le c\beta^{-p} \qquad \text{for } |\eta| \le a_2 - \epsilon_0, \tag{11.3.13}$$

i.e., for $\beta \in (0, \beta_0)$ with small enough $\epsilon_0 > 0, \beta_0 > 0$. Recall that $\beta = b_2^2 - a_2^2 b_2^2 \eta^2$. Therefore (11.3.12) holds for the η indicated here.

On the other hand, Φ is an analytic map for $\beta \ne b_1^2$ (the case when rays are tangent to the inner ellipse Y_1). Therefore function $f(\eta)$ is analytic for $|\eta| < \eta_1 = a_2^{-1} b_2^{-1}(b_2^2 - b_1^2)^{\frac{1}{2}}$ and for $\eta_1 < |\eta| \le a_2^{-1}$. Obviously, $f(\eta)$ is not constant in each of these three intervals. One can easily see that $f_\pm(\zeta) = f(\bar\eta \pm \zeta^2)$ is an analytic function for $|\zeta| \le \epsilon_0$ and $\bar\eta = +\eta_1, -\eta_1$.

However one can easily see that

$$\operatorname{mes}(\{\zeta : |f(\zeta)| \le \varepsilon\}) = O(\varepsilon^{\delta}) \tag{11.3.14}$$

holds for every $\mathscr{C}^K$ function on the closed segment provided

$$\sum_{1 \le j \le K-1} |f^{(j)}| \ge \epsilon_0. \tag{11.3.15}$$

Therefore (11.3.12) holds.

(b) Let us treat the zone $\{(x, \xi) : 0 > \beta > b_2^2 - a_2^2\}$. In this case every line $\{\eta = \mathsf{const}\}$ consists of two connected components and every component is equivalent to $\mathbb{S}^1$ and the x-projection maps every component to one of two segments on the ellipse Y_2 describing by the inequality $x_2^2 \ge -\beta b_2^2 (a_2^2 - b_2^2)^{-1}$. The inverse map is two-valued.

Let us introduce on this connected component the cyclic coordinate u' which can be done in the following way: let us consider the confocal hyperbola described by the equation $\frac{x_1^2}{a'^2} - \frac{x_2^2}{b'^2}$ with $a'^2 + b'^2 = a_2^2 - b_2^2$. This hyperbola intersects Y_2 in four points $(\bar{x}_1, \bar{x}_2)$ with

$$\bar{x}_1^2 = a_2^2(a_2^2 - b_2^2 + \beta)(a_2^2 - b_2^2)^{-1}, \quad \bar{x}_2^2 = -b_2^2\beta(a_2^2 - b_2^2)^{-1}.$$

Let us introduce parameter $u' \in \mathbb{S}^1$ on this hyperbola: $x_1 = a' \sec u'$, $x_2 = b' \tan u'$. Through every point of the hyperbola corresponding to this parameter there passes a tangent straight line hitting the corresponding segment of the ellipse. Then (x_1, x_2) is an intersection point and (ξ_1, ξ_2) is a unit vector along this tangent line, directed inside and outside for $\eta < 0$ and $\eta > 0$ respectively.

Let us introduce the parameter $w' \in \mathbb{R}/\mathbb{Z}$ in the same way as before. Again we take $dw' = m_\eta^{-1} d\nu_\eta$ with $m_\eta \asymp \log|\beta|$ as $\beta \to -0$ and $m_\eta \asymp 1$ as $\beta \to b_2^2 - a_2^2$. Moreover, in the latter case m_η is an analytic function of $\zeta' = (\beta + a_2^2 - b_2^2)^{\frac{1}{2}}$. One can easily see that if there is no inner ellipse Y_1 then Φ transforms one connected component of every line $\{\eta = \mathsf{const}\}$ to another and therefore for $\beta < -n^{-\sigma}$ the inequality $\mathsf{dist}(z', \Phi^k(z')) < \epsilon$ is impossible for odd k and the measure of the set $\{|\beta| \le n^{-\sigma}\}$ is $O(n^{-\delta})$.

On the other hand, if there is an inner ellipse Y_1 then the map Φ preserves both connected components of $\eta = \mathsf{const}$ but changes their orientations, i.e., the map is of the form $w_\eta \to \frac{1}{2} f(\eta) - w_\eta$ and therefore for odd k the $d\nu_\eta$-measure of points with $\mathsf{dist}(z', \Phi^k z') \le \varepsilon$ is $O(n^p \varepsilon)$. Therefore *in both cases only even k should be treated.*

Furthermore, on every connected component map Φ^2 is of the form $w_\eta \to w_\eta + f(\eta)$ where function f is again continuous and satisfies (11.3.13), this time for $|\eta| \ge a_2^{-1} + \epsilon_0$, i.e. $\beta \in (-\beta_0, 0)$ with small enough ϵ_0 and β_0.

This function is analytic for $a_2^{-1} < |\eta| < b_2^{-1}$, i.e., for $\beta \in (0, b_2^2 - a_2^2)$, and the functions $f(\pm(b_2^{-1} - \zeta^2))$ are analytic with respect to ζ with $|\zeta| \le \epsilon_0$. Therefore we obtain that (11.3.15) and thus (11.3.14) and (11.3.12) remain true in this case. □

Example 11.3.11. Let X be an elliptical domain or domain between two confocal ellipses. Then

(i) If the boundary problem is regular then asymptotics (11.3.7) holds for any $\vartheta \ge 0$.

(ii) We can allow singularities: transitions between different boundary value problems and cuts along some segments on semi-axis of the ellipses[22)]

Then the only condition to be checked is $\gamma^{-1-\vartheta} \in \mathscr{L}^1(X)$ where $\gamma = \mathsf{dist}(x, \mathcal{K})$ and $\mathcal{K}$ is the set of the cut ends and transition points. This condition again is equivalent to $\gamma^{-\vartheta} \in \mathscr{L}^1(\partial X)$ and $|\log \gamma| \in \mathscr{L}^1(\partial X)$ as $\vartheta = 0$.

Example 11.3.12. Let X be a ball or domain between two concentric spheres in $\mathbb{R}^d$.

Then

(i) If the boundary problem is regular then asymptotics (11.3.7) holds for any $\vartheta \geq 0$.

(ii) We can allow singularities: transitions between different boundary value problems. Then the only condition to be checked is $\gamma^{-1-\vartheta} \in \mathscr{L}^1(X)$ where $\gamma = \mathsf{dist}(x, \mathcal{K})$ and $\mathcal{K}$ is the set of the cut ends and transition points. Again, this condition is equivalent to $\gamma^{-\vartheta} \in \mathscr{L}^1(\partial X)$ for $\vartheta > 0$ and $|\log \gamma| \in \mathscr{L}^1(\partial X)$ as $\vartheta = 0$.

There are many open problems:

Problem 11.3.13. (i) Investigate domains bounded by one or two ellipses and one or two hyperbolas (all with the same foci).

(ii) Investigate domains whose boundary consists of segments of ellipses and hyperbolas (all with the same foci); investigate if one can add cuts along some segments on semi-axis.

(iii) Investigate ellipsoids of revolutions (i.e. with semi-axis $a_1 = a_2 = \cdots = a_p = a > a_{p+1} = \ldots = a_d = b$) or domains between two confocal ellipsoids of revolution (with the obvious meaning).

(iv) Investigate general ellipsoids.

Miscellaneous Remarks

Remark 11.3.14. Since the billiard trajectories of the principal symbol a^m coincide with those of a, all our discussion remains true for operators with the senior part Δ^m.

[22)] One can show easily that the reflection from cuts does not violate power non-periodicity.

Remark 11.3.15. (i) Billiard geodesic flow has power growth for certain Lie groups. One can easily see that in this case

$$\mu(\{z, \exists t \in [0, T] : \gamma(\Psi_t(z)) \le \varepsilon\}) \le C\varepsilon^{\sigma} T$$

and $\mu(\mathcal{Q}_{T,\varepsilon}) \le C\varepsilon T$ where $\mathcal{Q}_{T,\varepsilon} = \{z, \exists t \in [0, T]$ such that the distance from x to ∂X and the angle between $\frac{dx}{dt}$ and $T\partial X$ do not exceed $\epsilon\varepsilon$, and $\gamma(\Psi_t(z)) \ge \varepsilon\}$.

(ii) Billiard geodesic flow has power growth for manifolds with periodic geodesic flow (such manifolds can be either smooth or γ-admissible with $\gamma^{-1-\sigma} \in \mathscr{L}^1(X)$ for arbitrarily small $\sigma > 0$). We will analyze them in the next subsection.

(iii)

(iv) Recall that according to Theorem 7.4.10 in Euclidean strongly convex domains with $\mathscr{C}^3$-boundary geodesic billiard flow has (no more than) exponential growth. The same is true for compact Riemannian manifolds with geodesically strongly concave boundary and for compact closed Riemannian manifolds.

Remark 11.3.16. One can construct new manifolds of the corresponding type by the following methods:

(a) Taking a Cartesian product.

(b) Taking a factor manifold of X with respect to a finite group of its isometries.

(c) Considering reflections, i.e., metric preserving diffeomorphisms $S : \bar{X} \to \bar{X}$ such that $S^2 = Id$ and $S(x) = x \implies dS(x) \ne dx$ (X can have cuts so S does not necessarily preserve X) and removing from X some appropriate subset K of the set of fixed points of S. The condition on K is $\delta^{-1-\sigma} \in \mathscr{L}^1(X)$ for $\delta(x) = \mathsf{dist}(x, K)$.

Remark 11.3.17. (i) Let us note that $\mathsf{dist}(z, \Psi_t(z)) \asymp |t|$ for $|t| \le \epsilon\gamma(x)$, $z = (x, \xi)$. Therefore the non-periodicity condition should be checked only for $|t| \ge \epsilon\varepsilon(T)$.

(ii) Let us consider the Cartesian product X of two Riemannian manifolds X' and X'' with dimensions ≥ 1. Let us note that $\Psi_t(x', x'', \xi', \xi'') = (\Psi'_{t|\xi'|}(x', \xi'), \Psi''_{t|\xi''|}(x', \xi''))$ for $(x', x'', \xi', \xi'') \in S^*X$.

Then, if inequality $\mathsf{dist}(z', \Psi_t(z')) \geq \varepsilon|t|$ is fulfilled for all $t \in (0, T)$ for a point $z' = (x', \xi'_0) \in S^*X'$ then the same inequality with ε^2 instead of ε is fulfilled for a point $(x', x'', \xi', \xi'') \in S^*X$ with $\xi' = |\xi'|\xi'_0$ and $|\xi'| \geq \varepsilon$.

Note that $\mu\{(x, \xi) : |\xi'| \leq \varepsilon = O(\varepsilon)$. Therefore one should treat only points $(x', x'', \xi', \xi'') \in S^*X$ such that the non-periodicity condition is violated at both points $z' = (x', \xi'|\xi'|^{-1}) \in S^*X'$ and $z'' = (x'', \xi''|\xi''|^{-1}) \in S^*X''$. Let

$$T' = \min\bigl(t \geq \epsilon\varepsilon : \mathsf{dist}(z', \Psi'_t(z')) \leq \varepsilon\bigr)$$

and let define T'' in a similar way. Recall that we can assume that $|\xi'| \geq \varepsilon$ and $|\xi''| \geq \varepsilon$. Then taking into account condition on the Jacobi matrix we obtain that $\mathsf{dist}(\Psi'_t(z'), z') \geq \varepsilon^2$ provided $|\frac{t}{T'} - m| \geq \varepsilon\ \forall m \in \mathbb{Z}$ and $|t| \leq T$ (where $\varepsilon = \varepsilon(T)$. Therefore one needs to estimate the measure of the set

$$\begin{aligned} \mathcal{Q} = \bigl\{\phi \in \mathbb{S}^1 : \exists m', m'' \in \mathbb{Z} :& \\ &|m'| \leq \varepsilon^{-\sigma}, |m''| \leq \varepsilon^{-\sigma}, |m'T'\cos\phi + m''T''\sin\phi| \leq \varepsilon\bigr\} \end{aligned}$$

where ϕ is associated with ξ by the relations $|\xi'| = \cos\phi$, $|\xi''| = \sin\phi$ and one can take $\sigma > 0$ arbitrarily small. However, $\mathsf{mes}(\mathcal{Q}) = O(\varepsilon^\sigma)$ obviously.

Therefore for Cartesian products the non-periodicity condition always is fulfilled.

(iii) The above arguments also yield that $\mathsf{dist}(G^*z, \Psi_t(z)) \leq \varepsilon^2$ implies that $\mathsf{dist}(G^{*k}z, \Psi_{kt}(z)) \leq \varepsilon$ for constant k and the isometry $G^* : S^*\bar{X} \to S^*\bar{X}$ induced by the isometry $G : \bar{X} \to \bar{X}$. Therefore Definition 11.3.6 is stable under the above constructions of new manifolds in Remark 11.3.16(b) and (c) .

11.3.2 Operators with Periodic Flows

Regular Operators on Closed Manifolds

Let us now consider a closed manifold X and a scalar operator A on X with periodic Hamiltonian trajectories of the principal symbol. As before we assume that the manifold and operator are smooth enough; furthermore, as before we divide the operator A by τ, introduce $h=\tau^{-\frac{1}{2m}}$ and obtain an h-pseudodifferential operator. Then if all the Hamiltonian trajectories on the energy level $\Sigma=\{a(x,\xi)=1\}$ are periodic with period T_0 then every trajectory is T-periodic with $T(x,\xi)=T_0a^{\frac{1}{2m}-1}(x,\xi)$ and then the Hamiltonian flow of the symbol $a^{\frac{1}{2m}}$ is $2mT_0$-periodic.

Without any loss of the generality we assume that $2mT_0=1$. According to Section 6.2 we must replace operator A by operator $\bar{A}=(\tau^{-1}A)^{\frac{1}{2m}}$ (one should use this formula only in a neighborhood of Σ).

Then conditions (6.2.2), (6.2.4) and (6.2.7) with $T_0=1$ are fulfilled with $\Omega_0\supset\Sigma$. Calculating the action by formula (6.2.6) we obtain that $\alpha_1=0$ because symbol $\bar{a}$ is positively homogeneous of degree 1 on ξ. In what follows all the objects related to the operator $\bar{A}$ are denoted by the same letters as the corresponding objects related to A but equipped with a bar. Then according to (6.2.12)

$$b_0(x,\xi)=\frac{1}{2mT_0}\int_0^{T_0}(a^s\circ\Psi_t)(x,\xi)dt-\frac{\pi\alpha_0}{2}\qquad\text{on }\Sigma \tag{11.3.16}$$

and is positively homogeneous of degree 0 on ξ in a neighborhood of Σ. Here α_0 is the Maslov index.

Then results of Subsubsection *6.2.3.3 Spectral Gaps (the First Attempt)* and Subsection 6.4 yield

Theorem 11.3.18. *Let X be a closed compact $\mathscr{C}^K$ manifold and A an elliptic self-adjoint differential operator*[23] *of order $2m$. Let the Hamiltonian flow of $a(x,\xi)$ be T_0-periodic on Σ.*

Let $\mathcal{L}$ be the image of the map $\Sigma\ni(x,\xi)\to b(x,\xi)\xrightarrow{\iota}\mathbb{R}/\mu\mathbb{Z}$ where $\mu=\pi(mT_0)^{-1}$ and $\iota:\mathbb{R}\to\mathbb{R}/\mu\mathbb{Z}$ is a natural map. Assume $\mathcal{L}$ does not coincide with $\mathbb{R}/\mu\mathbb{Z}$; then $\iota^{-1}\mathcal{L}$ is a union of intervals $\{I_n\}$ of length $O(1)$ such that $\omega\in I_n\implies|\omega|\asymp n$ for $|n|\ge1$.

It follows that

$$\operatorname{Spec}(A^{\frac{1}{2m}})\subset\bigcup_{n\ge1}J_n\cup[-C,C] \tag{11.3.17}$$

where J_n are the $Cn^{\delta-1}$-neighborhoods of I_n, $\delta>0$ is arbitrarily small and $C=C_\delta$; without loss of generality we assume that $A\ge I$.

Moreover, for $b_0=\mathsf{const}$ one can take $\delta=0$.

[23] Or a classical pseudodifferential operator. In all cases we assume that $a\ge0$.

Theorems 6.3.6 and 6.3.12 yield then

Theorem 11.3.19. *Let the conditions of Theorem 11.3.18 be fulfilled and let $\omega_0 \in \mathcal{L}$,*

$$(x,\xi)\in\Sigma,\ b_0(x,\xi)=\omega_0 \implies \textit{either } \nabla_\Sigma b_0(x,\xi)\neq 0 \quad \textit{or } \operatorname{rank}\operatorname{Hess}_\Sigma b_0(x,\xi)\geq r. \tag{11.3.18}$$

Further, let us assume that either the Hamiltonian flow Ψ_t has no subperiodic trajectory on Σ or

$$\operatorname{dist}\bigl(\Psi_t(x,\xi),(x,\xi)\bigr)\geq t(T_0-t)\rho(x,\xi) \qquad \forall t\in(0,T_0)\ (x,\xi)\in\Sigma \tag{11.3.19}$$

where

$$|\nabla\rho|\leq c \quad \textit{and} \quad \upmu\{(x,\xi),\rho(x,\xi)\leq h\}=O(h^q) \quad \textit{as } h\to+0. \tag{11.3.20}$$

Then for $\tau\to+\infty$ such that $|\iota\tau^{\frac{1}{2m}}-\omega_0|\leq\epsilon_0$ (where $\epsilon_0>0$ is small enough) asymptotics

$$\mathsf{N}^-(\tau)=\varkappa_0\tau^{\frac{d}{2m}}+\bigl(\varkappa_1+F(\tau^{\frac{1}{2m}})\bigr)\tau^{\frac{d-1}{2m}}+O\bigl(\tau^{\frac{d-1-p}{2m}}\bigr) \tag{11.3.21}$$

holds with

$$F(z)=(2mT_0)^{-1}(2\pi)^{-d}\int_\Sigma \Upsilon\Bigl(2mT_0\bigl(z-b(x,\xi)\bigr)\Bigr)\,dxd\xi:da, \tag{11.3.22}$$

$p=\min(1,\frac{r}{2}-\delta,\frac{q}{2}-\delta)$ with an arbitrary small exponent $\delta>0$. Here and below $\Upsilon(z)$ is a 2π-periodic function on $\mathbb{R}$ equal to $\pi-z$ on $[0,2\pi)$.

The following should be rather easy:

Problem 11.3.20. (i) Show that for generic differential and pseudodifferential symbols a^s condition (11.3.18) is fulfilled with $r=2d-2$ for all ω_0. In fact the function $b_0(x,\xi)$ is given *on* $\Sigma'=\Sigma/\Psi_t$ (the factor of Σ with respect to the Hamiltonian flow) which is a $2d-2$-dimensional manifold *rather than on* Σ.

(ii) Derive more precise asymptotics are possible as well and improved asymptotics of $\mathsf{N}^-_{(\vartheta)}(\tau)$ with remainder estimates which could be as good as $O(\tau^{(d-2-2\vartheta)/2m})$ (under an appropriate non-degeneracy condition).

Surely, Theorem 11.3.19 is applicable to a Laplace-Beltrami operator on a Riemannian manifold with periodic geodesic flow which is perturbed by a first-order operator. For a generic vector $(V_1, \dots, V_d)$ the field corresponding to the perturbation condition (11.3.18) is fulfilled with $r = 2d - 2$.

In particular, for the Laplace-Beltrami operator on the standard sphere $\mathbb{S}^d$ one can see easily that $\alpha_0 \equiv 2(d-1) \mod 4\mathbb{Z}$ and $T_0 = \pi$. Let us consider this operator with perturbation more carefully. Let $j : \mathbb{S}^d \to \mathbb{S}^d$ be the antipodal map $x \to -x$ and let $j^* : S^*\mathbb{S}^d \to S^*\mathbb{S}^d$ be the adjoint map $(x, \xi) \to (-x, -\xi)$. Then (11.3.16) yields that b_0 depends only on the even part of a^s: $a^s_{\text{even}} = \frac{1}{2}(a^s + a^s \circ j^*)$. Let us now assume that a^s is skew-symmetric with respect to j^*:

$$a^s = -a^s \circ j^*. \tag{11.3.23}$$

Then $b_0 = 0$ and B is a (-1)-order operator (i.e., $\eta = h = \tau^{-\frac{1}{2}}$: see Section 6.2). Let us calculate then the principal symbol b_{-1} of the operator B. Given that the eigenvalues of unperturbed Laplace-Beltrami A_0 on $\mathbb{S}^d$ are $n(n+d-1)$ with $n \in \mathbb{Z}^+$ it will be more convenient for us to set

$$\bar{A} = h\Bigl(\bigl(A + \frac{1}{4}(d-1)^2\bigr)^{\frac{1}{2}} - \frac{1}{2}(d-1)\Bigr) \tag{11.3.24}$$

here instead of $\bar{A} = A^{\frac{1}{2}}$. Let

$$\bar{A}_0 = h\Bigl(\bigl(A_0 + \frac{1}{4}(d-1)^2\bigr)^{\frac{1}{2}} - \frac{1}{2}(d-1)\Bigr) \tag{11.3.25}$$

be the corresponding unperturbed operator.

Let $U(t) = \exp(ih^{-1}t\bar{A})$ and $U_0(t) = \exp(ih^{-1}t\bar{A}_0)$. Then

$$U_0(2\pi) = I, \qquad U_0(\pi) = J \quad \text{where } Jv = v \circ j. \tag{11.3.26}$$

So *we want to calculate* $U(2\pi)$ *modulo operators of order* (-2) because near Σ the orders in the sense of standard pseudodifferential operators and in the sense of h-pseudodifferential operators coincide. Let us calculate $U(\pi)J$ modulo operators of order (-1) first. To do it we consider $V(t) = U(t)U_0(-t)$ which is a standard (h-)pseudodifferential operator. Then

$$\frac{d}{dt}V(t) = ih^{-1}U(t)(\bar{A} - \bar{A}_0)U_0(-t) = iV(t)U_0(t)QU_0(-t)$$

where $Q = h^{-1}(\bar{A} - \bar{A}_0)$ is an operator of order 0. Considering principal symbols we obtain that $\partial_t v = iv(b \circ \Psi_t)$. Since $v|_{t=0} = 1$ we obtain that

$$v(t) = \exp(-i \int_0^t l \circ \Psi_{t'}\, dt')$$

where $q = \frac{1}{2} a^s a^{-\frac{1}{2}}$ is the principal symbol of Q. Therefore (11.3.26) yields that

(11.3.27) $U(\pi)J$ is an h-pseudodifferential operator with the principal symbol $\exp(i\varsigma \cdot \int_0^\pi q \circ \Psi_t\, dt)$

with $\varsigma = 1$.

Let us consider operator $\bar{A}_\varsigma = \bar{A}_0 + \varsigma hQ$ and let us consider the corresponding unitary group $U_\varsigma(t)$. Differentiating equalities

$$(hD_t - \bar{A}_\varsigma)U_\varsigma(t) = 0, \qquad U_\varsigma(0) = I$$

with respect to ς we obtain that

$$(hD_t - \bar{A}_\varsigma)U'_\varsigma(t) = hQU_\varsigma, \qquad U'_\varsigma(0) = 0$$

where the prime means the derivative with respect to ς. Then

$$U'_\varsigma(t) = i \int_0^t U_\varsigma(t - t')QU_\varsigma(t')\, dt'.$$

Let us set $t = 2\pi$ and divide interval $[0, 2\pi]$ of integration into subintervals $[0, \pi]$ and $[\pi, 2\pi]$ and replace t' by t in the first subinterval and by $t + \pi$ in the second one. We obtain that

$$U'_\varsigma(2\pi) = iU_\varsigma(2\pi) \int_0^\pi U_\varsigma(-t)\tilde{Q}U_\varsigma(t)\, dt \tag{11.3.28}$$

with $\tilde{Q} = Q + U_\varsigma(-\pi)QU_\varsigma(\pi)$. We know that $U_\varsigma(\pi) = V_\varsigma J$ and $U_\varsigma(-\pi) = JV^*_\varsigma$ where V_ς is a unitary operator. Therefore

$$\tilde{Q} = Q + JQJ + JV^*_\varsigma[Q, V_\varsigma]J.$$

Then (11.3.23) yields that $\tilde{Q}$ is an operator of order (-1) with the principal symbol

$$\tilde{q} = 2q_{\mathsf{even}} + \varsigma\{q, \int_0^\pi q \circ \Psi_{t'}\, dt'\}$$

where q_{even} is the even (with respect to j^*) part of q^s and the second term is also even. Therefore (11.3.28) yields that $-iU_\varsigma(-2\pi)U'_\varsigma(2\pi)$ is an operator of order (-1) with principal symbol

$$\int_0^{\pi} \tilde{q}\circ\Psi_{-t}\,dt = \int_0^{2\pi} q^s\circ\Psi_t\,dt + \varsigma\int_0^{\pi}\int_0^{\pi}\{q\circ\Psi_t, q\circ\Psi_{t+t'}\}\,dtdt'$$

where we changed variables in the integration in the obvious way. Integrating over $\varsigma\in[0,1]$ we obtain that

(11.3.29) $U(2\pi)-I$ is an operator of order (-1) with principal symbol $2\pi i b_{-1}$ where

$$b_{-1} = \frac{1}{2\pi}\Bigl(\int_0^{2\pi} q^s\Psi_t dt + \int_0^{\pi}\int_0^{\pi}\{q\circ\Psi_t, q\circ\Psi_{t+t'}\}\,dtdt'\Bigr). \tag{11.3.30}$$

Now theorems 6.3.6 and 6.3.12 yield

Theorem 11.3.21. *Let $A=\Delta+A'+A''$ where Δ is the (positive) Laplace-Beltrami operator on the standard sphere $\mathbb{S}^d$, $A'=A'(x,D)$ and $A''=A''(x,D)$ are symmetric operators of first and zero order respectively and A' anti-commutes with J (in which case a^s satisfies (11.3.23)). Let $b_{-1}(x,\xi)$ be defined by (11.3.30) with $q=\frac{1}{2}a'a^{-\frac{1}{2}}$, $q^s=\frac{1}{2}a''a^{-\frac{1}{2}}-\frac{1}{8}a'^2a^{-\frac{3}{2}}$ where a' and a'' are the principal symbols of A' and A'' respectively.*

Let $\mathcal{L}'=[\lambda_1,\lambda_2]$ be the image of the map $\Sigma\ni(x,\xi)\to b_{-1}(x,\xi)\in\mathbb{R}$ and let $f(\tau)=(\tau+\frac{1}{4}(d-1)^2)^{\frac{1}{2}}-\frac{1}{2}(d-1)$. Then

(i) $f(\operatorname{Spec} A)\subset\bigcup_{n\ge1}J'_n\cup[-C,C]$ where

$$J'_n=\bigl[n+\lambda_1 n^{-1}-Cn^{\delta-2}, n+\lambda_2 n^{-1}+Cn^{\delta-2}\bigr], \tag{11.3.31}$$

$C=C(\delta)$ is large enough, $\delta>0$ is arbitrarily small and for $\lambda_1=\lambda_2$ one can take $\delta=0$.

(ii) Let $\omega_1\in\mathcal{L}'$ and condition (11.3.18) be fulfilled with ω_1 and b_{-1} instead of ω_0 and b_0. Then for $\tau\to+\infty$ such that

$$f(\tau)\subset\bigcup_{n\ge1}[n+(\omega_1-\epsilon)\frac{1}{n}, n+(\omega_1+\epsilon)\frac{1}{n}]$$

with a sufficiently small constant $\epsilon > 0$ *asymptotics*

$$\mathsf{N}^-(\tau) = \varkappa_0\tau^{\frac{d}{2}} + \varkappa_1\tau^{\frac{d-1}{2}} + F(f(\tau),\tau)\tau^{\frac{d-1}{2}} + O(\tau^{\frac{d-1-p}{2}}) \tag{11.3.32}$$

holds with $p = \min(1, \frac{r}{2} - \delta)$ *with an arbitrarily small exponent* $\delta > 0$ *and with*

$$F(z,\tau) = 2(2\pi)^{-d-1}\int_\Sigma \Upsilon\Bigl(2\pi\bigl(z - \tau^{-\frac{1}{2}}b_{-1}(x,\xi)\bigr)\Bigr)\,dxd\xi : da. \tag{11.3.33}$$

The following seems to be easy:

Problem 11.3.22. (i) Prove that for an arbitrary fixed symbol a'' and for a generic differential symbol (i.e., vector field) as well as for a generic pseudodifferential symbol a' satisfying (11.3.23) the modified[24] condition (11.3.18) is fulfilled with $r = 2d - 2$ for all $\omega_1 \in \mathbb{R}$.

(ii) Prove that the same is true for an arbitrary fixed a'' satisfying (11.3.23) and for a generic differential symbol (i.e., potential) as well as for a generic pseudodifferential symbol a''.

Let us now assume that

$$a' = 0, \quad a'' \circ j^* = -a''. \tag{11.3.34}$$

Then $b_{-1} = 0$ and the operator $U(2\pi) - I$ obviously is an operator of order ≤ -2 (we will prove that its order is actually (-3)) and our goal is to calculate its principal symbol. Now the calculations are simpler because the perturbation is weaker. Let $\bar{A}$ and $\bar{A}_0$ be defined as above. Then $\bar{A} = \bar{A}_0 + Q' + Q''$[25] where Q' is an operator of order (-1) with principal symbol $q' = -\frac{1}{2}a^{-\frac{1}{2}}a''$ anti-commuting with J and Q'' is an operator of order (-3) with principal symbol $q'' = -\frac{1}{8}a^{-\frac{3}{2}}a''^2$[26]. Then the equalities

$$(D_t - \bar{A}_0)U(t) = QU(t), \qquad U(0) = I$$

[24] I.e., with ω_1, b_{-1} instead of ω_0, b_0.

[25] For the sake of simplicity we do not write the factor $\tau^{-\frac{1}{2}}$ which in any case is reduced in the final formula.

[26] In contrast with the above case we did not include a lower order operator commuting with J in A because we are mainly interested in differential operators. This generalization is left to the reader.

with $Q = Q' + Q''$ yield that

$$U(t) = U_0(t) + i\int_0^t U(t-t')QU(t')\,dt'$$

(this is similar to the calculation of V). After iteration we obtain $U(t)$ as an asymptotic series of operators with order decreasing by step 1. Let us set $t = 2\pi$. We then obtain that modulo operators of order (-4)

$$\text{(11.3.35)}\quad U(t) \equiv I + i\int_0^{2\pi} Q(t_1)\,dt_1 - \iint_{\Pi_2} Q(t_1)Q(t_2)\,dt_1dt_2 - \\ i\iiint_{\Pi_3} Q(t_1)Q(t_2)Q(t_3)\,dt_1dt_2dt_3$$

with $\Pi_k = \{t = (t_1,\dots,t_k) \in \mathbb{R}^k, 0 \le t_1 \le \cdots \le t_k \le 2\pi\}$ and $Q(t) = U_0(-t)QU_0(t)$. Moreover, in the third and fourth terms we can replace Q by Q'.

Dividing the interval of integration in the second term into subintervals $[0,\pi]$ and $[\pi, 2\pi]$ and replacing t_1 by t and $t+\pi$ on these subintervals respectively we obtain that this term is a pseudodifferential operator with principal symbol $i\int_0^{2\pi} q''\circ\Psi_t\,dt$.

The third term in the right-hand expression is equal to

$$-\frac{1}{2}\iint_{\Pi_2} (Q(t_1)Q(t_2) + Q(t_2)Q(t_1))\,dt_1dt_2 - \frac{1}{2}\iint_{\Pi_2} [Q(t_1), Q(t_2)]\,dt_1dt_2.$$

The first term here equals half of $-Q(t_1)Q(t_2)$ integrated over the square $(t_1,t_2) \in [0,2\pi]^2$; therefore it is equal to $-\frac{1}{2}(\int_0^{2\pi} Q(t)dt)^2$ and therefore is negligible as its principal symbol is 0.

The second term is an operator of order (-3) and after simple calculations we obtain that its principal symbol is

$$i\int_0^{\pi}\int_0^{\pi} \{q\circ\Psi_t, q\circ\Phi_{t+t'}\}\,dtdt'.$$

Finally, the third term in (11.3.35) is obviously an operator of order (-3). Making permutations of (t_1, t_2, t_3) and summing we obtain that modulo an operator of order (-4) this term is one-sixth of $Q(t_1)Q(t_2)Q(t_3)$ integrated over the cube $[0, 2\pi]^3$ and is therefore negligible.

Thus,

(11.3.36) $(U(2\pi)-I)$ is an operator of order (-3) with the principal symbol $2\pi i b_{-3}(x,\xi)$ where b_{-3} is given by (11.3.30) with $q'=\frac{1}{2}a''a^{-\frac{1}{2}}$ and $q''=-\frac{1}{8}a''^2a^{-\frac{3}{2}}$ instead of q and q^s respectively.

Therefore Theorem 6.2.3 yields

Theorem 11.3.23. *Suppose that conditions of Theorem 11.3.21 and condition* (11.3.34) *are fulfilled. Let* $\mathcal{L}''=[\lambda_1,\lambda_2]$ *be the image of the map* $\Sigma\ni(x,\xi)\to b_{-3}(x,\xi)\in\mathbb{R}$*. Then*

(i) $f(\operatorname{Spec} A)\subset\bigcup_{n\ge1}J''_n\cup[-C,C]$ *where*

$$J''_n=\bigl[n+\lambda_1 n^{-3}-Cn^{\delta-4},\,n+\lambda_2 n^{-3}+Cn^{\delta-4}\bigr], \tag{11.3.37}$$

$C=C(\delta)$ *is small enough,* $\delta>0$ *is arbitrarily small and for* $\lambda_1=\lambda_2$ *one can take* $\delta=0$.

(ii) Let $\omega_2\in\mathcal{L}''$ *and condition* (11.3.18) *be fulfilled with* ω_2 *and* b_{-3} *instead of* ω_0 *and* b_0*. Then for* $\tau\to+\infty$ *such that*

$$f(\tau)\subset\bigcup_{n\ge1}[n+(\omega_2-\epsilon)n^{-3},\,n+(\omega_2+\epsilon)n^{-3}]$$

with a sufficiently small constant $\epsilon>0$ *asymptotics* (11.3.32) *holds with the same* p *as above and with*

$$F(z,\tau)=2(2\pi)^{-d-1}\int_\Sigma\Upsilon\Bigl(2\pi\bigl(z-\tau^{-\frac{3}{4}}b_{-3}(x,\xi)\bigr)\Bigr)\,dxd\xi:da. \tag{11.3.38}$$

We leave to the reader

Problem 11.3.24. (i) Prove that for a generic differential symbol (i.e., potential) as well as for a generic pseudodifferential symbol a'' satisfying (11.3.34) the modified[27] condition (11.3.18) is fulfilled with $r=2d-2$ for all $\omega_2\in\mathbb{R}$.

(ii) Thus, in all the above cases for generic perturbations the remainder estimate is $O(\tau^{\frac{d-2}{2}})$ for $d\ge3$ and $O(\tau^{\frac{1}{2}(d-2+\delta)})$ with an arbitrarily small exponent $\delta>0$ for $d=2$. Applying the more precise results of Section 6.2 we obtain that for $d=2$ the remainder estimate is $O(1)$ if the principal symbol b has no saddle points at the corresponding level and $O(\log\tau)$ otherwise.

[27] I.e., with ω_2, b_{-3} instead of ω_0, b_0.

(iii) Derive the remainder estimate $O(1)$ in the latter case as well by including in the principal symbol the next symbol too (as we have seen it does not necessarily vanish even if the principal symbol vanishes identically).

(iv) Consider Riesz means asymptotics to the reader. Recall that the best possible remainder estimate is $O\bigl(\tau^{\frac{1}{2}(d-1-n\vartheta)}\bigr)$ where $n=0,1,3$ is the order of B taken with the opposite sign.

11.3.3 Singular Operators on Closed Manifolds

Let us now consider the singular case when not only an operator but also a manifold itself has singularities. Singularities are described by scaling function $\gamma(x)$:

Definition 11.3.25. (i) Let X be a Riemannian manifold (with or without boundary) such that $\bar{X}$ a closure of X in the Riemannian metrics is compact. Then we call X *singular compact Riemannian manifold.*

(ii) If the regular part of the boundary is empty, we call X *singular compact closed Riemannian manifold.*

(iii) Let γ be a scaling function on $\bar{X}$, such that $|\gamma(x,y)\le \frac{1}{2}\mathsf{dist}(x,y)$, $\partial X=\{x,\gamma(x)=0\}$, $X=\{x,\gamma(x)>0\}$ and let metrics on X be γ-admissible. Then we call X (singular) *γ-admissible compact Riemannian manifold.*

However we will need also a γ-admissible function $\zeta(x,\xi)$ describing how close trajectory $\Psi_t(x,\xi)$ can approach $\{\gamma=0\}$ (see condition (11.3.39) below):

Theorem 11.3.26. *Let X be a γ-admissible compact closed Riemannian manifold, A a scalar elliptic γ-admissible operator of order $2m$ with the lower order terms satisfying the standard conditions of this section. Let us assume that $\gamma^{-1-\delta}\in \mathscr{L}^1(X)$ with $\delta>0$.*

Further, let us assume that the Hamiltonian flow of $a(x,\xi)$ is T_0-periodic on Σ. Furthermore, let us assume that a γ-admissible bounded function ζ is given on Σ such that

$$|D\Psi_t(z)|\le C\zeta^{-n}(z),\quad \gamma(\Psi_t(z))\ge \epsilon\zeta^{n}(z)\qquad \forall z\in\Sigma\quad \forall t. \tag{11.3.39}$$

Finally, let us assume that conditions (11.3.18) *with* $r \ge 1$ *(for some fixed* ω_0*) and* (11.3.19)–(11.3.20) *(which is condition to subperiodic trajectories) with* $q > 0$ *are fulfilled.*

Then for $\tau \to +\infty$ *such that* $|\iota\tau^{\frac{1}{2m}} - \omega_0| \le \epsilon$ *asymptotics* (11.3.21) *holds with sufficiently small exponent* $p > 0$.

Proof. Dividing Σ into the zones $\{\zeta \ge \frac{1}{2}\tau^{-\delta}\}$ and $\{\zeta \le \tau^{-\delta}\}$ we can apply results of Section 6.2 in the former zone and the standard asymptotics in the latter one. □

Problem 11.3.27. We derived asymptotics which is marginally better than the standard one only as $\vartheta \le p$. The problem which is definitely rather difficult and very interesting is to derive asymptotics with the better remainder estimates; ideally one should derive asymptotics with the remainder estimates which are as good as in the regular case (surely one needs to impose more restricting conditions).

Example 11.3.28. (i) Consider *Zoll-Tannery manifold* with the rotational symmetry (see Example 8.3.9(iii). Recall that these manifolds are isometric (up to a factor) to a sphere $\mathbb{S}^d$ with the metrics $(r+h(\cos(s)))^2\, ds^2 + \sin^2(s)\, d\theta^2$ with $s \in [0,\pi]$, $r = p/q$, $p, q \in \mathbb{N}$, $\gcd(p,q) = 1$ and odd function $h : (-1,1) \to (-r,r)$. Then the length of equator is 2π, the length of meridian $2\pi r$ and the length of general geodesics $2\pi p$. See A. Besse [1], Theorem 4.13.

Then $T_0 = 2\pi q$. Therefore one can take $A = \Delta^m$ (plus lower order terms) where Δ is the Laplace-Beltrami operator; these lower order terms must satisfy the standard assumptions of this section and also provide condition (11.3.12) with $p \ge 1$. Singularities (if there are any) are located at $s = 0, \pi$.

However these singularities do not prevent us from recovering remainder estimate $O(\tau^{(d-2)/2m})$ as long as non-degeneracy condition to the lower order terms is fulfilled. This asymptotics contains non-Weyl correction term.

(ii) Consider the same example replace rational r by irrational r; then we are basically in the framework of the Subsection 11.3.1 and can recover remainder estimate $O(\tau^{(d-1)/2m-\delta})$ without any condition to the lower order terms.

Repeating the arguments of Theorems 11.3.21 and 11.3.26 we can generalize them to operators with the singular perturbations on $\mathbb{S}^d$.

Remark 11.3.29. Let us consider the Laplace-Beltrami operator of Example 11.3.28 perturbed by a potential $V \in \mathscr{L}^s$ such that $\|V\|_{\mathscr{L}_s} < \epsilon$ with $s > d = 2$ or $s = d \geq 3$. Then the standard functional-analytic arguments yield that the cluster character of the spectrum remains true. The size of a cluster is $O(\epsilon\tau^{\frac{d}{2s}})$ and the distance between clusters is $\asymp \tau^{\frac{1}{2}}$; here $\epsilon = \|V\|_{\mathscr{L}^s}$ is assumed to be small only for $s = d$.

11.3.4 Manifolds with the Boundary

The possibility of singularities increases the possibilities even if the boundary is smooth.

Example 11.3.30. Consider manifolds of Example 8.3.9 which have boundaries (and these boundaries may also have singularities). Therefore one can take $A = \Delta^m$ (plus lower order terms) where Δ is the Laplace-Beltrami operator with the appropriate boundary conditions.

Then

(i) If r is rational the lower order terms and boundary conditions must satisfy the standard assumptions of this section and also provide condition (11.3.12) with $p \geq 1$. Then asymptotics with remainder estimate $O(\tau^{(d-1)/2m-\delta})$ holds; this asymptotics contains non-Weyl correction term.

(ii) If r is irrational we can recover remainder estimate $O(\tau^{(d-1)/2m-\delta})$ without any condition to the lower order terms and boundary conditions.

11.3.5 The Indefinite Case

In this subsection we discuss the case when the principal symbol is neither positive nor negative definite. We assume that the lower order terms are appropriate: conditions on these terms will be discussed in the next Subsection 11.3.6.

All the results of Sections 11.2 and 11.3 remain true for operators with elliptic principal symbols which are neither positive nor negative definite. In this case $\mathsf{N}^-(\tau)$ is the number of eigenvalues in the interval $(0,\tau)$ for $\tau > 0$ and $\mathsf{N}^+(\tau)$ is the number of eigenvalues in the interval $(\tau, 0)$ for $\tau < 0$ and we consider asymptotics of $\mathsf{N}^\mp(\tau)$ for $\tau \to \pm\infty$.

Surely, for $\tau \to +\infty$ and for $\tau \to -\infty$ we may obtain different asymptotics and the conditions on Hamiltonian trajectories may be different;

however in two most interesting examples – Dirac and Maxwell operators (see Example 8.4.4) results should be the same modulo remainder estimate and equal to $\frac{1}{2}\mathsf{N}^-(A^2-\tau^2)$; this would make those results look trivial.

However it is not the case, at least not for the Maxwell operator which should be considered on the subspace defined by under-defined system and as domain is irregular (for example, if it has edges directed inside, like in $(\mathbb{R}^d \setminus \mathbb{R}^{+2}\times\mathbb{R}^{d-2})$) one should be really careful ; in particular formally Maxwell system and system of hydrodynamics[28] look the same but in such domains results are different. So, we suggest very difficult and interesting problem

Problem 11.3.31. Accurate analysis with the results of the same sharpness as in this Sections 11.2 and 11.3 for

(i) Maxwell system:

(a) isotropic (already difficult);

(b) anisotropic (much more difficult);

(ii) System of hydrodynamics[28].

Further, all the results for $\vartheta>0$ remain true if we set

$$\mathsf{N}^{\mp}_{(\vartheta)}(\tau) = \pm\int_0^\tau |\tau'|^{\vartheta-1}N(\tau')d\tau' \tag{11.3.40}$$

for $\pm\tau>0$.

Furthermore, the results of Subsection 11.2.3 should remain true. In this case (instead of the assumption that $A(x,D)$ is a scalar operator) we assume that $a(x,\xi)$ has only one positive (or negative) eigenvalue $\lambda(x,\xi)$ which is simple provided we treat asymptotics as $\tau\to+\infty$ ($\tau\to-\infty$ respectively). In this case all the conditions of these subsections are assumed to be fulfilled for the Hamiltonian flow generated by $\lambda(x,\xi)$.

Moreover, to obtain conditions on lower order terms one should find pseudodifferential operator Q such that

$$Q^*Q\equiv QQ^*\equiv I \qquad \text{and} \qquad Q^*AQ\equiv\begin{pmatrix}A_1 & 0\\ 0 & A_2\end{pmatrix}$$

[28] Without viscosity.

where A_1 is a scalar operator with principal symbol $\lambda(x,\xi)$ and A_2 an operator with principal symbol of the opposite sign. Then all the conditions should be imposed on the lower order terms of A_1.

Moreover, we can treat the Dirac operator; nevertheless the eigenvalues of its principal symbol are simple only for $D=2 \implies d=1,2,3$ (for $d=3$ one has a principally massless particle). However, one needs to check that all the symbols associated with A_1 and taken into account are scalar. Similar notes remain true for isotropic Maxwell and hydrodynamics systems and I suggest another very difficult and interesting problem

Problem 11.3.32. Accurate analysis with the results of the same sharpness as in this subsection 11.3.2 for

(i) Dirac operator;

(ii) Isotropic Maxwell system;

(iii) System of hydrodynamics[28)];

In order to prove the above statements one should apply the results of Parts II and III to the interval $(\frac{1}{2}\tau,\tau)$ (we consider the case $\tau>0$ for the sake of simplicity). Then the routine remainder estimate $O(\tau^\nu)$ or $O(\tau^\nu\log^\sigma\tau)$ is obtained by standard methods. Let us replace τ by $2^{-j}\tau$ (with $j=0,\dots,n$ such that $\tau\asymp 2^n$) and sum all these asymptotics; then we obtain asymptotics for the number of eigenvalues lying in (τ_1,τ) with the same remainder estimate for $\nu>0$ and with the remainder estimate $O(\log^{\sigma+1}\tau)$ for $\nu=0,\sigma\ge 0$. On the other hand it is relatively easy to prove that $\mathsf{N}(0,\tau_1)<\infty$.

Moreover, in order to improve the final remainder estimate in the case when it grows slowly, one can apply the following arguments to obtain the remainder estimate $O(\tau^{-s})$ with arbitrarily chosen s for mollifications of $\mathsf{N}(\tau',\tau'')$ (the number of eigenvalues in the interval (τ',τ'')) with respect to τ' and τ'' with mollifying functions $\tau^{-1}\phi_1(\tau'\tau^{-1})$ and $\tau^{-1}\phi_2(\tau''\tau^{-1})$ where ϕ_1 and ϕ_2 are supported near $\frac{1}{2}$ and 1 respectively. Replacing τ by $2^{-j}\tau$ and making a partition of unity one obtains the remainder estimate $O(1)$ for $\mathsf{N}(0,\tau')$ mollified near $\frac{1}{2}\tau$. On the other hand, for $N(\tau',\tau)$ (even without mollification) we have the appropriate remainder estimate provided $\tau'\asymp\tau$.

So, semiclassical arguments look rather straightforward but there is also analysis in the singular zone and for indefinite operators it may become rather non-trivial. For $\vartheta>0$ one can apply analysis of the same type.

11.3.6 Conditions to Lower Order Terms

Let us discuss conditions to the lower order terms starting from a series of remarks:

Remark 11.3.33. All our results were proven for (ρ, γ)-admissible lower order terms provided

$$\rho^{d+2m\vartheta} \in \mathscr{L}^1(X), \tag{11.3.41}$$

$$\rho^{2m} \in \mathscr{L}^1(X), \tag{11.3.42}$$

$$\rho^{2m} \in \mathscr{L}^2(X), \tag*{(11.3.42)*}$$

where (11.3.42) and (11.3.42)* are conditions in the semi-bounded and non-semibounded cases respectively. Here $2m$ is the order of the operator and $m \in \frac{1}{2}\mathbb{Z}$.

Recall that so far ρ was not mentioned at this section.

We can give a series of admissible lower order terms for all the types of operators treated above. However, it is very simple in the context of (ρ, γ)-admissibility. We leave this analysis to the reader and discuss how one can weaken these conditions which may be excessive (in contrast to Section 11.1 because now ρ does not enter in the assumptions on the principal part.

Example 11.3.34 (singularity at origin). (i) Let $\gamma = \epsilon|x|$ and $\rho = |x|^{\mu}$. Then conditions (11.3.42), (11.3.42) and (11.3.42)* mean respectively that

$$\mu > -d(d+2m\vartheta)^{-1}, \qquad \mu > -\frac{d}{2m}, \qquad \mu > -\frac{d}{4m}. \tag{11.3.43}$$

Let μ^* be the maximal forbidden value. Then for $\mu = \mu^*$ we can treat $\rho = |x|^{\mu}\bigl|\log|x|\bigr|^{\nu}$ with appropriate ν, etc.

However, arguments associated with estimates (11.1.66)–(11.1.67) imply that it is sufficient to assume that

$$\mu > \max(-1, -\frac{d}{2m}), \qquad \mu > \max(-1, -\frac{d}{4m}) \tag{11.3.44}$$

in the semi-bounded and non-semi-bounded cases respectively, i.e. that one should check (11.3.43) only with $\vartheta = 0$.

Really, in such vicinity operator A_0 or A_0^2 (where A_0 contains no lower order terms) is larger in operator sense than $r^{-2m+\delta}$, $r^{-4m+\delta}$ with an arbitrarily small exponent $\delta > 0$ respectively.

Moreover, for $d > 2m$, $d > 4m$ respectively in these cases one can even take $\mu = -1$ assuming that in a neighborhood of 0

$$|a_{\alpha',\alpha''}| \le \epsilon\rho^{2m-|\alpha'|-|\alpha''|} \qquad \forall \alpha', \alpha'' : |\alpha'| \le m, |\alpha''| \le m, |\alpha'| + |\alpha''| \ge 1, \tag{11.3.45}$$

$$|a_\alpha| \le \epsilon\rho^{2m-|\alpha|} \qquad \forall \alpha : 1 \le |\alpha| \le 2m \tag{11.3.45*}$$

respectively with small enough $\epsilon = \mathsf{const} > 0$.

(ii) Let us assume that in a neighborhood of 0 domain X coincides with the cone $\{(r, \theta) \in \mathbb{R}^+ \times U\}$ where $\partial U \in \mathscr{C}^K$. Further, let us assume that the Dirichlet boundary condition is given on a part of the boundary $\{(r, \theta) : r \le r_0, \theta \in U_1 \subset \partial U\}$ and that $r_0 > 0$ and $U_1 \ne \emptyset$ contains the ∂U-neighborhood of some point. Then one even does not need condition $\mu > -\frac{d}{2m}$ in the semi-bounded case.

Really, in such vicinity operator A_0 is larger in operator sense than ϵr^{-2m} with sufficiently small constant $\epsilon > 0$.

(iii) Moreover, for the Schrödinger operator with <u>either</u> $d \ge 3$ and with the Dirichlet boundary condition <u>or</u> with $d \ge 2$ and a domain and boundary conditions of the type described in (ii), variational arguments yield that one can assume that in a neighborhood of 0

$$V \ge -\epsilon\rho^2, \qquad |V_j| \le \epsilon\rho \tag{11.3.46}$$

with $\rho = |x|^{-1}$.

(iv) Let us treat the Schrödinger operator with $\mu = -1$, $d \ge 3$ assuming that either the Dirichlet boundary condition is given on ∂X or the domain and the boundary condition are of the type described in (ii). Then

$$(Au, u) \ge \sum_j \epsilon_0 \|(D_j - V_j)u\|^2 + (Vu, u). \tag{11.3.47}$$

Let us introduce spherical coordinates (r, θ) and remove the radial component of the magnetic potential by means of a gradient transform. We obtain that for functions supported in the vicinity of 0

$$(Au, u) \ge \epsilon_0 \|D_r u\|^2 + \epsilon_0 \sum_j \|r^{-1}\mathcal{L}_j u\|^2 + (Vu, u) - \varepsilon \|r^{-1}u\|^2 \tag{11.3.48}$$

where $\mathcal{L}_j$ are first-order operators with bounded coefficients on $\mathbb{S}^{d-1}$ and $\varepsilon > 0$ is arbitrarily small (the vicinity depends on ε). Let us consider

$$\|D_r u\|^2 = \int\int_0^{r_0} r^{d-1}|D_r u|^2\, drd\theta.$$

In order to do this we treat $\|r^\nu D_r u\|_1^2$ where $\|.\|_1$ and $(.,.)_1$ mean respectively $\mathscr{L}^2$-norm and inner product on $[0, r_0]$, $\nu = \frac{1}{2}(d-1)$. Substituting $u = r^{-\nu}v$ we obtain that

$$\begin{aligned} \|r^\nu \partial_r u\|_1^2 = \|(\partial_r - \nu r^{-1})v\|_1^2 = \|\partial_r v\|_1^2 + \nu^2 \|r^{-1}v\|_1^2 - \\ 2\nu \operatorname{Re}(\partial_r v, r^{-1}v)_1 = \|\partial_r v\|_1^2 + (\nu^2 - \nu)\|r^{-1}v\|_1^2. \end{aligned} \tag{11.3.49}$$

Let $d > 3$; then $\nu^2 > \nu$ and we conclude that in this case

$$\|r^\nu \partial_r u\|_1^2 \ge \epsilon_1 \|r^{\nu-1}u\|_1^2. \tag{11.3.50}$$

On the other hand, let $d = 3$. Then $\nu = 1$ in which case $\|r^\nu \partial_r u\|^2 = \|\partial_r v\|^2$ but then obviously (11.3.50) holds again. Therefore

(11.3.51) For $d \ge 3$ the second of conditions (11.3.46) is not necessary.

On the other hand, for $d = 2$, $\nu = \frac{1}{2}$ and $\nu^2 - \nu = -\frac{1}{4}$. In this case only one operator $\mathcal{L}_1 = D_\theta - w(r, \theta)$ is present and if for $\theta = \theta_1$ a Dirichlet boundary condition is given then $\|\mathcal{L}_1 u\|_2 \ge \epsilon_1 \|u\|_2$ where $\|.\|_2$ is $\mathscr{L}^2$-norm with respect to θ.

On the other hand, if $U = \{\theta, \theta_1 < \theta < \theta_2\}$ and if for both $\theta = \theta_1$ and $\theta = \theta_2$ the Neumann boundary condition is given then this estimate fails. However, if X is a complete neighborhood of 0 then this estimate remains true provided

$$|\int_0^{2\pi} w(\theta, r)d\theta - 2\pi m| \ge \epsilon_0 \qquad \forall m \in \mathbb{Z} \cdot \forall r \in (0, r_0). \tag{11.3.52}$$

Conversely, if (11.3.52) fails for every $\epsilon_0 > 0$ then the previous inequality also fails. We conclude that

(11.3.53) Let $d = 2$ and <u>either</u> $U = \{\theta_1 < \theta < \theta_2\}$ and for θ_1 or for θ_2 a Dirichlet boundary condition is given, <u>or</u> X contains a complete neighborhood of 0 and (11.3.52) holds. Then the second of conditions (11.3.46) is not necessary.

(v) All these arguments remain true in the case $\gamma = \epsilon|x''|$, $\rho = |x''|^{\mu}$ with d replaced by d'' in the conditions.

Example 11.3.35. Using the arguments of Remark 11.1.13 we see that in the context of Example 11.2.19 one can assume that $\rho f(|x''|) \le \epsilon$ (with small enough constant $\epsilon > 0$) instead of assumption (11.3.42).

Remark 11.3.36. Certain generalizations can be obtained from functional-analytic arguments. For example, consider Schrödinger operator and asymptotics of $\mathsf{N}^-(\tau)$. Then

(i) Asymptotics with the remainder estimate $O(\tau^{\frac{1}{2}(d-1)})$ remain true if one replaces potentials V and V_j by $V + V'$ and $V_j + V'_j$ with $V' \in \mathscr{L}^d(X)$, $V'_j \in \mathscr{L}^\infty(X)$.

(ii) Moreover, asymptotics with the remainder estimate $O(\tau^{\frac{1}{2}(d-1-\sigma)})$ remain true provided $V'_j = 0$, $V' \in \mathscr{L}^{d(1-\sigma)^{-1}}(X)$.

(iii) Therefore approximating $V' \in \mathscr{L}^d(X)$ by the potentials $V'_\varepsilon \in \mathscr{L}^\infty(X)$ we obtain that asymptotics with the remainder estimate $o(\tau^{\frac{1}{2}(d-1)})$ remain true for such perturbations.

(iv) Moreover, similar approximation arguments yield that asymptotics with the the remainder estimate $O(\tau^{\frac{1}{2}(d-1)}\log^{-\nu}\tau)$ remain true provided $V'|\log(|V'|+1)|^{\nu} \in \mathscr{L}^d(X)$.

(v) On the other hand, remainder estimates weaker than $O(\tau^{\frac{1}{2}(d-1)})$ remain true under weaker conditions. However, our weakly singular potentials (i.e., the potentials which we considered in this section) do not necessarily belong to the indicated classes. Thus, a combination of the two methods is the most effective.

11.4 Large Eigenvalues for Operators in Essentially Unbounded Domains

11.4.1 Coercivity Condition and Standard aAsymptotics

In this section, under the assumptions of the previous section concerning the domain X and the principal symbol, we consider operators with the lower order terms essential from the point of view of the domain of the operator and asymptotics[29]. These terms could be more singular than those studied before but they must satisfy *coercivity condition.*

We start with operators with positive principal symbol. In this case the coercivity condition is

$$\langle A(x,\xi)v,v\rangle \ge \epsilon_0(|\xi|+\rho)^{2m}\|v\|^2 \qquad \forall\xi \quad \forall v, \tag{11.4.1}$$

$$\sum_{\alpha,\beta:|\alpha|\le m,|\beta|\le m} \langle a_{\alpha,\beta}w_\alpha, w_\beta\rangle \ge \epsilon_0 \sum_{\alpha:|\alpha|\le m} \|w_\alpha\|^2\rho^{2m-2|\alpha|} \qquad \forall w=(w_\alpha)_{\{|\alpha|\le m\}} \tag{11.4.2}$$

where $\epsilon_0>0$,

$$A(x,\xi)=\sum_{\alpha,\beta:|\alpha|\le m,|\beta|\le m} a_{\alpha,\beta}\xi^{\alpha+\beta} \tag{11.4.3}$$

is the complete (symmetric) symbol of an operator A of the form (11.1.1) with $h=1$ and the first condition (11.4.1) is assumed to be fulfilled for all $x\in X$ while the second condition (11.4.2) (which is surely stronger) is assumed to be fulfilled only in $\{x\in X, \rho\gamma\ge\epsilon_0, \gamma\le\epsilon_0\}$. However, in the certain cases this condition is assumed to be fulfilled in a part of X which larger than this. Again we have assumed that

$$\varrho=\rho^{2m} \tag{11.4.4}$$

for the sake of simplicity. We leave to the reader to consider the more general case. In what follows we treat lower order terms which are sums of coercive

[29] Thus, it would be better to speak about senior symbols and junior terms.

and admissible indefinite (lower order) terms. For the Schrödinger operator (11.1.6) conditions (11.4.1) and (11.4.2) coincide with the condition

$$V \ge \epsilon_0 \rho^2 \qquad \forall x \tag{11.4.5}$$

provided condition (11.1.6) is fulfilled.

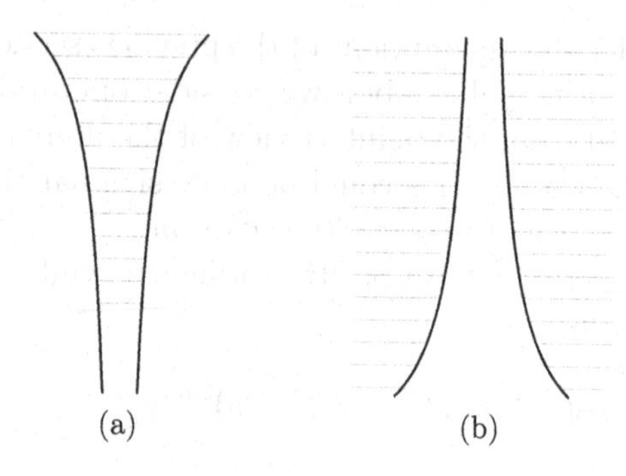

Figure 11.7: Non-coercive weakly singular potential (a) and coercive singular potential (b); blue lines show symbolically energy levels of the corresponding Schrödinger operator. For weakly singular potential the lowest energy level on (a) exists.

As before we set $\gamma_\tau = \gamma$ and $\rho_\tau = \rho + \tau^{\frac{1}{2}}$ and consider domain

$$X_{\mathsf{reg}} = \bigl\{ x \in X,\ \tau^{\frac{1}{2m}} \gamma(x) \ge 1,\ \exists \xi\, \exists v \ne 0 :\ \ \langle A(x,\xi)v, v\rangle \le (1+\epsilon_1)\tau \|v\|^2 \bigr\}$$

with an arbitrarily small constant $\epsilon_1 > 0$. Then condition (11.4.1) yields that $\rho \le C\tau^{\frac{1}{2m}}$.

As before we apply the results of Parts II and III in X_{reg} to obtain local semiclassical spectral asymptotics for $\mathsf{N}^-(\tau)$ (case $\vartheta > 0$ will be considered later) in the ball $\mathcal{B}(x, \gamma(x), X)$ with the remainder estimate $O\bigl(\tau^{\frac{1}{2m}(d-1)}\gamma^{d-1}\bigr)$ provided $\rho \le \epsilon_2 \tau^{\frac{1}{2m}}$, in which case operator is ξ-microhyperbolic.

Moreover, under condition to the Hamiltonian flow generated by the senior symbol this remainder estimate can be improved to $o\bigl(\tau^{\frac{1}{2m}(d-1)}\gamma^{d-1}\bigr)$ or sharper.

On the other hand, for balls with $\rho \ge \epsilon_2 \tau^{\frac{1}{2m}}$ the remainder estimate (as well as the principal part of the asymptotics) is $O\bigl(\tau^{\frac{d}{2m}}\gamma^d\bigr)$ in the general case; however it could be improved to $O\bigl(\tau^{\frac{1}{2m}(d-1+\delta)}\gamma^{d-1+\delta}\bigr)$ with an arbitrarily small exponent $\delta > 0$ for the scalar operators.

In order to obtain the remainder estimate $O\bigl(\tau^{\frac{1}{2m}(d-1)}\gamma^{d-1}\bigr)$ in the local spectral asymptotics in this zone one should check microhyperbolicity or another non-degeneracy condition there. In particular, for the Schrödinger operator with $d\geq 2$ we obtain this remainder estimate without any additional conditions. This improvement and condition will be important in the next sections but not now.

On the other hand, in the zone

$$X_+=\bigl\{x\in X,\ \rho\gamma\geq\epsilon_1, \forall\xi\ \forall v\ \langle A(x,\xi)v,v\rangle\geq(1+\epsilon_1)\tau\|v\|^2\bigr\}$$

we obtain the remainder estimate[30] $O(\rho^{-s}\gamma^{-s})$ with arbitrarily chosen exponent s provided either $B\bigl(x,\gamma(x)\bigr)$ does not intersect ∂X, or m-th order Dirichlet boundary condition is given on ∂X.

Moreover, one can obtain the same answer provided[31]

$$(Au,u)\geq\sum_{\alpha,\beta:|\alpha|,|\beta|\leq m}(a_{\alpha,\beta}D^\alpha u, D^\beta u)\quad\forall u\in\mathfrak{D}(A). \tag{11.4.6}$$

Let us note that $X_{\mathsf{reg}}\cup X_+\supset\bigl\{x\in X, \tau^{\frac{1}{2m}}\gamma\geq C_0\bigr\}$. Thus we include X_+ either into semiclassical zone X_{reg} or into variational zone X_{sing}.

Let us assume that condition (11.4.2) is fulfilled in the domain

$$\mathcal{X}\supset\{x\in X,\ \rho\gamma\leq\epsilon_1,\ \gamma\leq\epsilon_1\}.$$

Then for u supported in $\mathcal{X}$ the following estimate

$$\bigl((A-\tau I)u,u\bigr)\geq -C\tau\int_{\mathcal{X}_-}\|u\|^2dx$$

holds with $\mathcal{X}_-=\mathcal{X}\setminus\mathcal{X}_+$,

$$\mathcal{X}_+=\bigl\{x\in\mathcal{X},\ \sum_{\alpha,\beta:|\alpha|\leq m,|\beta|\leq m}\langle a_{\alpha,\beta}(x)w_\alpha, w_\beta\rangle\geq\tau\sum_{\alpha:|\alpha|\leq m}\|w_\alpha\|^2\quad\forall w_\alpha\bigr\}.$$

Surely $\mathcal{X}_+\supset\bigl\{x\in X,\rho\geq C\tau^{\frac{1}{2m}}\bigr\}$. Therefore we conclude that

(11.4.7) Contribution of $\mathcal{X}$ to the remainder does not exceed $C\tau^{\frac{d}{2m}}\operatorname{mes}(\mathcal{X}_-)$ provided either the Dirichlet boundary condition or any other condition providing (11.4.6) on $\partial X\cap\bar{\mathcal{X}}$ is given; in the latter case we assume that the cone condition is fulfilled.

---

[30] The principal part vanishes here.

[31] This inequality is a restriction on the boundary conditions.

Let us note that

Remark 11.4.1. $\mathcal{X}$ and X_+ overlap and therefore we can subtract from X_+ and add to $\mathcal{X}$ zone

$$\tilde{\mathcal{X}} = \{x \in X, \operatorname{dist}(x, X \setminus \mathcal{X}) \ge \epsilon_1 \gamma\}.$$

So we arrive to the estimate

$$|\mathsf{N}^-(\tau) - \mathsf{N}^{\mathsf{W}}(\tau)| \le C\bigl(R_1(\tau) + R_2(\tau) + R_3(\tau)\bigr) \tag{11.4.8}$$

with

$$\mathsf{N}^{\mathsf{W}}(\tau) = (2\pi)^{-d} \iint_{X_{\mathsf{reg}} \times \mathbb{R}^d} \boldsymbol{n}(x, \xi; \tau)\, dx d\xi \tag{11.4.9}$$

where $\boldsymbol{n}(x, \xi; \tau)$ is now the eigenvalue counting function for the complete symbol $A(x, \xi)$,

$$R_1(\tau) = \int_{X_{\mathsf{reg}}} \tau^{\frac{d-1}{2m}} \gamma^{-1}\, dx, \tag{11.4.10}$$

$$R_2(\tau) = \tau^{\frac{d}{2m}} \operatorname{mes}(\mathcal{X}_- \cup \tilde{\mathcal{X}}), \tag{11.4.11}$$

$$R_3(\tau) = \int_{X_+ \setminus \tilde{\mathcal{X}}} \rho^{d-s} \gamma^{-s}\, dx, \tag{11.4.12}$$

$\tilde{\mathcal{X}} = \{x \in X_{\mathsf{reg}},\ \rho \ge \epsilon_2 \tau^{\frac{1}{2m}}\}$ in the general case. This zone can be reduced (or even voided) under microhyperbolicity or other non-degeneracy condition; in particular $\tilde{\mathcal{X}} = \emptyset$ for the Schrödinger operator with $d \ge 2$.

Similarly, for $\vartheta > 0$ estimate

$$|\mathsf{N}^-_{(\vartheta)}(\tau) - \mathsf{N}^{\mathsf{W}}_{(\vartheta)}(\tau)| \le C\bigl(R_{1(\vartheta)}(\tau) + R_{2(\vartheta)}(\tau) + R_{3(\vartheta)}(\tau)\bigr) \tag{11.4.13}$$

holds with $\mathsf{N}^-_{(\vartheta)}(\tau) = \tau_+^{\vartheta-1} * \mathsf{N}^-(\tau)$, $\mathsf{N}^{\mathsf{W}}_{(\vartheta)}(\tau) = \tau_+^{\vartheta-1} * \mathsf{N}^{\mathsf{W}}(\tau)$, $\mathsf{N}^{\mathsf{W}}(\tau)$ containing now lower order terms as well,

$$R_{1(\vartheta)}(\tau) = \int_{X_{\mathsf{reg}}} \tau^{\frac{1}{2m}(d-1-\vartheta)+\vartheta} \gamma^{-1-\vartheta}\, dx, \tag{11.4.14}$$

$$R_{2(\vartheta)}(\tau) = \tau^{\frac{d}{2m}+\vartheta} \operatorname{mes}(\mathcal{X}_- \cup \tilde{\mathcal{X}}), \tag{11.4.15}$$

$$R_{3(\vartheta)}(\tau) = \int_{X_+ \setminus \tilde{\mathcal{X}}} \rho^{2m\vartheta+d-s} \gamma^{-s}\, dx. \tag{11.4.16}$$

In particular, we have proven

Proposition 11.4.2. *Let*

$$\gamma^{-1-\vartheta} \in \mathscr{L}^1(X), \tag{11.4.17}$$

$$\rho^{2m\vartheta+d-s}\gamma^{-s} \in \mathscr{L}^1(X \setminus \tilde{\mathcal{X}}) \tag{11.4.18}$$

and

$$\mathsf{mes}(\mathcal{X}_- \cup \tilde{\mathcal{X}}) = O\bigl(\tau^{-\frac{1}{2m}(\vartheta+1)}\bigr); \tag{11.4.19}$$

then

$$\mathsf{N}^-_{(\vartheta)}(\tau) = \mathsf{N}^{\mathsf{W}}_{(\vartheta)}(\tau) + O\bigl(\tau^{\frac{1}{2m}(d-1-\vartheta)+\vartheta}\bigr) \qquad \textit{as } \tau \to +\infty. \tag{11.4.20}$$

For the Schrödinger operator with $d \geq 2$ and other scalar operators with $\vartheta > 0$ condition (11.4.19) is fulfilled automatically, but in the general case this condition is very unpleasant and it is best to check the microhyperbolicity condition for $(A - \tau)$ in $\tilde{\mathcal{X}}_-$ and then reduce this zone.

Remark 11.4.3. Let us discuss the principal differences between weakly and strongly singular potentials:

(i) For strongly singular potentials coercivity condition is required; without it operator may be even not essentially self-adjoint;

(ii) For weakly singular potentials small vicinity of the singular point is a singular zone; for strongly singular potential it belongs to a classically forbidden (due to coercivity) part of regular zone;

(iii) Finally, in the general case the principal parts of the asymptotics $\mathsf{N}^{\mathsf{W}}(\tau)$ and $\mathsf{N}^{\mathsf{W}}_{(\vartheta)}(\tau)$ cannot be decomposed into a power series of $\tau^{\frac{1}{2m}}$ as in the previous section. Usually this decomposition leads to much worse remainder estimate than the original one.

11.4.2 Sharp Remainder Estimates

Let us discuss an improvement of the remainder estimate in the zone $X_{\mathsf{reg}} \cap \{\rho \leq C\tau^{\frac{1}{2m}}\}$; thus we consider the term $R_1(\tau)$ or $R_{1(\vartheta)}(\tau)$. Let us write our results as a series of remarks:

Remark 11.4.4. (i) Let the standard non-periodicity condition to the (billiard) Hamiltonian trajectories is fulfilled in X_{reg} for the symbol $a(x,\xi)$ on the energy level 1.

Then for any fixed $\bar{\gamma} > 0$ one can replace the contribution of zone $\{X, \gamma(x) \ge \bar{\gamma}\}$ to the remainder by $R^*_{1(\vartheta)}(\tau) = o(R_{1(\vartheta)}(\tau))$.

(ii) Further, let condition (11.4.17) be fulfilled; then $R_{1(\vartheta)}(\tau) \asymp \tau^{\frac{1}{2m}(d-1-\vartheta)+\vartheta})$ and we can replace it by $R^*_{1(\vartheta)}(\tau) = o(\tau^{\frac{1}{2m}(d-1-\vartheta)+\vartheta})$.

So, under condition (11.4.17) and improved assumption (11.4.19)

$$\operatorname{mes}(\mathcal{X}_- \cup \tilde{\mathcal{X}}) = o(\tau^{-\frac{1}{2m}(\vartheta+1)}) \tag{11.4.19*}$$

we can improve asymptotics (11.4.20) to

$$\mathsf{N}^-_{(\vartheta)}(\tau) = \mathsf{N}^{\mathsf{W}}_{(\vartheta)}(\tau) + o(\tau^{\frac{1}{2m}(d-1-\vartheta)+\vartheta}) \qquad \text{as } \tau \to +\infty. \tag{11.4.20*}$$

In particular, for $\vartheta = 0$ the two-term asymptotics holds with the remainder estimate $o(\tau^{(d-1)/2m})$.

(iii) All the examples of Subsection 11.3.1 remain true. However, the principal part of these asymptotics should be given by the formulas of the present section.

Remark 11.4.5. (i) Let the enhanced non-periodicity condition to the (billiard) Hamiltonian trajectories is fulfilled in X_{reg} for the symbol $a(x, \xi)$ on the energy level 1. Namely, let us assume that this flow is of either exponential or power type and that the corresponding non-periodicity condition is fulfilled. Then one can replace $R_{1(\vartheta)}(\tau)$ by

$$R^{**}_{1(\vartheta)}(\tau) = \int_{X_{\mathsf{reg}}} \tau^{\frac{1}{2m}(d-1-\vartheta)+\vartheta} \gamma^{-1-\vartheta} \bigl(\log(\tau^{1/2m}\gamma)\bigr)^{-1-\vartheta}\, dx, \tag{11.4.14**}$$

or

$$R^{***}_{1(\vartheta)}(\tau) = \int_{X_{\mathsf{reg}}} \tau^{\frac{1}{2m}(d-1-\vartheta)+\vartheta} \gamma^{-1-\vartheta} (\tau^{1/2m}\gamma)^{-\delta}\, dx \tag{11.4.14***}$$

respectively.

(ii) So, under condition (11.4.17) and further improved assumption (11.4.19)

$$\operatorname{mes}(\mathcal{X}_- \cup \tilde{\mathcal{X}}) = O(\tau^{-\frac{1}{2m}(\vartheta+1)} |\log \tau|^{-1-\vartheta}) \tag{11.4.19**}$$

or

$$\operatorname{mes}(\mathcal{X}_- \cup \tilde{\mathcal{X}}) = O(\tau^{-\frac{1}{2m}(\vartheta+1)-\delta}) \tag{11.4.19***}$$

we can improve asymptotics (11.4.20) to

$$\mathsf{N}^{-}_{(\vartheta)}(\tau) = \mathsf{N}^{\mathsf{W}}_{(\vartheta)}(\tau) + O\bigl(\tau^{\frac{1}{2m}(d-1-\vartheta)+\vartheta}|\log\tau|^{-1-\vartheta}\bigr) \tag{11.4.20**}$$

or

$$\mathsf{N}^{-}_{(\vartheta)}(\tau) = \mathsf{N}^{\mathsf{W}}_{(\vartheta)}(\tau) + O\bigl(\tau^{\frac{1}{2m}(d-1-\vartheta)+\vartheta-\delta}\bigr) \tag{11.4.20***}$$

respectively.

In particular, for $\vartheta = 0$ the two-term asymptotics holds with the remainder estimate $O(\tau^{(d-1)/2m}(\log\tau)^{-1})$ or $O(\tau^{(d-1)/2m-\delta})$ respectively.

Remark 11.4.6. In the context of Subsection 11.3.2 (periodic billiard Hamiltonian flow) under condition (11.4.19)*** asymptotics

$$\mathsf{N}^{-}(\tau) = \mathsf{N}^{\mathsf{W}}(\tau) + \tau^{\frac{d-1}{2m}}\bigl(\varkappa_1 + F(\tau^{\frac{1}{2m}})\bigr) + O\bigl(\tau^{\frac{d-1}{2m}-\delta}\bigr) \tag{11.4.21}$$

holds as $\tau \to +\infty$.

Remark 11.4.7. Non-periodicity condition may be useful even with weakened assumptions (11.4.17)–(11.4.19).

11.4.3 Examples

Let us consider some examples. We start from the Schrödinger operator $\Delta + V(x)$ with potential $V(x) = \kappa|x|^{2\mu}$, $\kappa = \mathsf{const} \ne 0$, $\mu \le -1$, singular at origin, gradually adding the complexity. Note that as $\mu < -1$ operator V is not Δ-bounded (see Definition 10.2.4) and as $\mu = -1$ it may be Δ-bounded[32] but it's Δ-bound is not 0.

So, as $\mu < -1$, we *must* assume that $0 \notin X$ and $\kappa > 0$, defining operator $\Delta + V$ through quadratic form; as $\mu = -1$ we may[32] allow $\kappa \ge \kappa_0$ with $\kappa_0 < 0$.

Then we can consider the case of singularity at origin when $\gamma(x) \asymp |x|$ and $\rho \asymp |x|^{\mu}$ with $\mu \le -1$ and to generalize it to singularity at $\{x, x'' = 0\}$ of codimension d''.

Example 11.4.8 (Power singularities). (i) Let all the conditions be fulfilled with $\gamma \asymp |x''|$, $\rho \asymp |x''|^{\mu}$, $\mu \le -1$. There one can take $\mathcal{X} = \emptyset$ for $\mu < -1$ and $\mathcal{X} = \{x \in X, |x''| \le \epsilon_0\}$ for $\mu = -1$.

[32] It is the case unless $d = 2$ and 0 is surrounded by X.

Then in the general case we should take the transitional zone $\tilde{\mathcal{X}} = \{\epsilon\tau \le |x''|^{2m\mu} \le C\tau\}$ arriving to

(a) For $d'' > \vartheta + 1$, $\mu \ge -d''(\vartheta+1)^{-1}$ asymptotics (11.4.20) holds.

Moreover, for $\mu > -d''(\vartheta+1)^{-1}$ under the standard non-periodicity condition to the Hamiltonian trajectories asymptotics (11.4.20)* holds.

Note that for $\vartheta = 0$ and $\mu > -d''$ one can replace full Weyl expression (including all terms of A) $\mathsf{N}^{\mathsf{W}}(\tau)$ by the standard two-term expression (one needs the second term only for sharp asymptotics) $\varkappa_0\tau^{\frac{d}{2m}} + \varkappa_1\tau^{\frac{d-1}{2m}}$ with the standard coefficients $\varkappa_0, \varkappa_1$ and with the same remainder estimate as above.

On the other hand, for $\mu = -d''$, $\mathsf{N}^{\mathsf{W}}(\tau) = \varkappa_0\tau^{\frac{d}{2m}} + O(\tau^{\frac{d-1}{2m}}\log\tau)$ in the general case and $\mathsf{N}^{\mathsf{W}}(\tau) = \varkappa_0\tau^{\frac{d}{2m}} + O(\tau^{\frac{d-1}{2m}})$ in the case of a scalar principal symbol.

(b) For $d'' = \vartheta + 1$, $\mu = -1$ asymptotics

$$\mathsf{N}^-_{(\vartheta)}(\tau) = \mathsf{N}^{\mathsf{W}}_{(\vartheta)}(\tau) + O\bigl(\tau^{\frac{1}{2m}(d-1-\vartheta)+\vartheta}\log\tau\bigr) \tag{11.4.22}$$

holds and for $\vartheta = 0$, $d'' = 1$, $\mu = -1$ one can replace $\mathsf{N}^{\mathsf{W}}(\tau)$ by $\varkappa_0\tau^{\frac{d}{2m}}$.

(c) If $\mu < -d''(\vartheta+1)^{-1}$ then asymptotics

$$\mathsf{N}^-_{(\vartheta)}(\tau) = \mathsf{N}^{\mathsf{W}}_{(\vartheta)}(\tau) + O\bigl(\tau^{\frac{d}{2m}+\frac{d''}{2m\mu}+\vartheta}\bigr) \tag{11.4.23}$$

holds.

Let us note that $\mu \ge -d''(\vartheta+1)^{-1}$, $\mu \le -1$ yield $d'' \le \vartheta + 1$ and therefore only the listed cases are possible.

(ii) One can slightly improve the above results in the number of cases. For the Schrödinger operator with $d \ge 2$ and general scalar operators with $\vartheta > 0$ one can extend X_{reg} and take $\tilde{\mathcal{X}} = \emptyset$. Moreover, one can easily see that if X is a conic with respect to the x'' domain in a neighborhood of $\{x'' = 0\}$ and

$$D^\alpha\bigl(a_{\beta',\beta''} - a^0_{\beta',\beta''}\bigr) = o\bigl(\rho^{2m-|\beta'|-|\beta''|}\gamma^{-|\alpha|}\bigr) \qquad \forall\beta',\beta''\quad \forall\alpha : |\alpha| \le 1 \tag{11.4.24}$$

as $|x''| \to 0$ with functions $a^0_{\beta',\beta''}$ which are positively homogeneous of degrees $(2m - |\beta'| - |\beta''|)\mu$ then[33] checking the microhyperbolicity condition with

$$\ell = \langle\xi, \partial_\xi\rangle + \mu^{-1}\langle x'', \partial_{x''}\rangle$$

[33] We call conditions similar to (11.4.24) *stabilization conditions.*

we see that we can take $\tilde{X}=\emptyset$ as well.

In all these we obtain that

(a) For $d''>\vartheta+1$ asymptotics (11.4.20) holds. Moreover, under the standard non-periodicity condition to Hamiltonian trajectories asymptotics (11.4.20)* holds.

(b) For $d''=\vartheta+1$ asymptotics (11.4.22) holds and for $d''<\vartheta+1$ asymptotics (11.4.23) holds.

Let us note that in both cases for $\mu\le -d''$, even for the Schrödinger operator, it is possible that

$$\mathsf{N}^{\mathsf{W}}(\tau)-\varkappa_0\tau^{\frac{d}{2m}}-\varkappa_1\tau^{\frac{d-1}{2m}}\asymp \tau^{\frac{d}{2m}+\frac{d''}{2m\mu}}$$

and therefore for $\mu<-d''$ we cannot replace $\mathsf{N}^{\mathsf{W}}(\tau)$ by $\varkappa_0\tau^{\frac{d}{2m}}$ and preserve the remainder estimate. Moreover, for $\mu=-d''$ we cannot replace $\mathsf{N}^{\mathsf{W}}(\tau)$ by $\varkappa_0\tau^{\frac{d}{2m}}+\varkappa_1\tau^{\frac{d-1}{2m}}$ and preserve the remainder estimate $o\bigl(\tau^{\frac{d-1}{2m}}\bigr)$.

Example 11.4.9 (Power singularities again). Let us consider Schrödinger operator with $d\ge 2$ [34] and let us assume that in a neighborhood of $\{x''=0\}$ domain X coincides with $X^0=\{x=(x',r,\theta)\in\mathbb{R}^{d''}\times\mathbb{R}^+\times U\}$ where $U\subset\mathbb{S}^{d''-1}$ is a $\tilde{\gamma}$-admissible domain with $\tilde{\gamma}=\tilde{\gamma}(x',\theta)$.

Let us assume that $g^{jk}=g^{jk}(x',\theta)$, $V_j=v_j(x',\theta)|x''|^{\mu}$, $V=v(x',\theta)|x''|^{2\mu}$ and $\mu\le -1$. Further, let us assume that $v\ge 0$ and the functions g^{jk}, v_j and v on $\mathbb{R}^{d'}\times\mathbb{S}^{d''-1}$ satisfy standard conditions (11.1.6)–(11.1.8) and (11.4.5) with $\tilde{\rho}=\sqrt{v}$ and with $\tilde{\gamma}$ instead of ρ and γ. Then

(i) Under additional assumption $|v_j|\le c\tilde{\rho}$ one can take $\rho=\tilde{\rho}|x''|^{\mu}$ and $\gamma=\epsilon_1\tilde{\gamma}|x''|$ with small enough constant $\epsilon_1>0$.

(a) Then we obtain that under assumption

$$\tilde{\gamma}^{-1-\vartheta}\in\mathscr{L}^1(\mathbb{R}^{d'}\times U) \tag{11.4.25}$$

asymptotics (11.4.20) holds for $d''>\vartheta+1$.

(b) Moreover, under the standard non-periodicity condition to the Hamiltonian trajectories asymptotics (11.4.20)* holds.

[34] One can consider general scalar operators for $\vartheta>0$.

(c) On the other hand, under condition (11.4.25) for $d'' = \vartheta + 1$ and $d'' < \vartheta + 1$ we obtain asymptotics (11.4.22) and (11.4.23) respectively.

(ii) Without condition $|v_j| \le c\tilde{\rho}$ we cannot use variational estimates. In this case we should assume that $\mu < -1$ and

$$\tilde{\rho}\tilde{\gamma} \ge \epsilon_0 > 0. \tag{11.4.26}$$

These conditions forbid the degeneration of v but permit certain singularities. Then all the statements of the previous assertion remain true.

(iii) Let $\mu < -1$, condition (11.4.26) be fulfilled and $V_j = v_j(x', \theta)|x''|^{\mu'}$ with $\mu' < \mu$ (the case $\mu' > \mu$ is trivial).

Then one can take $\rho = \tilde{\rho}|x'|^{\mu}$, $\gamma = \epsilon_1\tilde{\gamma}|x''|^{-\mu'+\mu+1}$. Then $\rho\gamma \ge \epsilon_2|x''|^{-\sigma}$ with $\sigma > 0$ provided $2\mu + 1 < \mu'$.

(a) Then, under assumption (11.4.25) we obtain asymptotics (11.4.20) for $d'' > (1+\vartheta)(1+\mu-\mu')^{-1}$.

(b) Moreover, under the standard non-periodicity condition to Hamiltonian trajectories we obtain asymptotics (11.4.20)*.

(c) On the other hand, for $d'' = (1+\vartheta)(1+\mu-\mu')^{-1}$ we obtain asymptotics (11.4.22) and for $d'' < (1+\vartheta)(1+\mu-\mu')^{-1}$ we obtain asymptotics

$$\mathsf{N}^{-}_{(\vartheta)}(\tau) = \mathsf{N}^{\mathsf{W}}_{(\vartheta)}(\tau) + O(\tau^{\frac{d-\eta}{2}+\vartheta}) \tag{11.4.27}$$

with $\eta = \mu^{-1}\bigl((1+\vartheta)(1+2\mu-\mu') - d''\bigr) \in (0, 1+\vartheta)$.

Here we cannot consider the case $\mu' \le 2\mu + 1$ because the magnetic field is too strong near the origin. This case is left to the next part.

Let us note that every time assertions with large enough s are used, K (in the conditions on coefficients) depends on μ, μ', etc. (via s).

Example 11.4.10 (Exponential singularities). Let X, $g^{jk} = g^{jk}(x', \theta)$, $v_j = v_j(x', \theta)$ and $v = v(x', \theta)$ be as in statements (i) or (ii) of the previous Example 11.4.9. Let $V_j = v_j \exp(|x''|^{-\nu})$, $V = v\exp(2|x''|^{-\nu})$ with $\nu > 0$.

Then all the conditions will be fulfilled with $\rho = \tilde{\rho}\exp(|x''|^{-\nu})$ and $\gamma = \epsilon_1\tilde{\gamma}|x''|^{\nu+1}$. We obtain that

(a) For $(1+\nu)(1+\vartheta) < d''$ asymptotics (11.4.20) holds and under the standard non-periodicity condition to Hamiltonian trajectories asymptotics (11.4.20)* holds.

(b) Further, for $(1+\nu)(1+\vartheta) = d''$ asymptotics

$$\mathsf{N}^-_{(\vartheta)}(\tau) = \mathsf{N}^{\mathsf{W}}_{(\vartheta)}(\tau) + O\bigl(\tau^{\frac{d-1+\vartheta}{2}} \log\log\tau\bigr) \tag{11.4.28}$$

holds and

(c) For $(1+\nu)(1+\vartheta) > d''$ asymptotics

$$\mathsf{N}^-_{(\vartheta)}(\tau) = \mathsf{N}^{\mathsf{W}}_{(\vartheta)}(\tau) + O\bigl(\tau^{\frac{d-1+\vartheta}{2}} \log^{\sigma}\tau\bigr) \tag{11.4.29}$$

holds with $\sigma = \nu^{-1}\bigl((1+\nu)(1+\vartheta) - d''\bigr)$.

We leave it to the reader the following series of rather easy exercises:

Problem 11.4.11. (i) Consider examples in the spirit of Example 11.4.9(iii) with exponential singularities, i.e., with $V_j = v_j|x''|^{-l}\exp(|x''|^{-\nu})$ or with $V_j = v_j\exp(k|x''|^{-\nu})$, $k\in(1,2)$.

(ii) For different examples estimate $\bigl(\mathsf{N}^{\mathsf{W}}(\tau) - \varkappa_0\tau^{\frac{d}{2}}\bigr)$ from above and below.

(iii) Consider quasiconical domains and domains with cusps for potentials which grow at 0 or infinity.

(iv) Consider general scalar operators with $\vartheta > 0$.

11.4.4 Indefinite Perturbations

Now we consider the case when a strongly singular coercive potential is perturbed by a less singular indefinite potential. Let us assume that conditions (11.1.3)–(11.1.4) are fulfilled with $\rho+\rho'$ instead of ρ and with $\varrho = (\rho+\rho')^{2m}$ where ρ and ρ' are γ-admissible weight functions. We denote such conditions by (11.1.3)′–(11.1.4)′.

Meanwhile we replace conditions (11.4.1) and (11.4.2) by conditions

$$\langle A(x,\xi)v,v\rangle \ge \epsilon_0\bigl((|\xi|+\rho)^{2m} - C\rho'^{2m}\bigr)\|v\|^2 \qquad \forall\xi\quad\forall v \tag{11.4.1$'$}$$

and

$$\sum_{\alpha,\beta:|\alpha|\le m,|\beta|\le m} \langle a_{\alpha,\beta}(x)v_\alpha, v_\beta\rangle \ge \epsilon_0 \sum_{\alpha:|\alpha|\le m} \rho^{2m-2|\alpha|}\|v_\alpha\|^2 - C\sum_{\alpha:|\alpha|\le m-1} \rho'^{2m-2|\alpha|}\|v_\alpha\|^2 \qquad \forall v=(v_\alpha)_{\{|\alpha|\le m\}} \tag{11.4.2$'$}$$

respectively.

Remark 11.4.12. Then the arguments of this and previous sections show that all the estimates of this section remain true with all domains defined by ρ if to their right-hand expressions we add

$$R'(\tau) = C\Bigl(\int_{\mathcal{Z}} (\rho'^{2m})_+^q\, dx\Bigr)^{\frac{d}{2mq}} \times \bigl(\mathrm{mes}(\mathcal{Z})\bigr)^{1-\frac{d}{2mq}} \tag{11.4.30}$$

for $\vartheta=0$, and

$$R'_{(\vartheta)}(\tau) = C\int_0^\infty t^{\vartheta-1}\Bigl(\int_{\mathcal{Z}} (\rho'^{2m}-t)_+^q\, dx\Bigr)^{\frac{d}{2mq}}\, dt \times \bigl(\mathrm{mes}(\mathcal{Z})\bigr)^{1-\frac{d}{2mq}} \tag{11.4.31}$$

for $\vartheta>0$ where $q\ge \frac{d}{2m}$, $q\ge 1$ and one of these inequalities is strict, and

$$\mathcal{Z}=\{x\in X, \rho\le C\rho'\}.$$

Example 11.4.13. Consider a power singularity under the assumptions of Example 11.4.9(i) with potentials V, V_j replaced by $V+V'$, $V_j+V'_j$ respectively where $V'=w(x',\theta)|x''|^{2\nu}$, $V'_j=w_j(x',\theta)|x''|^{\nu}$, w and w_j are also functions on $\mathbb{S}^{d''-1}\times\mathbb{R}^{d'}$ satisfying conditions (11.1.6)–(11.1.8) with $\tilde\rho'$, $\tilde\gamma$.

Here $\tilde\rho'$ is a $\tilde\gamma$-admissible weight function on $\mathbb{S}^{d''-1}\times\mathbb{R}^{d'}$ and $\nu>\mu$, $\mu\le -1$, $\nu<0$. Let us set $\gamma=\epsilon_1\tilde\gamma|x''|$, $\rho=\tilde\rho|x''|^{\mu}$, $\rho'=\tilde\rho'|x''|^{\nu}$ with $\tilde\rho=\sqrt{v}$. Then term (11.4.30) or (11.4.31) should be added to the right-hand expressions of our estimates, and one can see easily that this term does not exceed existing remainder estimate provided one of the following conditions is fulfilled:

$$\iint_{\{\tilde\rho'^2\ge\epsilon v\}} \tilde\rho'^{-a} v^{\frac{b}{2}}\, d\theta dx' <\infty \qquad \text{if } b<0,\ d\ge 3 \tag{11.4.32$_1$}$$

with $a=(d''+d\mu+2\vartheta\mu)(\nu-\mu)^{-1}$, $b=(d''+d\nu+2\vartheta\nu)(\nu-\mu)^{-1}$,

$$\iint_{\{\tilde\rho'^2\ge\epsilon v\}}\tilde\rho'^{\,d+2\vartheta}\bigl(\log\frac{\tilde\rho'^2}{v}\bigr)_+\,d\theta dx'<\infty \qquad \text{if}\ \ b=0,\ d\ge 3 \tag{11.4.32$_2$}$$

$$\iint_{\{\tilde\rho'^2\ge\epsilon v\}}\tilde\rho'^{\,d+2\vartheta}\,d\theta dx'<\infty \qquad \text{if}\ \ b>0,\ d\ge 3, \tag{11.4.32$_3$}$$

$$\iint_{\{\tilde\rho'^2\ge\epsilon v\}}\tilde\rho'^{\,-a-\delta\bar a}v^{\frac{b}{2}+\delta}\,d\theta dx'<\infty \qquad \text{if}\ \ b\le 0,\ d=2, \tag{11.4.32$_4$}$$

$$\iint_{\{\tilde\rho'^2\ge\epsilon v\}}\tilde\rho'^{\,d+2\vartheta+\delta}\,d\theta dx'<\infty \qquad \text{if}\ \ b>0,\ d=2, \tag{11.4.32$_5$}$$

where $\delta>0$ is arbitrarily small, $\bar a=\mu(\nu-\mu)^{-1}$, $\bar b=\nu(\nu-\mu)^{-1}$.

We leave to the reader

Problem 11.4.14. Modify and analyze the same way the other examples of this section.

11.4.5 Operators with Indefinite Principal Symbols

We leave to the reader

Problem 11.4.15. Analyze operators with indefinite principal symbols.

In this case one should replace conditions (11.4.1), (11.4.2) by

$$\|A(x,\xi)v\|\ge\epsilon_0(|\xi|+\rho)^{2m}\|v\| \qquad \forall\xi\quad\forall v \tag{11.4.1*}$$

and

$$\|\sum_{\alpha:|\alpha|\le 2m}a_\alpha(x)v_\alpha\|\ge\epsilon_0\sum_{\alpha:|\alpha|\le 2m}\rho^{2m-|\alpha|}\|v_\alpha\| \qquad \forall v=(v_\alpha)_{\{\alpha:|\alpha|\le 2m\}} \tag{11.4.2*}$$

respectively. Moreover, one should replace conditions (11.4.1)$'$, (11.4.2)$'$ by

$$\|A(x,\xi)v\|\ge\epsilon_0\bigl((|\xi|+\rho)^{2m}-C\rho'^{2m}\bigr)\|v\| \qquad \forall\xi\quad\forall v \tag{11.4.1$^{*\prime}$}$$

and

$$(11.4.2)^{*\prime} \quad \|\sum_{\alpha:|\alpha|\le 2m} a_\alpha(x)v_\alpha\| \ge \epsilon_0 \sum_{\alpha:|\alpha|\le 2m} \rho^{2m-|\alpha|}\|v_\alpha\| - C\sum_{\alpha:|\alpha|\le 2m-1} \rho'^{2m}\|v_\alpha\| \qquad \forall v=(v_\alpha)_{\{\alpha:|\alpha|\le 2m\}}$$

respectively.

Unfortunately we are not aware of any examples of this type which are interesting from the physical point of view. The reader can easily construct artificial examples (like generalized Dirac operator) and analyze them.

11.5 Large Eigenvalues for Operators in Essentially Unbounded Domains

Now we consider *essentially unbounded domains* which are domains containing massive[35] exits to infinity. In domains of this type the spectrum of the operator in question is discrete only because the operator contains lower degree terms which are coercive with the function ρ tending to ∞ at infinity (in almost all directions).

Therefore even the order of the asymptotics of $\mathsf{N}^{\mathsf{W}}(\tau)$ depends on the lower degree terms (which are not anymore "lower order terms" from our point of view).

We treat here only operators with positive definite senior symbols and leave the indefinite case to the reader.

11.5.1 Standard Results

We begin with potentials of the power growth at infinity. As the first example we use only the results of Parts II and III to obtain the following well known assertion:

Theorem 11.5.1. *Let X be an unbounded connected domain in $\mathbb{R}^d$ and let conditions* (11.1.1)–(11.1.3), (11.4.1) *and* (11.4.4) *be fulfilled with* $\gamma=\epsilon_1\langle x\rangle$, $\rho=\langle x\rangle^\mu$, $\mu>0$ *where here and below* $\langle x\rangle=(|x|^2+1)^{\frac{1}{2}}$.

[35] E.g., conical. Domains with cusps even having infinite volume do not qualify as essentially unbounded and will be treated in the next chapter.

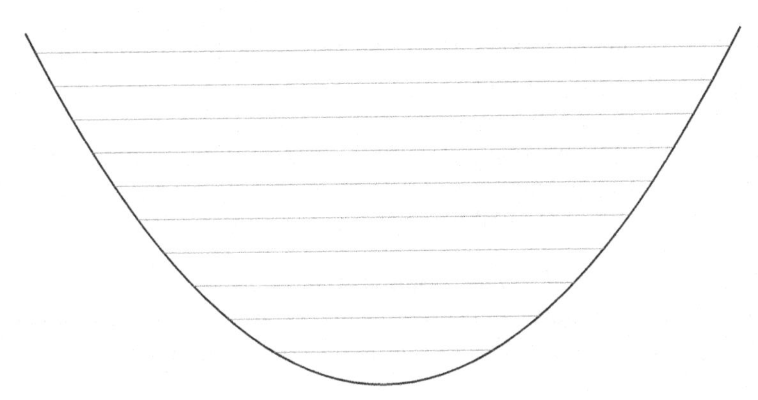

Figure 11.8: Coercive at infinity potential; blue lines show symbolically energy levels of the corresponding Schrödinger operator.

(i) Let us assume that for every $\tau > c$ *one of the non-degeneracy conditions is fulfilled after rescaling for the symbol* $A(x,\xi)\tau^{-1}$ *on the energy level* 1 *uniformly with respect to* τ, x, ξ *(this assumption is only restrictive for* $|x| \ge C_0$*). Then asymptotics*

$$\mathsf{N}^-(\tau) = \mathsf{N}^{\mathsf{W}}(\tau) + O(\tau^{l(d-1)}) \tag{11.5.1}$$

holds as $\tau \to +\infty$ *where* $\mathsf{N}^{\mathsf{W}}(\tau)$ *is given by* (11.4.9) *and* $\mathsf{N}^{\mathsf{W}} \asymp \tau^{ld}$, $l = \frac{1}{2m\mu}(\mu+1)$.

(ii) Let us assume that for every $\tau > c$ *one of the corresponding non-degeneracy conditions*[36] *is fulfilled for the symbol* $A(x,\xi)\tau^{-1}$ *on the energy level* 1 *uniformly with respect to* τ, x, ξ. *Then for* $\vartheta \ge 0$ *asymptotics*

$$\mathsf{N}^-_{(\vartheta)}(\tau) = \mathsf{N}^{\mathsf{W}}_{(\vartheta)}(\tau) + O(R_{(\vartheta)}(\tau)) \tag{11.5.2}$$

holds as $\tau \to +\infty$ *with*

$$R_{(\vartheta)}(\tau) = \begin{cases} \tau^{l(d-1)+\vartheta(1-l)} & \text{for } \vartheta < d-1, \\ \tau^{d-1}\log\tau & \text{for } \vartheta = d-1, \\ \tau^{\frac{1}{2m}(d-1-\vartheta)+\vartheta} & \text{for } \vartheta > d-1 \end{cases} \tag{11.5.3}$$

where $\mathsf{N}^{\mathsf{W}}_{(\vartheta)}(\tau) \asymp \tau^{ld+\vartheta}$.

[36] Recall that for operators with the scalar symbols some very weak non-degeneracy condition is needed only as $\vartheta = 0$.

Proof. Proof trivially follows from the corresponding results of Parts II and III (also see Introduction). □

Remark 11.5.2. (i) Obviously domain $\{|x| \asymp \tau^{1/(2m\mu)}\}$ provides the main contribution to the asymptotics. Further, as $\vartheta < \vartheta^* = (d-1)$ this domain also provides the main contribution to the remainder estimate.

(ii) However as $\vartheta > \vartheta^*$ inner domain $\{|x| \asymp 1\}$ provides the the main contribution to the remainder estimate while for $\vartheta = \vartheta^*$ intermediate domain $\{\epsilon^{-1} \le |x| \le \epsilon\tau^{1/(2m\mu)}\}$ provides the main contribution to the remainder estimate.

Therefore to improve remainder estimate one needs non-periodicity assumption for the corresponding Hamiltonian flow.

Theorem 11.5.3. *Let assumptions of Theorem 11.5.1 be fulfilled. Moreover, let $X \cap \{x \ge c_1\} = X^0 \cap \{x \ge c_1\}$ where X^0 is a conical (with respect to x) domain with $\partial X^0 \setminus 0 \in \mathscr{C}^K$ and let the stabilization conditions*

$$D^\alpha(a_\beta - a^0_\beta) = o(|x|^{(2m-|\beta|)\mu - |\alpha|}) \qquad \forall\beta : |\beta| \le 2m \quad \forall\alpha : |\alpha| \le 1 \tag{11.5.4}$$

be fulfilled as $|x| \to \infty$ with functions $a^0_\beta \in \mathscr{C}^K(\bar{X}^0 \setminus 0)$ which are positive homogeneous of degrees $(2m - |\beta|)\mu$ with respect to x.

(i) Further, let us assume that on the energy level $\tau = 1$ for the quasihomogeneous symbol

$$A^0(x,\xi) = \sum_{|\beta|\le 2m} a^0_\beta(x)\xi^\beta \tag{11.5.5}$$

the standard non-periodicity condition to the Hamiltonian trajectories in T^*X^0 is fulfilled and that these permitted trajectories[37] *never hit $0 \times \mathbb{R}^d$.*

Then for $\vartheta < \vartheta^* = (d-1)$ asymptotics

$$\mathsf{N}^-_{(\vartheta)}(\tau) = \mathsf{N}^{\mathsf{W}}_{(\vartheta)}(\tau) + o(R_{(\vartheta)}(\tau)) \tag{11.5.6}$$

holds $\tau \to +\infty$ with $R_{(\vartheta)}(\tau) = \tau^{l(d-1)+\vartheta(1-l)}$ according to (11.5.3).

[37] I.e. trajectories originated at $\{A^0(x,\xi) = 1\} \setminus \Lambda$ where Λ is an exceptional set.

(ii) On the other hand, let $\vartheta > \vartheta^*$, and the standard non-periodicity condition to Hamiltonian trajectories be fulfilled on the energy level $\tau = 1$ for the symbol

$$a(x,\xi) = \sum_{|\beta|=2m} a_\beta(x)\xi^\beta \tag{11.5.7}$$

in the domain X. Then asymptotics (11.5.6) also holds as $\tau \to +\infty$ but this time with $R_{(\vartheta)}(\tau) = \tau^{\frac{1}{2m}(d-1-\vartheta)+\vartheta}$ according to (11.5.3).

(iii) Finally, let $\vartheta = \vartheta^*$ and stabilization condition (11.5.4) and the standard non-periodicity condition to Hamiltonian trajectories is fulfilled on the energy level $\tau = 1$ for the symbol

$$a^0(x,\xi) = \sum_{|\beta|=2m} a^0_\beta(x)\xi^\beta \tag{11.5.8}$$

in the domain X^0 and these permitted[37)] trajectories never hit $0 \times \mathbb{R}^d$. Then asymptotics (11.5.6) holds as $\tau \to +\infty$ but this time with $R_{(\vartheta)}(\tau) = \tau^{d-1}\log\tau$ according to (11.5.3).

Remark 11.5.4. (i) If only stabilization condition (11.5.4) is fulfilled then the microhyperbolicity condition is automatically fulfilled for the symbol $A(x,\xi)\tau^{-1}$ on the energy level 1 and

$$\mathsf{N}^{\mathsf{W}}_{(\vartheta)}(\tau) = \varkappa_{0(\vartheta)}\tau^{ld}\bigl(1+o(1)\bigr) \tag{11.5.9}$$

as $\tau \to +\infty$.

(ii) As $\vartheta = 0$ asymptotics must include the second term $\varkappa_1\tau^{l(d-1)}$ where for the Schrödinger operator,

$$\varkappa_0 = (2\pi)^{-d}\omega_d \int_{X^0} (1-V^0)_+^{\frac{d}{2}}\sqrt{g^0}\,dx \tag{11.5.10}$$

and

$$\varkappa_1 = \mp\frac{1}{4}(2\pi)^{1-d}\omega_{d-1}\int_{\partial X^0} (1-V^0)_+^{\frac{d-1}{2}}\,dS^0 \tag{11.5.11}$$

for Dirichlet and Neumann boundary conditions respectively, dS^0 is the Riemannian density on ∂X^0 generated by the induced Riemannian metrics, $g^0 = \det(g^{jk0})^{-1}$.

Proof of Theorem 11.5.3. (i) One can see easily that as $\vartheta < \vartheta^*$ the contribution of zone $\{|x| \le \varepsilon\tau^{\frac{1}{2m\mu}}\}$ to the remainder does not exceed $\varepsilon^\delta R_{(\vartheta)}(\tau)$ and therefore only zone $\{\varepsilon\tau^{\frac{1}{2m\mu}} \le |x| \le C_0\tau^{\frac{1}{2m\mu}}\}$ should be considered: in all cases zone $\{|x| \ge C_0\tau^{\frac{1}{2m\mu}}\}$ is elliptic.

Let us apply the theorems of Parts II and III and we examine the behavior of the billiards generated by the symbol $A(x,\xi)$ on $\Sigma_\tau = \{(x,\xi), \tau \in \operatorname{Spec} A(x,\xi)\}$. We take

$$\Lambda_1 = \Lambda_{1,\tau,\epsilon,T} = \{(x,\xi) \in \Sigma_\tau, |x| \ge \varepsilon\tau^{\frac{1}{2m\mu}}, \\ |x\tau^{-\frac{1}{2m\mu}} - y| + |\xi\tau^{-\frac{1}{2m}} - \eta| \ge \epsilon \quad \forall (y,\eta) \in \Lambda_T^0\} \tag{11.5.12}$$

where $\Lambda_T^0 = \Sigma^0 \setminus \Lambda_T^1$, $\Sigma^0 = \{(y,\eta), 1 \in \operatorname{Spec} A^0(y,\eta)\}$ and for every point of Λ_T^1 the Hamiltonian frow of A^0 (with reflection from ∂X^0) for time T in an appropriate time direction does not enter ε-vicinity of $0 \times \mathbb{R}^d$ and periodicity is violated with at least divergence ε; we fix arbitrary small ε in advance.

By assumption $\operatorname{mes}(\Lambda_T^0) = 0$, on the other hand, Λ_T^0 is a relatively closed subset of $\Sigma^0 \setminus \{x = 0\}$ for every T. The remaining arguments repeat those in the proof of Theorem 11.1.1: we check that the conditions on billiards generated by $A^0(x,\xi)$ and $A(x,\xi)$ coincide provided $\tau \ge C_{\varepsilon,T}$.

(ii) On the other hand, for $\vartheta > \vartheta^*$ the principal contribution to the remainder estimate is given by the zone $\{x \in X, |x| \le r\}$ (with arbitrarily large fixed r) in the sense that the contribution of the remaining part does not exceed $r^{-\delta} R_{(\vartheta)}(\tau)$.

Therefore, to prove Statement (ii) one should treat Hamiltonian billiards in this zone, in which one can replace $A(x,\xi)$ by $a(x,\xi)$ for large τ.

(iii) Finally, for $\vartheta = \vartheta^*$ the principal contribution to the remainder estimate is given by the zone $\{r \le |x| \le \varepsilon\tau^{\frac{1}{2m\mu}}\}$ for every fixed ε and r in the sense that the contribution of the remaining part does not exceed $(\varepsilon^\delta + r^{-\delta})R_{(\vartheta)}(\tau)$.

Therefore, to prove Statement (iii) one should treat Hamiltonian billiards in this zone, in which one can replace $A(x,\xi)$ by $a^0(x,\xi)$ for large τ, $|x|$. □

Let us consider scalar operators. Observing that the microhyperbolicity condition is automatically fulfilled for $|x| \le \epsilon\tau^{\frac{1}{2m\mu}}$ and recalling the results of Chapter 5 we arrive immediately to

Theorem 11.5.5. *Let X be an unbounded connected domain in $\mathbb{R}^d$ and let conditions* (11.1.1)–(11.1.3), (11.4.1), (11.4.4) *be fulfilled with $\gamma = \epsilon_1\langle x\rangle$, $\rho = \langle x\rangle^\mu$, $\mu > 0$.*

Moreover, let A be a scalar operator. Then

(i) Asymptotics

$$\mathsf{N}^-(\tau) = \mathsf{N}^{\mathsf{W}}(\tau) + O\bigl(\tau^{l(d-1)+\delta}\bigr) \tag{11.5.13}$$

holds as $\tau \to +\infty$ with an arbitrarily small exponent $\delta > 0$ and asymptotics (11.5.2) *holds for $\vartheta > 0$.*

(ii) Moreover, for $0 < \vartheta < \vartheta^* = (d-1)$, $\vartheta > \vartheta^*$, $\vartheta = \vartheta^*$ respectively Statements (i)–(iii) of Theorem 11.5.3 remain true.

11.5.2 Strongly Non-Periodic Hamiltonian Flow

In what follows we consider homogeneous coefficients with singularities and (what is more essential here) with degenerations. However, we now improve the remainder estimates under stronger conditions to Hamiltonian trajectories.

Theorem 11.5.6. *Let the conditions of Theorem 11.5.1 be fulfilled. Let A be a scalar operator. Assume that*

(11.5.14) *Either $X^0 = \mathbb{R}^d$ or A is the Schrödinger operator or its power.*

Further, let the strong stabilization condition

$$D^\alpha(a_\beta - a^0_\beta) = O\bigl(|x|^{2m-|\beta|)\mu-|\alpha|-\delta}\bigr) \qquad \forall \beta : |\beta| \le 2m \quad \forall \alpha : |\alpha| \le 1 \tag{11.5.4*}$$

be fulfilled as $|x| \to \infty$ with some $\delta > 0$. Furthermore, let us assume that the Hamiltonian flow (with reflections at ∂X^0) is of non-periodic power type[38] *and that along trajectories $(x(t), \xi(t))$ with $(x(0), \xi(0)) \in \Lambda^0_{\varepsilon,T} \subset \Sigma_1$, $|x(0)| = 1$, $\varepsilon = T^{-M}$, condition $|x(t)| \ge T^{-M_1}$ is fulfilled.*

[38] We are aware of no example when the flow is of non-periodic exponential type but is not of non-periodic power type. We would appreciate construction of such example. We leave to the reader to formulate the statement in the case of non-periodic exponential type.

Then asymptotics

$$\mathsf{N}^-(\tau) = \mathsf{N}^{\mathsf{W}}(\tau) + \bigl(\varkappa_1 + O(\tau^{-\delta'})\bigr)\tau^{l(d-1)} \tag{11.5.15}$$

holds as $\tau \to +\infty$*. Moreover, for* $0 < \vartheta \le d - 1 - \delta$ *asymptotics*

$$\mathsf{N}^-_{(\vartheta)}(\tau) = \mathsf{N}^{\mathsf{W}}_{(\vartheta)}(\tau) + O\bigl(\tau^{l(d-1)+\vartheta(1-l)-\delta'}\bigr) \tag{11.5.16}$$

holds as $\tau \to +\infty$ *with* $\delta' = \delta'(d, m, \delta, l) > 0$.

Example 11.5.7. (i) Growth and non-periodicity conditions of Theorem 11.5.6 to the Hamiltonian flow are fulfilled for the Schrödinger operator with $X^0 = \mathbb{R}^d$ and $A^0(x,\xi) = \bigl(|\xi|^2 + |x|^{2\mu}\bigr)$, $0 < \mu \ne 1$.

(ii) Therefore the statement of Theorem 11.5.6 is true for this operator (provided all the other conditions of this theorem, including (11.5.4)*, are fulfilled) and for the $2m$-order operator with $A^0(x,\xi) = (|\xi|^2 + |x|^{2\mu})^m$, and one can replace $X^0 = \mathbb{R}^d$ by the X^0 obtained after (a sequence of) the following transformations:

(a) We take the factor manifold of X^0 with respect to a finite rotation group (with center at the origin),

(b) We consider the reflection $S : \bar{X}^0 \to \bar{X}^0$ with respect to a hyperplane and remove from X^0 some subset $\mathcal{K}$ of this plane provided $\tilde{\gamma}^{-1-\vartheta} \in \mathscr{L}^1(\mathbb{S}^{d-1})$ with $\tilde{\gamma} = \mathsf{dist}(x, \mathcal{K})$.

Proof. Obviously it is sufficient to prove Statement (i). Moreover, the potential $V^0 = |x|^{2\mu}$ is spherically symmetric and therefore an angular momentum is constant along every Hamiltonian trajectory and therefore the x-projection of this trajectory lies in some 2-dimensional plane. Therefore it is sufficient to treat only the case $d = 2$. We introduce polar coordinates (r, θ) and use the fact that the *angular momentum* $M = \dot{\theta} r^2$ is constant where $\dot{\theta} = \frac{d\theta}{dt}$, etc.

Therefore the energy is equal to

$$E = \frac{1}{2}\dot{r}^2 + \frac{1}{2}M^2 r^{-2} + r^{2\mu} = \frac{1}{2}\dot{r}^2 + W(r, M) \tag{11.5.17}$$

and is also constant and therefore r obeys equation of oscillations

$$\ddot{r} = -W'(r, M) \tag{11.5.18}$$

with the *effective potential* $W(r) = \frac{1}{2}M^2 r^{-2} + r^{2\mu}$.

Then a trajectory with $|M| \ge \varepsilon$ lies permanently in the zone $\{(\theta, r), r \ge M\sqrt{\frac{2}{E}}\}$. One can easily see that

$$|M| \le M_0(E) = (2\mu)^{\frac{1}{2}} (\frac{E}{\mu+1})^{\frac{1}{2\mu}(\mu+1)} \tag{11.5.19}$$

and every trajectory oscillates between the largest and the smallest values $r = r_2(E, M)$ and $r = r_1(E, M)$ respectively where $r_1 < r_2$ [39] are roots of the equation $W(r, M) = E$ (i.e. $\frac{1}{2}M^2 r^{-2} + r^{2\mu} = E$) and the period of the oscillation is

$$T_\sqcap = T_\sqcap(E, M) = 2\int_{r_1(E,M)}^{r_2(E,M)} \bigl(2E - 2W(r, M)\bigr)^{-\frac{1}{2}}\, dr \tag{11.5.20}$$

while the increment of θ during a period is

$$\theta_\sqcap = 2\int_{r_1(E,M)}^{r_2(E,M)} Mr^{-2}\bigl(2E - 2W(r, M)\bigr)^{-\frac{1}{2}}\, dr. \tag{11.5.21}$$

One can easily see that

(11.5.22) If $\varepsilon \le |M| \le M_0(E) - \varepsilon$ and $E \asymp 1$ then

$$|r_2(E, M) - r_1(E, M)| \ge \epsilon\varepsilon^p, \quad r_1(E, M) \ge \epsilon\varepsilon^p, \quad T_\sqcap(E, M) \ge \epsilon\varepsilon^p, \quad \theta_\sqcap \ge \epsilon\varepsilon^p$$

and the functions $f(E, M) = r_1(E, M), r_2(E, M), T_\sqcap(E, M), \theta_\sqcap(E, M)$ satisfy the inequalities

$$|D^\beta_{E,M} f| \le C_{|\beta|}\varepsilon^{-p|\beta|} \qquad \forall\beta$$

with appropriate constants $p, \epsilon > 0$ and $C_{|\beta|}$; $\varepsilon > 0$ is a small parameter here.

On the other hand,

(11.5.23) The $dxd\xi : dA^0$-measure of the set of points

$$\{(r, \theta) : E = 1, |M| \le \varepsilon \text{ or } E = 1, |M| \ge M_0(E) - \varepsilon\}$$

is less than $C\varepsilon^\delta$ with $\delta > 0$.

[39] And $r_1(E, M) = r_2(E, M)$ if and only if $|M| = M_0(E)$.

These assertions (11.5.22) and (11.5.2) yield that the Hamiltonian flow is of power type. Let us prove that as $\mu \neq 1$ this flow is of non-periodic power type.

The arguments above yield that it is sufficient to prove that

(11.5.24) For $k, n \in \mathbb{Z}^+, m \le \varepsilon^{-1}, n \le \varepsilon^{-1}$ the measure of the set of points (x,ξ) with $E(x,\xi) = 1$ such that

$$|\theta_\sqcap(1, M(x,\xi)) - 2\pi\frac{k}{n}| \le \varepsilon^L \tag{11.5.25}$$

does not exceed $O(\varepsilon^{\frac{L}{q}})$ where L is arbitrarily large.

Obviously, $\theta_\sqcap(1, M)$ is an analytic function of $M \in (0, M_0(1))$; further, $\theta_\sqcap(1, M_0(1) - \eta^2)$ is an analytic function of $\eta \in (-\epsilon_0, \epsilon_0)$.

Furthermore, it is well known (see, e.g., V. I. Arnold [2]) that

(11.5.26) For $\mu > 0, \mu \neq 1$ function $\theta_\sqcap(1, M)$ is not constant.

Therefore the measure of points (x,ξ) with $E(x,\xi) = 1$ and $|M(x,\xi)| \ge \epsilon_1$, satisfying (11.5.25) is $O(\varepsilon^{\frac{L}{q}})$ as $\varepsilon \to +0$. Here $\epsilon_1 > 0$ is an arbitrary constant.

On the other hand, one can easily see that $\theta_\sqcap$ is continuous at $[0, M_0(1)]$ and $\theta_\sqcap(+0) = \pi$. Therefore in the zone $\{(x,\xi), 0 < M(x,\xi) \le \epsilon\epsilon_0\}$ inequality (11.5.25) can be fulfilled only with $k \asymp n$.

Further, one can easily see that for $M \in (0, \epsilon_0)$, $\theta_\sqcap(1, M) = \mathcal{F}(M, M^{2\mu})$ where $\mathcal{F}$ is an analytic function of two variables in a neighborhood of $(0, 0)$. Furthermore, $\theta_\sqcap(1, M) \neq \mathsf{const}$ and therefore $|\nabla_M \theta_\sqcap(1, M)| \ge \epsilon_1 M^p$ for $M \in (0, \epsilon_1)$ with appropriate $\epsilon_1 > 0$ and p and therefore the measure of the set of points with $E(x,\xi) = 1$ and $M(x,\xi) \in (0, \epsilon_1)$ satisfying (11.5.25) is $O(\varepsilon^{\frac{L}{p}})$. □

Remark 11.5.8. (i) For $\vartheta > d - 1 + \delta$ one can easily improve the remainder estimate $O(\tau^{\frac{1}{2m}(d-1\vartheta)+\vartheta})$ to $O(\tau^{\frac{1}{2m}(d-1\vartheta)+\vartheta-\delta'})$ under a corresponding condition on the Hamiltonian flow of $a(x,\xi)$.

In order to reach uniformity one can assume that stabilization condition (11.5.4)* is fulfilled only for terms with $|\alpha| = 2m$. In this case only these terms should be scalar but if some junior terms are not scalar one should check that some non-degeneracy condition is fulfilled.

(ii) Moreover, one can consider the case $\vartheta = d-1$. It is easy to replace the factor $\log\tau$ by $\log^{1-\delta'}\tau$ but better improvements are possible only under much more complicated and restrictive conditions.

11.5.3 Periodic Hamiltonian Flow

Let A be a scalar operator and condition (11.5.14) be fulfilled. Further, let us assume that

(11.5.27) The Hamiltonian trajectories of $A^0(x,\xi)$ are periodic with period $T(A^0(x,\xi))$ and

$$A = A^{0\mathrm{w}}(x,D) + A'^{\mathrm{w}}(x,D), \tag{11.5.28}$$

where

$$D^\beta(a'_\alpha - a^1_\alpha) = O\bigl(|x|^{(2m-|\alpha|)\mu-\delta_1-\delta'-|\beta|}\bigr) \qquad \forall\alpha:|\alpha|\le 2m \quad \forall\beta:|\beta|\le K \qquad \text{as } |x|\to\infty \tag{11.5.29}$$

where $a^1_\alpha(x)$ are positively homogeneous of degrees $(2m-|\alpha|)\mu-\delta_1$ with respect to x and are smooth away from 0, $0<\delta_1<2(\mu+1)$, $\delta'>0$.

Let us introduce function

$$b(x,\xi) = \frac{l}{T_0}\int_0^{T_0} A^1\circ\Psi_t\,dt \qquad \text{with } A^1(x,\xi)=\sum_{|\alpha|\le 2m} a^1_\alpha(x)\xi^\alpha \tag{11.5.30}$$

on the surface $\Sigma^0=\bigl\{(x,\xi), A^0(x,\xi)=1\bigr\}$; recall that $l=\frac{1}{2m\mu}(\mu+1)$ and $T_0=T(1)$.

It follows from the quasihomogeneity of A^0 with respect to (x,ξ) that the action vanishes[40].

Then the results of Sections 6.2, 8.3 immediately yield the following theorem:

---

[40] In fact, the action is equal to

$$\frac{1}{T_0}\int_0^{T_0}\bigl(k\langle\xi,A^0_\xi\rangle + (1-k)\langle x,A^0_x\rangle - A\bigr)\circ\Psi_t\,dt$$

with an arbitrary constant k. Let us take $k=\mu(\mu+1)^{-1}$. Then the Euler identity yields that the integrand vanishes for $l=1$. On the other hand, we must assume that $l=1$: otherwise we should replace A by A^l in order to obtain $T=T_0$ for all (x,ξ).

Theorem 11.5.9. *Let conditions* (11.5.14) *and* (11.5.27)–(11.5.30) *be fulfilled. Further, let us assume that*

$$\operatorname{mes}_\Sigma\bigl(\{(x,\xi)\in\Sigma, |b(x,\xi)-\omega|\le\varepsilon\}\bigr)=O(\varepsilon^\delta)\qquad as\ \varepsilon\to+0 \tag{11.5.31}$$

with $\delta>0$ *and*

$$\begin{aligned}\operatorname{dist}(\Psi_t(x,\xi),(x,\xi))\ge t(T_0-t)\gamma(x,\xi)\\ \forall(x,\xi)\in\Sigma\quad\forall t\in(0,T_0)\end{aligned} \tag{11.5.32}$$

with $\gamma\ge0$ *such that* $\operatorname{mes}\bigl(\{(x,\xi),\gamma(x,\xi)\le\varepsilon\}\bigr)=O(\varepsilon^\delta)$ *as* $\varepsilon\to+0$.

Then

(i) If $\delta_1<\mu+1$ *and condition* (11.5.31) *is fulfilled for every* $\omega\in\mathbb{R}$ *then asymptotics*

$$\mathsf{N}^-(\tau)=\mathsf{N}^{\mathsf{W}}(\tau)+\kappa_1\tau^{l(d-1)}+O\bigl(\tau^{l(d-1)-\delta''}\bigr) \tag{11.5.33}$$

holds as $\tau\to+\infty$ *with* $\delta''>0$.

(ii) If $\delta_1\in[\mu+1,2\mu+2)$ *then for* $\tau\to+\infty$ *such that* $\tau^l\in\bigcup_{n\in\mathbb{Z}^+}J_n$ *asymptotics*

$$\mathsf{N}^-(\tau)=\mathsf{N}^{\mathsf{W}}(\tau)+\bigl(\kappa_1+F(\tau)\bigr)\tau^{l(d-1)}+O\bigl(\tau^{l(d-1)-\delta''}\bigr) \tag{11.5.34}$$

holds with $\delta''>0$ *where*

$$J_n=\bigl[\nu(n+\frac{1}{4}\omega_0)+(\omega-\epsilon_1)(\nu n)^{l-\frac{\delta_1}{2m\mu}},\nu(n+\frac{1}{4}\omega_0)+(\omega+\epsilon_1)(\nu n)^{l-\frac{\delta_1}{2m\mu}}\bigr],$$

ω_0 *is the Maslov index of the trajectory* $\Psi_t(x,\xi)$ *with* $t\in(0,T_0)$, $\nu=\frac{2\pi l}{T_0}$ *and* $\epsilon_1>0$ *is small enough,*

$$\begin{aligned}F(\tau)=(2\pi)^{-d}lT_0^{-1}\int_\Sigma\Upsilon\Bigl(T_0l^{-1}\bigl((\tau^l-\frac{1}{4}\pi\omega_0)-\tau^{l-\frac{\delta_1}{2m\mu}}b\bigr)\Bigr)\times\\ dxd\xi:dA^0.\end{aligned} \tag{11.5.35}$$

Let us consider the perturbed harmonic oscillator $A_0=\frac{1}{2}(|D|^2+|x|^2)$ in $\mathbb{R}^d$ more carefully. In this case $\mu=l=m=1$, $T_0=2\pi$, $\omega_0\equiv 2d\mod 4\mathbb{Z}$ and the x-projections of the Hamiltonian trajectories are ellipses centered at the origin.

Moreover,

(11.5.36) $\operatorname{Spec} A_0 = \{\tau_n = \frac{d}{2} + n, n \in \mathbb{Z}^+\}$. The multiplicity of τ_n is $P(n)$ where $P(.)$ is a polynomial of degree $d-1$.

Functional-analytic arguments yield that

(11.5.37) If the perturbing operator is a potential V with $\|V\|_{\mathscr{L}^\infty} \le \epsilon$ then the spectrum consists of clusters of length $\le \epsilon$.

We leave to the reader

Problem 11.5.10. Prove that for generic functions a^1_α ($|\alpha| \le k$) which are positively homogeneous of degree $(2-|\alpha|-\delta_1)$ condition (11.5.31) is fulfilled for every ω.

For such perturbations the theory of Section 6.2 yields that

(11.5.38) Asymptotics (11.5.33) holds for $0 < \delta_1 < 2$ and asymptotics (11.5.34)–(11.5.35) holds for $\delta_1 \ge 2$ where $\delta_k > 0$ ($k = 0, 1, 2$) are now arbitrary.

In particular, one can take $k = 0$ and a perturbing potential $V' = v_0 r^{2\mu-\delta_1}$ with arbitrary $v_0 \ne 0$ and $0 < \delta_1 \ne 2$.

Let us consider the case $b \equiv 0$. Then we have two possibilities.

The first (rather trivial) possibility is to assume that $A = A^{0\mathrm{w}} + A^{1\mathrm{w}} + A''^{\mathrm{w}}$ where A'' (instead of A') satisfies condition (11.5.29) with a^2_α, δ_2 instead of a^1_α, δ_1 respectively, $\delta_2 \in (\delta_1, 2\delta_1)$. Then one can construct the symbol $b(x, \xi)$ by formula (11.5.30) with A^2 instead of A^1, assume that condition (11.5.31) is fulfilled and obtain asymptotics (11.5.33) or (11.5.34)–(11.5.35) with δ_2 instead of δ_1.

However, for $\delta_2 = 2\delta_1$ the situation is more complicated because in this case b contains a term which is quadratic with respect to A^1.

So, let us assume that

$$A^1 \circ j^* = -A^1 \tag{11.5.39}$$

where j^* is the antipodal map $(x, \xi) \to (-x, -\xi)$. Repeating the arguments of the proof of Theorem 11.3.21 we again obtain formula (11.3.27):

$$b = \frac{1}{2\pi}\Bigl(\int_0^{2\pi} A^2 \circ \Psi_t \, dt + \int_0^{\pi}\int_0^{\pi} \{A^1 \circ \Psi_t, A^1 \circ \Psi_{t+t'}\}\, dtdt'\Bigr) \tag{11.5.40}$$

is a positive homogeneous symbol of degree $(2-2\delta_1)$ with respect to (x, ξ).

Therefore we have proven

Theorem 11.5.11. *Let $X^0 = \mathbb{R}^d$, $A = A^{0w} + A^{1w} + A''^{w}$ where A'' satisfies (11.5.29) with a^2_α and $\delta_2 > 0$ instead of a^1_α and δ_1. Here a^1_α and a^2_α are functions smooth for $x \ne 0$ and positively homogeneous of degrees $(2-|\alpha|-\delta_1)$, $(2-|\alpha|-\delta_2)$, and $\delta_2 = 2\delta_1$.*

Further, let condition (11.5.39) be fulfilled and let the symbol b be defined by (11.5.40). Then

(i) If $\delta_2 < 2$ and condition (11.5.31) is fulfilled for every $\omega \in \mathbb{R}$ then asymptotics (11.5.33) holds for $\tau \to +\infty$.

(ii) If $\delta_2 \ge 2$ and condition (11.5.31) is fulfilled for some $\omega \in \mathbb{R}$ then asymptotics (11.5.34)–(11.5.35) holds for $\tau \to +\infty$ and belonging to the intervals J_n indicated in Theorem 11.5.9(ii) with $\omega_0 = 2(d-1)$, $\nu = 1$. In these asymptotics as well as in the definition of the intervals J_n one should replace δ_1 with δ_2.

We leave to the reader

Problem 11.5.12. (i) Prove that for generic positively homogeneous functions a^2_α ($|\alpha| \le k$) with $k = 0, 1, 2$ condition (11.5.33) is fulfilled for all $\omega \in \mathbb{R}$.

(ii) Consider Riesz means with $\vartheta < \vartheta^* = (d-1)$.

Remark 11.5.13. It does not look right that in our results here $\delta'' > 0$ is unspecified small exponent. Later under certain conditions we will bring our results to be more sharp (up to being consistent with results of Section 6.2.

11.5.4 Singularities and Degenerations

Let us consider operators with homogeneous terms with singularities and (what is now more interesting) degenerations. In order to avoid non-degeneracy conditions we consider only the Schrödinger operator.

Theorem 11.5.14. *(i) Let $d \ge 3$, $X = \{x = (r, \theta) \in \mathbb{R}^+ \times U\}$ where $U \subset \mathbb{S}^{d-1}$. Let $g^{jk} = g^{jk}(\theta)$, $V_j = v_j(\theta) r^\mu$, $V = v(\theta) r^{2\mu}$ with $\mu > 0$.*

Further, let $v \ge 0$ and let U, g^{jk}, v_j, v and $\tilde\rho = \sqrt{v}$, $\tilde\gamma = \tilde\gamma(\theta)$ satisfy conditions (11.1.6)–(11.1.6) and (11.4.5).

Finally, let

$$V^{-\frac{d}{2\mu}},\, V^{-\frac{d-1}{2\mu}}\tilde{\gamma}^{-1} \in \mathscr{L}^1(U). \tag{11.5.41}$$

Then asymptotics (11.5.1) *holds and one can substitute* $\mathsf{N}^{\mathsf{W}}(\tau)$ *by* $\varkappa_0\tau^{ld}$ *with* $l = \frac{1}{2\mu}(\mu+1)$ *and with the coefficient* $\varkappa_0$ *given by* (11.5.10).

Moreover, if $0 < \vartheta < \vartheta^* = (d-1)$ *and*

$$V^{-\frac{1}{2\mu}(d-1-\vartheta)}\tilde{\gamma}^{-1-\vartheta} \in \mathscr{L}^1(U) \tag{11.5.42}$$

then asymptotics (11.5.2)–(11.5.3) *holds.*

(ii) Let us assume that there exists a set $\Xi \subset \Sigma = \{(x,\xi), a(x,\xi) = 1\}$ *such that* $\operatorname{mes}_\Sigma(\Xi) = 0$ *and through every point of* $\Sigma \setminus \Xi$ *there passes an infinitely long non-periodic Hamiltonian billiard generated by the Hamiltonian*

$$H^0 = \tilde{\gamma}|x|\bigl(g^{jk}(\xi_j - V_j)(\xi_k - V_k) + V - 1\bigr). \tag{11.5.43}$$

Then asymptotics (11.5.6) *holds and one can substitute* $\mathsf{N}^{\mathsf{W}}(\tau)$ *by* $\varkappa_0\tau^{ld}$.

Moreover, if $0 < \vartheta < \vartheta^*$ *and*

$$V^{-\frac{1}{2\mu}(d-1-\vartheta)}\tilde{\gamma}^{-1-\vartheta} \in \mathscr{L}^1(U) \tag{11.5.44}$$

then asymptotics (11.5.6), (11.5.3) *holds.*

Proof. One can easily prove this theorem by the same method as Theorem 11.5.1 but also with analysis in the zone $X_{\mathsf{sing}} = \{\gamma\cdot(\tau^{\frac{1}{2}} + \rho) \le C_0\}$ with $\gamma = \tilde{\gamma}r$, $\rho = \tilde{\rho}r^\mu$.

In the calculation of $\mathsf{N}^{\mathsf{W}}(\tau)$ we use spherical coordinates and the change of variable $r = \tau^{\frac{1}{2\mu}}r'$. □

Remark 11.5.15. It should be noted that if in (11.5.43) we skip the regularizing factor $\tilde{\gamma}r$ in H^0 (i.e., replace H^0 by H^0_{st}) then the conditions of the Statement (ii) do not change because it follows from the assumptions of the Statement (i) that the set of the points through which pass Hamiltonian billiards generated by H^0_{st} and reaching $\{\tilde{\gamma}r = 0\}$ in a finite time is a set of measure 0.

We leave to the reader the following rather easy

Problem 11.5.16. Generalize Theorem 11.5.14 to operators different from the Schrödinger operator.

Example 11.5.17. Let $X = \mathbb{R}^d$, $d \geq 3$ and as $|x| \geq 1$ $V \asymp |x''|^{2\nu}|x|^{2\mu-2\nu}$ where $x = (x'; x'') = (x_1, \dots, x_{d'}; x_{d'+1} \dots, x_d)$, $d'' = d - d'$, $0 < d'' < d$ (in which case one should try $\rho(x) = |x''|^\nu \cdot |x|^{\mu-\nu}$, $\gamma(x) = \frac{1}{2}|x''|$).

Let us assume that the conditions to g^{jk}, V_j, V are fulfilled with the indicated γ and ρ. Then

(i) $v^{-\frac{d}{2\mu}} \in \mathscr{L}^1(\mathbb{S}^{d-1})$ and $v^{-\frac{1}{2\mu}(d-1-\vartheta)}\tilde{\gamma}^{-1-\vartheta} \in \mathscr{L}^1(\mathbb{S}^{d-1})$ if and only if

$$\nu < \nu_0^* = \mu \frac{d''}{d} \tag{11.5.45}$$

and

$$\nu < \nu_{1+\vartheta}^* = \mu(d'' - 1 - \vartheta)(d - 1 - \vartheta)^{-1} \tag{11.5.46}$$

respectively.

Therefore for $\nu < \mu(d''-1)(d-1)^{-1}$ asymptotics (11.5.1)[41] holds, and for $0 < \vartheta < d'' - 1$ under assumption (11.5.46) asymptotics (11.5.2)–(11.5.3) [42] holds.

(ii) Let $\vartheta = 0$. In order to treat the case $\nu \in [\nu_1^*, \nu_0^*]$ when condition (11.5.45) holds (or reaches the borderline value) and (11.5.46) fails, recall that the original remainder estimate was

$$R(\tau) = \int_{\{V \leq C\tau, \sqrt{\tau}\gamma \geq 1\}} \tau^{\frac{d-1}{2}} \gamma^{-1}\, dx + \int_{\{V \leq C\tau, \sqrt{\tau}\gamma \leq 1\}} \tau^{\frac{d}{2}}\, dx.$$

Introducing spherical coordinates and integrating over r we obtain the remainder estimate

$$R(\tau) = \tau^{l(d-1)} \int_{\{v \leq C\tau^{\mu+1}\tilde{\gamma}^{2\mu}\}} v^{-\frac{d-1}{2\mu}} \tilde{\gamma}^{-1}\, d\theta + \tau^{ld} \int_{\{v \geq C\tau^{\mu+1}\tilde{\gamma}^{2\mu}\}} v^{-\frac{d}{2\mu}}\, d\theta.$$

We can replace v by $|x''|^{2\nu} \cdot |x|^{2\mu-2\nu}$ here in which case we arrive to asymptotics

$$\mathsf{N}^-(\tau) = \mathsf{N}^{\mathsf{W}}(\tau) + O(\tau^{l(d-1)} \log \tau) \tag{11.5.47}$$

[41] As well as asymptotics (11.5.6) under the standard condition on Hamiltonian trajectories.

[42] As well as asymptotics (11.5.6), (11.5.3) under the standard condition on Hamiltonian trajectories. Both these footnotes remain true below.

as $\tau \to +\infty$ provided $\nu = \nu_1^*$ and the asymptotics

$$\mathsf{N}^-(\tau) = \mathsf{N}^{\mathsf{W}}(\tau) + O\big(\tau^{l(d-\eta)}\big) \tag{11.5.48}$$

as $\tau \to +\infty$ provided $\nu \in (\nu_1^*, \nu_0^*)$ and $\eta = (-\nu d + \mu d'')(\mu - \nu)^{-1} \in (0, 1)$.

Here again one can replace $\mathsf{N}^{\mathsf{W}}(\tau)$ by $\varkappa_0 \tau^{ld}$ with $\varkappa_0$ given by (11.5.10).

(iii) Similarly, for $\nu = \nu_{\vartheta+1}^* = \mu(d'' - \vartheta - 1)(d - \vartheta - 1)^{-1}$ asymptotics

$$\mathsf{N}_{(\vartheta)}^-(\tau) = \mathsf{N}_{(\vartheta)}^{\mathsf{W}}(\tau) + O\big(\tau^{l(d-1-\vartheta)+\vartheta}\log \tau\big) \tag{11.5.49}$$

holds as $\tau \to +\infty$ and for $\nu \in (\nu_{\vartheta+1}^*, \nu_0^*)$ asymptotics

$$\mathsf{N}_{(\vartheta)}^-(\tau) = \mathsf{N}_{(\vartheta)}^{\mathsf{W}}(\tau) + O\big(\tau^{l(d-\eta)+\vartheta}\big) \tag{11.5.50}$$

holds as $\tau \to +\infty$ with $\eta = \big((d'' - \vartheta)\mu - (d - \vartheta)\nu\big)(\mu - \nu)^{-1} \in (0, \vartheta + 1)$.

(iv) Finally, for $\nu = \nu_0^*$ asymptotics

$$\mathsf{N}^-(\tau) = \mathsf{N}^{\mathsf{W}}(\tau) + O(\tau^{ld}) \tag{11.5.51}$$

holds as $\tau \to +\infty$ where

$$\mathsf{N}^{\mathsf{W}}(\tau) = (2\pi)^{-d}\omega_d \int_{\gamma \ge \epsilon\tau^{-\frac{1}{2}}} (\tau - V)_+^{\frac{d}{2}} \sqrt{g}\, dx \tag{11.5.52}$$

and $\mathsf{N}^{\mathsf{W}}(\tau) \asymp \tau^{ld} \log \tau$ [43]. The results for $\vartheta > 0$ obviously follow from this statement.

In order to prove this statement one needs only note that in the domain $X_{\mathsf{sing}} = \{x, |x''| \le \epsilon\tau^{-\frac{1}{2}}\}$ one can take $\tilde{\rho} = 0$ in the inequality

$$((Au - \tau), u) \ge \epsilon \|u\|_m^2 - C\|\tilde{\rho}^m u\|^2 \tag{11.5.54}$$

for all $u \in \mathfrak{D}(A)$ supported in $\bar{X}_{\mathsf{sing}}$. Moreover, this estimate yields that

(v) All the results of Statements (i)–(iv) remain true for $d = 2$.

---

[43] Moreover, under stabilization condition (11.5.4)

$$\mathsf{N}^{\mathsf{W}}(\tau) = (\bar{\varkappa} + o(1))\tau^{ld} \log \tau. \tag{11.5.53}$$

This statement remains true below.

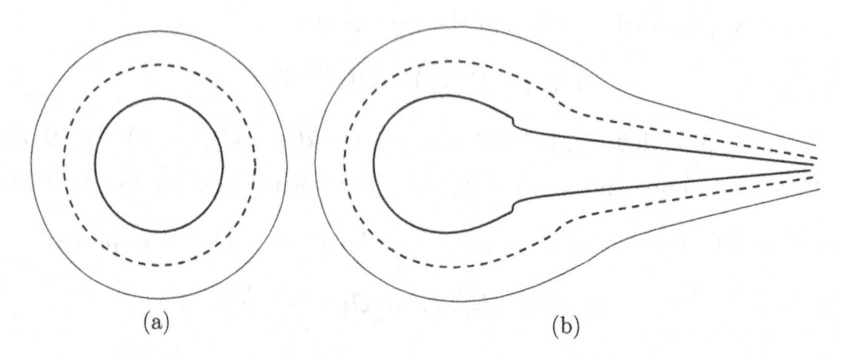

Figure 11.9: Level lines for non-degenerate (a) and degenerate in one direction (b) potential. In the latter case $V \le \tau$ forms a cusp but in contrast to cusps studied in Section 11.2 and 11.3 this cusp grows and fattens as $\tau \to +\infty$.

It is interesting to perturb the potentials treated above by potentials which have a lesser "average" growth at infinity bet does not have this degeneration. Namely, we can take perturbing potentials satisfying condition $(11.1.7)_3$ with the same γ as above and with $\rho' = (|x|+1)^{\mu'}$, $0 < \mu' < \mu$ instead of ρ:

$$|D^\alpha V'| \le c\rho'^2\gamma^{-|\alpha|} \tag{11.5.55}$$

Then two different situations may arise:

(a) The perturbing potential is also coercive i.e., satisfies (11.4.5) with ρ replaced by ρ'

$$V' \ge \epsilon_0\rho'^2 \qquad \forall x \tag{11.5.56}$$

in which case the perturbing potential "helps" to the unperturbed one. In this case the above analysis can be applied with no change.

(b) The perturbing potential is indefinite.

In the former case (a) we leave the detailed analysis to the reader and formulate the following statements:

Theorem 11.5.18 (Coercive perturbations). *(i) Let $d \geq 3$ and X, g^{jk}, V_j^0 and V^0 be as in Theorem 11.5.14. Let $V_j = V_j^0 + V_j'$, $V = V^0 + V'$ where V_j' and V' satisfy (11.5.55) and (11.5.56) with $\rho' = \langle x\rangle^{\mu'}$, $0 < \mu' < \mu$. Then asymptotics (11.5.1) holds with $\mathsf{N}^{\mathsf{W}}(\tau) = \varkappa_0 \tau^{ld}(1 + o(1))$ as $\tau \to +\infty$.*

(ii) Moreover, if the Hamiltonian H^0 given by (11.5.43) satisfies the conditions of the Statement (ii) of Theorem 11.5.14 then asymptotics (11.5.6), (11.5.3) holds as $\vartheta = 0$.

(iii) Finally, if $0 < \vartheta < (d - 1)$ and (11.5.42) is fulfilled then in the frameworks of (i) and (ii) asymptotics (11.5.2)–(11.5.3) and (11.5.6), (11.5.3) hold respectively.

Example 11.5.19. Let X, g^{jk}, V_j^0 and V^0 be as in Example 11.5.17. Let $V_j = V_j^0 + V_j'$, $V = V^0 + V'$ where V_j' and V' satisfy (11.5.55) and (11.5.56) with $\rho' = \langle x\rangle^{\mu'}$, $0 < \mu' < \mu$, and $\gamma = \epsilon|x''|$. Then

(i) If $\nu < \nu_{\vartheta+1}$ defined by (11.5.46), $0 \leq \vartheta < (d'' - 1)$ then asymptotics (11.5.2)–(11.5.3) holds.

(ii) In the case of threshold value $\nu = \nu_{\vartheta+1}$, $0 \leq \vartheta < (d'' - 1)$ asymptotics (11.5.47) holds.

(iii) In the case of above threshold value $\nu_{\vartheta+1} < \nu < \nu_0$, with ν_0 defined by (11.5.45), $0 \leq \vartheta < (d'' - 1)$ asymptotics (11.5.50) holds with η given in Example 11.5.17(ii)$'$ provided $\mu' \leq \mu'_* = (\mu - \nu)(\nu + 1)^{-1}$.

(iv) On the other hand, for $\mu' > \mu'_*$ the remainder estimate is $O(\tau^q)$ with

$$q = \frac{1}{2}(d - \vartheta - 1) + \frac{1}{2\mu'\nu}\bigl(d'\nu - (\mu - \nu + \mu')(d'' - \vartheta - 1)\bigr)$$

for $\vartheta < (d'' - 1)$ and with $q = \frac{d'}{2} + \frac{d'}{2\mu'} + \vartheta$ for $\vartheta > d'' - 1$.

For $\vartheta = (d'' - 1)$ the remainder estimate i $O(\tau^q \log \tau)$ with the same q.

(v) All these results remain true for $\nu = \nu_0$ and $\vartheta = 0$ with $\mathsf{N}^{\mathsf{W}}(\tau) \asymp \tau^{lq} \log \tau$.

(vi) Moreover, for $\nu > \nu_0$, $\vartheta = 0$ and $\mu' > \mu'_*$ these results remain true with $\mathsf{N}^{\mathsf{W}}(\tau) \asymp \tau^p$ with $p = \frac{1}{2}(d + \frac{d}{\mu'} - \frac{d''\mu}{\mu'\nu} + \frac{d''}{\nu})$. Then $p - q = \frac{1}{2\nu}(\nu + 1 - \frac{1}{\mu'}(\mu - \nu)) > 0$.

Let us now consider indefinite perturbations. In this case the perturbation cannot be too strong.

Theorem 11.5.20 (Indefinite perturbations). *(i) Let $d \ge 3$ and let X, g^{jk}, V_j^0 and V^0 be as in Theorem 11.5.14. Let $V_j = V_j^0 + V_j'$, $V = V^0 + V'$ where V_j' and V' satisfy* (11.5.55) *with $\rho' = \langle x\rangle^{\mu'}$, $0 < \mu' < \mu$ and $\gamma = \tilde{\gamma}|x|$. Then, if*

$$v^{-nd-\mu'\vartheta(\mu-\mu')^{-1}} \in \mathscr{L}^1(U) \tag{11.5.57}$$

and

$$v^{-n(d-\vartheta-1)-\mu'\vartheta(\mu-\mu')^{-1}}\tilde{\gamma}^{-\vartheta-1} \in \mathscr{L}^1(U) \tag{11.5.58}$$

with $n = \frac{1}{2}(\mu'+1)(\mu-\mu')^{-1}$, $0 \le \vartheta < (d-1)$ then asymptotics (11.5.2)–(11.5.3) *holds and for $\vartheta = 0$ $\mathsf{N}^{\mathsf{W}}(\tau) = \varkappa_0\tau^{ld} + o(\tau^{l(d-1)})$.*

(ii) Moreover, if the Hamiltonian H^0 given by (11.5.43) *satisfies the conditions of Theorem 11.5.14(ii), then asymptotics* (11.5.6), (11.5.3) *holds.*

Proof. The proof is similar to the proof of Theorem 11.5.14 but the domain $\{V \le \tau\}$ consists of two overlapping zones $\{V^0 \le C\tau, \rho' \le C\tau^{\frac{1}{2}}\}$ and $\{V^0 \le C\rho'^2, \tau \le C\rho'^2\}$.

Then $\tilde{\rho}(\tau) \asymp \tau^{\frac{1}{2}}$ and $\tilde{\rho}(\tau) \asymp \rho'$ in the first and second zones respectively.□

Example 11.5.21. Let X, g^{jk}, V_j^0 and V^0 be as in Example 11.5.17. Let $V_j = V_j^0 + V_j'$, $V = V^0 + V'$ where V_j' and V' satisfy (11.5.55) with $\rho' = \langle x\rangle^{\mu'}$, $0 < \mu' < \mu$ and $\gamma = \epsilon|x''|$.

Let $\mu > \mu' + \nu$ (otherwise the potential V can be semi-bounded from above or even tend to $-\infty$ in the pipe $\{|x''| \le \epsilon\}$). Moreover, let us assume that

$$\nu < \frac{\mu-\mu'}{\mu'+1}. \tag{11.5.59}$$

Then considering the zone

$$X'' = \bigl\{x : |x''| \le z(x,\tau) = \epsilon \min\bigl(\tau^{-\frac{1}{2}}, |x|^{-(\mu-\nu)(\nu+1)^{-1}}\bigr)\bigr\}$$

we see that we can take $\tilde{\rho}(\tau) = (\rho' - \epsilon_1 z^{-1})_+$ in condition (11.5.54) where $\epsilon_1 > 0$ does not depend on ϵ. Then (11.5.54) yields that we can take $\tilde{\rho}(\tau) = 0$ in X'' for $\tau \ge \epsilon^{-1}$ and small enough ϵ.

On the other hand, (11.5.59) yields that

$$(\tau^{\frac{1}{2}}+\rho)\gamma \ge \epsilon_2 > 0, \quad \rho' \le C(\tau^{\frac{1}{2}}+\rho)^{1-\delta}$$

outside of X'' where $\delta > 0$, $\rho = |x''|^{\nu}\cdot |x|^{\mu\mu-\nu}$, $\gamma = \epsilon_3|x''|$. Using these assertions one can easily prove that under condition (11.5.59) all the statements of Example 11.5.17 remain true.

We leave to the reader

Problem 11.5.22. (i) Consider the Schrödinger operator with coefficients satisfying conditions (11.1.6)–(11.1.8), (11.4.5) with $\rho = \langle x\rangle^{\mu}$, $\gamma = \epsilon\langle x\rangle^{1-\delta}$, $\mu > 0$, $\delta > 0$, $\mu + 1 > \delta$.

In this case the asymptotics $\mathsf{N}^-_{(\vartheta)}(\tau) = \mathsf{N}^{\mathsf{W}}_{(\vartheta)}(\tau) + O(\tau^q)$ appear with $\mathsf{N}^{\mathsf{W}}(\tau) \asymp \tau^p$ in domains conical at infinity.

(ii) Consider domains or Riemannian manifolds of type (11.5.69) (see below). Then asymptotics appear with other exponents p and q. In this example the magnetic field can grow strongly enough at infinity.

11.5.5 Quasihomogeneous Potentials

Let us now consider the quasihomogeneous case.

Example 11.5.23. (i) Let X be an L-quasiconical domain in $\mathbb{R}^d$ and let g^{jk}, V_j, V be L-quasihomogeneous on x of degrees 0, μ and 2μ respectively. Here $\mu > 0$ and $L = (L_1, \dots, L_d)$ with $L_j = 1$ for $1 \le j \le d'$ and $L_j > 1$ for $d' + 1 \le j \le d$, $0 < d' < d$.

Let us assume that $\partial X \setminus 0 \in \mathscr{C}^K$ and $g^{jk}, V_j, V \in \mathscr{C}^K(\bar{X} \setminus 0)$ satisfy (11.1.8) and $V(x) > 0$ for $x \ne 0$.

Then all of the conditions (11.1.6)–(11.1.8) and (11.4.5) are fulfilled with

$$\rho = [x]_L^{\mu}, \qquad \gamma = \epsilon[x]_L \tag{11.5.60}$$

for

$$[x]_L := \bigl(\sum_j |x|^{\frac{\bar{L}}{L_j}}\bigr)^{\frac{1}{\bar{L}}} \ge 1; \tag{11.5.61}$$

for $[x]_L < 1$ one should take $\gamma = [x]_L^{\bar{L}}$ with $\bar{L} = \max_j L_j$ here.

Then asymptotics

$$\mathsf{N}^-(\tau) = \varkappa_0 \tau^p + O(\tau^q) \tag{11.5.62}$$

holds with $p = \frac{d}{2} + \frac{L^*}{2\mu}$, $q = \frac{1}{2}(d-1) + \frac{1}{2\mu}(L^* - 1)$, $L^* = \sum_j L_j$ and with the coefficient

$$\varkappa_0 = (2\pi)^{-d}\omega_d \int (1-V)_+^{\frac{d}{2}} \sqrt{g}\, dx. \tag{11.5.63}$$

We leave the analysis for $\vartheta > 0$ to the reader.

(ii) In order to improve these asymptotics let us consider the Hamiltonian billiards generated by the Hamiltonian

$$H = [x]_L^{1-\mu}\bigl(\sum_{j,k} g^{jk}(\xi_j - V_j)(\xi_k - V_k) + V - \tau\bigr). \tag{11.5.64}$$

Obviously, the contribution of the zones $\{x \in X, |x_i| \le \epsilon \tau^{\frac{L_j}{2\mu}}\}$ to the remainder estimate does not exceed $\epsilon_1 \tau^q$ with $\epsilon_1 \to 0$ as $\epsilon \to +0$ $(i = 1, \dots, d)$. So we consider billiards starting from complements to these zones.

One can easily see that along Hamiltonian trajectories

$$\frac{d\xi_i}{dt} = o(\tau^{\frac{1}{2}}), \qquad \frac{dx_i}{dt} = o(\tau^{\frac{L_i}{2\mu}}) \tag{11.5.65}$$

as $\tau \to +\infty$ for $i \ge d' + 1$ provided $[x]_L \ge \epsilon \tau^{\frac{1}{2\mu}}$. Let us fix arbitrarily large T and arbitrarily small $\epsilon > 0$ and treat billiards of the length T.

Let us assume that $x' \ne 0$, $\nu' \ne 0$ on $\partial X \setminus 0$ where $x' = (x_1, \dots, x_{d'})$ and $\nu' = (\nu_1, \dots, \nu_{d'})$, $\nu = (\nu_1, \dots, \nu_d) = (\nu', \nu'')$ is a unit normal to ∂X. Then in the indicated zone $|\nu''| = o(1)$ and therefore the increment of ξ_i when a billiard reflects from ∂X is $o(\tau^{\frac{1}{2}})$ for $i \ge d' + 1$.

Therefore $\tau^{-\frac{1}{2}}\xi_i$ and $\tau^{-\frac{L_i}{2\mu}} x_i$ with $i \ge (d'+1)$ are almost constant along billiards (if the length and the number of reflections are bounded).

Therefore if $(\bar{x}', \bar{\xi}')$ is a good starting point for a billiard generated by the Hamiltonian $H_{(\bar{x}'', \bar{\xi}'')} = H(x', \bar{x}'', \xi', \bar{\xi}'')$ in the cross-section domain $X_{(\bar{x}'')} = \{x' : (x', \bar{x}'') \in X\}$ then $(\bar{x}, \bar{\xi})$ is a good starting point for the original Hamiltonian.

Therefore the phase space measure of points which cannot be starting points does not exceed $\epsilon_2\tau^q$ with arbitrarily small $\epsilon_2 > 0$. Here $\tau \ge \tau_0(\epsilon_2)$.

Let us treat the non-periodicity. Let us assume that

$$g^{jk} = 0, \quad V_k = 0 \quad \text{for } j \le d',\ k \ge d'+1. \tag{11.5.66}$$

Then one can easily show that $|x''(0)-x''(t)| \ge \varepsilon|t|\tau^{\frac{1}{2\mu}}$ provided $t \in [-T, T]$, $|\xi''| \ge C_{\epsilon,T}\varepsilon\tau^{\frac{1}{2}}$, $\tau \ge \tau_0(\epsilon, \varepsilon, T)$ where ϵ is defined by the above conditions including the transversality condition. Therefore the non-periodicity condition is also fulfilled if we exclude a subset of measure $\epsilon_2\tau^q$ from the starting point set.

Therefore asymptotics

$$\mathsf{N}^-(\tau) = \mathsf{N}^{\mathsf{W}}(\tau) + \bigl(\varkappa_1 + o(1)\bigr)\tau^q \tag{11.5.67}$$

holds with the coefficient

$$\varkappa_1 = \mp\frac{1}{4}(2\pi)^{1-d}\omega_{d-1}\int_{\partial X}(1-V)_+^{\frac{d-1}{2}}\, dS'\sqrt{g''}\, dx'' \tag{11.5.68}$$

where dS' and $\sqrt{g''}dx''$ are Riemannian densities on the projections $(\partial X)_{x''} = \{x' : (x', x'') \in \partial X\}$ and $X_{x'} = \{x'' : (x', x'') \in X\}$ generated by the metrics $(g^{jk})_{j,k\le d'}$ and $(g^{jk})_{j,k\ge d'+1}$ respectively (these metrics and densities depend on x). Here the sign "$\mp$" corresponds to Dirichlet and Neumann boundary condition respectively.

Remark 11.5.24. Surely, these statements remain true in the case when the operator tends to an L-quasihomogeneous operator or if assumptions are fulfilled with ρ and γ defined by (11.5.60)–(11.5.61); in the latter case only Statement (i) remains true. In the former case $\mathsf{N}^{\mathsf{W}}(\tau) = \bigl(\varkappa_0 + o(1)\bigr)\tau^p$ without exact equality.

11.5.6 Quickly Growing Potentials

In this subsection we consider potentials with more than power growth at infinity. Then the volume of the domain $\{x : V(x) \le \tau\}$ grows slowly as $\tau \to +\infty$ and one should expect that

$$N^{\mathsf{W}}_{(\vartheta)} \asymp \mathsf{mes}(\{x : V(x) \le \tau\}) \cdot \tau^{\frac{d}{2}} \tag{11.5.69}$$

have slightly larger magnitude than in the case of the essentially bounded domain.

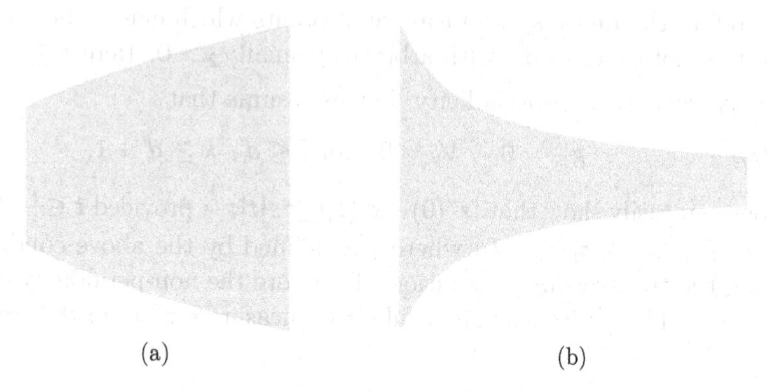

(a) (b)

Figure 11.10: Domains of the type (11.5.69)

Theorem 11.5.25. *(i) Let $d \geq 2$ and let X be a domain in $\mathbb{R}^d$ such that*

$$\mathsf{vol}(\{x \in X, |x| \asymp r\}) \asymp r^n \quad \textit{as} \quad r \to +\infty, \tag{11.5.70}$$

$n \leq d$.

Let conditions (11.1.6)–(11.1.8) *and* (11.4.5) *be fulfilled with* $\gamma = \epsilon_1 \langle x \rangle^{1-\mu}$, $\rho = \exp \langle x \rangle^{\mu}$, $\mu > 0$.

Then asymptotics (11.5.2) *holds with the principal part*

$$\mathsf{N}^{\mathsf{W}}_{(\vartheta)}(\tau) \asymp \tau^{\frac{d}{2}+\vartheta}\sigma(\tau), \tag{11.5.71}$$

$$\sigma(\tau) = \begin{cases} \log^n \tau & \textit{for } n > 0, \\ \log\log \tau & \textit{for } n = 0, \\ 1 & \textit{for } n < 0. \end{cases} \tag{11.5.72}$$

and remainder estimate

$$R_{(\vartheta)}(\tau) = \tau^{\frac{1}{2}(d+\vartheta-1)}\omega_{(\vartheta)}(\tau), \tag{11.5.73}$$

$$\omega_{(\vartheta)}(\tau) = \begin{cases} \log^q \tau & \textit{for } q := (n-(1-\mu)(1+\vartheta))/\mu > 0, \\ \log\log \tau & \textit{for } q = 0, \\ 1 & \textit{for } q < 0. \end{cases} \tag{11.5.74}$$

(ii) Moreover, let $q \geq 0$ and for every fixed $k > 0$ let condition

$$\mathsf{vol}\bigl\{x \in X, |x| \leq r,\ \mathsf{dist}(x, \partial X) \leq k\gamma(x)\bigr\} = o(r^n) \quad \textit{as} \quad r \to +\infty \tag{11.5.75}$$

be fulfilled. Furthermore, let

$$|D^\alpha g^{jk}| = o(\gamma^{-1}) \qquad \text{as } |x| \to +\infty \quad \forall j,k,\alpha : |\alpha| = 1, \tag{11.5.76}$$

and

$$|D^\alpha V_j| \le C\rho\gamma^{-|\alpha|} \qquad \forall j,\alpha : |\alpha| \le 1. \tag{11.5.77}$$

Then asymptotics (11.5.6) *holds with* $R_{(\vartheta)}(\tau)$ *defined above and with* $\varkappa_1 = 0$ *for* $\vartheta = 0$.

Proof. One can easily prove Statement (i) in the standard way.

In order to prove Statement (ii) let us note that if $q > 0$ then for $\tau \ge \tau_0(\varepsilon)$ the contributions of the zones $\{x \in X, |x| \le \varepsilon \log^{\frac{1}{\mu}} \tau^{\frac{1}{2}}\}$ and $\{x \in X, |x| \ge (1-\varepsilon) \log^{\frac{1}{\mu}} \tau^{\frac{1}{2}}\}$ to the remainder do not exceed $\varepsilon_1 R_{(\vartheta)}(\tau)$ with $\varepsilon_1 \to 0$ as $\varepsilon \to 0$.

Moreover, condition (11.5.75) yields that the contribution of the zone $\{X, \operatorname{dist}(x, \partial X) \le k\gamma(x)\}$ to the remainder is $o(R_{(\vartheta)}(\tau))$ and therefore one need consider only billiards generated by the improved Hamiltonian and starting in the zones

$$X_{k,\varepsilon,\tau} = \bigl\{X \in X, \varepsilon \le |x| \log^{-\frac{1}{\mu}} \tau^{\frac{1}{2}} \le 1-\varepsilon, \operatorname{dist}(x,\partial X) \ge k(\log \tau^{\frac{1}{2}})^{\frac{1}{\mu}-1}\bigr\}.$$

Then for $t \in [-T, T]$ and $\tau \ge \tau_0(T, k, \varepsilon)$, $k \ge k_0(T, \varepsilon)$ these billiards lie in $X_{\frac{k}{2},\frac{\varepsilon}{2},\tau}$, they do not hit ∂X, and along them

$$\begin{aligned}
\tau^{-\frac{1}{2}} V_j \equiv \tau^{-1} V \equiv 0, \quad g^{jk} \equiv \text{const}, \quad \gamma(\log\tau)^{1-\frac{1}{\mu}} \equiv \text{const},\\
\tau^{-\frac{1}{2}}\xi_j \equiv \text{const}, \quad \gamma^{-1}\frac{dx_j}{dt} \equiv 2\tau^{-\frac{1}{2}} \sum_k g^{jk}\xi_k, \quad \tau^{-1}\sum_{j,k} g^{jk}\xi_j\xi_k \equiv 1
\end{aligned}$$

modulo $o(1)$ as $\tau \to +\infty$.

The rest of the proof is similar to the proof of Theorems 11.2.7 and 11.2.8. The case $q = 0$ is treated in the same way with the removal of the zones $\{x \in X, |x| \le R\}$, $\{x \in X, |x| \ge \epsilon \log^{\frac{1}{\mu}} \tau\}$ and $\{x \in X, \operatorname{dist}(x, \partial X) \le k\gamma(x)\}$ with arbitrarily small $\epsilon > 0$ and arbitrarily large R. ∎

This theorem could be improved.

Theorem 11.5.26. *In the framework of the previous Theorem 11.5.25 assume that g^{jk} and the boundary satisfy smoothness assumptions with $\gamma(x) = \epsilon\langle x\rangle$ (then $n = d$). Then one can redefine* (11.5.74)

$$\omega_{(\vartheta)}(\tau) = \log^q \tau \qquad \textit{for } q := (d-1-(1-\mu)\vartheta)/\mu. \tag{11.5.78}$$

Proof. One can see easily that we can then replace $\gamma(x) = \langle x\rangle^{1-\mu}$ by

$$\gamma_1(x) = \epsilon \min\bigl((\tau/\rho^2)^\delta \gamma(x),\ \langle x\rangle\bigr). \tag{11.5.79}$$

Indeed, one can check easily that it is scaling function ($|\nabla\gamma| \le \epsilon_1$) and that $|\nabla^\alpha V| \le C\tau\gamma_1^{-|\alpha|}$, $|\nabla^\alpha V_j| \le C\tau^{1/2}\gamma_1^{-|\alpha|}$ for $|\alpha| \le K$.

Then contribution of the zone $\{x : V(x) \le 2\tau\}$ to the remainder does not exceed $C\int \tau^{(d-1-\theta)/2}\gamma_1^{-1-\theta}\,dx$. Plugging $\gamma_1(x) = (\tau/\rho^2)^\delta\gamma(x)$ and $dx = r^{d-1}\,dr$ we get (11.5.74) with $q = d-1-(1-\mu)\vartheta$ and plugging $\gamma_1 = \langle x\rangle$ we get even better remainder estimate. □

In the same way the following theorem is proven:

Theorem 11.5.27. *Let $d \ge 2$, $X = \mathbb{R}^+ \times X'$ where X' is a bounded domain in $\mathbb{R}^{d-1} \ni x'$ with $\mathscr{C}^K$ boundary. Let all the conditions* (11.1.6)–(11.1.8) *and* (11.4.5) *be fulfilled with $\rho = \exp x_1$ and $\gamma = \epsilon_0 > 0$ and let*

$$g^{jk} - g^{jk0} = o(1), \qquad D_1 g^{jk} = o(1) \qquad \textit{as } x_1 \to +\infty \tag{11.5.80}$$

where $g^{jk0} = g^{jk0}(x')$ and $g^{1k0} = 0$ for $k \ge 2$.

Further, let conditions (11.5.76)–(11.5.77) *be fulfilled. Then for every $\vartheta > 0$ asymptotics* (11.5.6) *holds with*

$$R_{(\vartheta)}(\tau) = \tau^{\frac{1}{2}(d+\vartheta-1)} \log \tau. \tag{11.5.81}$$

Further, for $\vartheta = 0$ the principal part is

$$\mathsf{N}^{\mathsf{W}}(\tau) + \varkappa_1 \tau^{\frac{1}{2}(d-1)} \log \tau \tag{11.5.82}$$

with coefficient

$$\varkappa_1 = \mp \frac{1}{8}(2\pi)^{1-d}\omega_{d-1} \int_{\partial X'} \frac{dS'}{\sqrt{g^{110}}} \tag{11.5.83}$$

where dS' is the Riemannian density on $\partial X'$ generated by the metric $(g^{jk0})_{j,k\ge 2}$.

Furthermore,

$$\mathsf{N}^{\mathsf{W}}(\tau) = \bigl(\bar{\varkappa}_0 + o(1)\bigr)\tau^{\frac{d}{2}}\log\tau \tag{11.5.84}$$

with the coefficient

$$\bar{\varkappa}_0 = \frac{1}{2}(2\pi)^{-d}\omega_d \int_{X'} \sqrt{g^{0\prime}} \frac{dx'}{\sqrt{g^{110}}} \tag{11.5.85}$$

where $\sqrt{g^{0\prime}}dx'$ *is a Riemannian density on* X' *generated by the same metric.*

The following problems look rather interesting but not very difficult:

Problem 11.5.28. (i) Consider more general operators than the Schrödinger operator.

(ii) Consider $X = \mathbb{R}^+ \times X'$, where X' is a bounded domain in $\mathbb{R}^{d-1}$ with $\mathscr{C}^K$-boundary, $\rho = exp(x_1^\mu)$, $\gamma \asymp x_1^\sigma$, $\sigma = \max(0, 1-\mu)$, $\mu > 0$.

(iii) Consider potentials with the degenerations and singularities (in the spirit of Section 11.5.4.

(iv) Consider other types of super-power growth.

11.5.7 Slowly Growing Potentials

Finally, let us consider potentials with slow growth at infinity. Here we should be very careful and specific since even formula (11.5.69) may be not valid anymore.

Namely, let $V = v(\theta)\log r$ in spherical coordinates (r,θ). We use $\rho_\tau = (\tau - V)_+^{\frac{1}{2}}$ and $\gamma_\tau = \epsilon_0 r\min(1, (\tau - V)_+ \log^{-1} r)$. Then we obtain

Theorem 11.5.29. *Let* $d \ge 3$ *and let* $g^{jk} = g^{jk}(\theta)$, $V_j = 0$, $V = v(\theta)\log r$, $X = \{\theta \in U\}$ *for* $r \ge c$ *where* (r,θ) *are spherical coordinates,* $g^{jk}, v, \partial U \in \mathscr{C}^K$ *and* (g^{jk}) *is a positive definite matrix. Let* $v_0 = \min_U v > 0$. *Then*

(i) Asymptotics (11.5.2) *holds with*

$$\mathsf{N}^{\mathsf{W}}_{(\vartheta)} \asymp \int_U \exp(\frac{\tau d}{v})\, d\theta, \tag{11.5.86}$$

$$R_{(\vartheta)}(\tau) = \tau^{1+\vartheta}\int_{U'} \exp(\frac{\tau(d-1-\vartheta)}{v})\, d\theta \qquad \textit{for } \vartheta < d-1. \tag{11.5.87}$$

Moreover,

$$\mathsf{N}^{\mathsf{W}}(\tau) = a\int_{U'} v^{\frac{d}{2}} \exp(\frac{\tau\, d}{v})\sqrt{g}\, d\theta \tag{11.5.88}$$

with $a = (2\pi)^{-d}\omega_d d^{-(1+\frac{d}{2})}\Gamma(1+\frac{d}{2}) = 2^{1-d}\pi^{-\frac{d}{2}} d^{-(1+\frac{d}{2})}$ *and* $U' = \{\theta\in U, v\le v_0+\varepsilon\}$ *with arbitrarily small* $\varepsilon>0$.

(ii) In particular, if $v = \mathrm{const} = v_0$ *then the principal part is* $\varkappa \exp(\frac{\tau d}{v_0})$ *and the remainder estimate is* $O\bigl(\tau^{1+\vartheta}\exp(\tau(d-\vartheta-1)v_0^{-1})\bigr)$ [44].

On the other hand, if $v(\theta) = v_0 \implies \theta\in\partial U, \nabla v(\theta)\ne 0$ *and* $\operatorname{rank}\operatorname{Hess} v|_{\partial U}(\theta) = d-2$ *then the principal part is* $\asymp \tau^{-\frac{d}{2}}\exp(\tau d v_0^{-1})$ *and the remainder estimate is* $O\bigl(\tau^{1+\vartheta-\frac{d}{2}}\exp(\tau(d-\vartheta-1)v_0^{-1}))\bigr)$.

In other cases the principal part of the asymptotics and the remainder estimate have intermediate orders.

We leave to the reader

Problem 11.5.30. (i) Investigate the case $\vartheta\ge(d-1)$. Note that one can take $\tilde\rho(\tau)=\tau^{\frac12}+(\log r)^{\frac12}$, $\tilde\gamma(\tau)=\epsilon_0 r$ and for $d\ge 2$ and smooth ∂X, g^{jk}, V_j and V we obtain the remainder estimate $R_{(\vartheta)}(\tau)=C\tau^d$ for $\vartheta=d-1$ and $R_{(\vartheta)}(\tau)=C\tau^{\frac12(d+\vartheta-1)}$ for $\vartheta>(d-1)$.

(ii) Consider potential $V=v(\theta)\log^\mu r$.

(iii) Consider potentials with degenerations and singularities. Obviously in this case the degeneration should be very weak, something like

$$(\log(|x|^2+2))^\mu\cdot(\log(|x''|^2+2))^{-\nu}.$$

11.5.8 Non-Dirichlet Boundary Conditions

All our results have been obtained for Dirichlet boundary conditions. Surely, they remain true for other boundary conditions provided these conditions are regular in the obvious sense and the domain satisfies the cone condition. In particular, for the Schrödinger operator Neumann conditions (and Robin

[44] One can take $\gamma_\tau=\epsilon_0 r\min(1,(\tau-V)_+)$ in this case and drop factor $\tau^{1+\vartheta}$ in the remainder estimate.

conditions provided the free term coefficient is regular in the obvious sense) are regular.

Then we obtain that if the domain satisfies the cone condition and if the asymptotics do not distinguish Dirichlet and Neumann boundary conditions then the asymptotics remain true for mixed boundary condition when Dirichlet and Neumann boundary conditions are given on $Y_0 \subset \partial X$ and $Y = \partial X \setminus Y_0$ respectively.

Let us consider more accurate asymptotics distinguishing Dirichlet and Neumann the boundary conditions. Then, repeating arguments of Subsection 11.2.3 we obtain that these asymptotics remain true for mixed boundary conditions with properly defined second coefficient provided either

(a) We are in the frameworks of Theorems 11.5.3, 11.5.14(ii)), 11.5.18(ii) and

$$\mathsf{mes}_{\partial X}\bigl((\bar{Y}_0 \cap \bar{Y}_1 \cap \{|x| \ge c\})_{\mathsf{cone}}\bigr) = 0 \tag{11.5.89}$$

or

(b) we are in the framework of Example 11.5.23(ii) and

$$\mathsf{mes}_{\Pi'\partial X}\bigl(\Pi'(\bar{Y}_0 \cap \bar{Y}_1 \cap \{|x| \ge c\})_{\mathsf{cone}}\bigr) = 0 \tag{11.5.90}$$

where Π' is the projection $\mathbb{R}^d \ni x \to x' \in \mathbb{R}^{d'}$ or

(c) We are in the framework of Theorem 11.5.27 and

$$\mathsf{mes}_{\partial X'}\bigl(\Pi'(\bar{Y}_0 \cap \bar{Y}_1 \cap \{|x| \ge c\})\bigr) = 0 \tag{11.5.91}$$

with the same Π'. Here M_{cone} is the conical hull of the set M.

Moreover, applying the arguments of Example 11.2.27 we obtain that asymptotics (11.5.16) remains true in the context of Example 11.5.7 provided

$$\mathsf{mes}_{\partial X}\{x \in \partial X, \mathsf{dist}(x, Y) \le \varepsilon\} = O(\varepsilon^{\delta}) \qquad \text{as } \varepsilon \to +0 \tag{11.5.92}$$

with $\delta > 0$ and $Y = (\bar{Y}_0 \cap \bar{Y}_1 \cap \{|x| \ge c\})_{\mathsf{cone}}$.

Finally, we would like to note that in almost all the above examples the remainder estimate is $O(\tau^{\frac{1}{2}(d+\vartheta-1)})$ for large enough ϑ (for the Schrödinger operator). In this case the principal contribution to the remainder estimate is provided by the zone $\{x \in X, |x| \le R\}$ with arbitrarily large R. In this case the standard non-periodicity condition to the Hamiltonian $\sum_{j,k} g^{jk}\xi_j\xi_k$ on the energy level $\tau = 0$ provides the remainder estimate $o(\tau^{\frac{1}{2}(d+\vartheta-1)})$ (see, e.g., Theorem 11.5.3).

The situation in which the principal contributions to the principal part of the asymptotics and to the remainder estimate are provided by different zones is not very exotic.

11.6 Asymptotics of Eigenvalues near -0

In this section we consider Schrödinger operators with potentials decaying at infinity. Under certain rather general assumptions its continuous spectrum is $[0,\infty)$ and the point -0 is the only possible limit point of the discrete spectrum. Moreover, there are no positive eigenvalues.

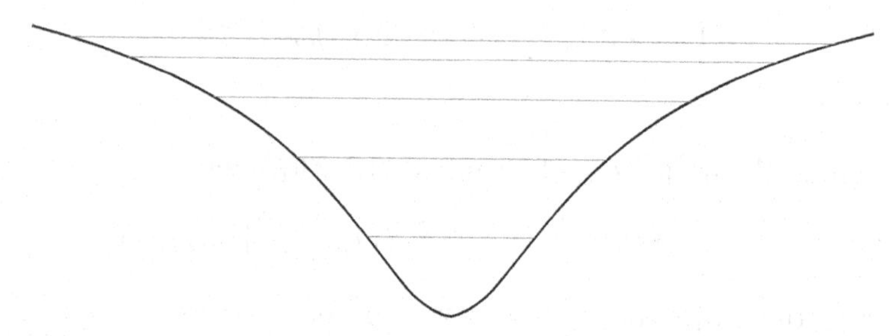

Figure 11.11: Slowly decaying at infinity potential; blue lines show symbolically energy levels of the corresponding Schrödinger operator.

Further, as potential is negative (at least in massive zones so that $\mathsf{mes}(\{x, V(x) < 0\}) = \infty$ and potential decays slowly ($|x|^{-2}$ is a threshold) -0 is a limit point of the discrete spectrum. In this case we obtain asymptotics of $\mathsf{N}^-(\tau)$ as $\tau \to -0$.

We treat more general higher-order operators and non-semi-bounded operators as well. In particular, we treat the Dirac operator with positive mass M where a spectral gap $(-M, M)$ appears in the case of $V = 0$ (while continuous spectrum occupies $(-\infty, -M] \cup [M, \infty)$ and in the general case of decaying potential there could be a discrete spectrum inside the spectral gap and if V decays slowly enough these eigenvalues accumulate to $-M + 0$ and/or $M - 0$. So we will treat *asymptotics of eigenvalues accumulating to the border of the essential spectrum*.

Later we consider much more subtle mechanism when the similar situation appears: Schrödinger and Dirac operators with magnetic field or operator with periodic coefficients. Essential spectrum may occupy spectral bands or separate points (separating spectral gaps). In both cases adding decaying potential one may create discrete spectrum inside the spectral gap and these eigenvalues may accumulate to its borders.

11.6.1 Standard Results

We start with semi-bounded operators. First of all the following statement holds:

Theorem 11.6.1 [45]. *Let X be an unbounded connected domain in $\mathbb{R}^d$ and let conditions* (11.1.1)–(11.1.3) *be fulfilled with $\gamma = \epsilon_1\langle x\rangle$, $\rho = \langle x\rangle^\mu$, $\mu \in (-1, 0)$.*

Let us assume that

$$\langle a(x,\xi)v, v\rangle \ge \epsilon_0|\xi|^{2m}\|v\|^2 \qquad \forall\xi\ \forall v \tag{11.6.1}$$

where $a(x,\xi)$ is the senior symbol of A.

Further, let us assume that for every $\tau \in (-\epsilon, 0]$ one of non-degeneracy conditions is fulfilled for the symbol $(A(x,\xi)\tau^{-1} - 1)$ uniformly with respect to (τ, x, ξ) (this condition is restrictive only for $|x| \ge C_0$). Then

(i) Asymptotics

$$\mathsf{N}^-(\tau) = \mathsf{N}^{\mathsf{W}}(\tau) + O\bigl(\tau^{l(d-1)}\bigr) \tag{11.6.2}$$

holds as $\tau \to -0$, where $\mathsf{N}^{\mathsf{W}}(\tau)$ is given by (11.4.9) *and $\mathsf{N}^{\mathsf{W}} = O(\tau^{ld})$ with $l = \frac{1}{2m\mu}(\mu + 1) < 0$.*

(ii) Moreover, for $0 < \vartheta \le \vartheta_0^* = -ld$ asymptotics

$$\mathsf{N}^-_{(\vartheta)}(\tau) = \mathsf{N}^{\mathsf{W}}_{(\vartheta)}(\tau) + O\bigl(R_{(\vartheta)}(\tau)\bigr) \tag{11.6.3}$$

holds as $\tau \to -0$ with $\mathsf{N}^{\mathsf{W}}_{(\vartheta)}(\tau) = \tau_+^{\vartheta-1} * \mathsf{N}^{\mathsf{W}}$ where in this case N^{W} should contain also lower order terms and

$$R_{(\vartheta)}(\tau) = \begin{cases} |\tau|^{l(d-1)+\vartheta(1-l)} & \text{for } \vartheta < \vartheta_1^*, \\ \log|\tau| & \text{for } \vartheta = \vartheta_1^*, \\ 1 & \text{for } \vartheta > \vartheta_1^* \end{cases} \tag{11.6.4}$$

[45] Cf. Theorem 11.5.1.

where $\vartheta_1^* = -l(d-1)(1-l)^{-1}$.

One can see easily that $0 < \vartheta_1^* < \vartheta_0^*$ *and that*

$$\mathsf{N}^{\mathsf{W}}_{(\vartheta)}(\tau) = \begin{cases} O(|\tau|^{ld+\vartheta}) & \text{for } \vartheta < \vartheta_0^* \\ O(\log|\tau|) & \text{for } \vartheta = \vartheta_0^* \end{cases} \tag{11.6.5}$$

as $\tau \to -0$ *while for* $\vartheta > \vartheta_0^*$ *we have estimate* $\mathsf{N}^{\mathsf{W}}_{(\vartheta)} = O(1)$.

(iii) Furthermore, if

$$\begin{aligned}\operatorname{mes}\bigl(\{x \in X : |x| \le r \ \ \&\ \exists \xi \in \mathbb{R}^d :\\ \operatorname{Spec} A(x,\xi) \cap (-\infty, -\epsilon_1\rho^{2m}] \ne \emptyset\}\bigr) \asymp r^d \qquad \text{as } r \to +\infty\end{aligned} \tag{11.6.6}$$

for some $\epsilon_1 > 0$ *then*

$$\mathsf{N}^{\mathsf{W}}_{(\vartheta)}(\tau) \asymp \begin{cases} |\tau|^{ld+\vartheta} & \text{for } \vartheta < \vartheta_0^* \\ \log|\tau| & \text{for } \vartheta = \vartheta_0^* \end{cases} \tag{11.6.7}$$

as $\tau \to -0$.

For the Schrödinger operator condition (11.6.7) *is equivalent to*

$$\operatorname{mes}\bigl(\{x \in X : |x| \le r,\, V \le -\epsilon_1\rho^2\}\bigr) \asymp r^d \qquad \text{as } r \to +\infty. \tag{11.6.8}$$

Theorem 11.6.2 [46]**.** *Let assumptions of Theorem 11.6.1 be fulfilled.*

Further, let us assume that $X \cap \{|x| \ge r_0\} = X^0 \cap \{|x| \ge r_0\}$ *where* X^0 *is a non-empty conical domain with* $\partial X^0 \setminus 0 \in \mathscr{C}^K$ *and that stabilization conditions* (11.5.4) *are fulfilled. Let us assume that as* $\tau = -1$ *the standard non-periodicity condition to Hamiltonian trajectories for the quasihomogeneous symbol*

$$A^0(x,\xi) = \sum_{|\beta| \le 2m} a^0_\beta(x)\xi^\beta \tag{11.6.9}$$

satisfies, and the permitted[37] *trajectories never hit* $0 \times \mathbb{R}^d$*. Then*

---

[46] Cf. Theorem 11.5.3.

(i) Asymptotics

$$\mathsf{N}^-(\tau) = \mathsf{N}^\mathsf{W}(\tau) + \bigl(\varkappa_1 + o(1)\bigr)|\tau|^{l(d-1)} \tag{11.6.10}$$

holds as $\tau \to -0$. *In particular, for the Schrödinger operator*

$$\varkappa_1 = \mp\frac{1}{4}(2\pi)^{1-d}\omega_{d-1}\int_{\partial X^0}(-1 - V^0)_+^{\frac{d-1}{2}}\,dS^0 \tag{11.6.11}$$

for Dirichlet or Neumann boundary conditions respectively where the notation of (11.5.10)–(11.5.11) *is used.*

(ii) Moreover, for $0 < \vartheta < \vartheta_1^*$ *asymptotics*

$$\mathsf{N}^-_{(\vartheta)}(\tau) = \mathsf{N}^\mathsf{W}_{(\vartheta)}(\tau) + o\bigl(R_{(\vartheta)}(\tau)\bigr) \qquad \text{as}\ \ \tau \to -0. \tag{11.6.12}$$

holds.

(iii) Finally, let $\vartheta = \vartheta_1^*$; *the condition on the domain and the stabilization condition remain true. Moreover, let the non-periodicity condition to Hamiltonian trajectories be fulfilled for the symbol* $A^0(x,\xi)$ *on the energy level* 0 *and let the permitted*[37)] *trajectories never hit* $0 \times \mathbb{R}^d$.

Then asymptotics (11.5.6) *also holds (so the remainder estimate is* $o(\log|\tau|)$ *rather than* $O(\log|\tau|)$*).*

Proof of Theorems 11.6.1 and 11.6.2. The proof is similar to the proof of theorems 11.5.1 and 11.5.3 but now in the zone X'_{reg}, $\rho_\tau \asymp \rho$ (instead of $\rho_\tau \asymp |\tau|^{\frac{1}{2m}}$) and $X'_{\mathsf{reg}} \subset \{\rho \ge \epsilon_2|\tau|^{\frac{1}{2m}}\}$ with appropriate $\epsilon_2 > 0$.

Further, every bounded domain gives a finite contribution to the remainder estimate (and therefore in the contrast to 11.5.3 there is no assertion about improved asymptotics as $\vartheta > \vartheta_1^*$), Furthermore, the reduced Hamiltonian is now $|x|^{1-2m\mu+\mu}A^0(x,\xi)$.

We would like to emphasize that for $\vartheta > \vartheta_0^*$ we obtain the estimate $\mathsf{N}^-_{(\vartheta)}(-0) < \infty$ instead of asymptotics. This remark remains true in many examples below.

In the framework of Theorem 11.6.2 the remainder estimate is $o(|\tau|^{l(d-1)+\vartheta(1-l)})$ and $o(\log|\tau|)$ in asymptotics (11.5.6) for $\vartheta < \vartheta_1^*$ and $\vartheta = \vartheta_1^*$ respectively.

Finally, the main contributions to the remainder estimate are made by the zones $\{x, |x| \ge \varepsilon|\tau|^{\frac{1}{2m}}\}$ and $\{x, R \le |x| \le \varepsilon|\tau|^{\frac{1}{2m}}\}$ for $\vartheta < \vartheta_1^*$ and $\vartheta = \vartheta_1^*$ respectively with arbitrarily small $\varepsilon > 0$ and arbitrarily large R. This difference leads to different non-periodicity conditions. □

Remark 11.6.3. If only stabilization conditions (11.5.4) (including the condition $X \cap \{|x| \ge r_0\} = X^0 \cap \{|x| \ge r_0\}$) are fulfilled then the non-degeneracy condition is automatically fulfilled and $\mathsf{N}^{\mathsf{W}}(\tau) = \varkappa_0 |\tau|^{ld}(1 + o(1))$ as $\tau \to -0$ where for the Schrödinger operator

$$\varkappa_0 = (2\pi)^{-d}\omega_d \int_{X^0} (-1 - V^0)_+^{\frac{d}{2}} \sqrt{g^0}\, dx \tag{11.6.13}$$

in the notation of (11.5.10)–(11.5.11).

For scalar operators we get the following

Theorem 11.6.4 [47]. *Let X be an unbounded connected domain in $\mathbb{R}^d$ and let conditions* (11.1.1)–(11.1.3) *be fulfilled with $\gamma = \epsilon_1 \langle x \rangle$, $\rho = \langle x \rangle^{\mu}$, $\mu \in (-1, 0)$.*

Let condition (11.6.1) *be fulfilled and let A be a scalar operator. Then asymptotics*

$$\mathsf{N}(\tau) = \mathsf{N}^{\mathsf{W}}(\tau) + O(\tau^{l(d-1)-\delta}) \tag{11.6.14}$$

hold as $\tau \to -0$, and for $\vartheta > 0$ asymptotics (11.6.3) *holds.*

11.6.2 Strongly Non-Periodic Hamiltonian Flow

In what follows we consider homogeneous potentials with singularities. Now we obtain a more precise remainder estimate under more restrictive conditions on the Hamiltonian flow. One can prove easily the following

Theorem 11.6.5 [48]. *Let the conditions of Theorem 11.6.1 be fulfilled and let A be a scalar operator. Moreover, let us assume* (11.5.14) *is fulfilled i.e. that <u>either</u> $X^0 = \mathbb{R}^d$ <u>or</u> A is the Schrödinger operator or its power.*

Further, let stabilization condition (11.5.4)* *be fulfilled. Furthermore, let us assume that the Hamiltonian flow (with reflections at ∂X^0) is of non-periodic power type*[49] *at negative energy levels and that along the trajectories $(x(t), \xi(t))$ with $(x(0), \xi(0)) \in \Lambda^0_{\varepsilon,T} \subset \Sigma_{-1}$, $|x(0)| = 1$, $\varepsilon = T^{-M}$ condition $|x(t)| \ge T^{-M_1}$ is fulfilled.*

Then asymptotics

$$\mathsf{N}^-(\tau) = \mathsf{N}^{\mathsf{W}}(\tau) + \bigl(\varkappa_1 + O(|\tau|^{\delta'})\bigr)\tau^{l(d-1)} \tag{11.6.15}$$

---

[47] Cf. Theorem 11.5.5.

[48] Cf. Theorem 11.5.6.

[49] Footnote [38] of the previous section remains true.

holds as $\tau \to -0$ *with* $\delta' > 0$. *Moreover, for* $0 < \vartheta \le \vartheta_1^* - \delta$ *with* $\delta > 0$ *asymptotics*

$$\mathsf{N}^-_{(\vartheta)}(\tau) = \mathsf{N}^{\mathsf{W}}_{(\vartheta)}(\tau) + O\bigl(|\tau|^{l(d-1)+\vartheta(1-l)+\delta'}\bigr) \tag{11.6.16}$$

holds as $\tau \to -0$ *with* $\delta' = \delta'(d, m, \delta, L) > 0$.

Let us consider the following

Example 11.6.6[50]. (i) The condition of Theorem 11.6.5 to the Hamiltonian flow is fulfilled for the Schrödinger operator with $X^0 = \mathbb{R}^d$ and $A^0(x,\xi) = |\xi|^2 - |x|^{2\mu}$ with $\mu \ne -\frac{1}{2}$. So we consider $\mu \in (-1, -\frac{1}{2}) \cup (-\frac{1}{2}, 0)$.

(ii) Therefore the statement of Theorem 11.6.5 is true for this operator (provided all the other conditions of this theorem, including (11.5.4)*, are fulfilled) and for any $2m$-order operator with $A^0(x,\xi) = (|\xi|^2 - |x|^{2\mu})^m$, and one can replace $X^0 = \mathbb{R}^d$ by the X^0 obtained after (a sequence of) the transformations (a),(b) of Example 11.5.7(ii).

11.6.3 Periodic Hamiltonian Flow

As in the previous section let us consider the case of periodic Hamiltonian flow:

Theorem 11.6.7[51]. *Let* $X^0 = \mathbb{R}^d$ *and conditions* (11.5.27)–(11.5.27), (11.5.32) *with* $\mu \in (-1, 0)$ *be fulfilled in the domain of negative energies. Let us introduce the symbol*

$$b(x,\xi) = -\frac{l}{T_0}\int_0^{T_0} A^1 \circ \Psi_t dt \tag{11.6.17}$$

where T_0 *is the period on the energy level* -1 *(compare with* (11.5.30)*; we have changed sign because* $A^0(x,\xi) = -1$ *now). Then*

(i) If $\delta_1 < (\mu + 1)$ *and condition* (11.5.31) *is fulfilled for all* ω *then asymptotics*

$$\mathsf{N}^-(\tau) = \mathsf{N}^{\mathsf{W}}(\tau) + O\bigl(|\tau|^{l(d-1)+\delta''}\bigr) \tag{11.6.18}$$

holds as $\tau \to -0$ *with* $\delta'' > 0$.

[50] Cf. Example 11.5.7.
[51] Cf. Theorem 11.5.9.

(ii) If $\delta_1 \in [\mu+1, 2\mu+2]$ then for $\tau \to -0$ such that $|\tau|^l \in \bigcup_{n\in\mathbb{Z}^+} J_n$ asymptotics

$$\mathsf{N}^-(\tau) = \mathsf{N}^\mathsf{W}(\tau) + F(\tau)|\tau|^{l(d-1)} + O\bigl(|\tau|^{l(d-1)+\delta''}\bigr) \tag{11.6.19}$$

holds with $\delta''>0$ and

$$F(\tau) = (2\pi)^{-d} l T_0^{-1} \int_{\Sigma_{-1}} \Upsilon\Bigl(T_0 l^{-1}\bigl((|\tau|^l - \frac{\pi}{4}\omega_0) - |\tau|^{l-\frac{\delta_1}{2m\mu}} b\bigr)\Bigr)\,dxd\xi : dA^0. \tag{11.6.20}$$

The intervals J_n are defined in Theorem 11.5.9(ii) with $\nu = -\frac{2\pi l}{T_0}$, ω_0 replaced by $-\omega_0$ where ω_0 is the Maslov index of the trajectory Ψ_t with $t \in (0, T_0)$.

Let us treat the perturbed Schrödinger operator with the Coulomb potential more carefully. In this case $A_0 = |D|^2 - |x|^{-1}$ and $\mu = -\frac{1}{2}$, $l = -\frac{1}{2}$, $T_0 = \frac{\pi}{2}$, $\nu = 2$ and the Hamiltonian trajectories are Keplerian ellipses (with one focus located at the origin). Moreover, $\omega_0 \equiv 2(d-1) \mod 4\mathbb{Z}$ and

$$\operatorname{Spec}(A_0) = \{\tau_n = -1(d+1+2n)^2,\ n\in\mathbb{Z}^+\} \tag{11.6.21}$$

and the multiplicity of τ_n is $P(n)$ where $P(n)$ is a polynomial of degree $(d-1)$ in n. Let us introduce the function $f(\tau) = \frac{1}{2}(-\tau)^{-\frac{1}{2}} - \frac{1}{2}$.

We leave to the reader

Problem 11.6.8. Prove that for generic positively homogeneous functions a^1_β (with $|\beta|\le k$) of degree $(-1+2|\beta|-\delta_1)$ condition (11.5.31) is fulfilled for all $\omega\in\mathbb{R}$ and therefore asymptotics (11.6.18) holds for $0<\delta_1<\frac{1}{2}$ while asymptotics (11.6.19)–(11.6.20) holds for $\delta_1 \ge \frac{1}{2}$, and now there is no restriction of the type "$\delta_1 \le 2\mu+2$" on δ_1 and $k=0,1,2$ is arbitrary.

Moreover, prove that one can take $V' = v_0|x|^{-1-\delta}$ with an arbitrary exponent $\delta>0$ and an arbitrary constant $v_0\ne 0$.

Let us assume that $b=0$. There are two possibilities in this case: the trivial possibility is to take $\delta_2 \in (\delta_1, 2\delta_1)$ and the non-trivial one is to take $\delta_2 = 2\delta_1$. In the first case all the previous statements remain true with δ_1 replaced by δ_2 etc.

In the second case we assume that condition (11.5.39) is fulfilled in which case the symbol $b(x,\xi)$ is given by formula (11.3.27) with b and b^s replaced by $-\frac{1}{2}A^1(-A^0)^{-\frac{1}{2}}$ and $-\frac{1}{2}A^2(-A^0)^{-\frac{1}{2}}+\frac{3}{4}(A^1)^2(-A^0)^{-\frac{3}{2}}$ respectively and with integration from 0 to π, $\frac{\pi}{2}$ instead of 2π, π respectively.

Further, the numerical factor on every integral is $\frac{1}{\pi}$ instead of $\frac{1}{2\pi}$. Furthermore, one can move A^0 out of the Poisson brackets and then take $A^0=-1$. Therefore

$$b(x,\xi)=\frac{1}{2\pi}\int_0^\pi \bigl(A^2-\frac{3}{2}(A^1)^2\bigr)\circ\Psi_t dt+ \frac{1}{2\pi}\int_0^{\frac{\pi}{2}}\int_0^{\frac{\pi}{2}}\{A^1\circ\Psi_t, A^1\circ\Psi_{t+t'}\}dtdt'. \tag{11.6.22}$$

Therefore the following assertion has been proven:

Theorem 11.6.9. *Let $x^0=\mathbb{R}^d$, $A=A^{0\mathrm{w}}+A^{1\mathrm{w}}+A''^{\mathrm{w}}$ where A'' satisfies condition (11.5.29) with a^2_β,δ_2 instead of a^1_β,δ_1, $\delta_2=2\delta_1$ and a^j_β are functions which are positively homogeneous of degrees $(-1+2|\beta|-\delta_j)$ with respect to x, smooth for $x\ne 0$. Let condition (11.5.39) be fulfilled and let us define the symbol b by (11.6.22). Then*

(i) If $\delta_2<\frac{1}{2}$ and condition (11.5.31) is fulfilled for every $\omega\in\mathbb{R}$ then asymptotics (11.6.18) hold as $\tau\to -0$.

(ii) If $\delta_2\ge\frac{1}{2}$ and condition (11.5.31) is fulfilled for some $\omega\in\mathbb{R}$ then asymptotics (11.6.19)–(11.6.20) holds as $\tau\to -0$ and $(-\tau)^{-\frac{1}{2}}\in\bigcup_{n\in\mathbb{Z}^+}$ where J_n are given in Theorem 11.5.9(ii) with $\omega_0=2(d-1),\mu=2$ and with δ_1 replaced by δ_2 everywhere.

One can easily prove that for fixed a^1_β with $|\beta|\le 2$ satisfying (11.5.39) and for generic a^2_β with $|\beta|\le k$ condition (11.5.33) is fulfilled for all $\omega\in\mathbb{R}$; here $k=0,1,2$ is arbitrary.

11.6.4 Singularities

Let us consider homogeneous lower order terms with singularities (for the Schrödinger operator only):

Theorem 11.6.10 [52]. *Let $d\ge 2$, $X=\{(r,\theta)\in\mathbb{R}^+\times U\}$ where $U\subset\mathbb{S}^{d-1}$ and let $g^{jk}=g^{jk}(\theta)$, $V_j=v_j(\theta)r^\mu$, $V=v(\theta)r^{2\mu}$, $\mu\in(-1,0)$. Further, let*

[52] Cf. Theorem 11.5.14.

$\tilde{\rho} = \tilde{\rho}(\theta)$ and $\tilde{\gamma} = \tilde{\gamma}(\theta)$ be functions on U and let conditions (11.1.6)–(11.1.8) *be fulfilled.*

Furthermore, assume that *__either__* $d \ge 3$ *__or__* $\tilde{\rho}\tilde{\gamma} \ge \epsilon_0 > 0$. *Finally, let*

$$\tilde{\rho}^{-\frac{d}{\mu}} \in \mathscr{L}^1(U), \tag{11.6.23}$$

$$\tilde{\rho}^{-\frac{1}{\mu}(d-1)}\tilde{\gamma}^{-1} \in \mathscr{L}^1(U). \tag{11.6.24}$$

Moreover, let $v(\bar{\theta}) < 0$ for some $\bar{\theta} \in U$.

Then asymptotics (11.6.2) *holds and $\mathsf{N}^{\mathsf{W}}(\tau) = \varkappa_0|\tau|^{ld}$ with the coefficient $\varkappa_0$ given by* (11.6.13).

Further, if $\vartheta \in (0, \vartheta_1^*)$, $\vartheta_1^* = -l(d-1)(1-l)^{-1} = (1+\mu)(d-1)(1-\mu)^{-1}$ and if

$$\tilde{\rho}^{-\frac{1}{\mu}(d-1-\vartheta)}\tilde{\gamma}^{-1-\vartheta} \in \mathscr{L}^1(U) \tag{11.6.25}$$

then asymptotics (11.6.3)–(11.6.4) *holds.*

Furthermore, let us assume that there exists a set $\Lambda \subset \Sigma = \Sigma_{-1}$ with $\operatorname{mes}_\Sigma(\Lambda) = 0$ such that through every point of $(\Sigma \setminus \Lambda)$ there passes an infinitely long non-periodic billiard generated by the Hamiltonian

$$H^0 = \tilde{\gamma} r^{1-\mu}(\tilde{\rho} + r^{-\mu})^{-1}\bigl(\sum_{j,k} g^{jk}(\xi_j - V_j)(\xi_k - V_k) + V + 1\bigr) \tag{11.6.26}$$

which never hits $0 \times \mathbb{R}^d \cup \Lambda$. Then asymptotics (11.6.10) *holds with the coefficient $\varkappa_0$ given by* (11.6.11).

Moreover, let $\vartheta \in (0, \vartheta_1^*)$ and (11.6.25) *be fulfilled. Then asymptotics* (11.6.12) *holds.*

Example 11.6.11[53]. Let $X = \mathbb{R}^d$, $d \ge 2$ and $\rho = |x''|^\nu \cdot |x|^{\mu-\nu}$, $\gamma = \frac{1}{2}|x''|$ with $\mu \in (-1, 0)$, $\nu \in (\mu, 0)$. Then

(i) Conditions (11.6.23) and (11.6.25) are fulfilled if and only if $\nu > \nu_0 = \mu d'' d^{-1}$ and $\nu > \nu_{\vartheta+1} = \mu(d'' - 1 - \vartheta)(d - 1 - \vartheta)^{-1}$ respectively.

Therefore for $\nu > \nu_1$ asymptotics (11.6.2) holds while for $\nu > \nu_{\vartheta+1}$, $\vartheta < \vartheta_1^*$ asymptotics (11.6.3) holds[54].

---

[53] Cf. Example 11.5.17.

[54] And under non-periodicity condition on the Hamiltonian flow asymptotics (11.6.10) holds. Further, asymptotics (11.6.12) (for $\nu > \nu_{\vartheta+1}$) holds.

(ii) The case $\nu \in [\nu_0, \nu_1)$ is treated in the same way as case (ii) of Example 11.5.17. We obtain that for $\nu = \nu_1$ asymptotics

$$\mathsf{N}^-(\tau) = \mathsf{N}^{\mathsf{W}}(\tau) + O\bigl(|\tau|^{l(d-1)} \log |\tau|\bigr) \tag{11.6.27}$$

holds as $\tau \to -0$ and for $\nu \in (\nu_0, \nu_1)$ asymptotics

$$\mathsf{N}(\tau) = \mathsf{N}^{\mathsf{W}}(\tau) + O\bigl(|\tau|^q\bigr) \tag{11.6.28}$$

holds as $\tau \to -0$ with $q = -\frac{1}{2}(\mu+1)d'(\nu-\mu)^{-1} = l(d-\eta)$, $\eta = (\mu d'' - \nu d)(\mu-\nu)^{-1} \in (0,1)$.

In these asymptotics $\mathsf{N}^{\mathsf{W}}(\tau) = \varkappa_0|\tau|^{ld}$.

(iii) Similarly, for $\nu = \nu_{\vartheta+1}$, $\vartheta \in (0, \vartheta_1^*)$ asymptotics

$$\mathsf{N}^-_{(\vartheta)}(\tau) = \mathsf{N}^{\mathsf{W}}_{(\vartheta)}(\tau) + O\bigl(|\tau|^{l(d-1-\vartheta)+\vartheta} \log |\tau|\bigr) \tag{11.6.29}$$

holds as $\tau \to -0$ and for $\nu \in (\nu_0, \nu_{\vartheta+1})$ asymptotics

$$\mathsf{N}^-_{(\vartheta)}(\tau) = \mathsf{N}^{\mathsf{W}}_{(\vartheta)}(\tau) + O\bigl(|\tau|^{q+\vartheta}\bigr) \tag{11.6.30}$$

holds as $\tau \to -0$ with the same q as before.

(iv) Finally, for $\nu = \mu d'' d^{-1}$ asymptotics

$$\mathsf{N}(\tau) = \mathsf{N}^{\mathsf{W}}(\tau) + O\bigl(|\tau|^{ld}\bigr) \tag{11.6.31}$$

holds as $\tau \to -0$ with $\mathsf{N}^{\mathsf{W}}(\tau)$ given by (11.5.52) and moreover $\mathsf{N}^{\mathsf{W}}(\tau) \asymp |\tau|^{ld} \log|\tau|$. Moreover, footnotes [41)], [42)] section hold.

We now have no need of the coercivity condition and therefore Theorems 11.5.18 and 11.5.20 and Examples 11.5.19 and 11.5.21 are almost senseless. However, the following problems make sense:

Problem 11.6.12. (i) Examine potentials of the type $V^0 + V'$ where V^0 is a slowly decreasing and not very singular potential and V' decreases faster and is more singular, etc.

(ii) Let us note that for $d \ge 3$ in the zone $\{V \ge 0\}$ no condition excluding $|V_j| \le c\sqrt{V}$ is necessary. Consider examples with $\rho = \langle x\rangle^\mu$, $\gamma = \langle x\rangle^{1-\sigma}$ with $\mu < 0$, $\sigma \in (0, 1+\mu)$, where asymptotics with a worse remainder estimate may be obtained. In this case the magnetic field can decrease more slowly at infinity.

11.6.5 Quasihomogeneous Potentials

Let us consider non-homogeneous operators. Repeating the arguments of Example 11.5.23 with obvious modifications we obtain

Example 11.6.13. (i) Let X be an L-quasiconical domain in $\mathbb{R}^d$ and let g^{jk}, V_j, V be positively L-quasihomogeneous with respect to x of degrees 0, μ, 2μ respectively where $\mu \in (-1,0)$, $L=(L_1,\dots,L_d)$ with $L_j=1$ for $1\le j\le d'$ and $L_j>1$ for $d'+1\le j\le d$, $1\le d'<d$.

Let us assume that g^{jk}, V_j, $V\in \mathscr{C}^K(\bar{X}\setminus 0)$ satisfy condition (11.1.8) and that $V(\bar\theta)<0$ for some $\bar\theta\in X$. Then conditions (11.1.7) are fulfilled with $\rho=[x]_L^\mu$, $\gamma=\epsilon[x]_L$ for $[x]_L\ge 1$; for $[x]_L\le 1$ one should take $\gamma=\epsilon[x]_L^{\bar L}$ with $\bar L=\max_j L_j$. Then the asymptotics

$$\mathsf{N}^-(\tau)=\varkappa_0|\tau|^p+O\bigl(|\tau|^q\bigr) \tag{11.6.32}$$

holds as $\tau\to -0$ with $p=\frac{d}{2}+\frac{L^*}{2\mu}$, $q=\frac{1}{2}(d-1)+\frac{1}{2\mu}(L^*-1)$, $L^*=\sum_j L_j$ and

$$\varkappa_0=(2\pi)^{-d}\omega_d\int (-1-V)_+^{\frac{d}{2}}\sqrt{g}\,dx. \tag{11.6.33}$$

(ii) Moreover, let us assume that $x'\ne 0$, $\nu'\ne 0$ on $\partial X\setminus 0$ where ν is a unit normal to ∂X and $x'=(x_1,\dots,x_{d'})$, $\nu'=(\nu_1,\dots,\nu_{d'})$. Let us assume that $g^{jk}=0$, $V_k=0$ for $j\le d'$, $k\ge d'+1$. Then asymptotics

$$\mathsf{N}^-(\tau)=\varkappa_0|\tau|^p+\bigl(\varkappa_1+o(1)\bigr)|\tau|^q \tag{11.6.34}$$

holds with

$$\varkappa_1=\mp\frac{1}{4}(2\pi)^{1-d}\omega_{1-d}\int (-1-V)_+^{\frac{d-1}{2}}\,dS'\sqrt{g''}dx'' \tag{11.6.35}$$

where dS' and $\sqrt{g''}dx''$ are the Riemannian densities introduced in Example 11.5.23.

11.6.6 Quickly Decreasing Potentials

Let us consider potentials which decrease rather quickly. We considered homogeneous potentials decreasing as $\langle x\rangle^{2\mu}$ with $\mu\in(-1,0)$ and $N^-(0)<\infty$ as either $\mu<-1$ or $\mu=-1$, $V\ge -\epsilon\langle x\rangle^{2\mu}$ with sufficiently small $\epsilon>0$. So "rather quickly" should mean "a bit slower than $\langle x\rangle^{-2}$" .

Theorem 11.6.14. *Let X be an unbounded domain in $\mathbb{R}^d$ and let conditions (11.1.1)–(11.1.3) and (11.6.1) be fulfilled with*

$$\gamma = \epsilon\langle x\rangle, \qquad \rho = \langle x\rangle^{-1}\log^{\nu}\langle x\rangle \tag{11.6.36}$$

where we now take $\langle x\rangle = (|x|^2+4)^{\frac{1}{2}}$, $\nu > 0$.

Further, let us assume that for every $\tau\in(-\epsilon,0]$ one of non-degeneracy conditions is fulfilled for the symbol $(A(x,\xi)\tau^{-1}-1)$ uniformly with respect to (τ,x,ξ) (this condition is restrictive only for $|x|\ge C_0$). Then

(i) Asymptotics

$$\mathsf{N}^-(\tau) = \mathsf{N}^{\mathsf{W}}(\tau) + O\bigl(\bigl|\log|\tau|\bigr|^{\nu(d-1)+1}\bigr) \tag{11.6.37}$$

holds as $\tau\to -0$ where $\mathsf{N}^{\mathsf{W}}(\tau) = O\bigl(\bigl|\log|\tau|\bigr|^{\nu d+1}\bigr)$ and under condition (11.6.8) $\mathsf{N}^{\mathsf{W}}(\tau)\asymp\bigl|\log|\tau|\bigr|^{\nu d+1}$.

(ii) Moreover, let $X\cap\{|x|\ge r_0\} = X^0\cap\{|x|\ge r_0\}$ where X^0 is a conical domain with $\partial X^0\setminus 0\in\mathscr{C}^K$ and let stabilization conditions (11.5.4) be fulfilled for functions $a_\beta\log^{-\nu(2m-|\beta|)}$ where the functions $a^0_\beta\in\mathscr{C}^K(\bar{X}^0\setminus 0)$ are positively homogeneous of degrees $-(2m-|\beta|)$ on x.

Then the microhyperbolicity condition is fulfilled automatically. Further, if the non-periodicity condition is fulfilled for the quasihomogeneous symbol

$$A^0(x,\xi) = |x|^{2m}\sum_{|\beta|\le 2m} a^0_\beta(x)\xi^\beta \tag{11.6.38}$$

on the energy level 0 then the asymptotics

$$\mathsf{N}^-(\tau) = \mathsf{N}^{\mathsf{W}}(\tau) + \bigl(\varkappa_1 + o(1)\bigr)\bigl|\log|\tau|\bigr|^{\nu(d-1)+1} \tag{11.6.39}$$

holds where, for the Schrödinger operator,

$$\varkappa_1 = \mp 2^{-\nu(d-1)-3}(2\pi)^{1-d}\omega_{d-1}\int_{\partial U}(V^0)_-^{\frac{d-1}{2}}\,dS, \tag{11.6.40}$$

$U = X^0\cap\mathbb{S}^{d-1}$ is a domain on $\mathbb{S}^{d-1}$, $dS = dS^0 : dr$ is the natural density on ∂U, dS^0 is the Riemannian density on ∂X^0 generated by the metric g^{jk0} (compare with ((11.6.12)).

(iii) On the other hand, if only the stabilization condition is fulfilled then $\mathsf{N}^{\mathsf{W}}(\tau) = (\varkappa_0 + o(1))\bigl|\log|\tau|\bigr|^{\nu d+1}$ *where, for the Schrödinger operator,*

$$\varkappa_0 = 2^{-\nu d-1}(2\pi)^{-d}\omega_d \int_U (V^0)_-^{\frac{d}{2}}\sqrt{g^0}\, d\theta, \tag{11.6.41}$$

$d\theta = dx : dr$, $g^0 = \det(g^{jk0})^{-1}$ *(compare with ((11.6.11)).*

Proof. Statement (i) follows immediately from our previous arguments; moreover, now $\rho\gamma = \log^{\nu}\langle x\rangle \to \infty$ as $|x| \to \infty$ and $X'' = \emptyset$.

In order to prove Statement (ii) let us note that the principal contribution to the remainder estimate is given by the zone $\{R \le |x| \le \epsilon\tau^{-\frac{1}{2m}}\}$ with arbitrary fixed R and $\epsilon > 0$. We need to examine the Hamiltonian trajectories starting in this zone and associated with the symbol

$$A_\tau(x,\xi) = \langle x\rangle^{2m}\log^{-\nu}\langle x\rangle\bigl(A(x,\xi) + \tau\bigr) \tag{11.6.42}$$

on $\Sigma_\tau = \{\det A_\tau(x,\xi) = 0\}$. Let $\Sigma^0 \supset \Lambda_T$ be a closed set of measure 0 such that through every point of $(\Sigma^0 \setminus \Lambda_T)$ there passes a non-periodic Hamiltonian trajectory (with reflections) of length $2T$ associated with the symbol A^0. Then one can take

$$\begin{aligned}\Lambda_{1,T,\epsilon,\varepsilon,\tau,R} = \bigl\{(x,\xi) \in \Sigma_\tau,\ & R \le |x| \le \epsilon\tau^{-\frac{1}{2m}},\\ &\operatorname{dist}(\frac{x}{|x|}, \partial X^0) > 0, \operatorname{dist}\bigl((\frac{x}{|x|}, \xi|x|\log^{-\nu}|x|), \Lambda_T\bigr) > \varepsilon\bigr\}\end{aligned}$$

with $R = R(\epsilon,\varepsilon,T)$. Moreover, one can see that $\log|x(t)|\log^{-1}|x(0)| = 1 + o(1)$ as $|x(0)| \to \infty$ uniformly with respect to τ and $t \in [-2T, 2T]$ with arbitrary fixed T. We leave the details to the reader. $\square$

We leave to the reader the following set of rather easy problems:

Problem 11.6.15. (i) Prove a statement similar to Theorem 11.6.10 with

$$g^{jk} = g^{jk}(\theta), \qquad V_j = v_j(\theta)r^{-1}\log^{\nu} r, \qquad V = v(\theta)r^{-2}\log^{2\nu} r,$$

$X = (1,\infty) \times U$.

(ii) Consider the case when

$$\rho = \langle x\rangle^{\mu}\log^{\nu}\langle x\rangle, \tag{11.6.43}$$

$\vartheta \geq 0$ with <u>either</u> $\mu > -d(d+2m\vartheta)^{-1}$ <u>or</u> $\mu = -d(d+2m\vartheta)^{-1}$, $\nu \geq -1(d+2m\vartheta)^{-1}$.

Prove that under these conditions the principal part of the asymptotics is $\asymp I$ with either $I = |\tau|^p$ (with $p < 0$) or $I = \bigl|\log|\tau|\bigr|^q$ (with $q > 0$) or $I = \log\bigl|\log|\tau|\bigr|$ and the remainder estimate is either from the same list or $O(1)$.

(iii) Moreover, consider the case of potentials decreasing even more quickly when

$$\rho = \langle x\rangle^{-1}\bigl(\underbrace{\log\ldots\log}_{r\text{ times}}\langle x\rangle\bigr)^{\nu} \tag{11.6.44}$$

when the principal part of the asymptotics and the remainder estimate are of the orders $\log|\tau| \times \zeta(\tau)^{\nu d}$ and $\log|\tau| \times \zeta(\tau)^{\nu(d-1)}$ respectively with $\zeta(\tau) = \underbrace{\log\ldots\log}_{r\text{ times}}|\tau|$.

(iv) Prove that in the framework of Theorem 11.6.14 $\mathsf{N}^-_{(\vartheta)}(-0) < \infty$ for $\vartheta > 0$.

11.6.7 Slowly Decreasing Potentials

Let us consider the slowly decreasing potential $V = v(\theta)\log^{-1} r$ in the spherical coordinate system (r,θ). Using

$$\tilde{\rho}_\tau = (\tau - V)_+^{\frac{1}{2}}, \qquad \tilde{\gamma}_\tau = \epsilon_0 r \min(1, (\tau - V)_+ \log r) \tag{11.6.45}$$

for $r \geq 2$ we easily prove

Theorem 11.6.16. *Let $d \geq 3$ and $X \cap \{r \geq r_0\} = \{r \geq r_0, \theta \in U\}$. For $|x| \geq r_0$ let*

$$g^{jk} = g^{jk}(\theta), \qquad V_j = 0, \qquad V = v(\theta)\log^{-1} r \tag{11.6.46}$$

where (r,θ) are spherical coordinates, $g^{jk}, v, \partial X \in \mathscr{C}^K$ and the matrix (g^{jk}) is positive definite. Let $v_0 = \min_U v < 0$. Then

(i) Asymptotics (11.6.3) *holds with*

$$\mathsf{N}^{\mathsf{W}}_{(\vartheta)}(\tau) \asymp |\tau|^{d+2\vartheta}\int_U \exp\bigl(\frac{vd}{\tau}\bigr)\,d\theta, \tag{11.6.47}$$

$$R_{(\vartheta)}(\tau) = |\tau|^{d-2}\int_U \exp\bigl(\frac{v}{\tau}(d-\vartheta-1)\bigr)\,d\theta \tag{11.6.48}$$

for $\vartheta < d-1$. *Moreover, one can take*

$$\mathsf{N}^{\mathsf{W}}(\tau) = a\int_{U'} |\tau|^d v_-^{-\frac{d}{2}} F_d(-\frac{\tau}{v_-})\exp(-\frac{vd}{\tau})\sqrt{g}\,d\theta \tag{11.6.49}$$

where $U' = U\cap\{v < v_0+\varepsilon\}$ *with arbitrarily small fixed* $\varepsilon > 0$, $a = 2^{1-d}\pi^{-\frac{d}{2}}d^{-\frac{d}{2}+1}$,

$$F_d(s) = \frac{1}{\Gamma(\frac{d}{2}+1)}\int_0^{\frac{d}{s}-\varepsilon} t^{\frac{d}{2}}\bigl(1-\frac{st}{d}\bigr)^{-\frac{d}{2}} e^{-t}dt = 1 + O(s) \tag{11.6.50}$$

as $s\to +0$.

(ii) In particular, for $v = v_0 = \mathrm{const}$ *the principal part of the asymptotics is* $\asymp |\tau|^{d+2\vartheta}\exp(\frac{v_0 d}{\tau})$ *and the remainder estimate is* $O(|\tau|^{d-2}\exp(\frac{v_0}{\tau}(d-1-\vartheta)))$.

Moreover, in this case one can take

$$\gamma_\tau = \epsilon_0 r\min\bigl(1, (\tau - V)_+\log^2 r\bigr)$$

and obtain a remainder estimate with the factor $|\tau|^{d+\vartheta-1}$ *instead of* $|\tau|^{d-2}$.

On the other hand, if $v = v(\theta) = v_0 \implies \theta\in\partial U, \nabla v(\theta)\neq 0$ *and* $\operatorname{rank}\operatorname{Hess} v|_{\partial U} = d-1$ *then the principal part is* $\asymp |\tau|^{\frac{3}{2}+2\vartheta}\exp(\frac{v_0 d}{\tau})$ *and the remainder estimate is* $O(|\tau|^{\frac{3}{2}d-3}\exp\bigl(\frac{v_0}{\tau}(d-1-\vartheta)\bigr))$. *In other cases the principal part and the remainder estimate are of intermediate orders.*

Problem 11.6.17. (i) Consider the case $\vartheta \geq (d-1)$. Then one can take $\rho_\tau = \log^{-\frac{1}{2}} r + |\tau|$ and $\gamma_\tau = \epsilon_0 r$ and for $d\geq 2$ one can derive the remainder estimates $O(|\tau|^d)$ for $\vartheta = (d-1)$ and $O(1)$ for $\vartheta > (d-1)$.

(ii) Consider potential $V = v(\theta)\log^{-\nu} r$ with arbitrary $\nu > 0$.

11.6.8 General Boundary Conditions

Remark 11.6.18. All the above results remain true for the Neumann boundary condition provided the domain satisfies the cone condition. Let us assume that this (cone) condition is fulfilled. Moreover, all the above results remain true for mixed boundary conditions (provided the asymptotics do not distinguish Dirichlet and Neumann boundary conditions) and also in the following cases:

(i) We are in the frameworks of Theorem 11.6.2 or 11.6.1011.6.10, and condition (11.5.89) is fulfilled.

(ii) We are in the frameworks of Example 11.6.13(ii), and condition (11.5.90) is fulfilled.

Moreover, in the context of Theorem 11.6.5 asymptotics (11.6.16) remain true if condition (11.5.91) is fulfilled.

11.6.9 Non-Semi-Bounded Operators

Finally, let us consider non-semi-bounded operators. We consider the Dirac operator with mass $M > 0$ (otherwise in the context of this section the spectrum coincides with $\mathbb{R}$). Moreover, let $V \to 0$, $V_j \to 0$ as $|x| \to \infty$.

Then $\mathsf{Spec}_{\mathsf{ess}}(A) = (-\infty, M] \cup [M, \infty)$ and the spectral gap $(-M, M)$ contains either a finite or infinite number of eigenvalues. In the second case these eigenvalues tend to $-M + 0$ and (or) to $M - 0$ and we can consider the corresponding eigenvalue asymptotics. Let us consider $M - 0$. In order to replace it by -0 we replace the operator A by $A - M \cdot I$. Let us fix $\tau_1 \in (-2M, 0)$ and let us consider

$$\mathsf{N}^-(\tau) := \mathsf{N}(\tau_1, \tau) \tag{11.6.51}$$

with $\tau \to -0$. Then

Remark 11.6.19. Under described modifications

(i) Theorems 11.6.1, 11.6.2, 11.6.14 and 11.6.5 and Examples 11.6.13, 11.6.6 remain true and one can check conditions on the Hamiltonian flow for the corresponding Schrödinger operator with the symbol

$$H = \sum_{j,k} g^{jk}(\xi_j - V_j)(\xi_k - V_k) + (V + M)^2 - M^2. \tag{11.6.52}$$

In this Hamiltonian one can skip the term V^2 which is negligible at infinity (as far as non-periodicity condition is concerned) in which case one can consider the Hamiltonian

$$H = \sum_{j,k} g^{jk}(\xi_j - V_j)(\xi_k - V_k) + 2MV. \tag{11.6.53}$$

Now, however, due to term V^2

$$\mathsf{N}^{\mathsf{W}}(\tau) = \varkappa_0 |\tau|^p (1 + O(|\tau|^\delta)) \tag{11.6.54}$$

with $\delta > 0$ instead of $\mathsf{N}^{\mathsf{W}}(\tau) = \varkappa_0 |\tau|^p$.

(ii) Theorem 11.6.10 (and Example 11.6.11(i) for $d \geq 3$) remain true. In order to prove this let us note that only the analysis in the zone $\{\tilde{\rho} r^\mu \geq 1\}$ should be changed. In this zone instead of $\rho = \tilde{\rho} r^\mu$ one should set $\rho = \tilde{\rho}^2 r^{2\mu}$.

We include this zone in X_{sing}. In this zone $\rho^d \in \mathscr{L}^1$ due to assumption $\tilde{\rho}^{-\frac{d}{2\mu}} \in \mathscr{L}^1(U)$. Therefore the contribution of this zone to the asymptotics and to the remainder estimate is $O(1)$.

(iii) Theorem 11.6.16 also remains true. However, $\mathsf{N}^{\mathsf{W}}(\tau)$ should be calculated for the Dirac operator, i.e., for the Schrödinger operator with the symbol (11.6.52) and with the spectral parameter equal to 0.

In the proof one should replace $\rho_\tau = (\tau - V)_+^{\frac{1}{2}}$ by

$$\rho_\tau = \bigl(\tau - V + C_1 r^{-1} \log^{-2} r\bigr)_+^{\frac{1}{2}} \tag{11.6.55}$$

with large enough C_1 in the zone $X_{\mathsf{sing}} \cap \{\tau - V \leq C|\tau|^{\frac{2}{3}} r^{-\frac{2}{3}}\}$. This does not increase the remainder estimate.

Let us consider Theorems 11.5.9 and 11.6.9. For simplicity we assume that

$$|v| \leq 2M - \epsilon_0 \qquad \forall x. \tag{11.6.56}$$

We leave to the reader the following easy

Problem 11.6.20. Conduct analysis below without assumption (11.6.56).

Under assumption (11.6.56) for appropriate τ_1,

$$\mathsf{N}^-(\tau) = \tilde{\mathsf{N}}^-(\tau) - \tilde{\mathsf{N}}^-(\tau_1) \tag{11.6.57}$$

where $\tilde{\mathsf{N}}^-(\tau)$ is the number of the negative eigenvalues for the operator

$$A(\tau) = \mathcal{L}^*(2M - V + \tau)^{-1}\mathcal{L} + V - \tau \tag{11.6.58}$$

with

$$\mathcal{L} = \sum_j \bigl(\sum_k \omega^{jk}(D_k - V_k) - \frac{i}{2}\omega^j\bigr)\sigma'_j \tag{11.6.59}$$

where we assume that

$$\sigma_0 = \begin{pmatrix} I & 0 \\ 0 & -1 \end{pmatrix}, \qquad \sigma_j = \begin{pmatrix} 0 & \sigma'^{*}_j \\ \sigma'_j & 0 \end{pmatrix} \quad (j \geq 1)$$

and use the notation $\omega^j = \sum_k \partial_k \omega^{jk}$.

If we could take $\rho = \langle x\rangle^{-\frac{1}{2}}$, $\gamma = \epsilon_0\langle x\rangle$ for the original Dirac operator and if

$$D^\alpha(\omega^{jk} - \delta^{jk}) = O(|x|^{-\delta_0 - |\alpha|}), \tag{11.6.60}$$

$$D^\alpha F_{jk} = O(|x|^{-\frac{3}{2} - \delta_3 - |\alpha|}) \qquad \text{as } |x| \to \infty \quad \forall \alpha : |\alpha| \leq K \tag{11.6.61}$$

(where $F_{jk} = i(\partial_j V_k - \partial_k V_j)$ is the tensor magnetic intensity) then the operator $A(\tau)$ is a scalar operator plus perturbation $\sum_j A'_j(D_j - V_j) + A'_0$ where

$$D^\alpha A'_j = O(|x|^{\max(-1-\delta_0, -2) - |\alpha|}), \tag{11.6.62}$$

$$D^\alpha A'_0 = O(|x|^{\max(-1-\delta_0, -2, -\frac{1}{2}-\delta_3) - |\alpha| - 1}) \quad \text{as } |x| \to \infty \tag{11.6.63}$$

and in the context of the previous subsections this error does not matter provided that under the assumptions of Theorems 11.6.7 and 11.6.9

$$\delta_\iota < \frac{3}{2}, \quad \delta_\iota < \frac{1}{2} + \delta_0, \quad \delta_\iota < \frac{1}{2} + \delta_3 \tag{11.6.64$_{1,2,3}$}$$

respectively with $\iota = 1, 2$ (because the "order" of the operator $D_j - V_j$ is $-\frac{1}{2} + \delta'$ with arbitrarily small $\delta' > 0$ if we take the "order" of $|x|$ equal to 1). Then the operator $A(\tau)$ differs by negligible terms from the Schrödinger operator

$$B(\tau) = \sum_{j,k} (D_j - V_j) g^{jk} (2M - V + \tau)^{-1} (D_k - V_k) + V - \tau \tag{11.6.65}$$

and moreover, for

$$\delta_\iota < 1 \qquad (\iota = 1, 2) \tag{11.6.66}$$

one can replace this operator by the operator

$$B'(\tau) = (2M)^{-1}\sum_{j,k}(D_j - V_j)g^{jk}(D_k - V_k) + V - \tau. \tag{11.6.67}$$

So we arrive to

Theorem 11.6.21. *Let conditions* (11.6.60)–(11.6.63), $(11.6.64)_{1,2,3}$ *and* (11.6.66) *be fulfilled.*

Furthermore, let assumptions of some statement of Theorem 11.6.7 or Theorem 11.6.9 be fulfilled for the Schrödinger operator $B'(\tau)$ defined by (11.6.65) *(or equivalently by* (11.6.67)*); then $\iota = 1, 2$ respectively.*

Then the corresponding asymptotics holds for the Dirac operator. Namely, asymptotics (11.6.18)*) holds for $\delta_\iota < \frac{1}{2}$ (then condition* (11.6.64)*) is fulfilled automatically) and asymptotics* (11.6.19)*) holds for $\delta_\iota \in (\frac{1}{2}, 1)$.*

In these asymptotics $\mathsf{N}^{\mathsf{W}}(\tau)$ is calculated for the Dirac operator and $F(\tau)$ is calculated for the Schrödinger operator and then it is multiplied by $\frac{1}{2}\mathsf{D}$.

On the other hand, let condition (11.6.66) be violated but condition $(11.6.63)_{1,2,3}$ be (almost) fulfilled. Then one should replace $B(\tau)$ by

$$B''(\tau) = (2M)^{-1}\sum_{j,k}(D_j - V_j)g^{jk}(D_k - V_k) + V - \tau - (2M)^{-1}(V - \tau)^2 \tag{11.6.68}$$

and in the last term one can replace V by V^0 where V^0 is a smooth function equal to $-1(2M|x|)^{-1}$ for $|x| \geq c$. Changing variables $x \to \zeta x$ and multiplying by $2M\zeta^2$ with $\zeta = 1 - \tau M^{-1}$ we obtain, modulo negligible terms, operator

$$B'''(\tau') = \sum_{j,k}(D_j - V_j)g^{jk}(D_k - V_k) + W - \tau' \tag{11.6.69}$$

with the new potential

$$W = 2MV - (V^0)^2 \tag{11.6.70}$$

and with the modified spectral parameter

$$\tau' = \tau\bigl(1 - \frac{3\tau}{2M}\bigr) \tag{11.6.71}$$

This yields that

Theorem 11.6.22. *Let assumptions of some statement of either Theorem 11.6.7 or Theorem 11.6.9 be fulfilled for the Schrödinger operator $B'''(0)$ defined by* (11.6.69).

Let conditions (11.6.60)–(11.6.63), $(11.6.63)_{1,2,3}$ *be fulfilled but $\delta_\iota \in [1, \frac{3}{2})$.*

Then asymptotics (11.6.19) *remains true where $\mathsf{N}^{\mathsf{W}}(\tau)$ is calculated for the Dirac operator and $F(\tau')$ is calculated for the Schrödinger operator $B'''(\tau')$ and is then multiplied by $\frac{1}{2}\mathsf{D}$, and τ' is defined by* (11.6.71).

Let us weaken condition $(11.6.64)_{1,2,3}$.

Remark 11.6.23. For $d = 2$ one can assume without any loss of the generality that $\mathsf{D} = 2$. Then $A(\tau)$ is surely a scalar operator but we have not taken into account some terms. Taking them into account and applying the above analysis we can discard condition $(11.6.64)_{1,2,3}$ completely (in which case we have additional corrections for W and τ').

In the general case we assume that conditions (11.6.60)–(11.6.62) are fulfilled. If $\delta_1 \ge 1$ then the non-scalar terms in $A(\tau)$ are

$$\frac{1}{2}(2M)^{-3}r^{-3}\sum_{l\neq k}\sigma_l'^{*}\sigma_k' F_{lk} \tag{11.6.72}$$

modulo operators of the form $\sum_j b_j' D_j + b_0'$ where b_j', b_0' satisfy condition (11.6.60), (11.6.61) with "-2" replaced by "-3." Here we used the skew-symmetry of the matrices $\sigma_l'^{*}\sigma_k'$ with $l \ne k$; recall that $F_{lk} = (\partial_k A_l - = \partial_l A_k)$.

This statement remains true for $\delta_2 \ge 1$ (i.e., for $\delta_1 \ge \frac{1}{2}$) in the framework of Theorem 11.6.7. However, in this case the operator $A(\tau)$ also contains terms of type $b_j'' D_j + b_0''$ with b_j'', b_0'' satisfying conditions (11.6.62)–(11.6.63) with "-2" replaced by "$-2 - \delta_1$" and this operator skew-commutes with the antipodal map. In both cases these terms are negligible provided

$$\delta_\iota < 2, \quad \delta_\iota < \frac{1}{2} + \delta_3 \quad (\iota = 1, 2). \tag*{$(11.6.73)_{1,2}$}$$

However, the second inequality is not necessary: for $\iota = 1$ it is fulfilled automatically and for $\iota = 2$ it is associated with terms of the type

$$-i\,\mathrm{const}\sum_{l\neq k}\sigma_l'^{*}\sigma_l' F_{lk}.$$

However, in this case $V_l \circ j^* = V_l$ and therefore $F_{lk} \circ j^* = -F_{lk}$ and the same is true for this extra term. Then one should check the condition $-2 - \delta_1 < -1 < -2\delta_1$ which follows from the first inequality.

Let us consider Schrödinger operator with a (perturbed) Coulomb potential. Let us consider

$$\int_0^{T_0} r^{-3}(x_j\xi_k - x_k\xi_j)\circ\Psi_t\,dt$$

where Ψ_t is the Hamiltonian flow generated by $|\xi|^2 - |x|^{-1}$. One can easily see that this integral is equal to $\pi\mu_{jk}\mu^{-3}$ where μ_{jk} are components of the angular momentum tensor

$$\mu_{jk} = (x_j\xi_k - x_k\xi_j) \qquad \text{and} \qquad \mu = \Bigl(\sum_{j<k}\mu_{jk}^2\Bigr)^{\frac{1}{2}}. \tag{11.6.74}$$

This calculation is easy for trajectories lying in the (x_1, x_2)-plane and then one can change coordinates.

Then we obtain that the contribution of the non-scalar terms to the operator $U(\pi,\tau)$ [55] is the multiplication by

$$I - i\pi\sum_{j,k}\sigma_j'^*\sigma_k'\mu_{jk}\mu^{-3} + O\bigl(|\tau|^{\frac{3}{2}-\delta'}\bigr)$$

with arbitrarily small $\delta' > 0$. Applying corollary 5.A.7 and calculating the eigenvalues of μ_{jk} we obtain that in order to take into account non-scalar terms for $d \ge 3$ we can treat the family of operators

$$B'''(\tau') + k\nu|\tau|^{\frac{3}{2}}M^{-2} \tag{11.6.75}$$

with an appropriate constant k and with $\nu \in \{-1, 0, 1\}$ of the corresponding multiplicities where M^s is an operator with the symbol μ^s with $\mu = \mu(x,\xi)$ defined by (11.6.74). One should check conditions for every such operator, calculate and sum $\mathsf{N}_\nu^-(\tau)$ with the corresponding multiplicities.

For $d = 2$ and $\mathsf{D} = 2$ we have $\nu = 0, 1$ and the operators

$$B'''(\tau') + 3\varsigma\nu|\tau'|^{\frac{3}{2}}\mathcal{M}M^{-3} \tag{11.6.76}$$

with $\mathcal{M} = x_1D_2 - x_2D_1$, $\varsigma = i\sigma_0\sigma_1\sigma_2 = \pm 1$.

We leave the following rather easy problem to the reader:

Problem 11.6.24. Analyze the Schrödinger operator with a (perturbed) Coulomb potential in details.

---

[55] Here $U(\pi,\tau) = \exp\bigl(i\pi(-A(\tau))^{-\frac{1}{2}}\bigr)$ and π is the period of the corresponding Hamiltonian flow for $\tau = 0$.

11.7 Multiparameter Asymptotics

11.7.1 Problem Set-up

In this section we consider operators depending on more than one parameter. First of all, we consider semiclassical asymptotics in the case when the scalar semiclassical parameter h is replaced by a vector $\boldsymbol{h} = (h_1, \dots, h_d)$ with $h_j \to +0$.

Next we assume that there are two parameters: the semiclassical parameter $h \to +0$ and the spectral parameter τ. Let us recall that each of the semiclassical asymptotics of Section 11.1 was uniform with respect to the energy level τ lying in the interval in which conditions of the corresponding theorem were fulfilled[56] uniformly. Therefore there are two principally different situations in which non-trivial results are possible. We will explain them for the Schrödinger operator.

In the first situation τ increases and tends to a value τ^* such that the conditions of the corresponding theorem of Section 11.1 are violated for $\tau = \tau^*$ and therefore the asymptotics are no longer uniform. There are two cases:

(a) $\tau \to +\infty$ as in Sections 11.2–11.5 or

(b) $\tau \to \tau^* - 0$ with $\tau^* < \infty$ as in Section 11.6 where in the this case $\tau^* = \liminf_{|x|\to\infty} V$. In this case τ^* is the bottom of the essential spectrum and $\tau^* - 0$ is either a limit point of the discrete spectrum for every fixed $h > 0$ or there is only a finite discrete spectrum for every fixed $h > 0$ and in the second case the number of eigenvalues and (or) the remainder estimate near $\tau^* - 0$ are unusually large ($\gg \mathsf{const} \cdot h^{-d}$ and $\gg \mathsf{const} \cdot h^{1-d}$ respectively).

In all these cases we obtain estimates for the number of eigenvalues and for the remainder which are *worse* than for "normal" τ.

In the second situation τ decreases to the special value $\tau^* = \inf V$ and we obtain estimates for the number of eigenvalues and for the remainder which are *better* than for "normal" τ. Again there are two different cases:

(c) either $\inf V$ is finite in which case $\tau \to \tau^* + 0$ or

[56] In this subsection for the sake of simplicity we consider only the Schrödinger and Dirac operators and we no not need to check non-degeneracy conditions.

(d) $\inf V = -\infty$ but $-\infty$ is not a limit point of the spectrum for every fixed $h > 0$. Then for $h \to +0$ the bottom eigenvalues of the operator "fall" to $-\infty$ and we consider this process. Only the last case is impossible for the Dirac operator.

11.7.2 Semiclassical Asymptotics with a Vector Semiclassical Parameter

Let us consider the Schrödinger operator

$$A = A_h = \sum_{j,k} P_j g^{jk} P_k + V, \qquad P_j = h_j - V_j \tag{11.7.1}$$

or the Dirac operator of the same type as before. Changing coordinates $x_{j\,\mathsf{new}} = x_{j\,\mathsf{old}} h_j^{-1}$ we obtain operator

$$A' = A'_h = \sum_{j,k} (D_j - V_{jh}) g^{jk}_h (D_k - V_{kh}) + V_h \tag{11.7.2}$$

in the domain $X_h = \{x, f_h(x) \in X\}$ where X is the original domain and $v_h = v \circ f_h$ for every function v where $f_h(x) = h \bullet x := (h_1x_1, \dots, h_dx_d)$, $h = (h_1, \dots, h_d)$. One can easily see that if (some of) the conditions (11.1.6)–(11.1.8), (11.4.5) were fulfilled for $X, \rho, \gamma, g^{jk}, V_j, V$ then the same conditions remain true for $X_h, \rho_h, h^{-1}\gamma_h, V_{jh}, V_h$ where $h = \max_j h_j$. Moreover, if the cone condition was fulfilled for the domain X then it remains true for X_h provided $h \le 1$. This immediately yields

Theorem 11.7.1. *(i) Let $d \ge 3$ and let conditions* (11.1.6)–(11.1.8) *be fulfilled for $g^{jk}, V_j, V, \rho, \gamma$. Let $\gamma > 0$ a.e. and $V_-^{\frac{d}{2}}, \rho^{d-1}\gamma^{-1} \in \mathscr{L}^1(X)$. Then asymptotics*

$$\mathsf{N}^-(A_h) = (h_1 \dots h_d)^{-1}(\mathsf{N}^{\mathsf{W}} + O(h)) \tag{11.7.3}$$

holds as $h = \max_j h_j \to +0$ where as usual

$$\mathsf{N}^{\mathsf{W}} = (2\pi)^{-d}\omega_d \int V_-^{\frac{d}{2}} \sqrt{g}\, dx. \tag{11.7.4}$$

(ii) Moreover, let us assume that

$$g^{jk} = V_j = 0 \qquad \textit{for } j \ge d'+1, k \le d', \tag{11.7.5}$$

$0 < d' < d$ and that

$$|\nu'| \ge \epsilon_0 > 0 \qquad \text{on } \partial X \cap \{V \le 0\} \tag{11.7.6}$$

where $\nu = (\nu_1, \dots, \nu_{d'}; \nu_{d'+1}, \dots, \nu_d) = (\nu'; \nu'')$ is a unit normal to ∂X. Then asymptotics

$$\mathsf{N}^-(A_{\boldsymbol{h}}) = (h_1 \dots h_d)^{-1}(\mathsf{N}^{\mathsf{W}} + \varkappa_1 h + o(h)) \tag{11.7.7}$$

holds as $h = \max_j h_j \to +0$, $h_j = o(h)$ $\forall j \ge d'+1$ where

$$\varkappa_1 = \mp\frac{1}{4}(2\pi)^{1-d}\omega_{d-1}\int_{\partial X} V_-^{\frac{d-1}{2}}\, dS' \sqrt{g''}dx'' \tag{11.7.8}$$

for Dirichlet and Neumann boundary conditions respectively and the Riemannian densities dS', $\sqrt{g''}dx''$ were introduced in Example 11.5.23(ii) (see (11.5.68)).

(iii) Furthermore, if $\vartheta > 0$, $V_-^{\frac{d}{2}+\vartheta}, \rho^{d-1+\vartheta} \in \mathscr{L}^1(X)$ we have, in the frameworks of Statements (i) and (ii), asymptotics

$$\mathsf{N}^-_{(\vartheta)}(A_{\boldsymbol{h}}) = (h_1 \dots h_d)^{-1}(\mathsf{N}^{\mathsf{W}}_{\vartheta,\boldsymbol{h}} + O(h^{1+\vartheta})) \tag{11.7.9}$$

and

$$\mathsf{N}^-_{(\vartheta)}(A_{\boldsymbol{h}}) = (h_1 \dots h_d)^{-1}(\mathsf{N}^{\mathsf{W}}_{\vartheta,\boldsymbol{h}} + o(h^{1+\vartheta})) \tag{11.7.10}$$

respectively.

(iv) The same results remain true for $d = 2$, $d' = d'' = 1$ under additional restrictions $\operatorname{mes}(Y) < \infty$ where $Y = \{X, \rho\gamma \le h_0\} < \infty$ for some $h_0 > 0$ and $V_-^{p+\vartheta} \in \mathscr{L}^1(X'')$ for some $p > 1$.

(v) Under the cone condition all these results remain true for Neumann boundary conditions and also for the Dirac operator (with the reasonable boundary conditions). In the case of Dirac operator we consider eigenvalues lying in the interval $[\tau_1, \tau_2]$ with arbitrary fixed τ_1, τ_2.

Proof. Statement (i) follows immediately from Theorem 9.1.7.

Statement (ii) follows from Theorem 9.1.7 with an improved remainder estimates under non-periodicity assumption. Analysis of the Hamiltonian billiards is based on the same arguments as in Example 11.5.23(ii); namely ξ_j

and h_jx_j with $j \ge d'+1$ are "almost constant" (constant modulo $o(1)$) along these billiards (which is not true for $j \le d'$). But then $\frac{d}{dt}x_j \equiv \sum_{k\ge d'+1} g^{jk}\xi_k \mod o(1)$ as $j \ge d'+1$ and for generic ξ non-periodicity condition is fulfilled automatically.

The other Statements (iii)–(v) are proven by a similar method. □

11.7.3 Asymptotics of Eigenvalues near $+\infty$

Here and below we consider the Schrödinger operator A_h and consider the number of its eigenvalues lesser than τ. For the Dirac operator A_h (we use the same notation) we consider the number of eigenvalues lying in the interval $[\tau_1,\tau)$ with fixed τ_1.

Applying the arguments of Sections 11.2–11.5 with $\rho_{\tau,h} = h^{-1}(\rho + \sqrt{\tau})$ and $\gamma_{\tau,h} = \gamma$ we obtain modifications of the statements of these sections. However, we cannot anymore replace $\mathsf{N}^{\mathsf{W}}(\tau)$ by $\varkappa_0\tau^{\frac{d}{2m}}$ in these assertions. In particular, we obtain the following statements:

Theorem 11.7.2. *(i) For the Schrödinger operator, let conditions* (11.1.6)–(11.1.8) *be fulfilled with auxiliary functions ρ,γ. Let $\vartheta \ge 0$ and*

$$\rho^{d+2\vartheta}, 1, \gamma^{-\vartheta-1} \in \mathscr{L}^1(X). \tag{11.7.11}$$

For $d=2$ let us assume also that

$$\rho^{2p+2\vartheta} \in \mathscr{L}^1(X) \qquad \text{with some } p>1. \tag{11.7.12}$$

Moreover, in the case of the Neumann condition let us in addition assume that the cone condition is also fulfilled[57]*. Then asymptotics*

$$\mathsf{N}^-_{(\vartheta)}(\tau,h) = \mathsf{N}^{\mathsf{W}}_{(\vartheta)}(\tau,h) + O\bigl(\tau^{\frac{1}{2}(d+\vartheta-1)}h^{1-d+\vartheta}\bigr) \tag{11.7.13}$$

holds as $\tau\to+\infty$, $h\to+0$. In particular, for $\vartheta=0$

$$\mathsf{N}^{\mathsf{W}}(\tau,h) = \mathsf{N}^{\mathsf{W}}(\tau)h^{-d}. \tag{11.7.14}$$

(ii) Further, if the set of all points which are periodic with respect to the geodesic flow with reflections is of measure 0 then asymptotics

$$\mathsf{N}^-_{(\vartheta)}(\tau,h) = \mathsf{N}^{\mathsf{W}}_{(\vartheta)}(\tau,h) + o_\tau\bigl(\tau^{\frac{1}{2}(d+\vartheta-1)}h^{1-d+\vartheta}\bigr) \tag{11.7.15}$$

[57] We leave the analysis of mixed boundary conditions to the reader.

holds as $\tau\to+\infty, h\to+0$ *where* $M'=o_\zeta(M)$ *means that* $M'M^{-1}\le\varepsilon(\zeta)$ *with* $\varepsilon(\zeta)=o(1)$ *as* ζ *tends to the described limit.*

In particular, for $\vartheta=0$ *asymptotics*

$$\mathsf{N}^-(\tau,h)=\mathsf{N}^{\mathsf{W}}(\tau)h^{-d}+\bigl(\varkappa_1+o_\tau(1)\bigr)\tau^{\frac{1}{2}(d-1)}h^{1-d} \tag{11.7.16}$$

holds as $\tau\to+\infty, h\to+0$ *with the coefficient*

$$\varkappa_1=\mp\frac{1}{4}(2\pi)^{1-d}\omega_{d-1}\,\mathsf{vol}'\bigl(\partial X\cap\{\gamma>0\}\bigr) \tag{11.7.17}$$

for Dirichlet and Neumann boundary conditions respectively.

(iii) Furthermore, let us assume that $\gamma^{-1-\theta}\in\mathscr{L}^1(X)$ *with* $\theta>\vartheta$ *and that the geodesic flow is* <u>*either*</u> *of non-periodic power type* <u>*or*</u> *of non-periodic exponential type. Then these asymptotics hold the remainder estimates*

$$O\bigl(\tau^{\frac{1}{2}(d+\vartheta-1-\delta')}h^{1-d+\vartheta}\bigr) \tag{11.7.18}$$

or

$$O\bigl(\tau^{\frac{1}{2}(d+\vartheta-1)}h^{1-d+\vartheta}\log^{-1-\vartheta}\tau\bigr) \tag{11.7.19}$$

respectively.

(iv) All these asymptotics remain true if we drop condition $\rho^{d+2\vartheta}\in\mathscr{L}^1(X)$ *replacing it by the coercivity condition* (11.4.5).

(v) Statements (i)–(iii) remain true for $\vartheta=0$ *for the Dirac operator*[58]. *In these assertions one should replace* $\mathsf{N}^-(\tau,h)$ *and* $\mathsf{N}^{\mathsf{W}}(\tau_2)$ *by* $\mathsf{N}(\tau_1,\tau_2,h)$ *and* $\mathsf{N}^{\mathsf{W}}(\tau_1,\tau_2)$ *respectively and take* $\tau=|\tau_1|+|\tau_2|\to\infty$. *One should assume that the boundary conditions are elliptic and the domain satisfies the cone condition.*

We leave to the reader the following easy set of problems

Problem 11.7.3. (i) Modify in the same way the complete set of Theorems 11.2.7, 11.2.8, 11.2.12 and Corollary 11.2.9.

(ii) Consider Examples 11.2.13, 11.2.14, 11.2.15, 11.2.17, 11.2.18, 11.2.19, 11.2.21, 11.2.22, 11.2.25, 11.2.26, 11.2.27, 11.3.34, 11.3.35, 11.4.8, 11.4.9, 11.4.10 and translate them into the framework of the present two-parameter theory (automatically or with minor modifications). The same for all other results of this type.

[58] For $\vartheta>0$ one should modify $\mathsf{N}_{(\vartheta)}(\tau_1,\tau_2,h)$ in the obvious manner.

(iii) Modify Theorems 11.3.7, 11.3.18, 11.3.19, 11.3.21, 11.3.23, 11.3.26 etc.

Here one should assume that $h \ge \tau^{-n}$, and the principal part of the asymptotics is of the form $\mathsf{N}^{\mathsf{W}}(\tau)h^{-d}$ and the second term is of the form

$$\bigl(\varkappa_1 + F(\tau h^{-m})\bigr)h^{1-d}\tau^{\frac{1}{2m}(d-1)} \tag{11.7.20}$$

and there is additional factor h^{1-d} in the remainder estimate.

Let us consider the statements of Section 11.5. For the sake of simplicity let us consider only the Schrödinger operator. Then the following statements hold:

Theorem 11.7.4. *(i) In the framework of Theorem 11.5.1 (with the additional assumption that A is the Schrödinger operator) asymptotics*

$$\mathsf{N}^-_{(\vartheta)}(\tau,h) = \mathsf{N}^{\mathsf{W}}_{(\vartheta)}(\tau,h) + O\bigl(R_{(\vartheta)}(\tau)h^{1-d+\vartheta}\bigr) \tag{11.7.21}$$

holds as $\tau\to+\infty, h\to+0$ where $R_{(\vartheta)}(\tau)$ is defined by (11.5.3).

(ii) Further, under the additional assumptions of Theorem 11.5.3(i) asymptotics

$$\mathsf{N}^-_{(\vartheta)}(\tau,h) = \mathsf{N}^{\mathsf{W}}_{(\vartheta)}(\tau,h) + o_\tau\bigl(R_{(\vartheta)}(\tau)h^{1-d+\vartheta}\bigr) \tag{11.7.22}$$

holds as $\tau\to+\infty, h\to+0$.

(iii) Furthermore, in the framework of Theorem 11.5.3(ii) asymptotics

$$\mathsf{N}^-_{(\vartheta)}(\tau,h) = \mathsf{N}^{\mathsf{W}}_{(\vartheta)}(\tau,h) + o_{\tau h^{-2}}\bigl(R_{(\vartheta)}(\tau)h^{1-d+\vartheta}\bigr) \tag{11.7.23}$$

holds as $\tau\to+\infty, h\to+0$. Moreover, the principal part in asymptotics (11.7.21) *with $\vartheta=0$ is $\mathsf{N}^{\mathsf{W}}(\tau)h^{-d}$ and in* (11.7.22), (11.7.23) *the principal part is $\mathsf{N}^{\mathsf{W}}(\tau)h^{-d}+\varkappa_1\tau^{l(d-1)}h^{1-d}$ with the coefficient $\varkappa_1$ given by* (11.5.11).

(iv) In the framework of Theorem 11.5.6 for $\vartheta\le d-1-\delta$ with $\delta>0$ asymptotics

$$\mathsf{N}^-_{(\vartheta)}(\tau,h) = \mathsf{N}^{\mathsf{W}}_{(\vartheta)}(\tau,h) + O\bigl(\tau^{l(d-1)+\vartheta(1-l)-\delta'}h^{1-d}\bigr) \tag{11.7.24}$$

holds as $\tau\to+\infty, h\to+0$. Here for $\vartheta=0$ the principal part is the same as in (11.7.22). *In particular, Example 11.5.7 applies.*

(v) In the frameworks of Theorems 11.5.14, 11.5.18 and 11.5.20 asymptotics (11.7.21) *remains true. Moreover, under the conditions of these theorems on the Hamiltonian trajectories with reflections asymptotics* (11.7.22) *remains true.*

(vi) In the frameworks of Example 11.5.23(i) and (ii) asymptotics

$$\mathsf{N}(\tau, h) = \mathsf{N}^{\mathsf{W}}(\tau)h^{-d} + O(\tau^q h^{1-d}), \tag{11.7.25}$$

and

$$\mathsf{N}(\tau, h) = \mathsf{N}^{\mathsf{W}}(\tau)h^{-d} + \bigl(\varkappa_1 + o_\tau(1)\bigr)\tau^q h^{1-d}, \tag{11.7.26}$$

hold respectively with the coefficient $\varkappa_1$ *given by* (11.5.68).

(vii) In the frameworks of Theorem 11.5.25(i) and (ii) the asymptotics

$$\mathsf{N}^-_{(\vartheta)}(\tau, h) = \mathsf{N}^{\mathsf{W}}_{(\vartheta)}(\tau)h^{-d} + O\bigl(\tau^{\frac{1}{2}(d+\vartheta-1)}h^{1-d+\vartheta}\omega(\tau)\bigr), \tag{11.7.27}$$

and

$$\mathsf{N}^-_{(\vartheta)}(\tau, h) = \mathsf{N}^{\mathsf{W}}_{(\vartheta)}(\tau)h^{-d} + o\bigl(\tau^{\frac{1}{2}(d+\vartheta-1)}h^{1-d+\vartheta}\omega(\tau)\bigr) \tag{11.7.28}$$

hold respectively with

$$N^{\mathsf{W}}_{(\vartheta)}(\tau, h) \asymp \tau^{\frac{d}{2}+\vartheta} \log^{\frac{n}{\mu}} \tau h^{-d} \tag{11.7.29}$$

and $\omega(\tau)$ *defined by* (11.5.74). *For* $\vartheta = 0$ *we have* $\mathsf{N}^{\mathsf{W}}(\tau, h) = \mathsf{N}^{\mathsf{W}}(\tau)h^{-d}$.

Moreover, asymptotics (11.7.28) *remains true in the framework of Theorem 11.5.27 and for* $\vartheta = 0$

$$\mathsf{N}^{\mathsf{W}}_{(0)}(\tau, h) = \mathsf{N}^{\mathsf{W}}(\tau)h^{-d} + \varkappa_1 \tau^{\frac{1}{2}(d-1)}h^{1-d}\log\tau \tag{11.7.30}$$

where the coefficient $\varkappa_1$ *is defined by* (11.5.83).

(viii) In the framework of Theorem 11.5.29(i) asymptotics with the principal part

$$\mathsf{N}^{\mathsf{W}}_{(\vartheta)}(\tau, h) \asymp \mathsf{N}^{\mathsf{W}}(\tau)h^{-d}, \qquad \mathsf{N}^{\mathsf{w}}_{(0)}(\tau, h) = \mathsf{N}^{\mathsf{W}}(\tau)h^{-d} \tag{11.7.31}$$

and with remainder estimate $h^{1-d+\vartheta}R_{(\vartheta)}(\tau)$ *holds where* $\mathsf{N}^{\mathsf{W}}(\tau)$ *and* $R_{(\vartheta)}(\tau)$ *are defined in the theorem.*

Moreover, the improvements made in Statement (ii) and in footnote [44] *remain true.*

We leave to the reader the following easy set of problems:

Problem 11.7.5. (i) Analyze Examples 11.5.17, 11.5.19 and 11.5.21.

(ii) Modify Theorems 11.5.5, 11.5.9 and 11.5.11.

Note that remark in Problem 11.7.3(iii) also remains true in this case.

11.7.4 Asymptotics of Eigenvalues near −0 and Related Asymptotics

Applying the arguments of Section 11.6 with $\rho_{\tau,h} = h^{-1}(\rho + \sqrt{|\tau|})$ and $\gamma_{\tau,h} = \gamma$ we obtain

Theorem 11.7.6. *Again, we treat only the Schrödinger operator.*

(i) In the framework of Theorem 11.6.1 asymptotics

$$\mathsf{N}^-_{(\vartheta)}(\tau, h) = \mathsf{N}^{\mathsf{W}}_{(\vartheta)}(\tau, h) + O\bigl(h^{1-d+\vartheta} R_{(\vartheta)}(\tau)\bigr) \tag{11.7.32}$$

holds as $\tau \to -0$, $h \to +0$ *with* $R_{(\vartheta)}(\tau)$ *defined by* (11.6.4).

(ii) Moreover, in the framework of Theorem 11.6.2(i) asymptotics

$$\mathsf{N}^-_{(\vartheta)}(\tau, h) = \mathsf{N}^{\mathsf{W}}_{(\vartheta)}(\tau, h) + o_\tau\bigl(h^{1-d+\vartheta} R_{(\vartheta)}(\tau)\bigr) \tag{11.7.33}$$

holds as $\tau \to -0$, $h \to +0$.

For $\vartheta = 0$ *the principal parts of these asymptotics are* $\mathsf{N}^{\mathsf{W}}(\tau)h^{1-d}$ *and*

$$\mathsf{N}^{\mathsf{W}}(\tau)h^{1-d} + \varkappa_1|\tau|^{l(d-1)}h^{1-d} \tag{11.7.34}$$

respectively with the coefficient $\varkappa_1$ *given by* (11.6.11).

(iii) In the framework of Theorem 11.6.5 with $\vartheta \le \vartheta_1 - \delta$, $\delta > 0$ *asymptotics*

$$\mathsf{N}^-_{(\vartheta)}(\tau, h) = \mathsf{N}^{\mathsf{W}}_{(\vartheta)}(\tau, h) + O(h^{1-d+\vartheta}|\tau|^{l(d-1)+\vartheta(1-l)+\delta'}) \tag{11.7.35}$$

holds with $\delta' > 0$. *In these asymptotics the principal part for* $\vartheta = 0$ *is* $\mathsf{N}^{\mathsf{W}}(\tau)h^{1-d} + \varkappa_1|\tau|^{l(d-1)}h^{1-d}$. *In particular, Example 11.6.6 applies.*

(iv) In the framework of Theorem 11.6.10 asymptotics (11.7.32) *remains true and under a condition on Hamiltonian trajectories with reflection asymptotics* (11.7.33) *also remains true.*

(v) In the frameworks of Example 11.6.13(i) and (ii) asymptotics

$$\mathsf{N}^-(\tau,h)=\mathsf{N}^{\mathsf{W}}(\tau)h^{-d}+O(|\tau|^q h^{1-d}), \tag{11.7.36}$$

and

$$\mathsf{N}^-(\tau,h)=\mathsf{N}^{\mathsf{W}}(\tau)h^{-d}+\bigl(\varkappa_1+o_\tau(1)\bigr)|\tau|^q h^{1-d} \tag{11.7.37}$$

hold with the coefficient $\varkappa_1$ given by (11.6.35).

(vi) In the frameworks of Theorems 11.6.14(i) and (ii) asymptotics

$$\mathsf{N}(\tau,h)=\mathsf{N}^{\mathsf{W}}(\tau)h^{-d}+O\bigl(\bigl|\log|\tau|\bigr|^{\nu(d-1)+1}h^{1-d}\bigr), \tag{11.7.38}$$

and

$$\mathsf{N}(\tau,h)=\mathsf{N}^{\mathsf{W}}(\tau)h^{-d}+\bigl(\varkappa_1+o_\tau(1)\bigr)\bigl|\log|\tau|\bigr|^{\nu(d-1)+1}h^{1-d} \tag{11.7.39}$$

respectively hold with the coefficient $\varkappa_1$ given by (11.6.40).

(vii) Moreover, Statement (vi) remains true for $\nu=0$ with $o_\tau(1)$ replaced by $o(1)$ in asymptotics (11.7.39).

Remark 11.7.7. Statement Statement (vii) dealing with potentials decaying as $|x|^{-2}$ has no analog in Section 11.6 because for $\nu=0$ for every small enough fixed $h>0$ the eigenvalues of A_h tend to -0 but in (11.7.38) both the principal part and the remainder estimate are of the same order $\log|\tau|$ and (11.7.38) is only an estimate (or a couple of estimates: from above and below), not asymptotics.

We leave to the reader

Problem 11.7.8. (i) Modify Example 11.6.11.

(ii) In the framework of Problem 11.6.15 derive asymptotics with the principal parts

$$N^{\mathsf{W}}_{(\vartheta)}(\tau,h)\asymp \mathsf{N}^{W}_{\vartheta,h}(\tau)h^{-d},\qquad N^{\mathsf{W}}_{(0)}(\tau,h)=\mathsf{N}^{\mathsf{W}}(\tau)h^{-d}$$

and remainder estimate $O\bigl(h^{1-d+\vartheta}R_{(\vartheta)}(\tau)\bigr)$ where $\mathsf{N}^{\mathsf{W}}(\tau)$, $R_{(\vartheta)}(\tau)$ are defined in that problem.

(iii) Modify Theorems 11.6.4, 11.6.7 and 11.6.9.

Note that remark in Problem 11.7.3(iii) also remains true in this case as $h\ge|\tau|^n$.

Let us consider the situation in which the conditions of Theorem 11.6.14 are fulfilled with $\nu < 0$ instead of $\nu > 0$.

In this case for every fixed $h > 0$ the discrete spectrum is finite. However, $V_-^{\frac{d}{2}} \notin \mathscr{L}^1(X)$ for $\nu \ge -d^{-1}$ provided condition (11.6.8) is fulfilled. In this case the above arguments lead us to the following

Theorem 11.7.9. *(i) Under the assumptions of Theorem 11.6.14(i) with* $\nu \in (-d^{-1}, 0)$ *asymptotics* (11.7.38) *holds for the Schrödinger operator for* $\tau \to -0$, $h \to +0$ *provided*

$$h = o\bigl(\bigl|\log|\tau|\bigr|^\nu\bigr). \tag{11.7.40}$$

In this asymptotics $\mathsf{N}^{\mathsf{W}}(\tau) = O\bigl(\bigl|\log|\tau|\bigr|^{\nu d+1}\bigr)$ *and under condition* (11.6.8) $\mathsf{N}^{\mathsf{W}}(\tau) \asymp \bigl|\log|\tau|\bigr|^{\nu d+1}$.

(ii) Under the assumptions of Theorem 11.6.14(i) with $\nu = -d^{-1}$ *asymptotics* (11.7.38) *holds for the Schrödinger operator for* $\tau \to -0$, $h \to +0$ *provided*

$$h = O\bigl(\bigl|\log|\tau|\bigr|^\nu\bigr). \tag{11.7.41}$$

In this asymptotics $\mathsf{N}^{\mathsf{W}}(\tau) = O\bigl(\log\bigl|\log|\tau|\bigr|\bigr)$ *and under condition* (11.6.8) $\mathsf{N}^{\mathsf{W}}(\tau) \asymp \log\bigl|\log|\tau|\bigr|$.

(iii) Under the assumptions of Theorem 11.6.14(ii) with $\nu \in [-d^{-1}, 0)$ *asymptotics* (11.7.39) *holds for the Schrödinger operator for* $\tau \to -0$, $h \to +0$ *satisfying* (11.7.40). *In this asymptotics coefficient* $\varkappa_1$ *is given by* (11.6.40).

Let us consider the case

$$h \ge \epsilon_0\bigl|\log|\tau|\bigr|^\nu. \tag{11.7.42}$$

Then the singular zone $X_{\mathsf{sing}} = \bigl\{(\rho + \sqrt{|\tau|})\gamma \le \epsilon h\bigr\}$ is not necessarily empty. Considering only the lower estimate we obtain that under condition (11.6.8)

$$\mathsf{N}^-(\tau, h) \ge \epsilon_1 h^{\frac{1}{\nu}} \qquad \text{for } \nu \in (-\frac{1}{d}, 0), \tag{11.7.43}$$

and

$$\mathsf{N}^-(\tau, h) \ge \epsilon_1 h^{\frac{1}{\nu}}|\log h| \qquad \text{for } \nu = -\frac{1}{d} \tag{11.7.44}$$

with $\epsilon_1 > 0$.

On the other hand, for $d \geq 3$ one can apply a variational estimate in X_{sing} (with $\rho = 0$ for $\log^\nu r \leq \epsilon_2 h$. Then we obtain estimate

$$\mathsf{N}^-(\tau, h) = O(h^{\frac{1}{\nu}}) \qquad \text{for } \nu \in (-\frac{1}{d}, 0), \tag{11.7.45}$$

and even asymptotics

$$\mathsf{N}^-(\tau, h) = \bigl(\mathsf{N}_h^{\mathsf{W}}(\tau) + O(1)\bigr)h^{-d} \qquad \text{for } \nu = -\frac{1}{d} \tag{11.7.46}$$

for $\tau \to -0$, $h \to +0$ with

$$\mathsf{N}_h^{\mathsf{W}}(\tau) = (2\pi)^{-d}\omega_d \int_{\{|\log|x||\leq Ch^{-\frac{1}{\nu}}\}} (-\tau - V)_+^{\frac{d}{2}} \sqrt{g}\, dx \tag{11.7.47}$$

with an arbitrarily chosen constant C. Under condition (11.6.8) this expression is $\asymp \min\bigl(\log\bigl|\log|\tau|\bigr|, |\log h|\bigr)$ for $\nu = -d^{-1}$.

Let us recall that the asymptotics and estimates (11.7.43)–(11.7.46) were obtained under the restriction (11.7.42).

Thus we arrive to

Remark 11.7.10. (i) If $d \geq 3$, $\nu \in (-d^{-1}, 0)$ then under condition (11.6.8) $\mathsf{N}^-(\tau, h) \asymp h^{\frac{1}{\nu}}$ for $\tau \to -0$, $h \to +0$ satisfying (11.7.42). One can easily obtain this result by purely variational arguments.

(ii) For $d \geq 3$, $\nu = -d^{-1}$ we combine Statement (ii) of Theorem 11.7.9 and (11.7.46) to obtain

$$\mathsf{N}^-(\tau, h) = \mathsf{N}_h^{\mathsf{W}}(\tau)h^{-d} + O\Bigl(\min\bigl(h^{-d}, h^{1-d}|\log|\tau||^{\frac{1}{d}}\bigr)\Bigr) \tag{11.7.48}$$

with $\mathsf{N}_h^{\mathsf{W}}(\tau)$ given by (11.7.47) and $\tau \to -0$, $h \to +0$.

We leave to the reader the following

Problem 11.7.11. (i) Treat the case $\nu \in [-(d-1)^{-1}, -d^{-1})$. In this case the principal part is standard and $\asymp h^{-d}$ but the remainder estimate is worse than usual. In this case we also obtain asymptotics for $\tau = 0$ in the context of Section 11.1.

(ii) rove that all the results of this subsection remain true for the Dirac operator with

$$\tau_1^0 = -M + \limsup_{|x|\to\infty} V < \tau_2^0 = M + \liminf_{|x|\to\infty} V. \tag{11.7.49}$$

One need only to replace $\mathsf{N}^-(\tau; h)$ by $\mathsf{N}(\tau_1, \tau; h)$ with fixed $\tau_1 \in (\tau_1^0, \tau_2^0)$ and respectively modify the formulas for N^{W}, $\mathsf{N}_h^{\mathsf{W}}$, $\varkappa_0$ and $\varkappa_1$; without any loss of the generality one can assume that $\tau_2^0 = 0$ and then $\tau \to -0$.

11.7.5 Asymptotics of Eigenvalues near $+0$

Arguments of the same type as above permit us also to treat the case $\tau \to +0$, $h \to +0$ and to obtain asymptotics of $\mathsf{N}^-(\tau, h)$ assuming that $\inf V = 0$ for the Schrödinger operator.

Similar results may be obtained for the Dirac operator for $\mathsf{N}(\tau_1, \tau; h)$ provided (11.7.49) holds, $\tau_1 \in (\tau_1^0, \tau_2^0)$ and

$$\tau_1^0 = -M + \sup V < \tau_1 \le \tau_2^0 = M + \inf V;$$

without any loss of the generality one can assume that $\tau_2^0 = 0$ and then $\tau \to +0$.

We formulate our statements only for the Schrödinger operator with Dirichlet boundary conditions. For the the Schrödinger operator with Neumann boundary conditions and for the Dirac operator with any self-adjoint boundary conditions these statements remain true as long as the domain X satisfies the cone condition.

It is easy to prove the following

Theorem 11.7.12. *(i) Let X be a bounded domain and let Y a $\mathscr{C}^1$ manifold of codimension d'' in $\mathbb{R}^d$, $2 \le d'' \le d$ such that $\bar{X} \cap Y \ne \emptyset$.*

Let conditions (11.1.6)–(11.1.8), (11.4.5) *be fulfilled with $\gamma = \epsilon r(x)$, $\rho(x) = r^\mu(x)$, $\epsilon > 0$, $\mu > 0$, $r(x) = \mathsf{dist}(x, Y)$. Then asymptotics*

$$\mathsf{N}^-(\tau, h) = \mathsf{N}^{\mathsf{W}}(\tau) h^{-d} + O(\tau^q h^{1-d}) \tag{11.7.50}$$

holds for $\tau \to +0$, $h \to +0$ with $\mathsf{N}^{\mathsf{W}}(\tau) = O(\tau^p)$, $p = \frac{1}{2}d + \frac{1}{2\mu}d''$ and $q = \frac{1}{2}(d-1) + \frac{1}{2\mu}(d''-1)$ [59),60)].

[59)] Surely we obtain asymptotics only for $h = o(\tau^l)$ with $l = \frac{1}{2\mu}(\mu+1)$, i.e., only if τ tends to $+0$ slowly enough. Otherwise we obtain only an upper estimate and for $h \ge \epsilon_0 \tau^l$ with a small enough constant ϵ_0 we obtain a lower estimate for $\mathsf{N}^-(\tau, h)$. This remark remains true for other asymptotics of this subsection.

[60)] We did not study the lowest eigenvalue but under rather general assumptions one can prove that it is $\tau_1 > 0 : \tau_1 \asymp h^{1/l}$.

Moreover, $\mathsf{N}^{\mathsf{W}}(\tau) \asymp \tau^{p}$ *provided*

$$\operatorname{vol}\bigl(X \cap \{r(x) < \bar{r}\}\bigr) \asymp r^{n} \quad \text{as } r \to 0 \tag{11.7.51}$$

with $n = d''$.

(ii) Further, let us assume that $Y = \{0\}$, $X \cap \{|x| \le r_0\} = X^0 \cap \{|x| \le r_0\}$ *where* $r_0 > 0$ *and* X^0 *is the non-empty open cone with* $\partial X^0 \setminus 0 \in \mathscr{C}^K$.

Let stabilization conditions (11.5.4) *be fulfilled for* $|x| \to 0$ *with functions* g^{jk0}, V_j^0 *and* V^0 *positively homogeneous of degrees* 0, μ *and* 2μ *respectively.*

Finally, let the standard non-periodicity condition be fulfilled for the domain X^0 *and the Hamiltonian*

$$H^0_{\mathsf{st}} = \sum_{j,k} g^{jk0}(\xi_j - V_j^0)(\xi_k - V_k^0) + V^0 - 1 \tag{11.7.52}$$

on the energy level 0. *Then the asymptotics*

$$\mathsf{N}^-(\tau, h) = \mathsf{N}^{\mathsf{W}}(\tau) h^{-d} + \bigl(\varkappa_1 + o(1)\bigr)\tau^{l(d-1)} h^{1-d} \tag{11.7.53}$$

holds as $\tau \to +0$, $h = o(\tau^{l})$ *with the coefficient* $\varkappa_1$ *given by* (11.5.11).

Note that under the stabilization conditions alone, $\mathsf{N}^{\mathsf{W}} = \bigl(\varkappa_0 + o(1)\bigr)\tau^{ld}$ *with the coefficient* $\varkappa_0$ *given by* (11.5.10).

(iii) Finally, let the stabilization conditions

$$D^{\alpha}(a_\beta - a^0_\beta) = O\bigl(|x|^{(2m-|\beta|)\mu - |\alpha| + \delta}\bigr) \qquad \forall \beta : |\beta| \le 2m \quad \forall \alpha : |\alpha| \le 1 \tag{11.7.54}$$

be fulfilled as $|x| \to 0$ *with some* $\delta > 0$. *Let the Hamiltonian flow generated by* H^0_{st} *near* 0 *be of non-periodic power type. Then the asymptotics*

$$\mathsf{N}^-(\tau, h) = \mathsf{N}^{\mathsf{W}}(\tau) h^{-d} + \bigl(\varkappa_1 + O(h^{\delta'}\tau^{-l\delta'})\bigr) h^{1-d}\tau^{l(d-1)} \tag{11.7.55}$$

with $\delta' > 0$ *holds for* $\tau \to +0$, $h = o(\tau^{l})$.

We leave to the reader the following easy Problem 11.7.13 and the proofs of Theorems 11.7.14–11.7.16 below.

Problem 11.7.13. (i) Generalize Statement (ii) of Theorem 11.7.12 to the case $d'' < d$. The arguments of Example 11.5.23(ii) should be applied here.

(ii) Consider the case of periodic Hamiltonian flow.

(iii) Consider the case $d''=1$ when the logarithmic factor appears in the remainder estimate.

(iv) Consider the general operators and the case $\vartheta>0$.

Theorem 11.7.14. *Let X be a bounded domain, $0\in X$ and*

$$X\cap\{|x|\le r_0\}=X^0\cap\{|x|\le r_0\} \tag{11.7.56}$$

where X^0 is an L-quasiconical domain in $\mathbb{R}^d$. Let g^{jk}, V_j and V be L-quasihomogeneous with respect to x of degrees 0, μ, 2μ respectively with $\mu>0$, $L=(L_1,\dots,L_d)$, $L_j=1$ for $j\le d'$ and $L_j<1$ for $j\ge d'+1$, $0<d'<d$.

Let us assume that $\partial X\setminus 0\in\mathscr{C}^K$ and $g^{jk},V_j,V\in\mathscr{C}^K(\bar{X}\setminus 0)$ satisfy condition (11.1.8)[61] *Finally, let us assume that*

$$V(x)>0\qquad\forall x\in\bar{X}\setminus 0. \tag{11.7.57}$$

Then

(i) Asymptotics

$$\mathsf{N}^-(\tau,h)=\varkappa_0\tau^ph^{-d}+O(\tau^qh^{1-d}) \tag{11.7.58}$$

holds for $\tau\to+0$, $h\to+0$ with $p=\frac{1}{2}d+\frac{1}{2\mu}L^*$, $q=\frac{1}{2}(d-1)+\frac{1}{2\mu}(L^*-1)$, $L^*=L_1+\dots+L_d$ and with the coefficient $\varkappa_0$ given by (11.5.63) *with X replaced by X^0.*

(ii) Under the additional conditions of Example 11.5.23(ii) asymptotics

$$\mathsf{N}^-(\tau,h)=\varkappa_0\tau^ph^{-d}+\bigl(\varkappa_1+o(1)\bigr)\tau^qh^{1-d} \tag{11.7.59}$$

holds for $\tau\to-0$, $h=o(\tau^{\frac{1}{2}+\frac{1}{2\mu}})$ with the coefficient $\varkappa_1$ given by (11.5.67).

Theorem 11.7.15. *Let X be a bounded domain and let Y be a $\mathscr{C}^1$ manifold of codimension d'' in $\mathbb{R}^d$, $2\le d''\le d$, $\bar{X}\cap Y\ne\emptyset$.*

Let conditions (11.1.6)–(11.1.8), (11.4.5) *be fulfilled with $\gamma=\epsilon r(x)$, $\rho=\exp(-r(x)^\nu)$ with $\nu<0$, $r(x)=\operatorname{dist}(x,Y)$.*

Let us assume that condition (11.7.51) *is fulfilled with an exponent n which is now not necessarily equal to d''. Then*

[61] Then one can take $\gamma=\epsilon[x]_L$, $\rho=\epsilon[x]_L^\mu$.

(i) Asymptotics

$$\mathsf{N}^-(\tau,h)=\mathsf{N}^{\mathsf{W}}(\tau)h^{-d}+O\bigl(\tau^{\frac{1}{2}(d-1)}|\log\tau|^{(\nu+n-1)\nu^{-1}}h^{1-d}\bigr) \tag{11.7.60}$$

holds for $\tau\to+0$, $h\to+0$ *and* $\mathsf{N}^{\mathsf{W}}(\tau)\asymp\tau^{\frac{1}{2}d}|\log\tau|^{n\nu^{-1}}$.

(ii) Moreover, let $Y=\{0\}$, $n=d$. *Let condition* (11.7.59) *be fulfilled for* $r\to+0$ *and every fixed* k *and let conditions* (11.5.76)–(11.5.77) *be fulfilled for* $|x|\to0$. *Then asymptotics*

$$\mathsf{N}^-(\tau,h)=\mathsf{N}^{\mathsf{W}}(\tau)h^{-d}+o\bigl(\tau^{\frac{1}{2}(d-1)}|\log\tau|^{(\nu+n-1)\nu^{-1}}h^{1-d}\bigr) \tag{11.7.61}$$

holds for $\tau\to+0$, $h=o\bigl(\tau^{\frac{1}{2}}|\log\tau|^{(1-\nu)\nu^{-1}}\bigr)$.

Theorem 11.7.16. *Let* $d\ge3$, $X=(0,\frac{1}{2})\times U$ *where* U *is a domain on* $\mathbb{S}^{d-1}$ *with* $\mathscr{C}^K$ *boundary,* $g^{jk}=g^{jk}(\theta)\in\mathscr{C}^K(\bar{U})$ *satisfy* (11.4.5). *Let* $V_j=0$, $V=v(\theta)|\log r|^{-1}$ *with* $v\in\mathscr{C}^K(\bar{U})$ *and* $\min_{\bar{U}}v=v_0>0$. *Then asymptotics*

$$\mathsf{N}^-(\tau,h)=\mathsf{N}^{\mathsf{W}}(\tau)h^{-d}+O\Bigl(h^{1-d}\tau^{d-2}\int_U\exp\bigl(-v(d-1)\tau^{-1}\bigr)\,d\theta\Bigr) \tag{11.7.62}$$

holds as $\tau\to+0$, $h\to+0$ *with*

$$\mathsf{N}^{\mathsf{W}}(\tau)=a\int_{U'}\tau^d v^{-\frac{d}{2}}F_d(\frac{\tau}{v})\exp(-vd\tau^{-1})\sqrt{g}\,d\theta \tag{11.7.63}$$

where $U'=U\cap\{v<v_0+\varepsilon\}$, $\varepsilon>0$ *is arbitrarily small, and* a *and* F_d *are introduced in Theorem 11.6.14 in which the orders of* $\mathsf{N}^{\mathsf{W}}(\tau)$ *and the remainder estimate are also discussed.*

11.7.6 Asymptotics of Eigenvalues near $-\infty$

Let us consider asymptotics of $\mathsf{N}(\tau,h)$ for $\tau\to-\infty$, $h\to+0$. Recall that we consider the Schrödinger operator and now we assume that

$$\inf V=-\infty; \tag{11.7.64}$$

thus it is still the case "$\tau\searrow\inf V$". The results of this subsection surely do not make any sense for the Dirac operator. The same arguments as above give us the following

Theorem 11.7.17. *(i) Let X be a bounded domain in $\mathbb{R}^d$, $0 \in \bar{X}$ and let conditions* (11.1.6)–(11.1.8) *be fulfilled with $\gamma = \epsilon|x|$, $\rho = |x|^\mu$, $\mu \in (-1, 0)$. Then asymptotics*[59),60)]

$$\mathsf{N}^-(\tau, h) = \mathsf{N}^{\mathsf{W}}(\tau)h^{-d} + O\bigl(|\tau|^{l(d-1)}h^{1-d}\bigr) \tag{11.7.65}$$

holds for $\tau \to -\infty$, $h \to +0$ and $\mathsf{N}^{\mathsf{W}}(\tau) = O(|\tau|^{ld})$.

Moreover, $\mathsf{N}^{\mathsf{W}}(\tau) \asymp |\tau|^{ld}$ provided

$$\operatorname{mes}\bigl(\{x \in X, |x| \le r, V \le -\epsilon_0\rho^2(x)\}\bigr) \asymp r^d \qquad \textit{for } r \to +0. \tag{11.7.66}$$

(ii) Further, let $X \cap \{|x| \le r_0\} = X^0 \cap \{|x| \le r_0\}$ with some $r_0 > 0$ and an open non-empty cone X^0 such that $\partial X^0 \setminus 0 \in \mathscr{C}^K$. Let stabilization conditions (11.5.4) *be fulfilled for $|x| \to 0$ with functions g^{jk0}, V_j^0, $V^0 \in \mathscr{C}^K(\bar{X}^0 \setminus 0)$.*

Let the standard non-periodicity condition to the Hamiltonian flow be fulfilled on the energy level 0 for the Hamiltonian

$$H^0(x,\xi) = |x|^{1-\mu}\bigl(\sum_{j,k} g^{jk0}(\xi_j - V_j^0)(\xi_k - V_k^0) + V^0 + 1\bigr). \tag{11.7.67}$$

Then asymptotics

$$\mathsf{N}^-(\tau, h) = \mathsf{N}^{\mathsf{W}}(\tau)h^{-d} + \bigl(\varkappa_1 + o(1)\bigr)|\tau|^{l(d-1)}h^{1-d} \tag{11.7.68}$$

holds for $\tau \to -\infty$, $h = o(|\tau|^l)$ with the coefficient $\varkappa_1$ given by (11.6.11).

(iii) Furthermore, let stabilization conditions (11.7.54) *with $\delta > 0$ be fulfilled for $x \to 0$ and let the Hamiltonian flow corresponding to the Hamiltonian*

$$H^0_{\mathsf{st}}(x,\xi) = \sum_{j,k} g^{jk0}(\xi_j - V_j^0)(\xi_k - V_k^0) + V^0 + 1 \tag{11.7.69}$$

be of power non-periodic type near the energy level 0. Then asymptotics

$$\mathsf{N}^-(\tau, h) = \mathsf{N}^{\mathsf{W}}(\tau)h^{-d} + \bigl(\varkappa_1 + O(h^{\delta'}|\tau|^{-l\delta'})\bigr)|\tau|^{l(d-1)}h^{1-d} \tag{11.7.70}$$

with $\delta' > 0$ holds for $\tau \to -\infty$, $h \to +0$.

We leave to the reader the following easy Problem 11.7.18 and Theorems 11.7.19, 11.7.20 and 11.7.22 below.

Problem 11.7.18. Treat the case of a singularity at a manifold of codimension less than d.

Theorem 11.7.19. *(i) Let conditions of Theorem 11.7.14 be fulfilled with $\mu \in (-1, 0)$. Then asymptotics*

$$\mathsf{N}^-(\tau, h) = \varkappa_0 |\tau|^p h^{-d} + O\big(|\tau|^q h^{1-d}\big) \tag{11.7.71}$$

holds as $\tau \to -\infty$, $h \to +0$ with the same exponents p, q and coefficient $\varkappa_0$ as in Theorem 11.7.14 (i.e. given by (11.5.63) *with X replaced by X^0).*

(ii) Further, under the additional restrictions of Theorem 11.7.14(ii) asymptotics

$$\mathsf{N}^-(\tau, h) = \varkappa_0 |\tau|^p h^{-d} + \big(\varkappa_1 + o(1)\big)|\tau|^q h^{1-d} \tag{11.7.72}$$

holds for $\tau \to -\infty$, $h = o(|\tau|^{\frac{1}{2\mu}(\mu+1)})$ with the coefficient $\varkappa_1$ given by (11.5.68).

Theorem 11.7.20. *Let X be a bounded domain in $\mathbb{R}^d$, $0 \in \bar{X}$ and let conditions* (11.1.6)–(11.1.8) *be fulfilled with $\gamma = \epsilon|x|$, $\rho = |x|^{-1}\big(|\log|x|| + 1\big)^\nu$, $\nu \le -d^{-1}$. Then*

(i) For $\nu < -(d-1)^{-1}$ asymptotics

$$\mathsf{N}^-(\tau, h) = \mathsf{N}^{\mathsf{W}}(\tau) h^{-d} + O\big((\log|\tau|)^{\nu(d-1)+1} h^{1-d}\big) \tag{11.7.73}$$

holds for $\tau \to -\infty$, $h \to +0$.

(ii) For $\nu = -(d-1)^{-1}$ the asymptotics

$$\mathsf{N}^-(\tau, h) = \mathsf{N}^{\mathsf{W}}(\tau) h^{-d} + O\big(\log(\log|\tau| \cdot h^{-\nu^{-1}}) h^{1-d}\big) \tag{11.7.74}$$

holds for $\tau \to -\infty$, $h = o(\log^\nu |\tau|)$.

(iii) Furthermore, let $X \cap \{|x| \le r_0\} = X^0 \cap \{|x| \le r_0\}$ with $r_0 > 0$ and a non-empty open cone X^0 with $\partial X^0 \setminus 0 \in \mathscr{C}^K$. Let the functions g^{jk}, $V_j \big|\log|x|\big|^{-\nu}$, $V\big|\log|x|\big|^{-2\nu}$ satisfy stabilization conditions (11.5.4) *for $|x| \to 0$ with $\mu = -1$ and with functions g^{jk0}, V_j^0, V^0 positively homogeneous of degrees 0, -1, -2 respectively.*

Moreover, let the standard non-periodicity condition be fulfilled with the Hamiltonian

$$H^0(x,\xi) = |x|^2\bigl(\sum_{j,k} g^{jk0}(\xi_j - V_j^0)(\xi_k - V_k^0) + V^0\bigr) \tag{11.7.75}$$

on the energy level 0*. Then in the framework of Statement (i) asymptotics*

$$\mathsf{N}^-(\tau,h) = \mathsf{N}^{\mathsf{W}}(\tau)h^{-d} + \bigl(\varkappa_1 + o(1)\bigr)(\log|\tau|)^{\nu(d-1)+1}h^{1-d} \tag{11.7.76}$$

holds for $\tau\to-\infty$, $h = o(\log^\nu|\tau|)$ *and in the framework of Statement (ii) asymptotics*

$$\mathsf{N}^-(\tau,h) = \mathsf{N}^{\mathsf{W}}(\tau)h^{-d} + \bigl(\varkappa_1 + o(1)\bigr)\log(\log|\tau|\cdot h^{-\nu^{-1}})h^{1-d} \tag{11.7.77}$$

holds for $\tau\to-\infty$, $h = o(\log^\nu|\tau|)$.

(iv) Furthermore, for $\nu\in(-(d-1)^{-1}, -d^{-1})$ *asymptotics*

$$\mathsf{N}^-(\tau,h) = \mathsf{N}^{\mathsf{W}}(\tau)h^{-d} + O\bigl(h^{\nu^{-1}}\bigr) \tag{11.7.78}$$

holds for $\tau\to-\infty$, $h = o(\log^\nu|\tau|)$.

In all these asymptotics $\mathsf{N}^{\mathsf{W}}(\tau) = O(\log^{\nu d+1}|\tau|)$ *and* $\mathsf{N}^{\mathsf{W}}(\tau)\asymp\log^{\nu d+1}|\tau|$ *under condition* (11.7.66).

(v) Let $\nu = -d^{-1}$*. Furthermore, suppose that* <u>*either*</u> $d\ge 3$ <u>*or*</u> $0\in\partial X$ *and a Dirichlet condition is given on* ∂X*. Then asymptotics*

$$\mathsf{N}^-(\tau,h) = \bigl(\mathsf{N}^{\mathsf{W}}(\tau,h) + O(1)\bigr)h^{-d} \tag{11.7.79}$$

holds for $\tau\to-\infty$, $h = o(\log^\nu|\tau|)$ *and* $\mathsf{N}^{\mathsf{W}}(\tau,h)$ *is give by* (11.7.47), $\mathsf{N}^{\mathsf{W}}(\tau,h) = O\bigl(\log(\log|\tau|\cdot h^{\nu^{-1}})\bigr)$.

Moreover, under condition (11.7.66) $\mathsf{N}^{\mathsf{W}}(\tau,h)\asymp\log(\log|\tau|\cdot h^{\nu^{-1}})$.

Remark 11.7.21. In the proof of the last statement of this theorem we also use the arguments of Remark 11.1.13(i).

Theorem 11.7.22. *Let* $d\ge 3$, X *be a conical domain with* $\partial X\setminus 0\in\mathscr{C}^K$*, and let* $g^{jk} = g^{jk}(\theta)\in\mathscr{C}^K(\mathbb{S}^{d-1})$ *satisfy condition* (11.4.5)*. Let* $V_j = 0$, $V = v(\theta)|\log r|$ *with* $v\in\mathscr{C}^K(\mathbb{S}^{d-1})$ *and* $\min_{\bar X} v = v_0 < 0$*. Then asymptotics*

$$\mathsf{N}^-(\tau,h) = \mathsf{N}^{\mathsf{W}}(\tau)h^{-d} + O\Bigl(h^{1-d}\tau\int_U \exp(-\tau(d-1)v^{-1})\,d\theta\Bigr) \tag{11.7.80}$$

holds as $\tau \to -\infty$, $h \to +0$ *with*

$$\mathsf{N}^{\mathsf{W}}(\tau) = a \int_U \exp(-\tau d v^{-1})\sqrt{g}\, d\theta \tag{11.7.81}$$

where $U = X \cap \mathbb{S}^{d-1}$ *and the coefficient* a *is introduced in Theorem 11.5.29. The orders of* $\mathsf{N}^{\mathsf{W}}(\tau)$ *and the remainder estimate and more accurate results are discussed in Statement (ii) of that theorem.*

11.8 Distributions of Spectra for Negative Order Operators

Let $A = A^{\mathsf{w}}(x, D)$ be a classical pseudodifferential operator of order $m < 0$ on a compact closed manifold X. We are interested in the asymptotics of the eigenvalue counting function

$$\mathsf{N}^{\pm}(\tau) = \#\{\lambda > \tau,\ \pm\lambda \text{ is an eigenvalue of } A\} \tag{11.8.1}$$

and

$$\mathsf{N}^{\pm}_{(\vartheta)}(\tau) = \int (\lambda - \tau)_+^{\vartheta}\, d_\lambda \mathsf{N}^{\pm}(\lambda) \tag{11.8.2}$$

as $\tau \to +0$. First of all, the following theorem holds:

Theorem 11.8.1. *(i) Let* $\mathsf{a}(x, \xi)$ *be a microhyperbolic symbol on the energy level* 0 *(in a direction depending on the point). Then asymptotics*

$$\mathsf{N}^{\pm}(\tau) = \varkappa^{\pm}\tau^{\frac{d}{m}} + O(\tau^{\frac{d-1}{m}}) \tag{11.8.3}$$

holds as $\tau \to +0$ *with*

$$\varkappa^{\pm} = (2\pi)^{-d} \int \boldsymbol{n}^{\pm}(x, \xi; 1)\, dx d\xi \tag{11.8.4}$$

where $\boldsymbol{n}^{\pm}(x, \xi; \tau)$ *is an eigenvalue counting function for* $\mathsf{a}(x, \xi)$ *i.e. the number of eigenvalues of the symbol* $\pm\mathsf{a}(x, \xi)$ *greater than* τ *and* a *is the principal symbol of* A.

(ii) Moreover, asymptotics

$$\mathsf{N}^{\pm}_{(\vartheta)}(\tau) = \mathsf{N}^{\mathsf{W},\pm}_{(\vartheta)}(\tau) + O(R_{(\vartheta)}(\tau)) \tag{11.8.5}$$

holds with

$$\mathsf{N}^{\mathsf{W},\pm}_{(\vartheta)}(\tau) = \sum_k \varkappa^{\pm}_{\vartheta k}\tau^{\frac{d-k}{m}} \tag{11.8.6}$$

and

$$R_{(\vartheta)}(\tau) = \begin{cases} \tau^{\frac{1}{m}(d-1-\vartheta)+\vartheta} & \text{for } \vartheta(1-m) < d-1,\\ |\log\tau| & \text{for } \vartheta(1-m) = d-1,\\ 1 & \text{for } \vartheta(1-m) > d-1. \end{cases} \tag{11.8.7}$$

(iii) Further, let $\vartheta(1-m) < d-1$ and let the standard non-periodicity condition be fulfilled for the Hamiltonian trajectories of the symbol $\pm a$ on the energy level 1. Then asymptotics

$$\mathsf{N}^{\pm}(\tau) = \varkappa^{\pm}\tau^{\frac{d}{m}} + \bigl(\varkappa_1 + o(1)\bigr)\tau^{\frac{d-1}{m}} \tag{11.8.8}$$

and asymptotics

$$\mathsf{N}^{\pm}_{(\vartheta)}(\tau) = \mathsf{N}^{\mathsf{W},\pm}_{(\vartheta)}(\tau) + o\bigl(R_{(\vartheta)}(\tau)\bigr) \tag{11.8.9}$$

hold as $\tau\to +0$.

(iv) Furthermore, let $\vartheta(1-m) = d-1$ and the standard non-periodicity condition be fulfilled for the Hamiltonian trajectories of the symbol $\pm a$ on the energy level 0. Then asymptotics (11.8.9) holds.

(v) On the other hand, if the symbol $\pm a(x,\xi)$ is negative-definite for every (x,ξ) (which under the microhyperbolicity condition means precisely that $\varkappa^{\pm}=0$) then

$$\mathsf{N}^{\pm}(\tau) = O(1). \tag{11.8.10}$$

Proof. We leave all the details of the proof to the reader and indicate only the rescaling which should be applied in this proof: namely, we apply the rescaling $(\epsilon, \epsilon r)$ with respect to (x,ξ) where $r=|\xi|$ and $h_{\mathsf{eff}} = r + |\tau|^{\frac{1}{m}}$ afterwards. □

Remark 11.8.2. Asymptotics (11.8.3) and (11.8.5) are standard results under ellipticity condition

$$|a(x,\xi)v| \ge \epsilon_0 |v|\cdot|\xi|^m \qquad \forall (x,\xi)\quad \forall v\in\mathbb{H} \tag{11.8.11}$$

and asymptotics (11.8.8) and (11.8.9) for $\vartheta(1-m) < d-1$ under this condition and non-periodicity condition on the energy level 1.

Let us get rid of the microhyperbolicity condition. In the general case we can do nothing better than introduce the rescaling

$$(x_1, x', \xi_1, \xi') \mapsto (\zeta^{-2}x_1, \zeta^{-1}x', r^{-1}\xi_1, \zeta^{-1}r^{-1}\xi') \qquad \text{with } r \asymp |\xi| \tag{11.8.12}$$

where the coordinate system is taken such that $\xi' = 0$ at some point of the element of the partition and

$$\zeta = \min_j (|\alpha_j| + \tau r^{-m} + r^{-\frac{1}{2}}) \qquad \text{for } \tau \le Cr^m \tag{11.8.13}$$

and $\alpha_j = \alpha_j(x, \xi)$ are eigenvalues of $a(x, \xi|\xi|^{-1})$. Further

(11.8.14) Here and below we set $\zeta = 1$ for $\tau \ge r^m$.

Then $h_{\mathsf{eff}} = r^{-1}\zeta^{-2}$ and the contribution of the zone $\{(x, \xi) : h_{\mathsf{eff}} \ge 1\}$ to the asymptotics does not exceed $CR'_{(\vartheta)}$ with

$$R'_{(\vartheta)} = \begin{cases} \tau^{l+\vartheta} & \text{for } \vartheta < -l, \\ |\log \tau| & \text{for } \vartheta = -l, \\ 1 & \text{for } \vartheta > -l, \end{cases} \qquad l = -\frac{2d}{1-2m}. \tag{11.8.15}$$

Then one can easily obtain the following

Theorem 11.8.3. *(i) In the general case the following asymptotics holds:*

$$\mathsf{N}^{\pm}_{(\vartheta)}(\tau) = \mathsf{N}^{\mathsf{W},\pm}_{(\vartheta)}(\tau) + O\bigl(R_{(\vartheta)}(\tau) + R'_{(\vartheta)}(\tau)\bigr) \tag{11.8.16}$$

where $R_{(\vartheta)}$ and $R'_{(\vartheta)}$ are defined by (11.8.7) *and* (11.8.16) *respectively.*

(ii) Moreover, suppose that $\vartheta < (d(1-2m)^{-1} - 1)$[62] *and let the standard non-periodicity condition be fulfilled for the Hamiltonian trajectories of the symbol $\pm a$ on the energy level* 1. *Then asymptotics* (11.8.8) *and* (11.8.9) *hold.*

(iii) On the other hand, if the symbol $\pm a(x, \xi)$ is non-positive definite for every $(x, \xi) (\Leftrightarrow \varkappa^{\pm} = 0)$ then estimate

$$\mathsf{N}^{\pm}(\tau) = O(\tau^{l}) \tag{11.8.17}$$

holds.

---

[62] Which implies that $d > 1 - 2m$ and $\vartheta(1-m) < (d-1)$ i.e. $R_{(\vartheta)} \gg R'_{(\vartheta)}$.

To improve results of Theorem 11.8.3 let us assume that

$$|D^\alpha_{(x,\xi)} a(x,\xi)| \le c\gamma^{\mu-|\alpha|} \qquad \forall \alpha: |\alpha|\le \mu \quad \forall (x,\xi): |\xi|=1, \tag{11.8.18}$$

$$|a(x,\xi)v| \ge \epsilon_0 \gamma^\mu |v| \quad \forall v\in\mathbb{H} \quad \forall (x,\xi): |\xi|=1, \tag{11.8.19}$$

where $\gamma(x,\xi)=\mathsf{dist}\big((x,\xi),\Lambda\big)$ and Λ is the union of a finite number of $\mathscr{C}^K$ manifolds of codimension $\ge p$. Then one can make rescalings

$$(x,\xi)\mapsto (\zeta^{-1}x, \zeta^{-1}r^{-1}\xi) \tag{11.8.20}$$

with

$$\zeta = \epsilon(\gamma + \tau^{\frac{1}{\mu}} r^{-\frac{m}{\mu}} + r^{-\frac{1}{2}}) \tag{11.8.21}$$

and obtain

Theorem 11.8.4. *Under assumptions* (11.8.18)–(11.8.19) *Statements (i)–(ii) of Theorem 11.8.3 hold with $R'_{(\vartheta)}$ defined by* (11.8.15) *but this time with* $l=-(2d-p)(\mu-2m)^{-1}$.

Let us now treat the scalar case. First of all, for $\vartheta>0$ one can apply exactly the construction of the proof of Theorem 11.8.1 (no microhyperbolicity-type condition is necessary at all). On the other hand, for $\vartheta=0$ one can apply rescaling (11.8.20) with

$$\zeta = \epsilon\big(\sum_{|\alpha|<M} |D^\alpha a|^{\frac{M}{M-|\alpha|}} + |\tau r^{-m}|\big)^{\frac{1}{M}} + \epsilon r^{-\frac{1}{2}} \tag{11.8.22}$$

with $M=2$. Then microhyperbolicity is fulfilled for $h_{\mathsf{eff}}\le h_0$. After calculations one gets the remainder estimate $O\big(\tau^{\frac{d-1}{m}}+\tau^{l}\big)$ where this time $l=-d(1-m)^{-1}$ which is good for $m<(1-d)$.

On the other hand, taking arbitrarily large M one can refer to the weakest possible non-degeneracy condition (5.2.126) and after easy calculations get the remainder estimate $O(\tau^{\frac{1}{m}(d-1+\delta)})$ with arbitrarily small $\delta>0$. Therefore the following statement holds:

Theorem 11.8.5. *Let A be a scalar operator. Then*

(i) For $\vartheta=0$ asymptotics (11.8.3) *hold if $m\le(1-d)$; further, asymptotics*

$$\mathsf{N}^\pm(\tau) = \varkappa^\pm \tau^{\frac{d}{m}} + O(\tau^{\frac{1}{m}(d-1+\delta)}) \tag{11.8.23}$$

holds with arbitrarily small exponent $\delta>0$ if $m>(1-d)$.

(ii) For $\vartheta > 0$ *asymptotics* (11.8.5) *holds.*

(iii) Suppose that either $\vartheta > 0$ *or* $m \le (1 - d)$ *and and that the standard non-periodicity condition is fulfilled for the Hamiltonian trajectories of the symbol* $\pm a$ *on the energy level* 1. *Then asymptotics* (11.8.8), (11.8.9) *hold.*

Comments

While some of the results in the simplest situations were established by different authors, all the results of Sections 11.1–11.7 were proven as the corollaries of the general theorems of Chapter 9 by the author during 1985–1986.

Results of Section 11.8 are brand new.

Chapter 12

Miscellaneous Asymptotics of Spectra

12.1 Introduction

In this chapter we apply the results and methods of the previous chapters in order to derive eigenvalue asymptotics for the miscellaneous problems, mostly to extend the results of the previous Chapter 11 to the cases when these results in their original forms are no longer valid.

In Section 12.2 we treat operators in the domains with the "thick" cusps. Let us recall that if the cusp is not thin enough then asymptotics

$$\mathsf{N}^-(\tau) = \mathsf{N}^{\mathsf{W}}(\tau) + O\bigl(\tau^{(d-1)l}\bigr) \tag{12.1.1}$$

(with $l = 1/(2m)$ where $2m$ is an order of operator) no longer follows from the results of Sections 11.2 and 11.3 and if the cusp is even thicker then even estimate

$$\mathsf{N}^-(\tau) = O\bigl(\tau^{dl}\bigr) \tag{12.1.2}$$

no more follows from these results. It occurs that these asymptotics and estimates are no longer valid and should be replaced by asymptotics either containing non-Weyl correction term or even having non-Weyl principal part.

Similarly, in Section 12.3 we consider operators in $\mathbb{R}^d$ with potentials growing at infinity as $|x|^{2m\mu}$ but degenerate in some directions (such as the Schrödinger operator with a positively homogeneous of degree 2μ potential

V. Ivrii, *Microlocal Analysis, Sharp Spectral Asymptotics and Applications II*, https://doi.org/10.1007/978-3-030-30541-3_12

$V_\infty \ge 0$, which vanishes on some manifold Θ). We proved in Section 11.5 asymptotics (12.1.1) with $l = (\mu+1)/(2m\mu)$ assuming that the degeneration is "mild" i.e. neither it is too strong nor $\dim\Theta$ is too large or, in other words, that the *valley* $\{x : V(x) \le \tau\}$ has only thin cusps for any fixed $\tau > 0$. If the degeneration is stronger then this asymptotics no longer follows from the results of Section 11.5 and if the degeneration is even stronger then even estimate (12.1.2) no more follows from these results.

Again it occurs that these asymptotics are no longer valid and should be replaced by asymptotics either containing non-Weyl correction term or even having non-Weyl principal part.

Further, in Section 12.4 we consider operators in $\mathbb{R}^d$ with potentials decaying at infinity slowly enough (as $|x|^{2m\mu}$ with $\mu \in (-1, 0)$) and consider asymptotics of $\mathsf{N}^-(\tau)$ as $\tau \to -\infty$. However these potentials are falling to $-\infty$ along some manifold Θ. We proved in Section 11.6 asymptotics (12.1.1) with $l = (\mu+1)/(2m\mu)$ assuming that the singularity is "mild" i.e. neither it is too strong nor $\dim\Theta$ is too large or, in other words, that the *gorge* $\{x : V(x) \le \tau\}$ has only thin cusps for any fixed $\tau < 0$. If the singularity is stronger then this asymptotics no longer follows from the results of Section 11.5 and if the singularity is even stronger then even estimate (12.1.2) no more follows from these results.

Again it occurs that these asymptotics are no longer valid and should be replaced by asymptotics either containing non-Weyl correction term or even having non-Weyl principal part.

Furthermore, Section 12.5 is devoted to maximally hypoelliptic operators with symplectic degeneration manifold Λ. In this case we derived no similar results in Chapter 11 but one can prove easily that if degeneration is mild then asymptotics (12.1.1) holds (with $l = 1/(2m)$) as if operator was elliptic; if degeneration is not that mild then this asymptotics holds with worse remainder estimate and if degeneration is even more severe then even estimate (12.1.2) is no longer valid. So, in Section 12.5 we derive correct asymptotics in the latter cases; again these asymptotics contain non-Weyl correction term or even non-Weyl principal part.

Moreover, Section 12.6 is devoted to asymptotics of Riesz means for operators with singularities at separate points: it may be conical singularities of the metrics and domain and weak singularities of lower order terms. Then while (12.1.1) with $l = 1/(2m)$ holds, asymptotics

$$\mathsf{N}^-_{(\vartheta)}(\tau) = \mathsf{N}^{\mathsf{W}}_{(\vartheta)}(\tau) + O\bigl(\tau^{(d-1-\vartheta)l+\vartheta}\bigr) \tag{12.1.3}$$

fails for $\theta \geq \bar{\vartheta}_1$ and for even larger ϑ even estimate $\mathsf{N}^-_{(\vartheta)}(\tau) = O(\tau^{dl+\vartheta})$ is no longer valid and again our goal is to derive correct asymptotics.

Further, in Section 12.7 we consider Neumann Laplacian in domain with cusps (and its certain generalizations). Then we need to assume that the cusps are really very thin: with the width decaying as $\exp(-|x|^\nu)$ with $\nu > 1$; otherwise essential spectrum appears. Still, only as $\nu > \bar{\nu}_1$ we derive asymptotics (12.1.1) and only as $\nu > \bar{\nu}_0$ estimate (12.1.2) is valid; our goal here again is to derive correct asymptotics.

Finally, in Section 12.8 we consider operator $A_t = A - tW$ where A is a periodic operator, W is a potential decaying at infinity. It is known that the spectrum of A consists of *spectral bands* (of essential, usually continuous) spectrum separated by *spectral gaps* (containing no spectrum of A). As $t \neq 0$ the essential spectrum of A_t coincide with one of A but discrete eigenvalues may appear inside spectral gaps. We fix a point E either inside the spectral gap or on its boundary and consider $\mathsf{N}_E(\tau)$ the number of eigenvalues of A_t crossing E as t grows from 0 to τ [1)] and consider asymptotics of $\mathsf{N}_E(\tau)$ as $\tau \to +\infty$.

We consider also the "lite" version of this theory: when A is not a periodic but operator with the constant coefficients.

All these problems have certain common features:

(a) We do not apply the results of Chapters 9 and 10;

(b) The asymptotics either contain non-Weyl terms or are completely non-Weyl (otherwise they would follow from the results of Chapters 9 and 10); however it is not necessarily true for Sections 12.5, 12.7 and 12.8.

(c) Most of the problems in question differ from problems which can be treated by means of results of Chapters 9 and 10 by some parameters (exponents, dimensions, orders of singularity, of degeneration, of Riesz means, etc.).

(d) In some sections we apply the theory of Chapters 2 and 4 for operators with the operator-valued symbols: namely, in Section 12.2 let $x' \in Z$ be the base of the cusp and $x'' \in \sigma(x')Y$ be the shape of the cross-section of the cusp respectively and let $d' = \dim Z$, $d'' = \dim Y$. After a coordinate

[1)] With eigenvalues crossing while increasing counted with the sign "+" and with eigenvalues crossing while decreasing counted with the sign "−".

transform $x'' \mapsto x''\sigma(x')^{-1}$ we obtain a cylindrical cusp $Z \times Y$ and treat operator $\boldsymbol{A}(x', D') = \boldsymbol{A}(x', D'; x'', D'')$ given on the base of the cusp for which the auxiliary Hilbert space is $\mathscr{L}^2(Y)$.

Under appropriate assumptions this is a perturbed operator

$$a(x', D', x'', \sigma^{-1}(x')D'') = \sigma^{-2m}(x')a^0(x', \sigma(x')D', x'', D'') \tag{12.1.4}$$

and it satisfies non-degeneracy condition on the energy level τ; here $\boldsymbol{a}$ is the principal symbol of A. In particular, as $A = \Delta' + \Delta''$ this unperturbed operator is $\Delta' + \sigma^{-2}(x')\Delta''$ and if we deal with unperturbed operator we can decompose it into the direct sum of operators $A_k = \Delta' + \sigma^{-2}(x')\lambda_k$ where $\lambda_1 \le \lambda_2 \le ...$ are eigenvalues of Δ'' (under inherited boundary conditions; we assume that those are Dirichlet conditions albeit one can consider more general boundary conditions as well).

As $\lambda_1 > 0$ all these operators are d'-dimensional Schrödinger operators and their asymptotics are given by Weyl formula with

$$\mathsf{N}_k^{\mathsf{W}}(\tau) = O\bigl(\tau^{d'l}\lambda_k^{-d'/(2m)}\bigr) \tag{12.1.5}$$

and with the remaining estimate

$$R_k^{\mathsf{W}}(\tau) = O\bigl(\tau^{(d'-1)l}\lambda_k^{-(d'-1)/(2m)}\bigr) \tag{12.1.6}$$

where $l = (\mu+1)/(2m\mu)$ as $\sigma(x') \asymp |x'|^{-\mu}$.

Now we need to sum all the main parts and we arrive to asymptotics with the principal part either $\asymp \tau^{d/(2m)}$ or $\asymp \tau^{d'l}$ whatever is greater but as degrees are equal we get principal part of magnitude $\asymp \tau^{d'l}|\log \tau|$.

Similarly, we need to sum all the remainder estimates and we arrive to $O(\tau^{(d-1)/(2m)})$ or $O(\tau^{d'l})$ whatever is greater but as degrees are equal we get the remainder $O(\tau^{(d'-1)l}|\log \tau|$.

Thus in Section 12.2 we extend this result to the case when operator in question does not decompose into direct sum and to the higher order operators. The main issue here is not only to express the principal part of the asymptotics (possibly containing more than one term) but also simplify it if possible invoking eigenvalue counting function $\boldsymbol{n}_\infty(\tau)$ of one axillary operator rather than the family $\boldsymbol{n}(x', \xi; \tau)$.

The similar approach works in Section 12.3 where in the case of the Schrödinger operator with potential $V(x)$, non-negative and positively homogeneous of degree $2\mu > 0$, x' are coordinates on $Z = \{x : V(x) = 0\}$

and x'' are coordinates in $Z^\perp$. Similarly in Section 12.4 in the case of the Schrödinger operator with potential $V(x)$, positively homogeneous of degree $2\mu \in (-2,0)$, x' are coordinates on $Z=\{x : V(x)=-\infty\}$ and x'' are coordinates in $Z^\perp$.

In Section 12.5 we consider pseudodifferential operators with the phase space Λ and with auxiliary Hilbert space $\mathscr{L}^2(\mathbb{R}^{d''})$ with $d''=\frac{1}{2}\operatorname{codim}\Lambda$.

In Sections 12.2–12.5 there are zones where we apply the standard Weyl asymptotics of Part II, zones where we can obtain the same asymptotics with improved remainder estimates (these enhancements are due to results on the propagation of singularities for operators with the operator-valued symbols and zones where results of Chapter 4 are applicable; the second and third zones are overlapping, and one can combine the asymptotics obtained.

In Section 12.6, instead of operators with the operator-valued symbols we treat model operators with positively quasihomogeneous symbols; for these operators

$$\int \Bigl(e_{(\vartheta)}(x,x,0)-e^{\mathsf{W}}_{(\vartheta)}(x,x,0)\Bigr)\,dx=\varkappa h^{-\kappa} \tag{12.1.7}$$

converges as $\vartheta>\bar{\vartheta}$ and the equality is due to quasihomogeneity; here $e(.,.,\tau)$ is the Schwartz kernel of the spectral projector, $e^{\mathsf{W}}(.,.,\tau)$ is its Weyl approximation, and $_{(\vartheta)}$ means a Riesz mean of order ϑ with respect to τ; the right-hand expression is *Scott correction term*. We consider also similar expressions with homogeneous factor in the integral and basically result is the same albeit in some cases we need to include cut-off in the left-hand expression and the right-hand expression becomes $(\varkappa|\log h|+\varkappa_1)h^{-\kappa}$. We cannot represent the left-hand expression in (12.1.7) as the difference as both integrals diverge at infinity.

We apply then Weyl method to the difference of the spectral functions of the original and the model operators and derive sharp asymptotics for $\int e_{(\vartheta)}(x,x,0)\psi(x)\,dx$ with a fixed supported near 0 function $\psi\in\mathscr{C}_0^\infty$; this asymptotics contains Weyl part and one or more Scott correction terms (which could be actually larger than the Weyl part).

Section 12.7 starts as Section 12.2, but $0=\lambda_1<\lambda_2$ (in more general situations $0=\lambda_1=\ldots=\lambda_k<\lambda_{k+1}$ and spectrum seems to become essential; however when we change variables $x''\mapsto\sigma^{-1}(x')x''$ we get a weighted $\mathscr{L}^2$-space $\mathscr{L}^2(Z,\mathscr{L}^2(Y),\sigma^d(x')dx')$; considering quadratic form and substituting $u=\sigma^{-d/2}v$ we obtain operator in the standard $\mathscr{L}^2(Z,\mathscr{L}^2(Y))$-space and this operator acquires an extra potential

$$V \sim \frac{d^2}{4} \sum_{j,k} g^{jk} (\partial_j \log \sigma)(\partial_k \log \sigma); \tag{12.1.8}$$

this potential grows at infinity as $|\nabla \log \sigma|^2 \asymp |x|^{2\mu}$ provided σ behaves like $\exp(-|x'|^\nu)$ with $\nu = \mu + 1$. Then the results similar to those of Section 12.2 are derived but since the Schrödinger operator corresponding to $\lambda_j = 0$ is so different from all others (which now have potentials with superexponential growth) logarithmic factor does not appear as two degrees collide.

In Section 12.8 we consider operator $A - tW$ where A is periodic operator and W is a decaying potential, and count eigenvalues crossing E which belongs either to spectral gap or its border. In this case we use Gelfand decomposition so A becomes $A(\xi)$ operator acting on $\mathscr{L}^2(\mathcal{Q})$ where $\mathcal{Q}$ is an elementary cell corresponding to periods lattice and $\xi \in \mathcal{Q}'$ is a quasimomentum where $\mathcal{Q}'$ is a dual cell. Meanwhile W becomes a negative order pseudodifferential operator with respect to ξ and we again apply the theory of operators with the operator-valued symbols. However in contrast to what we had before, dimension does not decrease.

12.2 Operators in Domains with Thick Cusps

12.2.1 Problem Set-up

Let us consider a uniformly elliptic operator of order $2m$

$$A = A^{\mathrm{w}}(x, D), \qquad A(x,\xi) = \sum_{|\alpha| \le 2m} a_\alpha(x) \xi^\alpha \tag{12.2.1}$$

where

(12.2.2) $a_\alpha(x)$ are Hermitian $\mathsf{D} \times \mathsf{D}$-matrix valued functions such that

$$|D^\beta a_\alpha| \le c \langle x \rangle^{-|\beta'| + \mu(2m - |\alpha|)} \qquad \forall \alpha : |\alpha| \le 2m \ \forall \beta : |\beta| \le K, \beta'' = 0, \tag{12.2.3}$$

$$D^\beta a_\alpha = o\bigl(|x'|^{-|\beta'|+\mu(2m-|\alpha|+|\beta''|)}\bigr) \qquad \text{as } |x'|\to\infty$$
$$\forall\alpha: |\alpha|\le 2m\ \forall\beta: |\beta|\le K, 2m-|\alpha|+|\beta''|>0, \tag{12.2.4}$$

$$\langle a(x,\xi)v,v\rangle \ge \epsilon|v|^2 \qquad \forall v\quad \forall(x,\xi) \tag{12.2.5}$$

where

$$a(x,\xi)=\sum_{|\alpha|=2m} a_\alpha(x)\xi^\alpha \tag{12.2.6}$$

is the principal symbol of A.

We consider this operator in the domain $X\subset\mathbb{R}^d$ such that

$$X = X_0\cup X_{\mathsf{cusp}},\ \partial X\in\mathscr{C}^K,\ X_0\Subset\mathbb{R}^d,$$
$$X_{\mathsf{cusp}}=\bigl\{x=(x',x''): x'\in Z,\ x''\in\sigma(x')Y\bigr\},\qquad Y\Subset\mathbb{R}^{d''},\ \partial Y\in\mathscr{C}^K \tag{12.2.7}$$

with

$$\sigma(x')\asymp\langle x'\rangle^{-\mu}\ \mu>0, \tag{12.2.8}$$

$$|D^{\beta'}\sigma|\le c\langle x'\rangle^{-\mu-|\beta'|}\quad \forall\beta': |\beta'|\le K \tag{12.2.9}$$

and

(12.2.10) Either $Z=\mathbb{R}^{d'}$ as $d'\ge 1$ or $Z=\mathbb{R}^\pm$ as $d'=1$[2].

Definition 12.2.1. We say that X_0 is the *compact part* of X, X_{cusp} is the *cusp*, Z and Y are the *cusp base and cusp cross-section* respectively and σ is the *cusp form defining function.*

Let us assume that

(12.2.11) A Dirichlet boundary condition (of order m) is given on ∂X.

One can prove easily that

(12.2.12) Under these conditions (12.2.1)–(12.2.11) A is a self-adjoint operator in $\mathscr{L}^2(X,\mathbb{H})$, $\mathbb{H}=\mathbb{C}^{\mathsf{D}}$ and has a compact resolvent.

---

[2] Analysis of more general cusps will be formulated as Problem 12.2.23.

Moreover, the following statement follows from the results of Sections 11.2 and 11.3:

Theorem 12.2.2. *Let conditions* (12.2.1)–(12.2.11) *be fulfilled. Then*

(i) As $\sigma^{d''-1} \in \mathscr{L}^2(Z)$ *which is equivalent to*

$$\mu > \mu_1 := d'(d''-1)^{-1}, \qquad \mu_0 := d'd''^{-1} \tag{12.2.13}$$

then Weyl asymptotics

$$\mathsf{N}^-(\tau) = \mathsf{N}^{\mathsf{W}}(\tau) + O(\tau^{\frac{1}{2m}(d-1)}) \tag{12.2.14}$$

holds as $\tau \to \infty$ *with*

$$\mathsf{N}^{\mathsf{W}}(\tau) \asymp \tau^{\frac{d}{2m}}. \tag{12.2.15}$$

Here $\mathsf{N}^{\mathsf{W}}(\tau)$ *is given by the standard Weyl formula, and moreover, one can replace it here by* $\kappa_0\tau^{\frac{d}{2m}}$ *where* κ_0 *and* κ_1 *are standard coefficients which are finite for* $\mu > \mu_0$, $\mu > \mu_1$ *respectively.*

(ii) Moreover, under the standard non-periodicity condition to the generalized Hamiltonian flow of $a(x,\xi)$ *with reflections at* ∂X *(and perhaps with branching at the points of reflection for* $m \geq 2$*) asymptotics*

$$\mathsf{N}^-(\tau) = \mathsf{N}^{\mathsf{W}}(\tau) + \kappa_1\tau^{\frac{1}{2m}(d-1)} + o\bigl(\tau^{\frac{1}{2m}(d-1)}\bigr) \tag{12.2.16}$$

holds as $\tau \to \infty$.

(iii) For $\mu = \mu_1$ *asymptotics*

$$\mathsf{N}^-(\tau) = \mathsf{N}^{\mathsf{W}}(\tau) + O\bigl(\tau^{\frac{1}{2m}(d-1)}\log\tau\bigr) \tag{12.2.17}$$

holds as $\tau \to \infty$ *and for* $\mu \in [\mu_0, \mu_1)$ *asymptotics*

$$\mathsf{N}^-(\tau) = \mathsf{N}^{\mathsf{W}}(\tau) + O\bigl(\tau^{\frac{1}{2m}(d-q)}\bigr) \tag{12.2.18}$$

holds as $\tau \to \infty$ *with* $q = d'\mu^{-1} + d'' \in [0,1)$*; here* $q = 0 \iff \mu = \mu_0$, $q = 1 \iff \mu = \mu_1$, *and for* $\mu > \mu_0$ (12.2.15) *holds. Further, for* $\mu = \mu_0$

$$\mathsf{N}^{\mathsf{W}}(\tau) \asymp \tau^{\frac{d}{2m}}\log\tau \tag{12.2.19}$$

is given by cut Weyl formula and one can replace it by $\kappa_0^*\tau^{d/2m}\log\tau$.

Remark 12.2.3. (i) Surely, our conditions here are more restrictive than necessary but the case $\mu > \mu_1$ is not our goal here. Our goal is to treat the case $\mu \le \mu_1$ and to derive asymptotics with better remainder estimates. Namely, we obtain asymptotics with the remainder estimate $O(\tau^{\frac{1}{2m}(d-1)})$ and even $o(\tau^{\frac{1}{2m}(d-1)})$ under the standard non-periodicity condition for $d'=1$ and for $d' \ge 2$, $\mu > \mu^*$ with $\mu^* < \mu_1$; otherwise we obtain asymptotics with a worse remainder estimate.

Surely, one should replace $\mathsf{N}^{\mathsf{W}}(\tau)$ by another expression and some additional conditions appear.

(ii) For Riesz means the crucial values in the Weyl asymptotics are

$$\mu_{\vartheta+1} = d'(d''-\vartheta-1)^{-1} \tag{12.2.13$'$}$$

and μ_0 respectively: namely, for $\mu > \mu_{\vartheta+1}$ we get the remainder estimate $O(\tau^{\frac{1}{2m}(d-\vartheta-1)+\vartheta})$ or even "o" instead of "O" under the standard non-periodicity condition. Further, for $\mu = \mu_{\vartheta+1}$ the remainder estimate is $O(\tau^{\frac{1}{2m}(d-\vartheta-1)+\vartheta}\log\tau)$ and for $\mu_0 < \mu < \mu_{\vartheta+1}$ it is $O(\tau^{\frac{d'(\mu+1)}{2m\mu}+\vartheta})$.

One can easily see that this exponent lies in $\bigl(\frac{1}{2m}(d-\vartheta-1)+\vartheta, \frac{d}{2m}+\vartheta\bigr)$. At the same time the principal Weyl part is of order $\tau^{\frac{d}{2m}+\vartheta}$ and $\tau^{\frac{d}{2m}+\vartheta}\log\tau$ for $\mu > \mu_0$ and $\mu = \mu_0$ respectively.

12.2.2 Sharp Asymptotics

Standard Remainder Estimate

First of all let us divide $\mathsf{N}^-(\tau)$ into the sum

$$\mathsf{N}^-(\tau) = \mathsf{N}^-_0(\tau) + \mathsf{N}^-_{\mathsf{cusp}}(\tau) \tag{12.2.20}$$

where

$$\mathsf{N}^-_\bullet(\tau) = \int \psi_\bullet(x')\,\operatorname{tr}(e(x,x,\tau))\,dx, \tag{12.2.21}$$

$e(x,y,\tau)$ is the Schwartz kernel of the spectral projector of A and $\psi_0, \psi_{\mathsf{cusp}} \in \mathscr{C}^K$ are supported in $\{|x'| \le 3c\}$, $\{|x'| \ge 2c\}$ respectively, $\psi_0 + \psi_{\mathsf{cusp}} = 1$. We assume that $X_0 \subset \{|x'| \le \frac{3}{2}c\}$.

Then

(12.2.22) Weyl asymptotics holds for $\mathsf{N}_0^-(\tau)$ with the proper remainder estimate $O\bigl(\tau^{\frac{1}{2m}(d-1)}\bigr)$ (and even $o\bigl(\tau^{\frac{1}{2m}(d-1)}\bigr)$ under the standard non-periodicity condition to the Hamiltonian billiards).

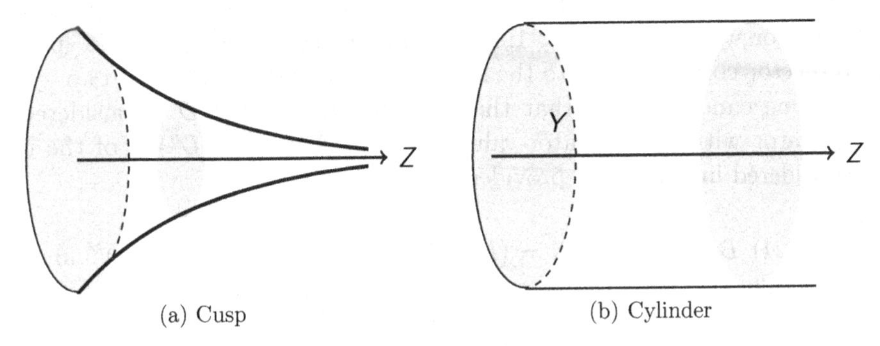

(a) Cusp (b) Cylinder

Figure 12.1: Cusp transforms to a cylinder

So we need to derive asymptotics of $\mathsf{N}^-_{\mathsf{cusp}}(\tau)$. Let us introduce the coordinates (x', x'') with $x''_{\mathsf{new}} = \sigma(x')^{-1}x''$. In these coordinates cusp X_1 transforms to the cylinder $Z\times Y$ and the differential expression A transforms to a differential expression which is symmetric with weight $\sigma(x')^{d''}$ provided our test functions vanish for $|x'|\le \frac{1}{3}c$. In order to get rid of this weight we multiply the differential expression by $\sigma(x')^{\frac{1}{2}d''}$ and $\sigma(x')^{-\frac{1}{2}d''}$ on the left and right respectively, obtaining

(12.2.23) $\tilde{A} = \tilde{A}^{\mathrm{w}}(x, D), \quad \tilde{A}(x,\xi) = \sum_{|\alpha|\le 2m} a_\alpha(\xi' - b'\beta)^{\alpha'}\xi''^{\alpha''}\sigma(x')^{-|\alpha''|},$

$$b_k = \partial_k(\log\sigma) \quad k = 1, \dots, d', \qquad \beta = \langle x'', \xi''\rangle$$

in the coordinates $x = (x', x'')$. We treat this operator as an operator in $\mathscr{L}^2(Z, \mathbb{H})$ with $\mathbb{H} = \mathscr{L}^2(Y, \mathbb{H})$. Surely, we cannot discuss self-adjointness now because we cannot guarantee that our functions vanish for $|x'| \le \frac{5}{3}c$.

Recall that we are interested in the spectral projector $E(\tau)$ of A corresponding to the spectral parameter τ. Let us consider the $\gamma(\bar{x}')$-neighborhood of the point $\bar{x}'$ with $\gamma(x') = \frac{1}{2}|x'|$ where we always assume that $|\bar{x}'| \ge c_1$ with a large enough constant c_1. Let us rescale $x' \mapsto (x' - \bar{x}')\bar{\gamma}^{-1}$ and the make a corresponding multiplication.

Let us first treat case of $\bar{\sigma} \le \epsilon\tau^{-\frac{1}{2m}}$ (i.e., $|x'| \ge \epsilon_1^{-1}\tau^{\frac{1}{2\mu m}}$) where here and below $\epsilon_j > 0$ are sufficiently small and $\bar{f} = f(\bar{x}')$ for any function f. Multiplying operator A by $\bar{\sigma}^{2m}$ after rescaling we obtain an h-differential operator with $h = \bar{\sigma}\bar{\gamma}^{-1} < 1$; we are interested in the operator's spectral projector corresponding to the spectral parameter $\tau' = \tau\bar{\sigma}^{2m} < \epsilon_2$.

One can easily see that this operator $\boldsymbol{A}(x', D'; x'', D'')$ considered as operator with the operator-valued symbol $\boldsymbol{A}(x', \xi'; x'', D'')$ is of the type considered in Chapter 4 provided

(12.2.24) $B = (L_{(m)\mathrm{D}})$, $B_1 = (L_{(M)\mathrm{N}})^{mM^{-1}}$ where $L_M = (\Delta'' + I)^M$ and the subscripts "D" and "N" mean m-th order Dirichlet and Neumann boundary conditions respectively (of the proper order), Δ'' is the positive Laplacian in Y and M is taken large in order to obtain large $l = l(M)$ [3].

Then for $j \in [0, l]$ we see that $\mathbb{H}_j = \mathfrak{D}(B_1^j) = \mathscr{H}^{2mj}(Y)$ are Sobolev spaces and for $j \in [-l, 0]$ $\mathbb{H}_j = B_1^{-j}(\mathbb{H})$ are the spaces dual to $\mathscr{H}^{-2mj}(Y)$.

In particular, one can easily check that conditions (4.2.7), $(4.2.8)_l$, $(4.2.13)_{\frac{1}{2}}$ (i.e. $(4.2.8)_l$ with $l = \frac{1}{2}$), $(4.2.14)_+$, (4.2.15), (4.2.17) and (4.2.21) are fulfilled.

Further, we see that

(12.2.25) Operator $\boldsymbol{A}(x', D'; x'', D'')$ [4] is elliptic on the energy level $s \le \epsilon_2$.

---

[3] See Chapter 4 for B, B_1 and l.

[4] Considered as an operator with the operator-valued symbol.

One can check this easily by dropping lower order terms (with $|\alpha| < 2m$) in $\tilde{A}$ and assuming that D'' and a_α commute and the energy level is non-positive. This procedure is justified by condition (12.2.4) and the appropriate choice of $\epsilon_2 > 0$.

Furthermore, since A is a self-adjoint operator, $\|\exp(itA)\| = 1$ and $\|E(\tau)\| \le 1$. On the other hand, $\exp(itA)$ and $E(\tau)$ generate operator-valued operators in $B(\bar{x}', \gamma(\bar{x}'))$ via their Schwartz kernels and one can apply the theory of Chapter 4 (namely, Propositions 4.2.8–4.2.10). This trick permits us to avoid the discussion of the self-adjointness of $\tilde{A}$, etc.

Proposition 12.2.4. *Let conditions* (12.2.1)–(12.2.11) *be fulfilled and let* c_0 *be large enough. Then*

(i) Estimate

$$|e(x,y,\tau)| \le C'_s \langle x'\rangle^{-s}\langle y'\rangle^{-s} \qquad \forall \tau \ge 1 \quad \forall x,y \in X : |x'| + |y'| \ge c_0\tau^{\frac{1}{2\mu m}} \tag{12.2.26}$$

holds.

(ii) Contribution of the far cusp zone $\{|x'| \ge c_0\tau^{1/2m\mu}\}$ *to the asymptotics is* $O(\tau^{-s})$.

Proof. Then the results of Chapter 4 immediately yield that

$$\|B_1^j e(x,y,\tau) B_1^k\| \le C'_{jks}\langle x'\rangle^{-s}\langle y'\rangle^{-s} \qquad \forall j,k \le l\ \forall s \tag{12.2.27}$$

where the norm in the left-hand expression is the operator norm in $\mathbb{H}$ and therefore as $l = l(d)$ (and thus $K = K(d)$) we arrive to estimate (12.2.26) thus proving Statement (i).

Statement (ii) follows from Statement (i) immediately. □

Let us now treat the *main cusp zone*

$$\{c_0 \le |x'| \le \bar{\gamma}(\tau) := c_0\tau^{\frac{1}{2m\mu}}\}. \tag{12.2.28}$$

Here we should multiply the operator by τ^{-1} and so we should consider the operator $\tau^{-1}A$ on the energy level 1. One can easily check that we are again in the framework of Chapter 4 with $h = \tau^{-\frac{1}{2m}}\bar{\gamma}^{-1}$ and

$$B = \tau^{-1}\bar{\sigma}^{-2m}(L_{(M)\mathrm{D}})^{mM^{-1}} + I, \quad B_1 = \tau^{-1}\bar{\sigma}^{-2m}(L_{(M)\mathrm{N}})^{mM^{-1}} + I \tag{12.2.29}$$

with the previous definition of $L_{(M)D}$ and $L_{(M)N}$.

One can easily check that this operator $A(x', D'; x'', D'')$ [4] is elliptic on the energy levels ≤ 1 for $|\xi'| \ge C_0$.

Moreover, one can easily check that this operator[4] is microhyperbolic on the energy level 1 in the direction $\ell \in T_{(x',\xi')}T^*Z$ in each of the following cases[5]:

(a) $m = 1$, $a_\alpha = o(1)$ for $|\alpha'| = |\alpha''| = 1$, $|\xi'| \ge \epsilon$, $\ell = \langle \xi', \partial_{\xi'}\rangle$;

(b) $m = 1$, $|\xi'| \le \varepsilon$ with a small enough constant $\varepsilon > 0$, $|\ell\sigma| \ge \epsilon\sigma|x'|^{-1}$, $\ell = \langle \ell', \partial_{x'}\rangle$ with an appropriate vector ℓ' such that $|\ell'| \le c$ and

$$\ell a_\alpha = o\bigl(|x'|^{-1+(2m-|\alpha|)}\bigr) \quad \text{as } |x'| \to \infty \quad \forall \alpha : |\alpha| < 2m \tag{12.2.30}$$

(here this condition should be checked for $\alpha' = 0$ only);

(c) $m \ge 1$, $\ell\sigma = \sigma\bigl(1 + o(1)\bigr)|x'|^{-1}$ as $|x'| \to \infty$ with an appropriate vector ℓ, $|\ell| \le c$, (12.2.30) holds and $\ell = \langle \xi', \partial_{\xi'}\rangle + \langle \ell', \partial_{x'}\rangle$.

All these conditions are given in the original coordinates and in Conditions (a) and (c) one can take ℓ' depending on x'.

Then we arrive to

Proposition 12.2.5. *The microhyperbolicity condition is fulfilled for any ξ' provided*[5] *one of the following conditions holds:*

(12.2.31) $m = 1$, $a_\alpha = o(1)$ *for* $|\alpha'| = |\alpha''| = 1$, $|\ell\sigma| \ge \epsilon\sigma|x|^{-1}$ *and* (12.2.30) *holds for* $\alpha' = 0$

(12.2.32) $a = \bar{a}^m + a'$ *where second-order symbol* $\bar{a}$ *satisfies* (12.2.31) *and all coefficients of* a' *satisfy*

$$a'_\alpha = o\bigl(|x'|^{\mu(2m-|\alpha|)}\bigr), \tag{12.2.33}$$

$$\ell a'_\alpha = o\bigl(|x'|^{-1+\mu(2m-|\alpha|)}\bigr) \tag{12.2.34}$$

(12.2.35) $m \ge 1$, $\ell\sigma = \sigma\bigl(1 + o(1)\bigr)|x'|^{-1}$ *and* (12.2.30) *holds.*

---

[5] These conditions are sufficient but not necessary.

In what follows we assume that operator $\boldsymbol{A}=\tau^{-1}\tilde{A}$ is microhyperbolic on the energy level 1 [4) 6)].

Then, applying Theorem 4.4.8 as this operator is ξ'-microhyperbolic and Theorem 4.4.9 as this operator is (x',ξ')-microhyperbolic (i.e. in the general case) we obtain that if $\psi=\psi(x')$ is a γ-admissible function supported in $\{|x'-\bar{x}'|\le \frac{1}{2}\bar{\gamma}\}$ then

$$
\begin{multline}
\tag{12.2.36} \|B_1^j\int \psi(x')\Bigl(e(x,y,\tau)_{x'=y'}-\\
(2\pi)^{-d'}\int \boldsymbol{e}(x',\xi',\tau)\,d\xi'\Bigr)\,dx'\cdot B_1^k\|\le C\tau^{\frac{1}{2m}(d'-1)}\gamma^{d'-1}\\
\forall j,k\le l
\end{multline}
$$

where the norm in the left-hand expression is the operator norm in $\mathbb{H}$ and

(12.2.37) $\boldsymbol{e}(x',\xi',\tau)$ is the spectral projector of the operator $\boldsymbol{a}(x',\xi'):=\boldsymbol{a}(x',\xi';x'',D'')$; the latter operator is an *operator-valued symbol* of $\tilde{A}(x,D)$ and we do not include factor τ^{-1} into operator or symbol.

Recall that $\tilde{A}$ is the operator $\sigma^{\frac{1}{2}d''}A\sigma^{-\frac{1}{2}d''}$ in the cylindrical coordinates (x',x'') (without division by τ)). Thus we obtain $\boldsymbol{a}(x',\xi';x'',D'')$ by replacing D' with ξ' in $\tilde{A}(x,D)$. So operator $\boldsymbol{a}(x',\xi')$ acts in $\mathbb{H}=\mathscr{L}^2(Y,\mathbb{H})$ and $\mathfrak{D}(\boldsymbol{a})=\mathscr{H}^{2m}(Y,\mathbb{H})\cap\mathscr{H}_0^m(Y,\mathbb{H})$.

Obviously, this operator is self-adjoint and therefore $\boldsymbol{e}(x',\xi',\tau)$ is well-defined.

Recalling the definition of B_1 one can easily see that

(12.2.38) As $l>\frac{1}{2m}d''$ B_1^{-l} is a trace operator in $\mathbb{H}$ with trace norm not exceeding $C\tau^{\frac{d''}{2m}}\sigma^{d''}$.

Therefore we immediately obtain the following estimate:

$$
\begin{multline}
\tag{12.2.39} |\int \psi(x')\operatorname{tr}(e(x,x,\tau))\,dx-(2\pi)^{-d'}\iint \psi(x')\boldsymbol{n}(x',\xi',\tau)\,dx'd\xi'|\le\\
C\tau^{\frac{1}{2m}(d-1)}\gamma^{d'-1-\mu d''}
\end{multline}
$$

where

6) Which is the case provided[5)] one of conditions (12.2.31), (12.2.32) or (12.2.35) is fulfilled.

(12.2.40) $n(x',\xi',\tau)=\operatorname{Tr}_{\mathbb{H}} e(x',\xi',\tau)$ is an eigenvalue counting function of the operator $a(x',\xi')$ in $\mathbb{H}$.

Really, due to (12.2.38) the trace norm (in $\mathbb{H}$) of the operator in the left-hand expression of (12.2.36) stripped of B_1^j and B_1^k does not exceed $C\tau^{\frac{1}{2m}(d'-1)}\gamma^{d'-1}\times\tau^{\frac{d''}{2m}}\sigma^{d''}$ which is exactly the right-hand expression of (12.2.39).

Summing over a partition of the unity in the main cusp zone (12.2.28) we obtain in the right-hand expression

$$C\tau^{\frac{1}{2m}(d-1)}\int_1^{\bar\gamma(\tau)}\gamma^{d'-1-\mu d''}\,d\gamma$$

which in turn is equal to

$$R^*(\tau)=\begin{cases}\tau^{\frac{1}{2m}(d-1)} & \text{for } d'<\mu d''+1,\\ \tau^{\frac{1}{2m}(d-1)}\log\tau & \text{for } d'=\mu d''+1,\\ \tau^{\frac{1}{2m\mu}(d'-1)(\mu+1)} & \text{for } d'>\mu d''+1.\end{cases}\tag{12.2.41}$$

Then, taking into account results in the zone $\{|x'|\le 2c_0\}$ (Weyl asymptotics) and far cusp zone $\{|x'|\ge\frac{1}{2}\bar\gamma(\tau)\}$ (Proposition 12.2.4) we obtain the following basic statement:

Theorem 12.2.6. *Let conditions* (12.2.1)–(12.2.11) *be fulfilled and let microhyperbolicity condition for* $a(x',\xi')$ *be fulfilled*[6]. *Then asymptotics*

$$N^-(\tau)=\mathcal{N}(\tau)+O\bigl(R^*(\tau)\bigr)\tag{12.2.42}$$

holds as $\tau\to+\infty$ *where* $R^*(\tau)$ *is given by* (12.2.41) *and*

$$\mathcal{N}(\tau)=(2\pi)^{-d}\iint\psi_0(x')n(x,\xi,\tau)\,dxd\xi+ \\ (2\pi)^{-d'}\iint\bigl(1-\psi_0(x')\bigr)n(x',\xi',\tau)\,dx'd\xi',\tag{12.2.43}$$

$n(x,\xi,\tau)$ *is the eigenvalue counting function for symbol* $a(x,\xi)$, $\psi_0\in\mathscr{C}^K$ *is supported in* $\{|x'|\le 2c_0\}$ *and* $\psi_0=1$ *for* $|x'|\le c_0$.

Moreover,

$$\mathcal{N}(\tau)\asymp\begin{cases}\tau^{\frac{d}{2m}} & \textit{for } d'<\mu d'',\\ \tau^{\frac{d}{2m}}\log\tau & \textit{for } d'=\mu d'',\\ \tau^{\frac{1}{2m\mu}d(\mu+1)} & \textit{for } d'>\mu d''.\end{cases}\tag{12.2.44}$$

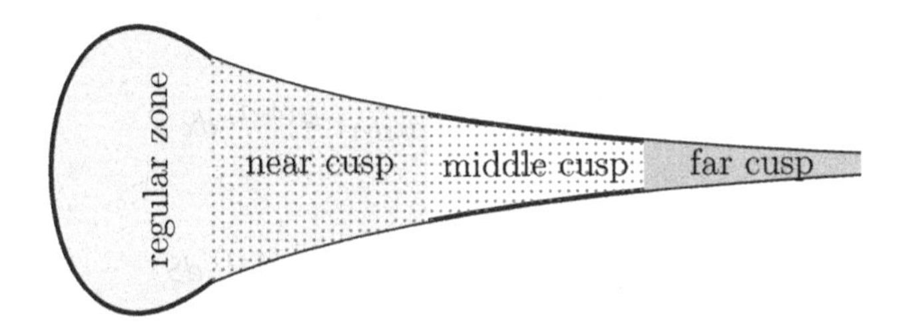

Figure 12.2: In both the *regular zone* and the *near cusp* we apply standard Weyl asymptotics (albeit in the near cusp we use propagation results for operators with the operator-valued symbols), in the *middle cusp* we use Weyl asymptotics for operators with the operator-valued symbols and *far cusp* is an elliptic zone for operator in question considered as operator with the operator-valued symbol; all zones overlap.

Improved Remainder Estimates

Assume first that $d' < 1+\mu d''$. Then the main contribution to the remainder estimate is given by the zone $\{|x'| \le t\}$ with an arbitrarily large constant t (the contribution of the complement does not exceed $\tau^{\frac{1}{2m}(d-1)} t^{-\delta}$ with $\delta > 0$); on the other hand, in this zone under the standard non-periodicity condition to the Hamiltonian billiards we have the remainder estimate $o\bigl(\tau^{\frac{1}{2m}(d-1)}\bigr)$.

Therefore the following statement holds:

Theorem 12.2.7. *Let conditions* (12.2.1)–(12.2.11) *be fulfilled and let microhyperbolicity condition for* $\mathsf{a}(x',\xi')$ *be fulfilled*[6].

Finally, let $d' < 1+\mu d''$ *and let the standard non-periodicity condition to the Hamiltonian billiards be fulfilled. Then asymptotics*

$$\mathsf{N}^-(\tau) = \mathcal{N}(\tau) + o\bigl(\tau^{\frac{1}{2m}(d-1)}\bigr) \tag{12.2.45}$$

holds as $\tau \to +\infty$ *where* <u>*now*</u>

$$\begin{aligned}\mathcal{N}(\tau) = (2\pi)^{-d} \iint \psi_t(x')\bigl(n(x,\xi,\tau) + \kappa_1 \tau^{\frac{d-1}{2m}}\bigr)\, dxd\xi + \\ (2\pi)^{1-d} \int \psi_t(x')\tilde{\kappa}_1 \tau^{\frac{d-1}{2m}}\, dS + \\ (2\pi)^{-d'} \iint \bigl(1-\psi_t(x')\bigr)\mathsf{n}(x',\xi',\tau)\, dx'd\xi',\end{aligned} \tag{12.2.46}$$

$$(2\pi)^{-d}\int \psi_t\kappa_1\tau^{\frac{1}{2m}(d-1)}\,dx \tag{12.2.47$_1$}$$

and

$$(2\pi)^{1-d}\int_{\partial X} \psi_t\tilde{\kappa}_1\tau^{\frac{1}{2m}(d-1)}\,dS \tag{12.2.47$_2$}$$

are the standard second terms in the Weyl asymptotics, $\psi_t(x') := \psi_0(\frac{x'}{t})$ *where* $t = t(\tau)$ *is an appropriate function such that* $t(\tau)\to+\infty$ *as* $\tau\to+\infty$ [7].

On the other hand, let $d' > 1 + d''\mu$. Then the main contribution to the remainder estimate in Theorem 12.2.6 is provided by the zone

$$\{\varepsilon \le \tau^{-\frac{1}{2m\mu}}|x'| \le c_0\} \tag{12.2.48}$$

with arbitrarily small $\varepsilon > 0$; contribution of zone $\{|x'| \le 2\varepsilon\tau^{\frac{1}{2m\mu}}\}$ does not exceed $\varepsilon^\delta R^*(\tau)$.

However, in zone (12.2.48) $\boldsymbol{n}(x',\xi',\tau)$ does not exceed $\nu(\varepsilon) = \varepsilon^M$ and therefore we have effectively here matrix- rather than operator-valued operators.

Applying Theorems 4.4.11 and 4.4.12 and replacing $\boldsymbol{a}$ if necessary by another operator in $\mathbb{H}$ with eigenvalues asymptotically close to those of $\boldsymbol{a}$ we obtain the following statement:

Theorem 12.2.8. *Let conditions* (12.2.1)–(12.2.11) *be fulfilled and let microhyperbolicity condition for* $\boldsymbol{a}(x',\xi')$ *be fulfilled* [6].

Further, let $d' > 1 + \mu d''$ *and let the stabilization conditions*

$$D^{\beta'}(a_\alpha - a_{\infty,\alpha}) = o(|x'|^{-|\beta'|}) \qquad \forall\alpha : |\alpha| = 2m, \forall\beta' : |\beta'| \le K, \tag{12.2.49$_1$}$$

$$D^{\beta'}(\sigma - \sigma_\infty) = o(|x'|^{-\mu-|\beta'|}) \qquad \forall\beta' : |\beta'| \le K \tag{12.2.49$_2$}$$

be fulfilled as $|x'|\to\infty$ *where* $a_{\infty,\alpha} = a_{\infty,\alpha}(x')$, $\sigma_\infty(x')$ *are positively homogeneous functions of* x' *of degrees* 0, $-\mu$ *respectively (surely,* $a_{\infty,\alpha}$ *can be matrix-valued).*

---

[7] If $t(\tau)$ works then every function $t'(\tau) \le t(\tau)$ such that $t'(\tau)\to+\infty$ as $\tau\to+\infty$ also works.

Let us consider operator-valued symbol

$$\mathsf{a}_\infty(x',\xi';D'') = \sum_{|\alpha|=2m} a_{\infty,\alpha}\sigma_\infty^{-|\alpha''|}\xi'^{\alpha'}D''^{\alpha''} \tag{12.2.50}$$

in $\mathbb{H}$; *let* $\zeta_j(x',\xi')$ *be its (non-decreasing) eigenvalues*[8].

Finally, let us consider $\mathsf{a}_{\infty,j}(x',\xi') = \mathsf{a}_\infty(x',\xi')\mathsf{e}_\infty(x',\xi',\zeta_j(x',\xi'))$ *where* $\mathsf{e}_\infty(x',\xi',\tau)$ *is the spectral projector for* $\mathsf{a}_\infty(x',\xi')$ *and assume that*

(12.2.51) *For each* j *the standard non-periodicity condition is fulfilled for matrix-valued symbol* $\mathsf{a}_{\infty,j}$ *which means that with the exception of measure* 0 *on* $\Sigma_j = \{1 \in \operatorname{Spec} \mathsf{a}_{\infty,j}(z)\}$ $\Psi_{j,t}(z) \not\ni z$ *and* $\Psi_{j,t}(z) \cap 0 \times \mathbb{R}^{d'} = \emptyset$ *for all* t *of the proper sign (depending on* $z = (x',\xi')$*) where* $\Psi_{j,t}$ *is a generalized Hamiltonian flow for* $\mathsf{a}_{\infty,j}$.

Then asymptotics

$$\mathsf{N}^-(\tau) = \mathcal{N}(\tau) + \bigl(\kappa_1^* + o(1)\bigr)\tau^{\frac{1}{2m\mu}(d'-1)(\mu+1)} \tag{12.2.52}$$

hold as $\tau \to +\infty$ *with an appropriate coefficient* κ_1^* *(given in Remark 4.4.13).*

Condition (12.2.51) looks extremely cumbersome and impossible to check. However in the following rather special example this condition becomes much simpler:

Example 12.2.9. Let A satisfy (12.2.31) with $\bar{a} = |\xi|^2$; then eigenvalues of a_∞ are $\zeta_j(x',\xi') = (|\xi'|^2 + \sigma^{0-2}(x')\lambda_j)$ and they have constant multiplicities, λ_j are eigenvalues of $\Delta_{Y,\mathsf{D}}$.

Then one needs to check the standard non-periodicity condition for symbols $\zeta_j(x',\xi')$ and all of them are equivalent to the same standard non-periodicity condition for a scalar symbol

$$\zeta(x',\xi') = |\xi'|^2 + \sigma_\infty(x')^{-2} \tag{12.2.53}$$

(on the energy level 1).

In this case we need to check only check the standard non-periodicity condition for (12.2.53). Note that

(12.2.54) As $d' \geq 2$ the generic trajectory does not hit $0 \times \mathbb{R}^{d'}$.

---

[8] Then obviously $\zeta_j(tx', t^\mu\xi') = t^{2m\mu}\zeta_j(x',\xi')$.

Really, $\upmu(\Sigma\cap\{|x'|\le 2\varepsilon\})=O(\varepsilon^{d'})$ where $\upmu=dx'd\xi':d\zeta$ is the standard measure on $\Sigma=\{(x',\xi'):\zeta(x',\xi')=1\}$ and trajectory intersecting $\{|x'|\le\varepsilon\}$ needs at least $\epsilon_0\varepsilon$-time to leave $\{|x'|\le 2\varepsilon\}$; therefore the $\upmu$-measure of points such that Ψ_t touches $\{|x'|\le\varepsilon\}$ for some $t:|t|\le T$ does not exceed $CT\varepsilon^{d'-1}$.

In particular, the standard non-periodicity condition for $\zeta(x',\xi')$ is fulfilled as $\sigma_\infty=|x'|^{-\mu}$, $\mu>0$ and $\mu\ne 1$; recall that $d'\ge 2$.

Finally, let us consider the case $d'=1+d''\mu$. Considering diagonal operator with diagonal elements $\zeta_j^{\mathrm{w}}(x',D')$ one can prove easily the remainder estimate $o(\tau^{(d-1)/2m}\log\tau)$ (not subject to any non-periodicity condition). This leads us to conjecture which we were unable to prove

Conjecture 12.2.10. *As $d'=1+d''\mu$ and stabilization conditions* (12.2.49)$_{1,2}$ *are fulfilled, remainder estimate is in fact $o(\tau^{(d-1)/2m}\log\tau)$.*

We leave to the reader the following easy

Problem 12.2.11. (i) Generalize Theorem 12.2.8 to the case when a_∞ depends on x''; then one should assume that $a_\infty(x',\sigma_\infty(x')x'')$ is positively homogeneous of degree 0 with respect to x'; then $\boldsymbol{a}_\infty=\boldsymbol{a}_\infty(x',\xi';x'',D'')$ and one should take Weyl quantization in the right-hand expression of (12.2.50).

(ii) Treat Riesz means and derive asymptotics with the remainder estimates

$$R^*_{(\vartheta)}(\tau)=\begin{cases}\tau^{\frac{1}{2m}(d-1-\vartheta)+\vartheta} & \text{for } d'<\mu d''+1+\vartheta,\\ \tau^{\frac{1}{2m}(d-1-\vartheta)+\vartheta}\log\tau & \text{for } d'=\mu d''+1+\vartheta,\\ \tau^{\frac{1}{2m\mu}(d'-1-\vartheta)(\mu+1)+\vartheta} & \text{for } d'>\mu d''+1+\vartheta.\end{cases}\tag{12.2.55}$$

and in the analogs of Theorems 12.2.7 and 12.2.8 one should assume that $d'<\mu d''+\vartheta+1$, $d'>\mu d''+\vartheta+1$ respectively.

12.2.3 Effective Asymptotics

Scheme of Effective Asymptotics

The main defect of Theorems 12.2.6–12.2.8 is that the expression for $\mathcal{N}(\tau)$ is not effective and includes the eigenvalue counting function for the *family* of auxiliary operators $\boldsymbol{a}(x',\xi')=\boldsymbol{a}(x',\xi';x'',D'')$. We would like to derive a formula including the eigenvalue counting function for *one* auxiliary operator or at least for a family of auxiliary operators with a smaller number of parameters (under certain conditions). We are aware that for a thick cusp (i.e., with $\sigma^{d''-1-\vartheta}\notin \mathscr{L}^1(Z)$) we cannot completely get rid of the auxiliary operators in the final answer.

Let us start with a more accurate Weyl analysis. Let $U(x,y,t)$ be the Schwartz kernel of the operator $\exp(it\tau^{-1}A)$. Then

$$\tau D_t U = A_x U,\qquad U|_{t=0}=\delta(x-y)I_{\mathbb{K}} \tag{12.2.56}$$

with $\mathbb{K}=\mathbb{C}^{\mathsf{D}}$.

Let us fix a point $\bar{x}\in X$ and consider its $\bar{\sigma}$-neighborhood. As usual, let us rescale $x_{\mathsf{new}}=(x_{\mathsf{old}}-\bar{x})\bar{\sigma}^{-1}$ and divide A by τ; then we transform A into an h_1-differential operator A_1 with $h_1=\bar{\sigma}^{-1}\tau^{-\frac{1}{2m}}$. Introducing a new time scale $t_1=th_1$ we obtain from (12.2.56) the scaled problem

$$h_1 D_{t_1} U = A_{1x} U,\qquad U|_{t_1=0}=\bar{\sigma}^{-d}\delta(x-y)I_{\mathbb{K}} \tag{12.2.56$_1$}$$

where we skip the subscript "new" at x and y. Then one can apply the results of Section 4.3; moreover, on the energy level 1 the microhyperbolicity condition is fulfilled provided h_1 is small enough.

Let $\bar{\chi}\in\mathscr{C}_0^K([-1,1])$, $\bar{\chi}=1$ on $[-\frac{1}{2},\frac{1}{2}]$ and let ψ' be $\bar{\sigma}$-admissible and be supported in the $\bar{\sigma}$-neighborhood of $\bar{x}$. Further, let $h_1\le\varepsilon$ (i.e., $\bar{\sigma}\ge c_0\tau^{-\frac{1}{2m}}$), and let $T_1=\varepsilon h_1^{-1}$ where here and below $\varepsilon>0$ is a small enough constant. Then according to Section 4.3 we obtain the following estimate:

$$\begin{aligned} &\Bigl|\int\Bigl(F_{t\to\lambda}\bar{\chi}_{T_1}(t)\Gamma_x U-\sum_{0\le j\le J-1}\kappa'_j(x,\lambda\tau)^{\frac{1}{2m}(d-j)-1}\tau\Bigr)\psi'\,dx-\\ &\qquad\int_{\partial X}\sum_{1\le j\le J-1}\tilde{\kappa}'_j(x,\lambda\tau)^{\frac{1}{2m}(d-j)-1}\tau\psi'\,dS\Bigr|\le C\tau^{\frac{1}{2m}(d-J)}\bar{\sigma}^{d-J}\end{aligned} \tag{12.2.57}$$

where $\lambda\in[\epsilon_0,c]$ here and below, dS is the standard Riemannian density on ∂X, κ'_j, $\tilde{\kappa}'_j$ are the standard Weyl coefficients and

$$|\kappa'_j|\le C\bar{\sigma}^{-j},\qquad |\tilde{\kappa}'_j|\le C\bar{\sigma}^{1-j}. \tag{12.2.58}$$

Recall that one can obtain these coefficients in the following way:

$$\mathsf{Tr}\bigl(\mathsf{Res}_{\mathbb{R}}\,\psi'(\tau-A)^{-1}\bigr)\sim\int\sum_j\psi'\kappa'_j\tau^{\frac{1}{2m}(d-j)-1}\,dx \tag{12.2.59}$$

and

$$\mathsf{Tr}\bigl(\mathsf{Res}_{\mathbb{R}}\,\psi'((\tau - A)_D^{-1} - (\tau - A)^{-1})\bigr) \sim \sum_j \int_{\partial X} \psi' \tilde{\kappa}'_j \tau^{\frac{1}{2m}(d-j)-1}\,dS \tag{12.2.60}$$

where $(\tau - A)^{-1}$ and $(\tau - A)_D^{-1}$ are parametrices of the pseudodifferential operator $(\tau - A)$ in terms of the formal calculus of pseudodifferential operators and the Dirichlet problem for the same operator in terms of the formal calculus of pseudodifferential operators (with the transmission property) and Green operators (see, e.g., Section 1.4), and for the sake of simplicity the ordinary matrix trace tr is included in the operation Γ_x in (12.2.57).

Therefore under the same conditions for the $\bar{\gamma}$-admissible function $\psi = \psi(x')$ supported in the $\bar{\gamma}$-neighborhood of $\bar{x}'$ estimate

$$\begin{aligned} &|\int \Bigl(F_{t\to\lambda}\bar{\chi}_{T_1}(t)\Gamma_x U - \sum_{0\le j\le J-1} \kappa'_j(x,\lambda\tau)^{\frac{1}{2m}(d-j)-1}\tau\Bigr)\psi\,dx - \\ &\qquad \int_{\partial X} \sum_{1\le j\le J-1} \tilde{\kappa}'_j(x,\lambda\tau)^{\frac{1}{2m}(d-j)-1}\tau\psi'\,dS| \le C\tau^{\frac{1}{2m}(d-J)}\bar{\sigma}^{d''-J}\bar{\gamma}^{d'} \end{aligned} \tag{12.2.61}$$

holds where the right-hand expression acquired factor $\bar{\gamma}^{d'}\bar{\sigma}^{-d'}$.

So far there is nothing beyond analysis of Sections 11.2 and 11.3. However, let us take another point of view. Namely, let us treat the $\bar{\gamma}$-neighborhood of $\bar{x}'$, rescale $x'_{\mathsf{new}} = (x'_{\mathsf{old}} - \bar{x}')\bar{\gamma}^{-1}$ and divide A by τ; earlier we transformed X_{cusp} to a cylinder in the routine way.

Then A is transformed to a h_2-differential operator (with respect to x') A_2 with the operator-valued symbol; surely $h_2 = \bar{\gamma}^{-1}\tau^{-\frac{1}{2m}} \ll 1$ here. Introducing the new time scale $t_2 = th_2\tau^{-1}$ we obtain from (12.2.56) another scaled problem

$$\tau D_{t_2} U = A_{2x} U, \qquad U|_{t_2=0} = \bar{\gamma}^{-d'}\delta(x'-y')I_{\mathbb{H}}. \tag{12.2.56$_2$}$$

Assume that A_2 is a microhyperbolic operator with the operator-valued symbol[6)]. Let us apply the strongest version of Theorem 2.1.19 to estimate

$$|\int F_{t\to\lambda}\chi_T(t)\Gamma_x U\psi'\,dx| \tag{12.2.62}$$

with $\chi \in \mathscr{C}_0^K([-1,1])$ vanishing on $[-\frac{1}{2},\frac{1}{2}]$.

Namely, formula (4.3.52) yields that in the right-hand expressions in (2.1.58) and (2.1.59) one can add the factor $h = h_2$, i.e., one can take the right-hand expressions to be $h^{1-d-|\alpha|}\bigl(\frac{h}{T}\bigr)^s$.

Therefore scaling we obtain that

Proposition 12.2.12. *Let A_2 be a microhyperbolic operator with the operator-valued symbol, $\chi \in \mathscr{C}_0^K([-1,1])$ vanishing on $[-\frac{1}{2},\frac{1}{2}]$ and ψ $\bar\gamma$-admissible and supported in the $\bar\gamma$-neighborhood of $\bar x'$.*

Then estimate

$$\|F_{t\to\lambda} B_1^j \chi'_T \Gamma' \psi U B_1^k\| \le C h_2^{-d'} T^{-s} \tag{12.2.63}$$

holds for any $T\in[\epsilon_0, T_2]$, $T_2 = \epsilon h_2^{-1}$; here Γ' means that we take $x'=y'$ and integrate over x', and we take the operator norm in $\mathbb{H}$ in the left-hand expression.

Applying Sobolev embedding theorem we take the trace norm in the left-hand expression with the price of additional factor $h_1^{-d''}$ in the right-hand expression and shedding of some factors B_1^j and B_1^k; we still assume that $\bar\sigma \ge c_0 T^{-\frac{1}{2m}}$.

Therefore we conclude that

(12.2.64) Expression (12.2.62) does not exceed $C h_2^{-d'} T^{-s} h_1^{-d''}$.

Taking an appropriate partition of unity in the interval $[T'_1, T'_2]$ we obtain estimate

$$\bigl|F_{t\to\lambda}\bigl(\bar\chi_{T'_2}(t) - \bar\chi_{T'_1}(t)\bigr)\Gamma\psi U\bigr| \le C h_2^{-d'} h_1^{-d''} T_1'^{-s} \tag{12.2.65}$$

for $\epsilon \le T'_1 < T'_2 \le T_2$.

Let us set $T'_j = T_j$ $(=\varepsilon h_j^{-1})$. Then estimates (12.2.65) and (12.2.61) yield that

(12.2.66) If A_2 is a microhyperbolic operator with the operator-valued symbol then estimate (12.2.61) remains true with T_1 replaced by T_2.

Applying the routine Tauberian technique we then obtain then the following estimate

$$\Bigl|\Bigl(\int \psi\bigl(\mathsf{tr}(e(x,x,\tau)\bigr) - \sum_{0\le j\le J-1} \kappa_j(x)\tau^{\frac{1}{2m}(d-j)}\Bigr)\,dx - \int_{\partial X} \psi \sum_{1\le j\le J-1} \tilde\kappa_j(x)\tau^{\frac{1}{2m}(d-j)}\,dS\Bigr)\Bigr|_{\tau=\tau'}^{\tau=\tau}\Bigr| \le C\bar\sigma^{d''}\bar\gamma^{d'-1}\tau^{\frac{1}{2m}(d-1)} + C\bar\sigma^{d''-J}\bar\gamma^{d'}\tau^{\frac{1}{2m}(d-J)} \tag{12.2.67}$$

provided $\tau \ge \epsilon\bar\sigma^{-2m}$, $\tau' \asymp \tau$ and $J \le d$.

For $J > d$ one should replace in the left-hand expression the terms with $j = d$ by $\kappa_d \log\tau$ and $\tilde\kappa_d \log\tau$. Here and below the Weyl coefficients κ_j and $\tilde\kappa_j$ satisfy (12.2.58) [9].

Thus, *using propagation of singularities for operators with the operator-valued symbols we have improved the Weyl asymptotics in the cusp zone* $\{x : \sigma(x') \ge \tau^{-\frac{1}{2m}}\}$.

However now we need to take into account many terms in the decomposition which undermines our intention to get an effective asymptotics. To remedy this we consider A as a perturbation of some other operator A^0. So let A^0 be another operator in the domain X^0 with the same cusp part X_{cusp} also satisfying all the conditions above with the same exponents. Moreover, let us assume that the operators A and A^0 are close at infinity in the sense that

$$|D^\beta(a_\alpha - a^0_\alpha)| \le C\eta\langle x'\rangle^{-|\beta'|+\mu(2m-|\alpha|+|\beta''|)} \qquad \forall\alpha : |\alpha| \le 2m \quad \forall\beta : |\beta| < K \tag{12.2.68}$$

with

$$\eta = \eta(x') \asymp \langle x'\rangle^{-\nu}, \qquad \nu > 0. \tag{12.2.69}$$

Remark 12.2.13. In most of the arguments below one need only assume that η is a γ-admissible weight function and $\eta = o(1)$ as $|x'| \to \infty$.

All the objects associated with A^0 have an additional superscript "0".

Let us introduce the formal differential expression $A^\theta = A^0 + \theta(A - A^0)$ with $\theta \in [0,1]$; then $A^1 = A$. We also assign Dirichlet boundary conditions to

[9] A similar estimate also holds for Riesz means. Namely, the first term in the right-hand expression of (12.2.67) should be multiplied by $\tau^\vartheta\bar\sigma^\vartheta\bar\gamma^{-\vartheta}$ and the second term should be multiplied by τ^ϑ.

A^θ. Below all the objects associated with A^θ have the additional superscript "θ". Then applying estimate (12.2.67) to both operators we obtain estimate

$$\begin{aligned}(12.2.70)\quad &|\Bigl(\int \psi\Bigl(e(x,x,\tau)-e^0(x,x,\tau)-\\ &\qquad\sum_{0\le j\le J-1}\bigl(\kappa_j(x)-\kappa_j^0(x)\bigr)\tau^{\frac{1}{2m}(d-j)}\Bigr)\,dx-\\ &\int_{\partial X}\psi\sum_{1\le j\le J-1}\bigl(\tilde\kappa_j(x)-\tilde\kappa_j^0(x)\bigr)\tau^{\frac{1}{2m}(d-j)}\,dS\Bigr)\Bigr|_{\tau=\tau'}^{\tau=\tau}|\le\\ &\qquad C\bar\sigma^{d''}\bar\gamma^{d'-1}\tau^{\frac{1}{2m}(d-1)}+C\bar\sigma^{d''-J}\bar\gamma^{d'}\tau^{\frac{1}{2m}(d-J)}.\end{aligned}$$

provided $J\le d$; otherwise terms with $j=d$ contain the additional factor $\log\tau$.

Moreover, differentiating representations (12.2.59), (12.2.60) with respect to θ one can easily see that

$$(12.2.71)\qquad |\partial_\theta\kappa_j^\theta(x)|\le c_1\bar\eta\bar\sigma^{-j},\qquad |\partial_\theta\tilde\kappa_j^\theta(x)|\le c_1\bar\eta\bar\sigma^{1-j}$$

in the same notations as before; we first obtain these estimates for the coefficients $\kappa_j'^\theta$, $\tilde\kappa_j'^\theta$.

Integrating on θ from 0 to 1 and passing from ψ' to ψ we obtain that

$$(12.2.72)\qquad |\kappa_j(x)-\kappa_j^0(x)|\le c_1\bar\eta\bar\sigma^{1-j},\qquad |\tilde\kappa_j(x)-\tilde\kappa_j^0(x)|\le c_1\bar\eta\bar\sigma^{1-j}$$

These estimates and (12.2.70) yield that

$$\begin{aligned}(12.2.73)\quad &|\Bigl(\int \psi\Bigl(e(x,x,\tau)-e^0(x,x,\tau)-\\ &\qquad\sum_{0\le j\le J-1}\bigl(\kappa_j(x)-\kappa_j^0(x)\bigr)\tau^{\frac{1}{2m}(d-j)}\Bigr)\,dx-\\ &\int_{\partial X}\psi\Bigl(\sum_{1\le j\le J-1}\bigl(\tilde\kappa_j(x)-\tilde\kappa_j^0(x)\bigr)\tau^{\frac{1}{2m}(d-j)}\,dS\Bigr)\Bigr|_{\tau=\tau'}^{\tau=\tau}|\le\\ &C\bar\sigma^{d''}\bar\gamma^{d'-1}\tau^{\frac{1}{2m}(d-1)}+C\bar\sigma^{d''-s}\bar\gamma^{d'}\tau^{\frac{1}{2m}(d-s)}+C\bar\eta\bar\sigma^{d''-J}\bar\gamma^{d'}\tau^{\frac{1}{2m}(d-J)}.\end{aligned}$$

with arbitrarily large exponent s where K surely depends on s.

This estimate is accurate enough to obtain the final theorems under certain restrictions on ν from below (we leave this easy task to the reader).

However, in order to prove these theorems under weaker conditions one needs a more refined analysis. Namely, *we want to shed of the second term in the right-hand expression of* (12.2.73) which is not nice for h_1 close to 1 (i.e., for $\bar{\gamma}$ close to $\tau^{\frac{1}{2m\mu}}$).

Remark 12.2.14. Without any loss of the generality we can assume that $X = X^0$. Really, one can arrange this by replacing both X and X^0 with $X^1 := X_{\mathsf{cusp}} \cap \{|x'| \ge c_0\}$ (or possibly $X_{\mathsf{cusp}} \cap \{x_1 \ge c_0\}$ in the case $d' = 1$) and taking Dirichlet boundary conditions on ∂X; surely the operators remain self-adjoint; moreover as $\mathfrak{D}(A^0) = \mathfrak{D}(A^1)$.

Surely, as we do this function $e^\theta(x, y, \tau)$ changes because we changed the geometrical domains. However the routine Tauberian arguments yield that in the frameworks of Theorems 12.2.6–12.2.8 the errors appearing in $\int \psi e^\theta(x, x, \tau)\, dx$ (with ψ supported in $\{|x'| \ge 2c_0\}$) do not exceed the remainder estimates in the asymptotics of these terms in the corresponding theorems and since we are not going to improve these remainder estimates but to provide a better principal part everything is fine.

Then operators $A^\theta = A^0 + \theta(A - A^0)$ with $\mathfrak{D}(A^\theta) = \mathfrak{D}(A^0)$ are self-adjoint and all the functions $e^\theta(x, y, \tau)$ and $U^\theta(x, y, t)$ do not lose their meanings as Schwartz kernels of $E^\theta(\tau)$ and $\exp(i\tau^{-1}tA^\theta)$ respectively as $\theta \in [0, 1]$; $E^\theta(.)$ is the spectral projector of A^θ.

Then

$$\partial_\theta \exp\bigl(it\tau^{-1}A^\theta\bigr) = i\tau^{-1}\int_0^t \exp\bigl(it'\tau^{-1}A^\theta\bigr)\cdot\bigl(A^1 - A\bigr)\cdot\exp\bigl(i\tau^{-1}(t-t')A^\theta\bigr)\, dt' \tag{12.2.74}$$

and one can easily prove that for $\lambda \in [\epsilon, c]$

$$\bigl|F_{t\to\lambda}\bar{\chi}_T(t)\partial_\theta U^\theta(x, y, t)\bigr| \le C\bar{\eta}h_1^{-p}T \tag{12.2.75}$$

for any $T > 0$ with large enough p. One can improve this estimate and calculate p but this improvement provides no advantage for arguments below.

We want to calculate (12.2.74) after we multiply it by $\bar{\chi}_T(t)\psi(x)$ and calculate the trace. Let us do it first as T is small enough: applying routine arguments associated with the successive approximation method and increasing T_1 from $h_1^{\frac{1}{2}+\delta}$ (*only here,* the subscript 1 at T means that we use the t_1-scale) to ϵ in Sections 4.3 and 7.2 we conclude that

$$(12.2.76)\quad |\partial_\theta\Bigl(\int\Bigl(F_{t\to\lambda}\bar{\chi}_T(t)\Gamma_x U^\theta - \sum_{0\le j\le J-1}\kappa_j'^\theta(\lambda\tau)^{\frac{1}{2m}(d-j)-1}{}_T\Bigr)\psi'\,dx - \int_{\partial X}\sum_{1\le j\le J-1}\tilde{\kappa}_j'^\theta(\lambda\tau)^{\frac{1}{2m}(d-j)-1}\psi'\,dS\Bigr)| \le C\bar{\eta}\tau^{\frac{1}{2m}(p-J)}\bar{\sigma}^{p-J}$$

where $T = T_1 = \epsilon h_1^{-1}$ and p is large enough[10].

Therefore estimate (12.2.76) holds with $\psi(x')$ (which is γ-admissible) instead of ψ' (which was σ-admissible) and with the additional factor $h_2^{-d'}h_1^{d'}$ in the right-hand expression.

Moreover, applying arguments associated with operators the with operator-valued symbols one can easily obtain from (12.2.75) that

$$(12.2.77)\qquad \bigl|F_{t\to\lambda}\chi_T(t)\Gamma\bigl(\psi\partial_\theta U^\theta\bigr)\bigr| \le \bar{\eta}h_2^{-d'}h_1^{-p}T^{-s}$$

for $T\in[T_1, T_2]$ and therefore

$$(12.2.78)\qquad \bigl|F_{t\to\lambda}\bigl(\bar{\chi}_{T_2'} - \bar{\chi}_{T_1'}\bigr)\Gamma\bigl(\psi\partial_\theta U^\theta\bigr)\bigr| \le C\bar{\eta}h_2^{-d'}h_1^{-p}T_1'^{-s}$$

with the same T_1, T_2, T_1', T_2', χ and $\bar{\chi}$ as in (12.2.65) and (12.2.65).

Therefore estimate (12.2.76) holds with $T = T_2$ instead of $T = T_1$. Integrating over θ from 0 to 1 we see that

(12.2.79) Estimate (12.2.76) holds with $T = T_2$ and with the derivative on θ replaced by the difference operator; so $\partial_\theta f^\theta$ is replaced by $(f^1 - f^0)$.

Let us apply the Tauberian technique. Surely, $\Gamma\bigl(\psi\bigl(e(.,.,\tau) - e^0(.,.,\tau)\bigr)\bigr)$ is not necessarily a monotone function of τ and one should apply the estimate

$$(12.2.80)\quad \Bigl|\Gamma\Bigl(\psi\bigl(e^\theta(.,.,\tau) - e^\theta(.,.,\tau')\bigr)\Bigr)\Bigr| \le \bar{\gamma}^{d'}\bar{\sigma}^{d''}\tau^{\frac{d}{2m}}h_2 = \bar{\gamma}^{d'-1}\bar{\sigma}^{d''}\tau^{\frac{1}{2m}(d-1)}$$

for $|\tau'-\tau|\le h_2$. This estimate was obtained by the Tauberian method (see Sections 4.2 and 4.4) for functions $\Gamma\bigl(\psi e^\theta(.,.,\tau)\bigr)$ which are monotone of τ.

[10] However, considering this estimate with large $J' = J + p$ instead of J and applying estimates (12.2.58) one can reach $p = d$.

This gives the improved estimate (12.2.73):

$$
\begin{aligned}
(12.2.81)\quad &|\Bigl(\int \Bigl(\bigl(e(x,x,\tau)-e^0(x,x,\tau)\bigr)\\
&\qquad - \sum_{0\le j\le J-1} \bigl(\kappa_j(x)-\kappa^0_j(x)\bigr)\tau^{\frac{1}{2m}(d-j)}\Bigr)\psi\,dx-\\
&\int_{\partial X}\sum_{1\le j\le J-1} \bigl(\tilde\kappa_j(x)-\tilde\kappa^0_j(x)\bigr)\tau^{\frac{1}{2m}(d-j)}\psi\,dS\Bigr)\Bigr|_{\tau=\tau'}^{\tau=\tau}| \le\\
&\qquad\qquad C\bar\sigma^{d''}\bar\gamma^{d'-1}\tau^{\frac{1}{2m}(d-1)} + C\bar\eta\bar\sigma^{d''-J}\bar\gamma^{d'}\tau^{\frac{1}{2m}(d-J)}.
\end{aligned}
$$

Recall that $\tau'\in[\epsilon\tau,\tau]$, $\tau\bar\sigma^{2m}\ge\epsilon$ here and terms with $j=d$ contain the factor $\log\tau$ [11].

Therefore, the following statement has been proven:

Proposition 12.2.15. *(i) In the framework of Theorem 12.2.6 estimate* (12.2.67) *holds.*

(ii) Moreover, let both operators A and A^0 satisfy assumptions of Theorem 12.2.6 and let (12.2.68) *be fulfilled (in $X_{\mathsf{cusp}}\cap|x'-\bar x'|\le\bar\gamma$). Then estimate* (12.2.81) *holds.*

Recall that ψ is γ-admissible and supported in the $\bar\gamma$-neighborhood of $\bar x'$. Let us replace τ and τ' by $2^{-n}\tau$ and $2^{-n-1}\tau$ and sum over $n=0,\dots,n_0=c_0\lfloor\log(\tau\bar\sigma^{2m})\rfloor+c_1$. Applying Proposition 12.2.4 we obtain

Corollary 12.2.16. *(i) In the framework of Theorem 12.2.6 estimate*

$$
\begin{aligned}
(12.2.82)\quad \mathsf{R}_1 := &|\int \Bigl(\operatorname{tr}(e(x,x,\tau)) - \sum_{0\le j\le J-1}\kappa_j(x)\tau^{\frac{1}{2m}(d-j)}\Bigr)\psi\,dx-\\
&\int_{\partial X}\sum_{1\le j\le J-1}\tilde\kappa_j(x)\tau^{\frac{1}{2m}(d-j)})\psi\,dS| \le\\
&\qquad C\bar\sigma^{d''}\bar\gamma^{d'-1}\tau^{\frac{1}{2m}(d-1)} + C\bar\sigma^{d''-J}\bar\gamma^{d'}\tau^{\frac{1}{2m}(d-J)}
\end{aligned}
$$

holds for $J<d$; for $J=d$ estimate

$$
(12.2.83)\qquad \mathsf{R}_1\le C\bar\sigma^{d''}\bar\gamma^{d'-1}\tau^{\frac{1}{2m}(d-1)} + C\bar\sigma^{-d'}\bar\gamma^{d'}\log(\tau\bar\sigma^{2m})
$$

[11] One can see easily that for Riesz means both terms in the right-hand expression should be modified in the way indicated in [9].

holds and for $J = d+1$ *the estimate*

$$R_1 \le C\bar{\sigma}^{d''}\bar{\gamma}^{d'-1}\tau^{\frac{1}{2m}(d-1)} + C\bar{\sigma}^{-d'}\bar{\gamma}^{d'} \tag{12.2.84}$$

holds, and in the latter case the terms with $j = d$ *are* $\kappa_d \log(\tau\bar{\sigma}^{2m})$ *and* $\tilde{\kappa}_d \log(\tau\bar{\sigma}^{2m})$.

(ii) In the framework of Proposition 12.2.15(ii) estimates

$$\begin{multline}
R_2 := \Bigl|\Bigl(\int \Bigl(\operatorname{tr}(e^\theta(x,x,\tau) - \sum_{0\le j\le J-1} \kappa_j^\theta(x)\tau^{\frac{1}{2m}(d-j)}\Bigr)\psi\, dx - \\
\int_{\partial X}\sum_{1\le j\le J-1} \tilde{\kappa}_j^\theta(x)\tau^{\frac{1}{2m}(d-j)}\psi\, dS\Bigr)\Bigr|_{\theta=0}^{\theta=1}\Bigr| \le \\
C\bar{\sigma}^{d''}\bar{\gamma}^{d'-1}\tau^{\frac{1}{2m}(d-1)} + C\bar{\eta}\bar{\sigma}^{d''-J}\bar{\gamma}^{d'}\tau^{\frac{1}{2m}(d-J)},
\end{multline}\tag{12.2.85}$$

$$R_2 \le C\bar{\sigma}^{d''}\bar{\gamma}^{d'-1}\tau^{\frac{1}{2m}(d-1)} + C\bar{\eta}\bar{\sigma}^{-d'}\bar{\gamma}^{d'}\log(\tau\bar{\sigma}^{2m}) \tag{12.2.86}$$

and

$$R_2 \le C\bar{\sigma}^{d''}\bar{\gamma}^{d'-1}\tau^{\frac{1}{2m}(d-1)} + C\bar{\eta}\bar{\sigma}^{-d'}\bar{\gamma}^{d'} \tag{12.2.87}$$

hold for $J < d$, $J = d$, $J = d+1$ *respectively (we remind the reader that terms with* $j = d$ *contain a logarithmic factor).*

Then after summation over a γ-admissible partition of unity in Z we can obtain miscellaneous formulas (later under certain conditions we will calculate the terms associated with the operator A^0):

(i) Let us sum the estimates for R_2 in the domain $\{t_1 \le |x'| \le c_0\tau^{\frac{1}{2m\mu}}\}$ and apply Theorem 12.2.2 in the domain $\{|x'| \le 2t_1\}$. Then we derive the following formula

$$\begin{multline}
N^-(\tau) = \\
\sum_{j=0,1}\int \phi(\frac{x'}{t_1})\kappa_j(x)\tau^{\frac{1}{2m}(d-j)}\, dx + \int \phi(\frac{x'}{t_1})\tilde{\kappa}_1(x)\tau^{\frac{1}{2m}(d-1)}\, dS +
\end{multline}\tag{12.2.88}$$

$$\sum_{j=0,1}\int \bigl(1-\phi(\frac{x'}{t_1})\bigr)\bigl(\kappa_j(x)-\kappa_j^0(x)\bigr)\tau^{\frac{1}{2m}(d-j)}\,dx+$$
$$\int \phi(\frac{x'}{t_1})\bigl(\tilde\kappa_1(x)-\tilde\kappa_1^0(x)\bigr)\tau^{\frac{1}{2m}(d-1)}\,dS+$$
$$\int \bigl(1-\phi(\frac{x'}{t_1})\bigr)\,\mathrm{tr}(e^0(x,x,\tau))\,dx+$$
$$O\bigl(\mathsf{R}^*(\tau)+\tau^{\frac{1}{2m\mu}(d'(\mu+1)-\nu)}\bigr)$$

where $R^*(\tau)$ was introduced by (12.2.41) and here and below $\phi\in\mathscr{C}_0^K(\mathbb{R}^{d'})$, $\phi=1$ in a neighborhood of the origin.

Remark 12.2.17. In formula (12.2.88) we have taken $J=2$ because we gain nothing by increasing J and so we have dropped all the terms with $j\geq 2$.

Further, in (12.2.88) t_1 is either a large enough constant or a slowly increasing function of τ. Moreover, one can skip terms with $j=1$ in the third and fourth items in the right-hand expression in all cases excluding $d'=(d''-1)\mu+\nu$, $\nu<\mu+1$ when the remainder estimate is $O\bigl(\tau^{\frac{1}{2m}(d-1)}\bigr)$, and the contribution of this term is $O\bigl(\tau^{\frac{1}{2m}(d-1)}\log\tau\bigr)$ (and this estimate is the best possible).

Furthermore, for $t_1=\mathsf{const}$ one can skip terms with $j=1$ in the first and second items in the right hand expression. In any case, one can effectively calculate all terms with $j=0,1$.

(ii) Consider the case $d'<1+\mu d''$, $\nu>\mu(1-d'')+d'$. In this case the remainder estimate in (12.2.88) is $O\bigl(\tau^{\frac{1}{2m}(d-1)}\bigr)$.

Moreover, the contribution of the zone $\{t_1\leq |x'|\}$ to the remainder estimate is $o\bigl(\tau^{\frac{1}{2m}(d-1)}\bigr)$ for $t_1=t_1(\tau)\to+\infty$ as $\tau\to+\infty$. Then, applying the improved Weyl estimate in the zone $\{|x'|\leq t_1\}$ under the standard non-periodicity condition to the Hamiltonian billiards we obtain asymptotics

$$\mathsf{N}^-(\tau)= \tag{12.2.89}$$
$$\sum_{j=0,1}\int \phi(\frac{x'}{t_1})\kappa_j(x)\tau^{\frac{1}{2m}(d-j)}\,dx+\int \phi(\frac{x'}{t_1})\tilde\kappa_1(x)\tau^{\frac{1}{2m}(d-1)}\,dS+$$
$$\int \bigl(1-\phi(\frac{x'}{t_1})\bigr)\bigl(\kappa_0(x)-\kappa_0^0(x)\bigr)\tau^{\frac{d}{2m}}\,dx+$$
$$\int \bigl(1-\phi(\frac{x'}{t_1})\bigr)\,\mathrm{tr}(e^0(x,x,\tau))\,dx+o\bigl(\tau^{\frac{1}{2m}(d-1)}\bigr)$$

where we have dropped terms with $j=1$ in the third and fourth items in the right-hand expression of (12.2.89).

(iii) Consider now an opposite case $d'>1+\mu d''$, $\nu>\mu+1$; then the remainder estimate in (12.2.88) is $O\bigl(\tau^{\frac{1}{2m\mu}(d'-1)(\mu+1)}\bigr)$. Moreover, in this case one can drop all terms with $j=1$.

Furthermore, the contribution of the zone $\{|x'|\le t_2\}$, $t_2=\varepsilon\tau^{\frac{1}{2m\mu}}$ to the remainder estimate is $\varepsilon'\tau^{\frac{1}{2m\mu}(d'-1)(\mu+1)}$ with $\varepsilon'=\varepsilon'(\varepsilon)\to 0$ as $\varepsilon\to 0$.

On the other hand, for every fixed $\varepsilon>0$ in the zone $t_2(\tau)\le |x'|\le c_0\tau^{\frac{1}{2m\mu}}$ in the context of Theorem 12.2.7 one can improve the remainder estimate. Namely, taking a standard partition of unity in (x',ξ')-space and applying the standard Tauberian arguments we obtain that the contributions of both elements of this partition do not exceed $\varepsilon''\tau^{\frac{1}{2m\mu}(d'-1)(\mu+1)}$ with arbitrarily small $\varepsilon''>0$ because the support of the first element is small enough and in the second element one can take $T_2=(h_2\varepsilon''')^{-1}$ with arbitrarily small $\varepsilon'''>0$ in the Tauberian arguments.

Remark 12.2.18. The only element of these arguments which is not completely standard is that we apply the propagation results derived from the theory of operators with the operator-valued symbols to Weyl asymptotics.

This approach will be heavily used in forthcoming sections and chapters.

Then we obtain the following asymptotics

$$
\begin{aligned}
\text{(12.2.90)}\quad \mathsf{N}^-(\tau) = &\int \phi\bigl(\frac{x'}{t_1}\bigr)\kappa_j(x)\tau^{\frac{d}{2m}}\,dx+\\
&\int \bigl(1-\phi\bigl(\frac{x'}{t_1}\bigr)\bigr)\bigl(\kappa_0(x)-\kappa_0^0(x)\bigr)\tau^{\frac{d}{2m}}\,dx+\\
&\int \bigl(1-\phi\bigl(\frac{x'}{t_1}\bigr)\bigr)\,\mathrm{tr}\bigl(e^0(x,x,\tau)\bigr)\,dx + o\bigl(\tau^{\frac{1}{2m\mu}(d'-1)(\mu+1)}\bigr);
\end{aligned}
$$

here we have dropped all terms with $j=1$ in (12.2.88). Here as above t_1 is either a large enough constant or a slowly increasing function of τ.

(iv) If one replaces t_1 by $t_1'\ge t_1$ in (12.2.88)–(12.2.90) then the left-hand expression does not change, and in order to estimate the increment of the right-hand expression one can apply one of estimates (12.2.82)–(12.2.84) for

R_1. Then one can easily see that for $t_1' \le \tau^\delta$ with a small enough exponent $\delta > 0$ this increment does not exceed the remainder estimate and therefore in (i)–(iii) one can formulate the condition of slow growth of t_1 more exactly: $t_1 \le \tau^\delta$. We leave the details and the calculation of this exponent $\delta > 0$ to the reader.

(v) One could apply formula (12.2.82) for $|x'| \le 2t_2$ and formula (12.2.85) for $t_2 \le |x'| \le c_0\tau^{\frac{1}{2m\mu}}$ where t_2 is now less than $c_0\tau^{\frac{1}{2m\mu}}$. Therefore the following formula holds:

$$\begin{multlined}\mathsf{N}^-(\tau) = \sum_{0\le j\le J} \int \phi(\frac{x'}{t_2})\kappa_j(x)\tau^{\frac{1}{2m}(d-j)}\,dx + \\ \sum_{1\le j\le J}\int_{\partial X} \phi(\frac{x'}{t_2})\tilde\kappa_j(x)\tau^{\frac{1}{2m}(d-j)}\,dS + \int \bigl(1-\phi(\frac{x'}{t_2})\bigr)e^0(x,x,\tau)\,dx + \\ O(R^* + R_2)\end{multlined} \tag{12.2.91}$$

with some remainder estimate expression $R_2 = R_{2,J}(\tau, t_2)$. Taking $t_2 = t_{2,J}(\tau)$ as providing the best possible estimate one can obtain certain asymptotics, maybe with worse remainder estimates. Improvements in the spirit of (ii) and (iii) are possible.

(vi) Let us pick two cases in which the formulas based on (12.2.85)–(12.2.87) and (12.2.82)–(12.2.84) well match and in (12.2.91) we can take $J = 2$ and $t_2 = c_0\tau^{\frac{1}{2m\mu}}$ and the same remainder estimate as in formulas (12.2.88)–(12.2.90). This is the case when $\kappa_j^0 = \kappa_j$ ($j = 0, 1$), $\tilde\kappa_1^0 = \tilde\kappa_1$. And these equalities surely hold in the following cases:

(a) A and A^0 differ only in lower order terms. The only coefficient in the Weyl formula depending on lower order terms is κ_1 but even this coefficient surely does not depend on lower order terms because A is a differential operator of even order; moreover, this coefficient vanishes provided the principal symbol a is not only Hermitian but also real-valued (in which case it is symmetric).

(b) The principal symbol of A can be obtained from the principal symbol of A^0 by the transformation $\xi' \to \xi' - T(x, \xi'')$ with scalar function $T(x, \xi'')$ linear in ξ''. In fact, under the completely Weyl procedure this transform surely preserves κ_0, κ_1 and some part of κ_1 which does not depend on the lower order terms; however, the other part surely vanishes.

Simplifying the Answer

Now let us consider the term

$$\int \bigl(1-\phi(\frac{x'}{t})\bigr)\,\mathrm{tr}(e^0(x,x,\tau))\,dx \tag{12.2.92}$$

with $t \ge c_0$ and apply Theorems 12.2.6–12.2.8. Then the remainder estimate is preserved.

Moreover, applying Theorem 12.2.7 one can replace the non-periodicity condition to billiards (which is now unnatural because it is associated with the operator A^0 in the truncated domain X^1) with the condition "$t \to +\infty$ as $\tau \to +\infty$". Let us investigate when the formulae for (12.2.92) associated with A^0 are simpler than in the general case.

(i) Let the principal operator-valued symbol of A^0 [12] be of the form

$$a^0(x',\xi') = \sigma(x')^{-2m} a_\infty(x'',\sigma(x')\xi';D'') \tag{12.2.93}$$

in which case the answer is

$$(2\pi)^{-d'}\iint \bigl(1-\phi(\frac{x'}{t})\bigr)\, n_\infty(\xi',\tau\sigma(x')^{2m})\sigma(x')^{d'}\,dx'd\xi' \tag{12.2.94}$$

where $n_\infty(\xi',\tau)$ is the eigenvalue counting function for the operator $a(\xi';x'',D'')$ in $\mathbb{H} = \mathscr{L}^2(Y,\mathbb{K})$. In this case the final answer uses the eigenvalue counting function for the operator family depending only on ξ'.

Moreover, one can consider the case when a_∞ depends on $x'|x'|^{-1}$.

(ii) On the other hand, let the principal operator-valued symbol of A^0 [12]

$$a^0(x',\xi') = \sigma(x')^{-2m} F(x',\sigma(x')\xi';\Lambda(x')) \tag{12.2.95}$$

where $\Lambda = \Lambda(x';x'',D'')$ is a self-adjoint operator in $\mathbb{H}$ (an elliptic differential operator of order $2m$ with Dirichlet boundary conditions) and $F(x',\xi',\lambda)$ is a monotone increasing function of λ.

Then the answer is

$$(2\pi)^{-d'}\iint \nu\bigl(x',f(x',\xi',\tau\sigma(x')^{2m})\bigr)\sigma(x')^{-d'}\,dx'd\xi' \tag{12.2.96}$$

where $\nu(x',\lambda)$ is the eigenvalue counting function for $\Lambda(x')$ and $\lambda = f(x',\xi',\tau)$ is solution of the equation $F(x',\xi',\lambda)=\tau$.

---

[12] After mapping the cusp to the cylinder.

(iii) In particular, let $m=1$ and let the principal operator-valued symbol of A^0 [12] be

$$a^0(x',\xi') = b(x',\xi') + \sigma(x')^{-2}\Lambda(x';x'',D'') \tag{12.2.97}$$

where $b(x',\xi') = \sum_{jk} b_{jk}(x')\xi'_j\xi'_k$ is a positive scalar quadratic form in ξ'. Then the final answer is

$$\frac{1}{2}(2\pi)^{-d'}d'\omega_{d'}\iint \nu(x',\lambda)\sigma(x')^{-d'}\beta(x')^{-\frac{1}{2}}\bigl(\tau\sigma(x')^2-\lambda\bigr)_+^{\frac{1}{2}(d'-2)}\,d\lambda dx' \tag{12.2.98}$$

where $\beta(x')$ is the determinant of the matrix of the quadratic form $b(x',.)$.

Main Results

Let us formulate the principal theorems proved in this section. They are due to the results above. Moreover, we need to assume only that A^0 is a symmetric operator given in the cusp and we should not assume that A^0 is self-adjoint. Indeed, otherwise it could be replaced by $\frac{1}{2}(\mathcal{A}+\mathcal{A}^*)$ with

$$\mathcal{A} = \phi(\frac{x'}{t_0})A + (1-\phi(\frac{x'}{t_0}))A^0.$$

Then we obtain the following statements:

Theorem 12.2.19. *(i) Let conditions* (12.2.1)–(12.2.11) *and stabilization conditions* (12.2.68), (12.2.69) *be fulfilled. Further, let microhyperbolicity condition for* $\mathbf{a}(x',\xi')$ *be fulfilled*[6].

Then asymptotics

$$\mathsf{N}^-(\tau) = \mathcal{N}_1(\tau) + O\bigl(R^*(\tau) + \tau^{\frac{1}{2m\mu}(d'(\mu+1)-\nu)}\bigr) \tag{12.2.99}$$

holds where $R^*(\tau)$ *is given by* (12.2.41) *and*

$$\begin{aligned}\mathcal{N}_1(\tau) = &\sum_{j=0,1}\int \phi(\frac{x'}{t_1})\kappa_j(x)\tau^{\frac{1}{2m}(d-j)}\,dx + \int_{\partial X}\phi(\frac{x'}{t_1})\tilde\kappa_1(x)\tau^{\frac{1}{2m}(d-1)}\,dS +\\ &\sum_{j=0,1}\int (1-\phi(\frac{x'}{t_1}))\bigl(\kappa_j(x)-\kappa_j^0(x)\bigr)\tau^{\frac{1}{2m}(d-j)}\,dx +\\ &\int_{\partial X}\phi(\frac{x'}{t_1})\bigl(\tilde\kappa_1(x)-\tilde\kappa_1^0(x)\bigr)\tau^{\frac{1}{2m}(d-1)}\,dS +\\ &(2\pi)^{-d'}\iint (1-\phi(\frac{x'}{t_1}))\,\boldsymbol{n}^0(x',\xi',\tau)\,dx'd\xi'\end{aligned} \tag{12.2.100}$$

where n^0 *is the eigenvalue counting function for* $\mathsf{a}^0(x',\xi';x'',D'')$ *in* $\mathbb{H}$, ϕ *is introduced above,* $c_0 \le t_1 \le \tau^\delta$ *with a small enough exponent* $\delta > 0$.

Moreover, (12.2.44) *remains true for* $\mathcal{N}_1(\tau)$.

(ii) Further, let $d' < 1+\mu d''$, $\nu > d' - \mu(d''-1)$ [13]. *Moreover, let the standard non-periodicity condition be fulfilled for the Hamiltonian billiards of* $a(x',\xi')$.

Then asymptotics

$$\mathsf{N}^-(\tau) = \mathcal{N}_1(\tau) + o\bigl(\tau^{\frac{d-1}{2m}}\bigr) \tag{12.2.101}$$

holds as $\tau\to+\infty$ *provided* $t_1 = t_1(\tau)\to+\infty$ *(we still assume that* $t_1 \le \tau^\delta$*).*

(iii) On the other hand, let $d' > 1+\mu d''$, $\nu > \mu+1$ [14] *Moreover, let stabilization conditions* $(12.2.49)_{1,2}$ *and condition* (12.2.51) *be fulfilled*[15].

Then the asymptotics

$$\mathsf{N}^-(\tau) = \mathcal{N}_1(\tau) + \varkappa_1^*\tau^{\frac{1}{2m\mu}(d'-1)(\mu+1)} + o\bigl(\tau^{\frac{1}{2m\mu}(d'-1)(\mu+1)}\bigr) \tag{12.2.102}$$

holds with the coefficient $\varkappa_1^*$ *given in Remark 4.4.13.*

We leave to the reader the following series of rather easy problems:

Problem 12.2.20. Derive miscellaneous results in the spirit of (12.2.91) (it would be better to take $J=2$ because we do not know nice formulas for κ_j, $\tilde\kappa_j$ with $j\ge 2$):

(i) Consider cases when σ and η are not purely power functions (for example, if these functions contain logarithmic factors), which is significant for the orders of the principal part of asymptotics and remainder estimate only for $d' \ge \mu d''$, $d' \ge 1+\mu d''$ respectively.

(ii) Consider slowly shrinking cusps with f.e. with $\sigma \asymp \log^{-\nu}|x'|$.

---

[13] Then the remainder estimate in (12.2.99) is $O\bigl(\tau^{\frac{1}{2m}(d-1)}\bigr)$.

[14] Then the remainder estimate in (12.2.99) is $O\bigl(\tau^{\frac{1}{2m\mu}(d'-1)(\mu+1)}\bigr)$.

[15] In these conditions $a_{\infty,\alpha}$ should not necessarily coincide with the coefficients in (12.2.68) and in the definition of $\mathcal{N}_1$ as here we need only "$o(.)$" in the right-hand expression.

(iii) Formulate and prove all results of this section for a domain with more than one cusp (with different characteristics)

$$X = X_0 \cup X_{\mathsf{cusp}(1)} \cup \ldots \cup X_{\mathsf{cusp}(N)}, \qquad X_{\mathsf{cusp}(j)} \cap X_{\mathsf{cusp}(k)} = \emptyset \text{ as } 0 < j < k.$$

This problem is trivial even if the dimensions of these cusps (d'_j) and the exponents of shrinking (μ_j) differ. We only note that in some situations the bounded part of the domain and even some cusps can be ignored in the final answer).

(iv) Derive asymptotics for the Riesz means. We mention only that $R^*_{(\vartheta)}(\tau)$ in this case is defined by (12.2.55) and the additional error due to perturbation is $O\bigl(R^*_{(\vartheta)}(\tau) + \tau^{\frac{1}{2m\mu}(d'(\mu+1)-\nu)+\vartheta}\bigr)$ and so the remainder estimate is $O(R^*_{(\vartheta)})$ provided $\nu \ge \min\bigl((\mu+1)(1+\vartheta), d' - d''\mu + (1+\vartheta)\mu\bigr)$.

Further, we leave to the reader the following series of problems:

Problem 12.2.21. (i) Generalize to the cases of mixed boundary conditions which are Dirichlet for $|x'| \ge c$, or Dirichlet on the upper boundary of the cusp and Neumann on the lower boundary of the cusp for $|x'| \ge c$ and $d'' = 1$.

Further, consider generalizations for $d'' \ge 2$. Domain must satisfy cone condition which does not contradict shrinking as cone is truncated by the part of the boundary on which Dirichlet condition is imposed.

(ii) Consider more general boundary conditions under assumptions that all the eigenvalues of $a(x', \xi')$ including the lowest ones have the same power growth at infinity. This is rather difficult and interesting modification as we cannot apply results of Chapter 2 directly because $\mathfrak{D}(a(x', \xi')$ depends on (x', ξ').

Probably, the easiest way to overcome this obstacle would be to introduce an unitary operator $Q(x', \xi')$ in $\mathbb{H}$ such that $Q(x', \xi')\,\mathfrak{D}(a(x', \xi')$ does not depend on (x', ξ').

(iii) Derive more precise remainder estimates in the framework of Theorem 12.2.7 as Hamiltonian flow associated with $a(x, \xi)$ is of power or exponential non-periodic type.

(iv) Derive more precise remainder estimate in the frameworks of Theorem 12.2.8 and Example 12.2.9 assuming that the Hamiltonian flow associated with $\zeta(x',\xi')$ is of power or exponential non-periodic type; in particular, consider Schrödinger operator in Euclidean metrics in domains of with $\sigma(x')=|x|^{-\mu}$, $\mu>0$, $\mu\neq 1$ (with appropriate d' and d'').

(v) In the frameworks of Theorem 12.2.8 and Example 12.2.9 investigate the case when the Hamiltonian flow associated with $\zeta(x',\xi')$ is periodic; in particular, consider Schrödinger operator in Euclidean metrics in domains of with $\sigma(x')=|x|^{-1}$ (with appropriate d' and d'').

(vi) Consider combs and trees as in Examples 11.2.25 and 11.2.26 respectively, but thicker than there.

Remark 12.2.22. If we wish to improve the remainder estimate in the final theorem in the case when the stabilization exponent ν is not large enough then the following algorithm is suggested. In the domain $|x'|\le t$ we apply the previous analysis. So we need to treat

$$\int \psi(x')\operatorname{tr}(\mathsf{e}^{\theta}(x',\xi',\tau))\,dx'd\xi'$$

with $\theta=1$ and a γ-admissible function ψ supported in $B(\bar{x}',\bar{\gamma})$ (and later sum on a partition of unity in the zone $|x|\ge t_2$). This expression is smooth with respect to θ because of the microhyperbolicity property.

Then one can apply Taylor's formula with N terms. The remainder estimate is $O\bigl(\bar{\gamma}^{d'}\bar{\sigma}^{d''}\bar{\eta}^{N}\tau^{d'}\bigr)$ (for $|\bar{x}''|\le \tau^{\frac{1}{2m\mu}}$) and after summation on the partition of unity we obtain $O\bigl(R^*(\tau)+R_3(t_2,\tau)\bigr)$ with R_3 depending on N and ν.

So, the total remainder estimate is $O\bigl(R^*(\tau)+R_2(\tau,t_2)+R_3(\tau,t_2)\bigr)$ where R_2 is an upper estimate for R_2. Picking $t_2=t_2(\tau)$ as providing the best possible remainder estimate we obtain the final formula. However, this formula is rather complicated even in the Case (iii) of the previous Subsubsection 12.2.3.2. Namely, the additional term with $n=1$ in the Taylor expansion is

$$\text{const}\iint \bigl(\tau\sigma(x')^{2m}-\lambda\bigr)^{\frac{d-2}{2}}\,d_\lambda\bigl(\operatorname{tr}(a'E(\lambda))\bigr)_+\,dx' \tag{12.2.103}$$

provided $a(x,D)-a^0(x,D)=a'(x,D'')$ where $E(\lambda)$ is the spectral projector for Λ in $\mathbb{H}$, and this term is slightly more complicated for more general perturbations.

However, the next terms (with $n \geq 2$) are very complicated. Still if we know that a_α and σ can be decomposed into cut-off series of positive homogeneous terms (modulo terms decaying fast enough) we can find that these complicated terms in $\mathcal{N}(\tau)$ are of the power (may be multiplied by a logarithm) type.

We leave this task to the reader as well as the generalization to the Riesz means.

It would be very interesting and under certain assumptions not very difficult to

Problem 12.2.23. Consider cusps with Z of more general type than here:

(i) for example, let Z be an open cone in $\mathbb{R}^{d'}$, $Z_0 = Z \times \{0\} \subset \mathbb{R}^d$ with the smooth boundary and

$$\sigma(x') \leq \langle x' \rangle^{-\mu} \quad \mu > 0, \tag{12.2.104}$$

$$|D^{\beta'}\sigma| \leq c\sigma\gamma^{-|\beta'|} \quad \forall \beta' : |\beta'| \leq K \tag{12.2.105}$$

with

$$\gamma = \mathsf{dist}(x', \partial Z) \tag{12.2.106}$$

Then provided

$$\int \sigma^{d''}\gamma^{-1}\, dx' \leq C|x'|^{-\mu d'' + d' - 1}. \tag{12.2.107}$$

we can recover the same remainder estimate as for $Z = \mathbb{R}^{d''}$.

(ii) Alternatively $\sigma \asymp \langle x' \rangle^{-\mu}$ but near ∂Z a rounding up of the radius $\asymp \sigma$ is added. Then (12.2.107) holds with the logarithmic factor in the right-hand expression and we can recover almost the same remainder estimate as in (i)–with $R^*_{(\vartheta)}(\tau)$ replaced by $R^*_{(\vartheta)}(\tau)\log\tau$.

(iii) In (i) and (ii) consider quasiconical base Z.

12.3 Operators with Degenerate Potentials Growing at Infinity

12.3.1 Problem Set-up

Let us start from

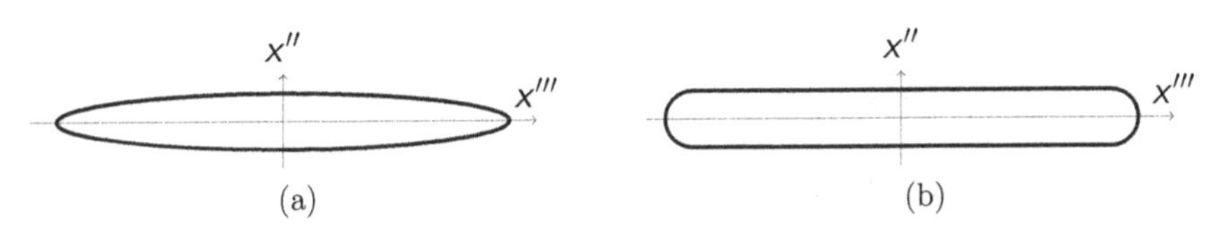

Figure 12.3: As $x' = (x_1, x'')$ we show cross-section with $x_1 = \mathsf{const}$ of the cusp from (i) on (a) and from (ii) on (b)

Example 12.3.1. Consider self-adjoint (matrix) operator on $\mathbb{R}^d$

$$A = \sum_{\alpha:|\alpha|\le 2m} a_\alpha(x) D^\alpha \tag{12.3.1}$$

(possibly in the divergent form) with coefficients such that

$$|D^\beta a_\alpha| \le c\langle x\rangle^{(2m-|\alpha|)\mu-|\beta|} \qquad \forall \beta : |\beta| \le K \tag{12.3.2}$$

with $\mu > 0$.

As usual we assume that operator is elliptic and positive

$$a^0(x,\xi) = \sum_{\alpha:|\alpha|=2m} a_\alpha(x)\xi^\alpha \ge \epsilon_0 |\xi|^{2m} \tag{12.3.3}$$

and that A is a perturbation of operator $A_{(0)}$ with the coefficients $a_{\alpha(0)}(x)$ such that for $j=0$

$$D^\beta a_{\alpha(j)}|_\Lambda = 0 \qquad \mathsf{as} \qquad |\beta| < (2m-|\alpha|)\nu - j \tag{12.3.4}$$

with $0 < \nu < \mu$ where $\nu \in \mathbb{Z}$ and

(12.3.5) Λ is a conic $\mathscr{C}^K$-manifold of codimension d'', $\zeta(x) = \mathsf{dist}(x, \Lambda)$

and

$$\sum_{\alpha:|\alpha|\le 2m} a_{\alpha,0}(x)\xi^\alpha \ge \epsilon_0 \bigl(|\xi|^2 + |x|^{2\mu-2\nu}\cdot \zeta(x)^{2\nu}\bigr)^m. \tag{12.3.6}$$

We will show that "admissible perturbations" are

$$a_\alpha(x) = \sum_{j\le (2m-|\alpha|)\nu} a_{\alpha(j)}(x) \tag{12.3.7}$$

where $a_{\alpha(j)}(x)$ satisfy (12.3.4) and also

$$|D^\beta a_{\alpha(j)}| \le c\langle x\rangle^{(2m-|\alpha|)\mu-|\beta|-(\mu+1)(\nu+1)^{-1}j} \tag{12.3.8}$$

and also condition (12.3.19) below.

Without any loss of the generality we can assume that (at least locally) $\Lambda = \{x : x'' = 0\}$. Then we need to introduce

$$\zeta(x) = \frac{1}{2}|x''|, \qquad \gamma(x) = \frac{1}{2}|x'|, \qquad \rho(x) = \zeta(x)^\nu \cdot |x'|^{\mu-\nu} \tag{12.3.9}$$

scales with respect to x'', x' and ξ respectively.

One can consider more general operators. Assume that $\nu \in (0,\mu)$ is not necessary integer anymore, ζ, γ and ρ are defined by (12.3.9), but ρ, ζ are then redefined by

$$\rho^*(x) = \max\bigl(\rho(x), \sigma_0(x')^{-1}\bigr), \qquad \zeta^*(x) = \max\bigl(\frac{1}{2}|x''|, \sigma_0(x')\bigr) \tag{12.3.10}$$

with

$$\sigma_0(x') := c_0\langle x'\rangle^{-(\mu-\nu)/(\nu+1)}. \tag{12.3.11}$$

Assume that

$$|D^\beta a_\alpha| \le C\rho^{*\,2m-|\alpha|}\gamma^{-|\beta'|}\zeta^{*\,-|\beta''|}. \tag{12.3.12}$$

Potential surfaces are shown on figure 12.4.

Consider this operator first in the framework of the theory of Section 11.5; after scaling $x_{\mathsf{new}} = x/\sigma$, $\xi_{\mathsf{new}} = \tau^{-1/(2m)}\xi$ we conclude that operator $\tau^{-1}A$ is elliptic on the energy levels 1 and below provided

$$|x''| \ge \sigma(x',\tau) := c_0|x'|^{-(\mu-\nu)/\nu}\tau^{1/(2m\nu)}, \tag{12.3.13}$$

(we assume that $\tau \gg 1$, $|x'| \gg |x''|$ and $|x'| \gg 1$) and the uncertainty principle $\rho(x)\zeta(x) \ge 1$ is fulfilled provided

$$|x''| \ge \sigma_0(x') := c_0\langle x'\rangle^{-(\mu-\nu)/(\nu+1)} \tag{12.3.14}$$

as in (12.3.11).

Then we arrive immediately to the following

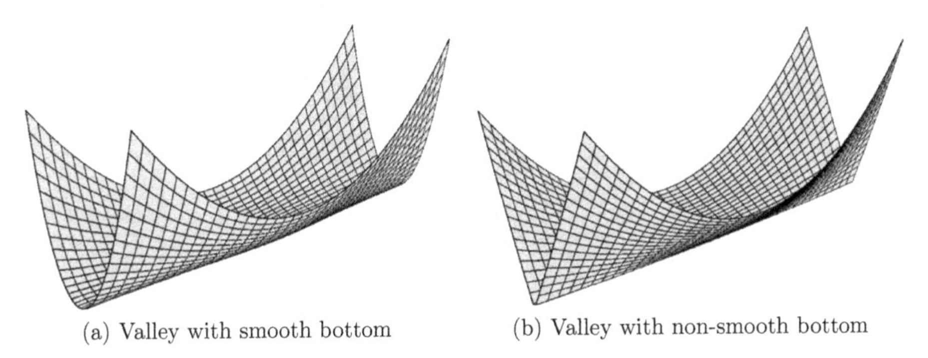

(a) Valley with smooth bottom (b) Valley with non-smooth bottom

Figure 12.4: Potential graphs

Proposition 12.3.2. *Let conditions* (12.3.1), (12.3.3), (12.3.6) *and* (12.3.9)–(12.3.12) *be fulfilled at the point* $\bar{x} \in \mathbb{R}^d$.

(i) Then

$$|e(x,y,\tau)| \le C\langle x\rangle^{(\mu-\nu)d/(\nu+1)}\zeta(x)^{-s}\sigma^s(x',\tau) \qquad \forall x,y\in B(\bar{x},\frac{1}{2}\zeta(\bar{x})) \tag{12.3.15}$$

with an arbitrarily large exponent s.

(ii) Moreover, if $U(x,y,t)$ *is the Schwartz kernel of the operator* $\tau^{-1}A$ *then*

$$\bigl|F_{t\to\lambda}\bar{\chi}_{T,T'}(t)U(x,y,t)\bigr| \le C\langle x\rangle^{(\mu-\nu)d/(\nu+1)}\zeta(x)^{-s}\sigma(x',\tau)^s \qquad \forall x,y\in B(\bar{x},\frac{1}{2}\zeta(\bar{x}))\quad \forall T'\in\mathbb{R}, T=\sigma(\bar{x})\quad \forall\lambda\le 1+\epsilon \tag{12.3.16}$$

with a small enough constant $\epsilon>0$.

Remark 12.3.3. Consider "valley" $\mathcal{X}(\tau)=\{x, |x''|\le\sigma(x',\tau)\}$. It consists of the *regular part* where $|x|\le c\tau^{1/(2m\mu)}$ and the *cusp* where $|x|\ge c\tau^{1/(2m\mu)}$; this cusp stretches along $Z=\{x,x''=0\}$ and shrinks at infinity (on $|x|$) for any $\tau>0$.

Let us apply now the theory of operators with the operator-valued symbols (in rather limited manner at this time); to do this we need to analyze operator $\boldsymbol{a}(x',\xi')=A(x',\xi';x'',D'')$:

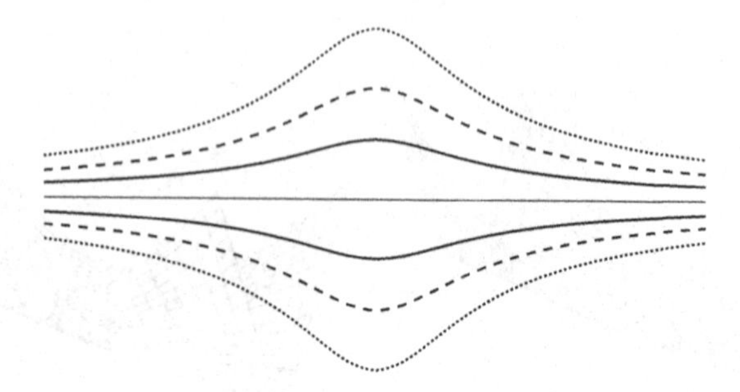

Figure 12.5: Cusps defined for different τ

Proposition 12.3.4. *Let conditions* (12.3.1), (12.3.3), (12.3.6) *and* (12.3.9)–(12.3.12) *be fulfilled. Let* $\lambda_k(x',\xi')$ *be eigenvalues of* $\mathbf{a}(x',\xi')$ *in* $\mathbb{H}=\mathscr{L}^2(\mathbb{R}^{d''})$. *Then*

$$|\lambda_k(x',\xi)|\le c_k\Bigl(|\xi'|+\frac{1}{\sigma_0(x')}\Bigr)^{2m} \tag{12.3.17}$$

and

$$\lambda_k(x',\xi')\ge \epsilon_0\Bigl(|\xi'|+\frac{1}{\sigma_0(x')}\Bigr)^{2m}\qquad \textit{as}\ \ k\ge\bar{k}. \tag{12.3.18}$$

where $\sigma_0(x')$ *is defined by* (12.3.11).

Proof. Scaling $x''\mapsto \sigma_0(x')^{-1}x''$, $D''\mapsto\sigma_0(x')D''$ we arrive to elliptic operator with a parameter in $\mathbb{R}^{d''}$ and both statements (12.3.17) and (12.3.18) are well-known. □

Finally we assume that

(12.3.19) Inequality (12.3.18) is fulfilled for all k.

Due to our assumptions $A(x',D';x'',D'')$ is elliptic as an operator with operator valued symbol on the energy level τ and below unless $|\xi'|\le c\tau^{\frac{1}{2m}}$ and $|x'|\le r^*(\tau)$ with

$$r^*(\tau):=c_0|\tau|^{(\nu+1)/2m(\mu-\nu)}. \tag{12.3.20}$$

Proposition 12.3.5. *Let conditions* (11.1.6), (11.1.8) *and* (12.3.1), (12.3.3), (12.3.6), (12.3.9)–(12.3.12) *and* (12.3.19) *be fulfilled. Then estimate*

$$|e(x,y,\tau)| \le C\langle x\rangle^{-s}\langle y\rangle^{-s}\tau^{-s} \qquad \forall x,y : |x|\ge r^*(\tau), |y|\ge r_*(\tau) \tag{12.3.21}$$

holds.

Proof. Our statement is due to Proposition 4.2.8 as we take

$$B_1 = \bigl(\Delta + |x''|^{2\nu}|x''|^{2(\mu-\nu)}\bigr)^m \tag{12.3.22}$$

where

$$\bigl(\Delta + |x''|^{2\nu}|x''|^{2(\mu-\nu)}\bigr) \ge \epsilon_0 |x'|^{2(\mu-\nu)/(\nu+1)} \tag{12.3.23}$$

in the operator sense. Recall that $\mathbb{H} = \mathscr{L}^2(\mathbb{R}^{d''}, \mathbb{K})$, $\mathbb{K} = \mathbb{C}^{\mathsf{D}}$. ☐

Remark 12.3.6. (i) $\sigma_0(x') \lesssim \sigma(x',\tau)$ if and only if $|x'| \lesssim r^*(\tau)$.

(ii) Note that $\operatorname{mes}(\mathcal{X}(\tau)) = O(\tau^{d/(2m\mu)})$ if and only if $\mu d'' > \nu d$. Under this condition the principal part of the Weyl formula is $\asymp \tau^{\frac{1}{2m\mu}d(\mu+1)}$.

(iii) Let us drop condition $\mu d'' > \nu d'$ and let us restrict integration to domain $\{x, |x| \le r^*(\tau)\}$ in the Weyl formula due Proposition 12.3.5. Then the principal part of the Weyl asymptotics (modified in this way) is

$$\asymp \begin{cases} |\tau|^{d(\mu+1)/(2m\mu)} & \text{for } d\frac{\nu}{\mu} < d'', \\ |\tau|^{d(\mu+1)/(2m\mu)} |\log|\tau|| & \text{for } d\frac{\nu}{\mu} = d'', \\ |\tau|^{d'(\mu+1)/2m(\mu-\nu)} & \text{for } d\frac{\nu}{\mu} > d''. \end{cases} \tag{12.3.24}$$

where in this section we can replace $|\tau|$ by τ and $|\log\tau|$ by $\log\tau$.

(iv) Let us consider the remainder estimate in the Weyl formula. We can integrate only over domain $\{x, |x| \le r^*(\tau)\}$ because of Proposition 12.3.5. Then the remainder estimate in the modified Weyl asymptotics is $O(R(\tau))$ with

(12.3.25) $R(\tau) =$

$$\begin{cases} |\tau|^{(d-1)(\mu+1)/(2m\mu)} & \text{for } (d-1)\frac{\nu}{\mu} < (d''-1), \\ |\tau|^{(d-1)(\mu+1)/(2m\mu)}|\log|\tau|| & \text{for } (d-1)\frac{\nu}{\mu} = (d''-1), \\ |\tau|^{d'(\mu+1)/2m(\mu-\nu)} & \text{for } (d-1)\frac{\nu}{\mu} > (d''-1). \end{cases}$$

(v) Let $\mu(d''-1) > \nu(d-1)$. Then under standard stabilization conditions of a_α as $|x| \to \infty$ to positively homogeneous $a_{\infty,\alpha}$ and standard restrictions to the Hamiltonian flow generated by the corresponding Hamiltonian $a_\infty(x,\xi)$ one can obtain the remainder estimate $o(\tau^{\frac{1}{2m\mu}(d-1)(\mu+1)})$.

In this remark the role of the "operator-valued theory" is very limited: it only provides a nice cut-off in the Weyl formulas.

We leave to the reader the following easy

Problem 12.3.7. Calculate the principal part of the asymptotics and remainder estimate for Riesz means in Remark 12.3.6.

12.3.2 Sharp Asymptotics

So, as we are interested in the asymptotics of $\mathsf{N}^-(\tau)$ and remainder estimate (12.3.25) is sharp as $(d-1)\nu/\mu < (d''-1)$, our analysis should be done as

$$\mu(d''-1) \le \nu(d-1). \tag{12.3.26}$$

Here we should treat only the cusp zone

$$\mathcal{X}_{\mathsf{cusp}}(\tau) = \{c_1\tau^{\frac{1}{2m}} := r_*(\tau) \le |x'| \le r^*(\tau),\ |x''| \le c_2\sigma(x',\tau)\tau^\delta\} \tag{12.3.27}$$

with $r^*(\tau)$ and $\sigma(x',\tau)$ defined by (12.3.20) and (12.3.13) and an arbitrarily small exponent $\delta > 0$.

In fact, estimate (12.3.15) and Proposition 12.3.5 yield that for $|x'| \ge c_1\tau^{\frac{1}{2m\mu}}$ outside of $\mathcal{X}(\tau)$ function $e(x,x,\tau)$ is negligible; on the other hand, we have

Proposition 12.3.8. *(i) Contribution of the regular zone*

$$\mathcal{X}_{\mathsf{reg}}(\tau) = \{x,\ |x'| \le cr_*(\tau)\} \tag{12.3.28}$$

(with arbitrarily large constant c) to the principal part of the Weyl asymptotics and to the remainder estimate are $O(\tau^{\frac{d(\mu+1)}{2m\mu}})$ and $O(\tau^{\frac{(d-1)(\mu+1)}{2m\mu}})$ respectively[16].

(ii) Moreover, under standard stabilization condition (11.5.4) to quasihomogeneous symbol $a_\infty(x,\xi)$ and the standard non-periodicity condition to the Hamiltonian flow generated by symbol $a_\infty(x,\xi)$ the contribution of the zone $\mathcal{X}_{\mathsf{reg}}(\tau)$ to the remainder estimate is $o(\tau^{\frac{1}{2m\mu}(d-1)(\mu+1)})$.

Proof. The proof is absolutely standard in the framework of Section 11.5.

To deal with the possible irregularity near $\{x''=0\}$ note that in the zone $\{x:\ |x'|\le cr_*(\tau),\ |x''|\le \epsilon r_*(\tau)\}$ with sufficiently small constant $\varepsilon>0$ operator $\tau^{-1}A$ after rescaling is ξ-microhyperbolic on the energy level 1 and $h_{\mathsf{eff}}=\tau^{-1/(2m)}\zeta^{*-1}\le \tau^{-\delta''}$ with $\delta''>0$; so propagation theory of Chapter 2 works. □

Furthermore, the same arguments yield that treating zone $\{|x'|\le c_0 r^*(\tau)\tau^{-\delta'}\}$ with an arbitrarily small exponent $\delta'>0$ one can set $\delta=0$ in the definition of $\mathcal{X}(\tau)$; in this case constant c_2 should be large enough.

To treat the cusp zone (12.3.27) let us apply theory of Section 4.4. For this we need to check if operator in question is (x',ξ') microhyperbolic on the energy level τ in (γ,ρ_τ)-scale with $\gamma(x')=\frac{1}{2}|x'|$ and $\rho_\tau=\tau^{1/2m}$.

Consider first the pilot-model operator with the coefficients

$$a_{\alpha(j)}(x)=\sum_{\beta:|\beta|=(2m-|\alpha|)\nu-j} b_{\alpha,\beta}(x')x''^{\beta} \tag{12.3.29}$$

where

(12.3.30) Coefficients $b_{\alpha,\beta}(x')$ in the decompositions (12.3.29) are positively homogeneous of degrees $(\mu-\nu)(2m-|\alpha|-j(\nu+1)^{-1})$

which is fulfilled as we take terms with $|\beta|=(2m-|\alpha|)\nu-j$ of Taylor decomposition of $a_{\alpha(j)}(x)$ provided

(12.3.31) Coefficients $a_{\alpha(j)}(x)$ are positively homogeneous of degrees $(2m-|\alpha|)-(\mu+1)(\nu+1)^{-1}j$.

[16] Subject to conditions of Section 11.5. For example, as either operator is the Schrödinger operator or its perturbed power or stabilization condition (11.5.4) is fulfilled.

Then rescaling $x'' \mapsto x''\sigma_0(x')$, $D'' \mapsto D''\sigma_0^{-1}(x')$ we arrive to operator

$$a(x', D') = \sum_{\alpha':|\alpha'|\le 2m} a_{\alpha'}(x')D'^{\alpha'} \tag{12.3.32}$$

where operator coefficients $a_{\alpha'}(x') = a_{\alpha'}(x'; x'', D'')$ are positively homogeneous of degrees $(2m - |\alpha|)(\mu - \nu)(\nu + 1)^{-1}$.

Therefore under assumption (12.3.29)–(12.3.30) microhyperbolicity condition is fulfilled. Then this condition is fulfilled under assumption (12.3.31) as well and therefore we arrive to the Statement (i) of Proposition 12.3.9 below; Statement (ii) follows from the similar arguments.

Proposition 12.3.9. *(i) In the framework of Example 12.3.1 operator $a(x', D')$ is (x', ξ')-microhyperbolic under stabilization condition*

$$D^\beta\bigl(a_{\alpha(j)} - a^\infty_{\alpha(j)}\bigr) = o\bigl(|x'|^{(2m-|\alpha|)\mu-|\beta|-(\mu+1)(\nu+1)^{-1}j}\bigr) \qquad \text{as } |x'| \to \infty, x'' = 0, |\beta| \le (2m - |\alpha|)\nu + 1 \tag{12.3.33}$$

(ii) In the general case operator $a(x', D')$ is (x', ξ')-microhyperbolic under stabilization condition

$$D^\beta_{x'}\bigl(a_\alpha - a^\infty_\alpha\bigr)(x', \sigma_0(x')y'') = o\bigl(|x'|^{(2m-|\alpha|)(\mu-\nu)/(\nu+1)-|\beta|}\bigr) \qquad \text{as } |x'| \to \infty, |y''| \le c_3, |\beta| \le 1. \tag{12.3.34}$$

We leave to the reader the following easy

Problem 12.3.10. Establish other conditions providing (x', ξ')-microhyperbolicity of $a(x', D')$.

So, assume that microhyperbolicity condition is fulfilled. Then (after all rescalings) one can take

$$B = B_1 = \theta\bigl(\Delta + \langle x''\rangle^{2\nu}\bigr)^m, \qquad \theta = \tau^{-1}\gamma^{2m(\mu-\nu)/(\nu+1)} \tag{12.3.35}$$

in $\mathscr{L}^2(\mathbb{R}^{d''})$.

We will need to take the trace norm; one can easily see that

(12.3.36) For $k > d''(\nu + 1)/(2m\nu)$ and $r_*(\tau) \le |x'| \le r^*(\tau)$ the trace norm of B^{-k} does not exceed $\theta^{-d''(\nu+1)/2m\nu}$ which in turn is equal to $\bigl(\gamma/r^*(\tau)\bigr)^{-d''(\mu-\nu)/\nu}$.

Then, applying Theorem 4.4.8 as this operator is ξ'-microhyperbolic and Theorem 4.4.9 as this operator is (x',ξ')-microhyperbolic (i.e. in the general case) we obtain that if $\psi=\psi(x')$ is a γ-admissible function supported in $\{|x'-\bar{x}'|\le \frac{1}{2}\bar{\gamma}\}$ with $\bar{\gamma}=\gamma(\bar{x})$ then

$$\begin{gathered}|\int \psi(x')e(x,x,\tau)\,dx-(2\pi)^{-d'}\iint \psi(x')\mathbf{n}(x',\xi',\tau)\,dx'd\xi'|\le\\ C\hbar^{1-d'}\bigl(\bar{\gamma}/r^*(\tau)\bigr)^{-d''(\mu-\nu)/\nu}=C(\tau^{1/(2m)}\bar{\gamma})^{(d'-1)}\bigl(\bar{\gamma}/r^*(\tau)\bigr)^{-d''(\mu-\nu)/\nu}\end{gathered}\tag{12.3.37}$$

where $\mathbf{n}(x',\xi',\tau)$ is the eigenvalue counting function for operator $\mathbf{a}(x',\xi')=\mathbf{a}(x',\xi';x'',D'')$ which is the operator-valued symbol of A (after a coordinate transform), $\hbar=\tau^{-1/(2m)}\bar{\gamma}^{-1}$.

Let us sum over a γ-admissible partition of unity in the zone

$$\{t_1r_*(\tau)\le |x'|\le r^*(\tau)\};\tag{12.3.38}$$

the right-hand expression then sums to

$$R^*(\tau)=\begin{cases}|\tau|^{(d-1)(\mu+1)/(2m\mu)} & \text{for } (d-1)\frac{\nu}{\mu}<d'',\\ |\tau|^{(d-1)(\mu+1)/(2m\mu)}|\log|\tau|| & \text{for } (d-1)\frac{\nu}{\mu}=d'',\\ |\tau|^{(d'-1)(\mu+1)/2m(\mu-\nu)} & \text{for } (d-1)\frac{\nu}{\mu}>d''.\end{cases}\tag{12.3.39}$$

Let us apply Weyl asymptotics in the regular zone $\{|x'|\le t_1\tau^{1/(2m\mu)}\}$. To ensure (x,ξ)-microhyperbolicity there we assume that the standard stabilization condition (11.5.4) is fulfilled[17].

Then we arrive to the following assertion

Theorem 12.3.11 [18]**.** *Let conditions* (12.3.1), (12.3.3), (12.3.6), (12.3.9)–(12.3.12) *and stabilization conditions* (11.5.4) [16),17)], (12.3.34) *be fulfilled.*

Let $\psi\in\mathscr{C}_0^K(\mathbb{R}^{d'})$, $\psi(x')=1$ *as* $|x'|\le c_0$. *Then*

(i) Asymptotics

$$\mathsf{N}^-(\tau)=\mathcal{N}(\tau)+O\bigl(R^*(\tau)\bigr)\tag{12.3.40}$$

[17] But for some operators this condition may be weakened or skipped completely. Also the total contribution of the regular zone to the asymptotics may be smaller than the contribution of the cusp to the remainder estimate.

[18] Cf. Theorem 12.2.6.

holds with $\mathcal{N}(\tau)$ *defined by* (12.2.43) *with* $\psi_t(x') = \psi(x'/t)$ *where* $t \asymp r_*(\tau)$ [19] *and with* $R^*(\tau)$ *defined by* (12.3.39).

(ii) Moreover, $\mathcal{N}(\tau)$ *has a magnitude described in* (12.3.24).

Let us note that for $(d-1)\nu < d''\mu$ the main contribution to the remainder estimate is provided by the zone

$$\{x:\ |x'| \le t\tau^{\frac{1}{2m}}\} \tag{12.3.41}$$

with arbitrarily large t (the contribution of the complement does not exceed $\tau^{(d-1)(\mu+1)/(2m\mu)}t^{-\delta}$ with $\delta > 0$); on the other hand, in this zone under the standard non-periodicity condition to the Hamiltonian billiards we have the remainder estimate $o\bigl(\tau^{\frac{1}{2m}(d-1)}\bigr)$. Then applying improved Weyl asymptotics we arrive to the following assertion:

Theorem 12.3.12 [20]. *Let the conditions of Theorem 12.3.11 be fulfilled with* $(d-1)\nu < d''\mu$.

Moreover, let the standard stabilization condition to quasihomogeneous symbol $a_\infty(x,\xi)$ *and the standard non-periodicity condition to the Hamiltonian trajectories generated by* $a_\infty(x,\xi)$ *be fulfilled. Then asymptotics*

$$\mathsf{N}^-(\tau) = \mathcal{N}(\tau) + o\bigl(\tau^{\frac{1}{2m}(d-1)(m+1)}\bigr) \tag{12.3.42}$$

holds as $\tau \to +\infty$ *where* $\mathcal{N}(\tau)$ *is defined by* (12.2.46),

$$(2\pi)^{-d}\int \psi\kappa_1\tau^{\frac{1}{2m}(d-1)}\,dx \quad \textit{and} \quad (2\pi)^{1-d}\int_{\partial X}\psi\tilde{\kappa}_1\tau^{\frac{1}{2m}(d-1)}\,dS \tag{12.3.43}$$

are the standard second terms[21] *in the Weyl asymptotics,* $\psi_t(x') = \psi_0(x'/t)$ *where* $t = t(\tau)$ *is an appropriate function such that* $t(\tau)/r_*(\tau) \to +\infty$ *as* $\tau \to +\infty$ [22].

---

[19] One can assume that $t = t(\tau)$ and $t(\tau)/r_*(\tau)$ is either bounded or tends slowly to $+\infty$. Recall that $r_*(\tau) = \tau^{1/(2m\mu)}$.

[20] Cf. Theorem 12.2.7.

[21] We will discuss domains with the boundaries later.

[22] If $t(\tau)$ works then every function $t'(\tau) \le t(\tau)$ such that $t'(\tau)/r_*(\tau) \to +\infty$ as $\tau \to +\infty$ also works.

On the other hand, as $d' > 1 + d''\mu$, the main contribution to the remainder estimate in Theorem 12.3.11 is provided by the zone

$$\{\varepsilon r^*(\tau) \le |x'| \le c_0 r^*(\tau)\} \tag{12.3.44}$$

with arbitrarily small $\varepsilon > 0$; contribution of zone $\{|x'| \le 2\varepsilon r^*(\tau)\}$ does not exceed $\varepsilon^\delta R^*(\tau)$.

However, in zone (12.3.44) $\boldsymbol{n}(x', \xi', \tau)$ does not exceed $\nu(\varepsilon) = \varepsilon^M$ and therefore we have effectively here matrix- rather than operator-valued operators.

Applying Theorems 4.4.11 and 4.4.12 and replacing $\boldsymbol{a}$ if necessary by another operator $\boldsymbol{a}_\infty$ in $\mathbb{H}$ with eigenvalues asymptotically close to those of $\boldsymbol{a}$ we obtain the following statement:

Theorem 12.3.13[23]. *Let the conditions of Theorem 12.3.11 be fulfilled with $(d-1)\nu < d''\mu$.*

Let us consider operator-valued symbol $\boldsymbol{a}_\infty(x', \xi'; D'')$ in $\mathbb{H}$; let $\lambda_j(x', \xi')$ be its (non-decreasing) eigenvalues[24]. *Let us consider*

$$\boldsymbol{a}_{\infty,j}(x', \xi') = \boldsymbol{a}_\infty(x', \xi')\boldsymbol{e}_\infty(x', \xi', \zeta_j(x', \xi')) \tag{12.3.45}$$

where $\boldsymbol{e}_\infty(x', \xi', \tau)$ is the spectral projector for $\boldsymbol{a}_\infty(x', \xi')$ and let assume that non-periodicity condition (12.2.51) *is fulfilled.*

Then asymptotics

$$\mathsf{N}^-(\tau) = \mathcal{N}(\tau) + \bigl(\kappa_1^* + o(1)\bigr)\tau^{(d'-1)(\mu+1)/2m(\mu-\nu)} \tag{12.3.46}$$

holds as $\tau \to +\infty$ with an appropriate coefficient κ_1^* (given in Remark 4.4.13).

Condition (12.2.51) looks extremely cumbersome and impossible to check. However in the following rather special example this condition becomes much simpler:

Example 12.3.14[25]. Let $d'' = 1$ and A be a Schrödinger operator stabilizing at infinity to $|\xi^2| + |x_d|^2 \cdot \omega(x')^2$ with ω, positively homogeneous of degree $2(\mu - \nu)$.

---

[23] Cf. Theorem 12.2.8.

[24] Then obviously $\lambda_j(tx', t^k\xi') = t^{2mk}\omega_j(x', \xi')$ with $k = (\mu - \nu)/(\nu + 1)$.

[25] Cf. Example 12.2.9.

Then one needs to check the standard non-periodicity condition for symbols $\lambda_j(x',\xi') = |\xi'|^2 + (2j-1)\omega(x')$ and all of them are equivalent to the same standard non-periodicity condition for a scalar symbol

$$\lambda(x,\xi) = |\xi|^2 + \omega(x')^{2/(\nu+1)} \tag{12.3.47}$$

(on the energy level 1). This conclusion holds as A is a Schrödinger operator stabilizing at infinity to $|\xi^2| + |x_d|^{2\nu}\cdot\omega(x')^2$.

In this case we need to check only check the standard non-periodicity condition for (12.3.47). Note that

(12.3.48) As $d'\geq 2$ the generic trajectory does not hit $0\times\mathbb{R}^{d'}$.

Really, $\upmu(\Sigma\cap\{|x'|\leq 2\varepsilon\}) = O(\varepsilon^{d'})$ where $\upmu = dx'd\xi' : d\lambda$ is the standard measure on the surface $\Sigma = \bigl\{(x',\xi') : \lambda(x',\xi') = 1\bigr\}$ and any trajectory intersecting $\{|x'|\leq\varepsilon\}$ needs at least $\epsilon_0\varepsilon$-time to leave $\{|x'|\leq 2\varepsilon\}$; therefore the $\upmu$-measure of points such that Ψ_t touches $\{|x'|\leq\varepsilon\}$ for some $t : |t|\leq T$ does not exceed $CT\varepsilon^{d'-1}$.

In particular, the standard non-periodicity condition for $\lambda(x',\xi')$ is fulfilled as $\sigma\sim|x'|^{-\eta}$, $\eta>0$ and $\eta\neq 1$; recall that $d'\geq 2$.

Finally, let us consider the case $d' = 1 + d''\mu$. Considering diagonal operator with diagonal elements $\lambda_j^{\mathrm{w}}(x',D')$ one can prove easily the remainder estimate $o(\tau^{(d-1)/2m}\log\tau)$ (not subject to any non-periodicity condition). This leads us to conjecture which we were unable to prove

Conjecture 12.3.15 [26]**.** *As $(d-1)\nu = d''\mu$ remainder estimate is in fact $o(\tau^{(d-1)/(2m\mu)}\log\tau)$.*

We leave to the reader the following easy

Problem 12.3.16[27]. Treat Riesz means and derive asymptotics with

$$R^*_{(\vartheta)}(\tau) = \begin{cases} |\tau|^{(d-1-\vartheta)(\mu+1)/(2m\mu)+\vartheta} & \text{for } (d-1-\vartheta)\dfrac{\nu}{\mu} < d'' \\ |\tau|^{(d-1-\vartheta)(\mu+1)/(2m\mu)+\vartheta}\,|\log|\tau|| & \text{for } (d-1-\vartheta)\dfrac{\nu}{\mu} = d'' \\ |\tau|^{(d'-1-\vartheta)(\mu+1)/2m(\mu-\nu)+\vartheta} & \text{for } (d-1-\vartheta)\dfrac{\nu}{\mu} > d'' \end{cases} \tag{12.3.49}$$

[26] Cf. Conjecture 12.2.10.
[27] Cf. Problem 12.2.11.

and in the analogs of Theorems 12.3.12 and 12.3.13 one should assume that $d' < \mu d'' + \vartheta + 1$, $d' > \mu d'' + \vartheta + 1$ respectively.

12.3.3 Effective Asymptotics

Simplifying the Answer

The main defect of Theorems 12.3.11–12.3.13 again is that the expression for $\mathcal{N}(\tau)$ is not effective and includes the eigenvalue counting function for the *family* of auxiliary operators $a(x',\xi')$. Exactly like in the previous Section 12.2 we would like to derive a formula including the eigenvalue counting function for *one* auxiliary operator or at least for a family of auxiliary operators with a smaller number of parameters; however in contrast to the previous section we want also to simplify these axillary operators.

Again, let us start with a more accurate Weyl analysis. Repeating arguments of the Subsection 12.2.3 one can easily prove in the framework of Example 12.3.1 the following estimate similar to (12.2.67):

$$|\int \psi(x')\bigl(\operatorname{tr}(e(x,x,\tau)) - \sum_{0\le j\le J-1} \kappa_j(x,\tau)\bigr)_{\tau=\tau'}^{\tau=\tau}\, dx| \le \\ C\bar{\gamma}^{d'-1-\frac{\mu d''}{\nu}}\tau^{\frac{d-1}{2m}+\frac{d''}{2m\nu}} + C\tau^{\frac{d-J}{2m}}\bar{\sigma}^{d''-J}\bar{\gamma}^{d'} \tag{12.3.50}$$

where $\psi = \psi(x')$ is γ-admissible and supported in $B(\bar{x}',\frac{1}{2}\bar{\gamma})$, $r_*(\tau)\le\bar{\gamma} = |\bar{x}'| \le r^*(\tau)$, $\tau'\in[\epsilon\tau,\tau)$, $\bar{\sigma} = \bar{\gamma}^{-(\mu-\nu)/\nu}\tau^{1/(2m\nu)}$ and the Weyl coefficients κ_j (which now include also powers of effective semiclassical parameter) satisfy the inequalities

$$|\int \psi(x')\kappa_j(x,\tau)\, dx| \le C\tau^{\frac{d-j}{2m}}\bar{\sigma}^{d''-j}\bar{\gamma}^{d'} \tag{12.3.51}$$

and

$$\kappa_0 = (2\pi)^{-d}\int n(x,\xi,\tau)\, d\xi \tag{12.3.52}$$

where as before $n(x,\xi,\tau)$ is an eigenvalue counting function for symbol $a(x,\xi)$.

However in the general case one should add to the right-hand expressions of (12.3.50) and (12.3.51) terms reflecting irregularity as $x''=0$; one can think about last terms in (12.3.50) and (12.3.51) with $\bar\sigma$ replaced by $\bar\sigma_0 = \bar\gamma^{-(\mu-\nu)/(\nu+1)}$ as $J > d''$ or $j > d''$ and with extra factor $\log(\bar\sigma/\bar\sigma_0)$ as $J = d''$ or $j = d''$ but in fact these terms are better.

Thus, as in the previous section, using propagation of singularities for operators with operator-valued symbols we have improved the Weyl asymptotics in the cusp zone. However again we need to take into account many terms in the decomposition and in contrast to the previous section singularity as $x'' = 0$ may prevent us from doing this. To remedy this we consider A as a perturbation of some other operator A^0. So let A^0 be another operator also satisfying all the conditions above with the same exponents.

Let us take $J = 1$ and assume that

$$|D^\beta(a_\alpha - a^0_\alpha)| \le \sum_{q\in Q} \varepsilon_q(x')|x''|^{q-|\beta|}\rho^{*\,2m-|\alpha|} \qquad \forall \beta: |\beta|\le 1 \tag{12.3.53}$$

with the finite set $Q \subset \mathbb{R}^+ \cup 0$. Note that effectively $(A - A^0)$ contains a factor $|x''|^q$ which due to Proposition 12.3.2 in the estimates translates into factor $\sigma(\gamma,\tau)^q$. Then the following estimate similar to (12.2.82) can be proven in the same way:

$$\mathsf{R}_1 := \Bigl|\int \Bigl(\bigl(e(x,x,\tau) - e^0(x,x,\tau)\bigr) - \psi(x')\bigl(\kappa_0 - \kappa^0_0\bigr)\Bigr)_{\tau=\tau'}^{\tau=\tau}\,dx\Bigr| \le \\ C\bar\gamma^{d'-1-\frac{\mu d''}{\nu}}\tau^{\frac{d-1}{2m}+\frac{d''}{2m\nu}} + C\tau^{\frac{d-1}{2m}}\bar\sigma^{d''-1}\bar\gamma^{d'}\bar\eta \tag{12.3.54}$$

with

$$\Bigl|\int \psi(x')\bigl(\kappa_0(x,\tau) - \kappa^0_0(x,\tau)\bigr)\,dx\Bigr| \le C\tau^{\frac{d}{2m}}\bar\sigma^{d''}\bar\gamma^{d'}\bar\eta \tag{12.3.55}$$

where A^0 satisfy the same conditions as A, all the objects associated with A^0 have an additional superscript "0" and in comparison with (12.3.50)–(12.3.51) we gain factor

$$\bar\eta = \sum_{q\in Q} \sigma(\bar\gamma,\tau)^q\varepsilon_q(\bar\gamma). \tag{12.3.56}$$

We leave to the reader to prove easily that the logarithmic factor does not appear in the last term (12.3.54) even if $d'' = 1$, $q = 0$; this proof is based on representation of κ_1 and assumption $\nu > 0$.

So we have proved the following assertion:

Proposition 12.3.17. *Let both operators A and A^0 satisfy assumptions of Theorem 12.3.11. Moreover, let condition* (12.3.53) *be fulfilled. Then estimates* (12.3.54)–(12.3.56) *hold.*

Now our arguments basically repeat those of Subsection 12.2.3.

Let us replace τ and τ' by $2^{-n}\tau$ and $2^{-n-1}\tau$ respectively and sum over $n = 0, \dots, n_0 = c_0\lfloor\log(\tau\bar{\sigma}^{2m})\rfloor + c_1$. Applying Proposition 12.3.2 we can get rid of τ' in all these estimates arriving to the statement similar to Corollary 12.2.16. We leave easy details to the reader.

Assume that

(12.3.57) ε_q are γ-admissible with no more than a polynomial growth and $\eta = O(\gamma^{-\varkappa})$ as $\gamma \to \infty$.

Recall that ψ is γ-admissible and supported in the $\bar{\gamma}$-neighborhood of $\bar{x}'$. Then after summation over a γ-admissible partition of unity in Z we can obtain miscellaneous formulae (later under certain conditions we will calculate the terms associated with the operator A^0):

First, we obtain formula (12.2.88); then immediately we arrive to

Proposition 12.3.18. *(i) Let both operators A and A^0 satisfy assumptions of Theorem 12.3.11. Moreover, let condition* (12.3.53) *be fulfilled with $Q = \{0, 1\}$, $\varepsilon_1 = \gamma^{-1}$, $\varepsilon_0 = \sigma_0\gamma^{-1}$. Then estimate* (12.3.54) *holds with the right hand expression $O\bigl(R^*(\tau)\bigr)$.*

(ii) Further, in the framework of Theorem 12.3.12 we obtain formula (12.2.89) *with the remainder estimate $o\bigl(\tau^{(d-1)(\mu+1)/(2m\mu)}\bigr)$ rather than $o\bigl(\tau^{(d-1)/(2m)}\bigr)$.*

(iii) Furthermore, in framework of Theorem 12.3.13 we obtain formula (12.2.90) *with the remainder estimate $o\bigl(\tau^{(d'-1)(\mu+1)/2m(\mu-\nu)}\bigr)$ as $\varepsilon_1 = o(\gamma^{-1})$, $\varepsilon_0 = o(\sigma_0\gamma^{-1})$.*

(iv) In particular, in the framework of Example 12.3.1 in Statements (i)–(ii) one can take

$$a^0_{\alpha(j)}(x) = \sum_{\beta'':|\beta''|=\mu(2m-|\alpha|)-j} \bigl(\partial_{x''}^{\beta''} a_{\alpha(j)}(x)\bigr)_{x''=0}\frac{x''^{\beta''}}{\beta''!} \tag{12.3.58}$$

while in Statement (iii) one should add a correction term $\varkappa'_1\tau^{(d'-1)(\mu+1)/2m(\mu-\nu)}$.

Main Theorems

Let us formulate the principal theorems proved in this section. They are due to the results above. Then we obtain the following statements:

Theorem 12.3.19 [28]. *Let conditions* (12.3.1), (12.3.3), (12.3.6), (12.3.9)–(12.3.12), (12.3.19) *and stabilization conditions* (11.5.4), (12.3.34) *and* (12.3.56)–(12.3.57) *be fulfilled.*

(i) Then asymptotics

$$\mathsf{N}^-(\tau) = \mathcal{N}_1(\tau) + O\bigl(R^*(\tau) + \tau^b\bigr),\qquad b = \frac{(\mu+1)}{2m(\mu-\nu)}\Bigl(d - 1 - \frac{(\nu+1)}{\nu}(d''-1) - \varkappa\Bigr) \tag{12.3.59}$$

holds where $R^*(\tau)$ *is given by* (12.3.39) *and* $\mathcal{N}_1(\tau)$ *is defined by* (12.2.100), $n^0(x',\xi',\tau)$ *is the eigenvalue counting function for operator* $\mathsf{a}^0(x',\xi';x'',D'')$ *in* $\mathbb{H}$ *and* $c_0 \le t_1\tau^{-1/(2m\mu)} \le \tau^\delta$ *with a small enough exponent* $\delta > 0$.

Moreover, (12.3.24) *remains true for* $\mathcal{N}_1$.

(ii) Further, let $(d-1)\nu < \mu d''$, $\kappa > \nu^{-1}\bigl(d-1-(\nu+1)(d''-1)\bigr)$ [29]. *Moreover, let the standard non-periodicity condition be fulfilled for the Hamiltonian flow of* $a(x',\xi')$.

Then asymptotics

$$\mathsf{N}^-(\tau) = \mathcal{N}_1(\tau) + o\bigl(\tau^{\frac{(d-1)(\mu+1)}{2m\mu}}\bigr) \tag{12.3.60}$$

holds as $\tau\to+\infty$ *provided* $t_1\to+\infty$ *(we assume that* $t_1\le\tau^\delta$*).*

(iii) Furthermore, let $(d-1)\nu < \mu d''$, $\kappa > (d'-1)\mu(\mu+1)^{-1} - (d''-1)\nu^{-1}$ [30].

Moreover, let non-periodicity condition (12.2.51) *be fulfilled. Then the asymptotics*

$$\mathsf{N}^-(\tau) = \mathcal{N}_1(\tau) + \varkappa_1^*\tau^{\frac{1}{2m\mu}(d'-1)(\mu+1)} + o\bigl(\tau^{\frac{1}{2m(\mu-\nu)}(d'-1)}\bigr) \tag{12.3.61}$$

holds with the coefficient $\varkappa_1^*$ *given in Remark 4.4.13.*

---

[28] Cf. Theorem 12.2.19.

[29] Then the remainder estimate in (12.3.59) is $O\bigl(\tau^{\frac{1}{2m\mu}(d-1)}\bigr)$.

[30] Then the remainder estimate in (12.3.59) is $O\bigl(\tau^{\frac{1}{2m(\mu-\nu)}(d'-1)}\bigr)$.

Discussion

We leave to the reader the following series of problems:

Problem 12.3.20. (i) Formulate and prove all results of this section for a domain where Λ (the manifold of the degeneration) consists of more than one connected components (with different characteristics) $\Lambda = \Lambda_1 \cup \cdots \cup \Lambda_N$, $\Lambda_j \cap \Lambda_k = \emptyset$ as $0 \ne j \ne k \ne 0$. This problem is trivial even if the dimensions of these cusps (d'_k) and the exponents of degenerations (μ_k) differ.

(ii) Consider the case when $X \ne \mathbb{R}^d$ but as $|x| \ge c$ it coincides with the conic domain assuming that

(12.3.62) each connected component of degeneration manifold Λ either does not intersect ∂X or is contained there.

Then coefficients $\tilde{\kappa}_j$ will appear here as well.

(iii) Investigate different boundary conditions.

(iv) In the framework of (ii) get rid of assumption (12.3.62).

(v) Write explicit expressions as $m = 2$, $\nu = 1$.

In this case operator $a_\infty(x', x'; x'', D'')$ is a quadratic Hamiltonian and due to ellipticity it is after metaplectic transformation (see Subsection 1.2.5) it becomes a direct sum of harmonic oscillators $\sum \omega_j (D_j^2 + x_j^2)$ with very well known spectrum and eigenvalue counting function

$$n_\infty(\tau) = \#\{\alpha \in \mathbb{Z}^{+k}, \sum_k \omega_k (2\alpha_k + 1) < \tau\}. \tag{12.3.63}$$

(vi) Conduct analysis similar to one in the cases (iv), (v) of Subsubsection *12.2.3.1 Scheme of Effective Asymptotics.*

(vii) Repeat with trivial modifications arguments of Subsubsection 12.2.3.

(viii) Derive more precise remainder estimates in the framework of Theorem 12.3.12 as Hamiltonian flow associated with $a(x, \xi)$ is of power or exponential non-periodic type.

(ix) Derive more precise remainder estimate in the framework of Theorem 12.3.13 and Example 12.3.14 assuming that the Hamiltonian flow associated with $\zeta(x',\xi')$ is of power or exponential non-periodic type.

(x) Investigate the case in the framework of Theorem 12.3.13 and Example 12.3.14 assuming that the Hamiltonian flow associated with $\zeta(x',\xi')$ is periodic.

(xi) Derive asymptotics for Riesz means.

Obviously Remark 12.2.22 stays.

12.4 Operators with Singular Potentials Decaying at Infinity

12.4.1 Problem Set-up

Let us start from the following example:

Example 12.4.1. Consider self-adjoint (matrix) operator on $\mathbb{R}^d$ (12.3.1) (possibly in the divergent form) with the positively homogeneous of degrees $(2m-|\alpha|)\mu$ (as $|x|\ge c$) coefficients a_α. In Section 11.6 we investigated the case when (12.3.2), (12.3.3) hold, $0>\mu>-1$ and established asymptotics for $\mathsf{N}^-(\tau)$ as $\tau\to -0$.

Now we want to consider the case when coefficients are singular at Λ, behaving basically as $\rho^{2m-|\alpha|}$ with $\rho(x)=\zeta(x)^\nu |x|^{\mu-\nu}$, $\zeta(x)=\mathsf{dist}(x,\Lambda)$ where Λ satisfies (12.3.5) and $0>\nu>\mu>-1$.

Without any loss of the generality we can assume that (at least locally) $\Lambda=\{x: x''=0\}$. Then we need to introduce $\zeta(x)$, $\gamma(x)$ and $\rho(x)$ by (12.3.9). However we do not want to have true singularities as $x''=0$: one can manage to investigate them too but it will be a pain with no gain - extra analysis without new effects; so instead we assume that (12.3.12) holds with ρ, ζ redefined as ρ^*, ζ^* by (12.3.10)–(12.3.11).

Remark 12.4.2. (i) We need condition $\nu>\mu$ for the cusp to shrink at infinity. As $0>\nu>-1$ the gorge would be too narrow to generate eigenvalues tending to $-\infty$ even if we did not "regularized" coefficients, departing from Example 12.4.1.

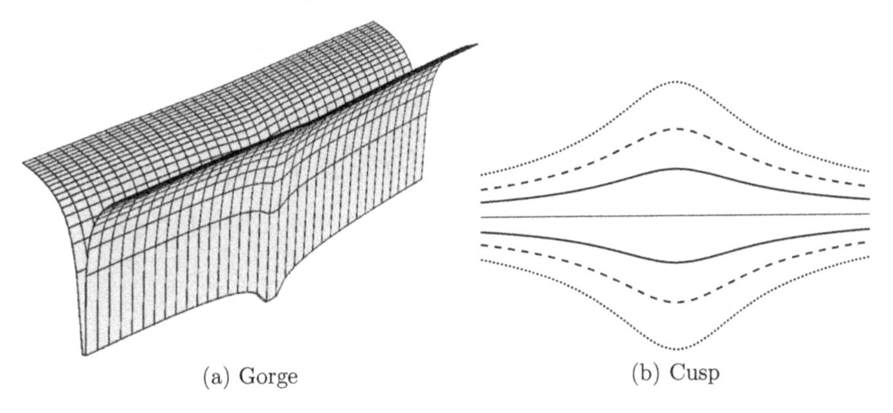

(a) Gorge (b) Cusp

Figure 12.6: Potential graph and level lines

(ii) Our results here are very similar to those of the previous Section 12.3 and could be formally obtained from them by replacing "$\tau \to +\infty$" by "$\tau \to -0$" and assumption $\mu > \nu > 0$ by $0 > \nu > \mu > -1$.

Consider operator first in the framework of the theory of Section 11.6; after scaling $x_{\mathsf{new}} = x\bar{\sigma}$, $\xi_{\mathsf{new}} = |\tau|^{-1/(2m)}\xi$ we conclude that operator $|\tau|^{-1}A$ is elliptic on the energy levels (-1) and below provided

$$|x''| \ge \sigma(x',\tau) := c_0|x'|^{-(\mu-\nu)/\nu}\tau^{\frac{1}{2m\nu}}, \tag{12.4.1}$$

(we assume that $|x'| \gg |x''|$ and $|x'| \gg 1$) and the uncertainty principle $\rho(x)\zeta(x) \ge 1$ is fulfilled provided

$$|x''| \ge \sigma_1(\tau) := |\tau|^{-\frac{1}{2m}}. \tag{12.4.2}$$

Then we arrive immediately to the following assertion:

Proposition 12.4.3 [31]. *Let conditions* (12.3.1), (12.3.3) *and* (12.3.9)–(12.3.12) *be fulfilled at the point* $\bar{x} \in \mathbb{R}^d$.

(i) Then

$$|e(x,y,\tau)| \le C|\tau|^{\frac{d}{2m}}\zeta(x)^{-s}\sigma_1(\tau)^s \qquad \forall x,y \in B(\bar{x},\frac{1}{2}\zeta(\bar{x})) \tag{12.4.3}$$

with an arbitrarily large exponent s.

[31] Cf. Proposition 12.3.2.

(ii) Moreover, if $U(x,y,t)$ *is the Schwartz kernel of the operator* $\tau^{-1}A$ *then*

$$|F_{t\to\lambda}\bar{\chi}_{T,T'}(t)U(x,y,t)| \le C|\tau|^{d/(2m)}\zeta(x)^{-s}\sigma_1(\tau)^s \qquad \forall x,y\in B(\bar{x},\frac{1}{2}\zeta(\bar{x}))\quad \forall T'\in\mathbb{R}, T=\sigma(\bar{x})\quad \forall\lambda\le 1+\epsilon \tag{12.4.4}$$

with a small enough constant $\epsilon > 0$.

Remark 12.4.4[32]. Consider the "gorge"

$$\mathcal{X}(\tau) = \{x, |x''| \le \sigma(x',\tau)\}. \tag{12.4.5}$$

It consists of the *regular part* where $|x| \le c|\tau|^{1/(2m\mu)}$ and the *cusp* where $|x| \ge c|\tau|^{1/(2m\mu)}$; this cusp stretches along $X' = \{x, x''=0\}$ and shrinks at infinity (on $|x|$) for any $\tau > 0$.

Let us apply now the theory of the operators with the operator-valued symbols (in rather limited manner at this time); to do this we need to analyze operator $\mathsf{a}(x',\xi') = A(x',\xi';x'',D'')$:

Proposition 12.4.5[33]**.** *Let conditions* (12.3.1), (12.3.3) *and* (12.3.9)–(12.3.12) *be fulfilled. Let* $\lambda_k(x',\xi')$ *be eigenvalues of* $\mathsf{a}(x',\xi')$ *in* $\mathbb{H}=\mathscr{L}^2(\mathbb{R}^{d''})$. *Then*

$$\lambda_k(x',\xi) \ge \epsilon_0|\xi'|^{2m} - \varepsilon_k\sigma_0(x')^{-2m} \tag{12.4.6}$$

where $\sigma_0(x')$ *is defined by* (12.4.2).

Proof. Coincides with the proof of Proposition 12.3.4. □

Due to our assumptions $|\tau|^{-1}\mathsf{a}(x',\xi')$ is elliptic as an operator with an operator-valued symbol on the energy level (-1) unless $|\xi'| \le c\tau^{1/(2m)}$ and $|x'| \le r^*(\tau)$ with $r^*(\tau)$ defined by (12.3.20).

Proposition 12.4.6[34]**.** *Let conditions* (11.1.6), (11.1.8), (12.3.1), (12.3.3) *and* (12.3.9)–(12.3.12) *be fulfilled. Then estimate* (12.3.21) *holds.*

---

[32] Cf. Remark 12.3.3.

[33] Cf. Proposition 12.3.4.

[34] Cf. Proposition 12.3.5.

Proof. Coincides with the proof of Proposition 12.3.5. □

Remark 12.4.7[35].

(i) Note that $|\tau|^{1/(2m)} \lesssim \sigma(x',\tau)$ if and only if $|x'| \lesssim r^*(\tau)$.

(ii) Note that $\operatorname{mes}(\mathcal{X}(\tau)) = O(|\tau|^{d/(2m\mu)})$ if and only if $\mu d'' < \nu d$. Under this condition the principal part of the Weyl formula is $\asymp |\tau|^{d(\mu+1)/(2m\mu)}$.

(iii) Let us drop condition $\mu d'' < \nu d'$ and let us restrict integration to the domain $\{x, |x| \le r^*(\tau)\}$ in the Weyl formula due Proposition 12.4.6. Then the principal part of the Weyl asymptotics (modified in this way) has the magnitude of (12.3.24).

(iv) Let us consider the remainder estimate in the Weyl formula. We can integrate only over domain $\{x, |x| \le r^*(\tau)\}$ due to Proposition 12.4.6. Then the remainder estimate in the modified Weyl asymptotics is (12.3.25)

(v) Let $\mu(d''-1) < \nu(d-1)$. Then under standard stabilization conditions of a_α as $|x| \to \infty$ to positively homogeneous $a_{\infty,\alpha}$ and standard restrictions to the Hamiltonian flow generated by the corresponding Hamiltonian $a_\infty(x,\xi)$ one can obtain the remainder estimate $o(|\tau|^{(d-1)(\mu+1)/(2m\mu)})$.

Again, in this remark the role of the "operator-valued theory" is very limited: it only provides a nice cut-off in the Weyl formulas.

We leave to the reader the following easy problem:

Problem 12.4.8[36]. Calculate the principal part of the asymptotics and remainder estimate for Riesz means in the Remark 12.4.7.

12.4.2 Sharp Asymptotics

So, as we are interested in the asymptotics of $\mathsf{N}^-(\tau)$, our analysis should be done as

$$\mu(d''-1) \le \nu(d-1). \tag{12.4.7}$$

[35] Cf. Remark 12.3.6.

[36] Cf. Problem 12.3.7.

Here we should treat only the cusp zone

$$\mathcal{X}_{\mathsf{cusp}}(\tau) = \bigl\{c_1\tau^{\frac{1}{2m}} := r_*(\tau) \le |x'| \le r^*(\tau),\ |x''| \le c_2\sigma(x',\tau)|\tau|^{-\delta}\bigr\} \tag{12.4.8}$$

with $r^*(\tau)$ and $\sigma(x',\tau)$ defined by (12.3.20) and (12.4.1) and an arbitrarily small exponent $\delta > 0$.

Really, estimate (12.4.3) and Proposition 12.4.6 yield that for $|x'| \ge c_1\tau^{1/(2m\mu)}$ outside of $\mathcal{X}$ function $e(x,x,\tau)$ is negligible; on the other hand, we have

Proposition 12.4.9 [37]**.** *(i) Contribution of the regular zone* (12.3.28) *(with arbitrarily large constant* c*) to the principal part of the Weyl asymptotics and to the remainder estimate are* $O\bigl(|\tau|^{d(\mu+1)/(2m\mu)}\bigr)$ *and* $O\bigl(|\tau|^{(d-1)(\mu+1)/(2m\mu)}\bigr)$ *respectively* [38].

(ii) Moreover, under standard stabilization conditions and the standard non-periodicity condition to the Hamiltonian flow the contribution of the zone $\mathcal{X}_{\mathsf{reg}}(\tau)$ *to the remainder estimate is* $o\bigl(|\tau|^{(d-1)(\mu+1)/(2m\mu)}\bigr)$.

Proof. Coincides with the proof of Proposition 12.3.8. □

Furthermore, the same arguments yield that treating zone $\bigl\{|x'| \le c_0 r^*(\tau)|\tau|^{\delta'}\bigr\}$ with an arbitrarily small exponent $\delta' > 0$ one can set $\delta = 0$ in the definition of $\mathcal{X}(\tau)$; in this case constant c_2 should be large enough.

To treat the cusp zone (12.4.8) let us apply theory of Section 4.4. For this we need to check if operator in question is (x',ξ') microhyperbolic on the energy level τ in (γ,ρ_τ)-scale with $\gamma(x') = \frac{1}{2}|x'|$ and $\rho(x') = \sigma_0(x')^{-1}$.

Proposition 12.4.10 [39]**.** *Operator* $a(x',D')$ *is* (x',ξ')*-microhyperbolic under stabilization condition* (12.3.34).

Proof. Coincides with the proof of Proposition 12.3.9. □

We leave to the reader the following easy problem:

---

[37] Cf. Proposition 12.3.8.

[38] Subject to conditions of Section 11.5. For example, as either operator is the Schrödinger operator or its perturbed power or stabilization condition (11.5.4) is fulfilled.

[39] Cf. Proposition 12.3.9.

Problem 12.4.11[40]. Establish other conditions providing (x',ξ')-microhyperbolicity of $a(x', D')$.

So, assume that microhyperbolicity condition is fulfilled. Then we can apply Theorem 4.4.8 as this operator is ξ'-microhyperbolic and Theorem 4.4.9 as this operator is (x',ξ')-microhyperbolic (i.e. in the general case) but there is a problem: operators B_1^{-k} is not of the trace norm. So we need to modify our arguments.

First of consider trace only with respect to x' and consider $\mathbb{H}\to\mathbb{H}$ operator norm; then the contribution of γ-element to the remainder does not exceed

$$C\hbar^{1-d'} = C\rho^{d'-1}\gamma^{d'-1} \tag{12.4.9}$$

as $\hbar=\rho^{-1}\gamma^1$. Then due to embedding theorem if we consider just pointwise norm (on (x'',y'')), the the contribution of γ-element to the remainder does not exceed expression (12.4.9) multiplied by $\rho^{d''/(2m)}$, i.e. $C\rho^{d-1}\gamma^{d'-1}$. Then after cut-off as $\zeta(x)\le\sigma(x',\tau)$ leads to the the remainder estimate

$$\int_{\sigma_0}^{\sigma} (\zeta^\nu\gamma^{\mu-\nu})^{d-1}\gamma^{d'-1}\zeta^{d''-1}\,d\zeta \tag{12.4.10}$$

on the other hand as $|x''|\ge\sigma(x',\tau)$ the pointwise norm does not exceed

$$C|\tau|^{\frac{d}{2m}}\gamma^{d'-1}\times|x''|^{-s}\sigma(x',\tau)^s$$

where we used the standard ellipticity of our operator here and the contribution of this zone to the remainder estimate does not exceed (12.4.10).

So, the total remainder estimate is

$$\int_{r_*(\tau)}^{r^*(\tau)}\int_{\sigma_0}^{\sigma} (\zeta^\nu\gamma^{\mu-\nu})^{d-1}\gamma^{d'-2}\zeta^{d''-1}\,d\zeta d\gamma \tag{12.4.11}$$

and one can prove easily that it is $\asymp R^*(\tau)$ defined by (12.3.39).

Let us apply Weyl asymptotics in the zone $\{|x'|\le t_1|\tau|^{1/(2m\mu)}\}$. To ensure (x,ξ)-microhyperbolicity there we assume that the standard stabilization condition (11.5.4) is fulfilled[17].

Then we obtain the following theorem:

[40] Cf. Problem 12.3.10.

Theorem 12.4.12 [41]. *Let conditions* (12.3.1), (12.3.3), (12.3.9)–(12.3.12) *and stabilization conditions* (11.5.4) [38),17)], (12.3.34) *be fulfilled.*

Let $\psi \in \mathscr{C}_0^K(\mathbb{R}^{d'})$, $\psi(x') = 1$ *as* $|x'| \le c_0$. *Then*

(i) Asymptotics (12.3.40) *holds as* $\tau \to -0$ *with* $\mathcal{N}(\tau)$ *defined by* (12.2.43) *with* $\psi_t(x') = \psi(x'/t)$ *where* $t \asymp r_*(\tau)$ [19] *and with* $R^*(\tau)$ *defined by* (12.3.39).

(ii) Moreover, $\mathcal{N}(\tau)$ *has a magnitude described in* (12.3.24).

Repeating corresponding arguments of Section 12.3 we arrive to the following theorem:

Theorem 12.4.13 [42]. *Let the conditions of Theorem 12.4.12 be fulfilled with* $(d-1)\nu > d''\mu$.

Moreover, let the standard non-periodicity condition to the Hamiltonian trajectories generated by $a_\infty(x,\xi)$ *be fulfilled. Then asymptotics* (12.3.42) *holds as* $\tau \to -0$ *where* $\mathcal{N}(\tau)$ *is defined by* (12.2.46),

$$(2\pi)^{-d}\int \psi\kappa_1\tau^{\frac{1}{2m}(d-1)}\,dx \quad \text{and} \quad (2\pi)^{1-d}\int_{\partial X}\psi\tilde{\kappa}_1\tau^{\frac{1}{2m}(d-1)}\,dS \tag{12.4.12}$$

are the standard second terms [29] *in the Weyl asymptotics,* $\psi_t(x') = \psi_0(x'/t)$ *where* $t = t(\tau)$ *is an appropriate function such that* $t(\tau)/r_*(\tau) \to +\infty$ *as* $\tau \to -0$ [30]

Theorem 12.4.14 [43]. *Let the conditions of Theorem 12.4.12 be fulfilled with* $(d-1)\nu < d''\mu$.

Let us consider operator-valued symbol $\mathbf{a}_\infty(x',\xi')$ *in* $\mathbb{H}$; *let* $\lambda_j(x',\xi')$ *be its (non-decreasing) eigenvalues* [24]. *Let us consider*

$$\mathbf{a}_{\infty j}(x',\xi') = \mathbf{a}_\infty(x',\xi')\mathbf{e}_\infty(x',\xi',\zeta_j(x',\xi')) \tag{12.4.13}$$

where $\mathbf{e}_\infty(x',\xi',\tau)$ *is the spectral projector for* $\mathbf{a}_\infty(x',\xi')$ *and assume that non-periodicity condition* (12.2.51) *is fulfilled.*

Then asymptotics (12.3.46) *holds as* $\tau \to -0$ *with an appropriate coefficient* κ_1^* *(given in Remark 4.4.13).*

---

[41] Cf. Theorems 12.2.6 and 12.3.11.

[42] Cf. Theorems 12.2.7 and 12.3.12.

[43] Cf. Theorems 12.2.8 and 12.3.13.

Condition (12.2.51) looks extremely cumbersome and impossible to check. However in the following rather special example this condition becomes much simpler:

Example 12.4.15[44]. Let $d''=1$ and A be a Schrödinger operator stabilizing at infinity to $|\xi^2|+|x_d|^{2\nu}\cdot\omega(x')^2$ with ω, positively homogeneous of degree $2(\mu-\nu)$.

Then one needs to check the standard non-periodicity condition for symbol In particular, the standard non-periodicity condition for $\lambda(x',\xi')$ is fulfilled as $\sigma^0=|x'|^{-\eta}$, $\eta<0$ and $\eta\ne\frac{1}{2}$; recall that $d'\ge 2$.

Finally, let us consider the case $d'=1+d''\mu$. Considering diagonal operator with diagonal elements $\lambda_j^{\mathrm{w}}(x',D')$ one can prove easily the remainder estimate $o(\tau^{(d-1)/2m}|\log|\tau||)$ (not subject to any non-periodicity condition). This leads us to conjecture which we were unable to prove

Conjecture 12.4.16[45]. *As* $(d-1)\nu<d''\mu$ *remainder estimate is in fact* $o(|\tau|^{(d-1)/(2m\mu)}|\log|\tau||)$.

We leave to the reader the following easy albeit interesting problem:

Problem 12.4.17[46]. (i) Treat Riesz means and derive asymptotics with $R^*_{(\vartheta)}(\tau)$ defined by (12.3.49).

(ii) Here one needs also to improve results of Section 11.6. To do it one can consider propagation and extend time from $T\asymp|x|^{1-\mu}$ to $T\asymp|\tau|^{(1-\mu)/(m\mu)}$.

12.4.3 Effective Asymptotics

Simplifying the Answer

Again, the main defect of Theorems 12.4.12–12.4.14 again is that the expression for $\mathcal{N}(\tau)$ is not effective and includes the eigenvalue counting function for the *family* of auxiliary operators $\boldsymbol{a}(x',\xi')$. Exactly like in the previous Section 12.3 we would like to derive a formula including the eigenvalue counting function for *one* auxiliary operator or at least for a family of auxiliary

[44] Cf. Examples 12.2.9 and 12.3.14.

[45] Cf. Conjectures 12.2.10 and 12.3.15.

[46] Cf. Problems 12.2.11 and 12.3.16.

operators with a smaller number of parameters; however in contrast to the previous section we want also to simplify these axillary operators.

Let us assume that (12.3.53) holds with the finite set $Q \subset \mathbb{R}^+ \cup 0$. Note that effectively $(A-A^0)$ contains a factor $|x''|^q$ which due to Proposition 12.4.3 in the estimates translates into factor $\sigma(\gamma,\tau)^q$. Then arguments leading to (12.2.82) (or (12.3.54) imply that

$$\mathsf{R}_1 \le R + \varepsilon_q(\gamma) \int_{\sigma_0}^{\sigma} (\zeta^\nu \gamma^{\mu-\nu})^{d-1} \gamma^{d'} \zeta^{d''-2+q}\, d\zeta \tag{12.4.14}$$

where R_1 is defined by (12.3.54) and the right-hand expression is (12.4.10) modified in the obvious way. Further, one can estimate the left-hand expression of (12.3.55) in the same way.

Recall that A^0 satisfy the same conditions as A, all the objects associated with A^0 have an additional superscript "0".

Then after summation with the respect to partition we conclude that redefined R_1 does not exceed

$$\int_{r_*(\tau)}^{r*(\tau)} \int_{\sigma_0}^{\sigma} (\zeta^\nu \gamma^{\mu-\nu})^{d-1} \gamma^{d'-1+(\mu-\nu)q/(\nu+1)} \zeta^{d''+q-2}\, d\zeta d\gamma \tag{12.4.15}$$

as $\varepsilon_q(\gamma) = \gamma^{-p+(\mu-\nu)q/(\nu+1)}$. One can see easily that this expression does not exceed $CR^* + C|\tau|^b$ because $(\zeta^\nu\gamma^{\mu-\nu})^{d-1}\gamma^{d'+(\mu-\nu)q/(\nu+1)}\zeta^{d''+q-1}$ calculated as $\gamma = r^*(\tau)$, $\zeta = \sigma_0(r^*(\tau))$ is exactly the second term here.

Now our arguments basically repeat those of Subsections 12.2.3 and 12.3.3. First, we obtain formula (12.2.88); then immediately we arrive to asymptotics similar to (12.3.54)–(12.3.56). Further, in the framework of Theorem 12.4.13 we obtain formula (12.2.89) with the remainder estimate $o(R^*) + O(|\tau|^b)$. Furthermore, in the framework of Theorem 12.4.13 we obtain formula (12.2.90) with the remainder estimate $o(R^*) + O(|\tau|^b)$.

Problem 12.4.18[47]. (i) Consider domains with the boundary (with appropriate boundary conditions), assuming that (12.3.62) holds. Then coefficients $\tilde{\varkappa}_j$ will appear here as well.

(ii) Conduct analysis similar to one in the Cases (iv) and (v) of Subsubsection *12.2.3.1 Scheme of Effective Asymptotics.*

(iii) Repeat with trivial modifications arguments of Subsubsection *12.2.3.2 Scheme of Effective Asymptotics.*

[47] Cf. Problem 12.3.20.

Main Theorems

Let us formulate the principal theorems proved in this section. They are due to the results above.

Then we obtain the following statements:

Theorem 12.4.19 [48]. *Let conditions* (12.3.1), (12.3.3), (12.3.9)–(12.3.12) *and stabilization conditions* (11.5.4), (12.3.34) *and* (12.3.56)–(12.3.57) *be fulfilled.*

(a) Then asymptotics (12.3.59) *holds where* $R^*(\tau)$ *is given by* (12.3.39), $\mathcal{N}_1(\tau)$ *is defined by* (12.2.100), $n^0(x',\xi',\tau)$ *is the eigenvalue counting function for operator* $a^0(x',\xi';x'',D'')$ *in* $\mathbb{H}$ *and* $c_0 \le t_1|\tau|^{-1/(2m\mu)} \le |\tau|^{-\delta}$ *with a small enough exponent* $\delta > 0$.

Moreover, (12.3.24) *remains true for* $\mathcal{N}_1$.

(b) Further, let $(d-1)\nu < \mu d''$ *and* $b > (d-1)\nu$. *Moreover, let the standard non-periodicity condition be fulfilled for the Hamiltonian flow of* $a(x',\xi')$.

Then asymptotics (12.3.60) *holds as* $\tau \to -0$ *provided* $t_1 \to +\infty$ *(we assume that* $t_1 \le |\tau|^{-\delta}$*).*

(c) Furthermore, let $(d-1)\nu < \mu d''$ *and* $R_1^* = o(R^*)$. *Moreover, let non-periodicity condition* (12.2.51) *be fulfilled.*

Then the asymptotics (12.3.61) *holds with the coefficient* $\varkappa_1^*$ *given in Remark 4.4.13.*

We leave to the reader the following easy series of problems:

Problem 12.4.20 [49]. (i) Formulate and prove all results of this section for a domain where Λ (the manifold of the degeneration) consists of more than one connected components (with different characteristics) $\Lambda = \Lambda_1 \cup \ldots \cup \Lambda_N$, $\Lambda_j \cap \Lambda_k = \emptyset$ as $0 \ne j \ne k \ne 0$. This problem is trivial even if the dimensions of these cusps (d'_k) and the exponents of degenerations (μ_k) differ.

(ii) Consider the case when $X \ne \mathbb{R}^d$ but as $|x| \ge c$ it coincides with the conic domain assuming that each component of Λ either does not intersect ∂X or is contained in ∂X. Investigate different boundary conditions.

[48] Cf. Theorems 12.2.19 and 12.3.19.

[49] Cf. Problem 12.4.20.

(iii) Derive more precise remainder estimates in the framework of Theorem 12.4.13 as Hamiltonian flow associated with $a(x,\xi)$ is of power or exponential non-periodic type.

(iv) Derive more precise remainder estimate in the framework of Theorem 12.4.14 assuming that the Hamiltonian flow associated with $\zeta(x',\xi')$ is of power or exponential non-periodic type.

(v) Derive asymptotics for Riesz means.

Obviously Remark 12.2.22 stays.

12.5 Spectral Asymptotics for Maximally Hypoelliptic Operators

12.5.1 Maximally Hypoelliptic Operators

In this section we consider a class of pseudodifferential operators on a compact closed $\mathscr{C}^{K'}$ manifold X degenerating on a symplectic manifold $\Lambda \subset T^*X \setminus 0$ conical in ξ. The same results hold with no change for operators acting in Hermitian bundles. Moreover, similar results hold after an obvious modification for differential operators on manifolds with boundary provided $\Lambda|_{\partial X} = \emptyset$. We leave these modification to the reader. So, let us assume that

(12.5.1) X is a compact $\mathscr{C}^{K'}$ manifold, $\Lambda \subset T^*X \setminus 0$ is a symplectic manifold conic in ξ, $d = \dim X$, $2d' = \dim \Lambda$, $1 \le d' < d$, $d'' = d - d'$.

Definition 12.5.1. Let $\Psi^{m,\mu}(X,\Lambda,\mathbb{K})$ with $m \in \mathbb{R}$, $\mu \in \mathbb{Z}^+$ denote the class of pseudodifferential operators from $\mathscr{H}^s(X,\mathbb{K})$ to $\mathscr{H}^{s-m}(X,\mathbb{K})$ [50] with symbols in every local chart of the form

$$a = \sum_{0 \le n \le K-1} a_{\frac{n}{2}} + a'_{\frac{K}{2}} \tag{12.5.2}$$

[50] With $s \in (-\bar{s},\bar{s})$ and with $\bar{s} = \bar{s}(K',d,m) \to +\infty$ as $K' \to +\infty$ and d,m are fixed. This footnote holds until the end of this section.

where $a_{\frac{n}{2}}$ are $\mathscr{C}^{K'-n}$ symbols positively homogeneous of degrees $(m-\frac{n}{2})$ in ξ and such that

$$a^{(\alpha)}_{\frac{n}{2}(\beta)}(z)=0 \qquad \forall z\in\Lambda \quad \forall\alpha,\beta:|\alpha|+|\beta|+n<\mu \tag{12.5.3}$$

and the remainder $a'_{\frac{K}{2}}$ is assumed to be smooth enough and decrease quickly enough as $|\xi|\to\infty$. Here and below $K'>2K$ and K are large enough.

It is well known (see also Section 1.2) that these classes are microlocally invariant. In an appropriate microlocal symplectic coordinate system on $T^*X\setminus 0$ we have

$$\Lambda=\bigl\{(x,\xi)\in T^*X:\ x''=\xi''=0\bigr\} \tag{12.5.4}$$

where $x=(x';x'')=(x_1,\dots,x_{d'};x_{d'+1},\dots,x_d)$ and similar notation is applied to ξ. Then for the operator $A\in\boldsymbol{\Psi}^{m,\mu}(X,\Lambda,\mathbb{K})$ and for $z\in\Lambda$ let us introduce the operator

$$A_z(x'',D''):=\sum_{|\alpha|+|\beta|+n=\mu}\frac{1}{\alpha!\beta!}a^{(\alpha)}_{\frac{n}{2}(\beta)}(z)x''^{\beta}D''^{\alpha} \tag{12.5.5}$$

acting in the space $\mathscr{L}^2(\mathbb{R}^{d''},\mathbb{K})$[51]. Recall that $f^{(\alpha)}_{\beta}=\partial^{\alpha}_{\xi}\partial^{\beta}_{x}f$.

Definition 12.5.2. This operator $A_z(x'',D'')$ is called the *microlocalization* (of order μ) of the operator A at the point $z\in\Lambda$.

It is well known (see Subsection 1.2.5) that in another microlocal coordinate system with the same form of Λ we obtain the microlocalization $A'_z=Q^*_zA_zQ_z$ where Q_z is a unitary metaplectic (see Definition 1.2.39) operator in $\mathscr{L}^2(\mathbb{R}^{d''})$ also acting in $\mathscr{L}^2(\mathbb{R}^{d''},\mathbb{K})=\mathscr{L}^2(\mathbb{R}^{d''})\otimes\mathbb{K}$.

Definition 12.5.3. One says that the operator $A\in\boldsymbol{\Psi}^{m,\mu}(X,\Lambda,\mathbb{K})$ is *maximally hypoelliptic* if and only if the following two conditions are fulfilled: the inequality

$$\|a_0(x,\xi)v\|\ge\epsilon_0\sigma^{\mu}(x,\xi)|\xi|^m\|v\| \qquad \forall(x,\xi)\in T^*X \quad \forall v\in\mathbb{K} \tag{12.5.6}$$

holds with a constant $\epsilon_0>0$ and with

$$\sigma(x,\xi)=\mathsf{dist}\bigl((x,\xi|\xi|^{-1}),\Lambda\bigr)\asymp|x''|+|\xi''|\cdot|\xi'|^{-1} \tag{12.5.7}$$

where the last relation holds as Λ is given by (12.5.4) and only in the conical vicinity of Λ and

[51] A Hermitian bundle $\mathbb{K}$ should be replaced by its fiber $\mathbb{K}_z$ over z in this definition.

(12.5.8) Operator A_z is injective in $\mathbb{H} = \mathscr{L}^2(\mathbb{R}^{d''}, \mathbb{K})$ at every point $z \in \Lambda$. In this case

$$\|A_z v\|_{\mathbb{H}} \ge \epsilon_0 \|v\|_{\mathbb{H}} \qquad \forall v \in \mathbb{H} \quad \forall z \in \Lambda \cap \{|\xi'| = 1\} \tag{12.5.9}$$

with an appropriate constant $\epsilon_0 > 0$.

Here and below $\|.\|_{\mathbb{K}}$, $\|.\|_{\mathbb{H}}$ and $\|.\|$ are the norms in $\mathbb{K}$, $\mathbb{H}$ and $\mathscr{L}^2(X, \mathbb{K})$ respectively and $\langle .,.\rangle_{\mathbb{K}}$, $\langle .,.\rangle_{\mathbb{H}}$ and $(.,.)$ are the corresponding inner products.

It is well known L. Boutet de Monvel [2] that

Theorem 12.5.4. *(i) Every maximally hypoelliptic operator is microlocally hypoelliptic; more precisely, operator $A \in \boldsymbol{\Psi}^{m,\mu}(X, \Lambda, \mathbb{K})$ satisfies*

$$\mathsf{WF}^{s+l}(u) \subset \mathsf{WF}^{s}(Au) \qquad \forall u \in \mathscr{D}'(X, \mathbb{K}) \tag{12.5.10}$$

with $l = m - \frac{\mu}{2}$ if and only if it is maximally hypoelliptic.

(ii) Moreover, in this case (and only in this case) the resolvent of A is a bounded operator from $\mathscr{H}^s(X, \mathbb{K})$ to $\mathscr{H}^{s+l}(X, \mathbb{K})$.

In this section we treat a maximally hypoelliptic operator A with

$$l = m - \frac{\mu}{2} > 0. \tag{12.5.11}$$

Corollary 12.5.5. *If $A \in \boldsymbol{\Psi}^{m,\mu}(X, \Lambda, \mathbb{K})$ and (12.5.11) holds then the resolvent of A is a compact operator in $\mathscr{L}^2(X, \mathbb{K})$ and the spectrum of A is discrete with no accumulation point (provided A is self-adjoint).*

Remark 12.5.6. One can easily prove that this is wrong for $l \le 0$; moreover, for $l < 0, m > 0$ the continuous spectrum fills either the semiaxis or axis, while for $l = 0, m > 0$ and $l < 0, m = 0$ the spectrum is more complicated: there is an infinite number of accumulation points or points of the infinite multiplicity point spectrum (provided the operator is self-adjoint); for $m < 0$ the structure of the spectrum is transparent.

We leave to the reader the following easy problem

Problem 12.5.7. Prove that

(i) If Weyl symbol of A is Hermitian at every point $z \in T^*X$ then the Weyl symbol of A_z is Hermitian at every point $z'' \in T^*(\mathbb{R}^{d''})$ and therefore A_z is obviously symmetric in $\mathbb{H}$.

(ii) Moreover, under condition (12.5.8) A_z is an elliptic global operator (see Section 1.1.4) and is therefore self-adjoint.

(iii) A symmetric maximally hypoelliptic operator satisfying (12.5.11) is self-adjoint. Moreover, a self-adjoint maximally hypoelliptic operator satisfying (12.5.11) is semibounded from above if and only if both of the following two conditions are fulfilled:

(12.5.12) $a_0(z)$ is a negative definite matrix in $\mathbb{K}$ at every point $z \in (T^*X \setminus 0) \setminus \Lambda$

and

(12.5.13) A_z is a negative definite operator in $\mathbb{H}$ at every point $z \in \Lambda$.

12.5.2 Getting Started

So, let $A \in \boldsymbol{\Psi}^{m,\mu}(X, \Lambda, \mathbb{K})$ be a maximally hypoelliptic and self-adjoint operator and let (12.5.11) be fulfilled. Let $\mathsf{N}^{\pm}(\tau)$ be the number of eigenvalues of A lying in the intervals $[0, \tau)$ and $(-\tau, 0]$ respectively. We are interested in the asymptotics of $\mathsf{N}^{+}(\tau)$ as $\tau \to +\infty$. One can obtain the asymptotics of $\mathsf{N}^{-}(\tau)$ by replacing A by $-A$.

First of all, let us rewrite $\mathsf{N}^{\pm}(\tau)$ in the form

$$\mathsf{N}^{\pm}(\tau) = \mathsf{N}_0^{\pm}(\tau) + \mathsf{N}_1^{\pm}(\tau), \qquad \mathsf{N}_j^{\pm}(\tau) = \int \psi_j(x)\operatorname{tr}(e(x, x,))\, dx \tag{12.5.14}$$

where $e(x, y, \tau)$ is the Schwartz kernel of the spectral projector of A and ψ_j are classical pseudodifferential operators of order 0,

$$\operatorname{supp}(\psi_0) \cap \Lambda = \emptyset, \qquad \operatorname{supp}(\psi_1) \subset \mathcal{Z} = \bigl\{(x, \xi) :\ \sigma(x, \xi) < \epsilon\bigr\} \tag{12.5.15}$$

where $\epsilon > 0$ is an arbitrarily small constant. Recall that $\sigma(x, \xi)$ is defined by (12.5.7). Then

Proposition 12.5.8. *Let A be a self-adjoint operator. Then for $\mathsf{N}_0^{\mp}(\tau)$ the standard Weyl asymptotics holds with principal part*

$$\kappa_0\tau^{\frac{1}{m}(d-1)} + \kappa_{\frac{1}{2}}\tau^{\frac{1}{m}(d-\frac{1}{2})} + \kappa_1\tau^{\frac{1}{m}(d-1)} \tag{12.5.16}$$

and with remainder estimate $O\bigl(\tau^{\frac{1}{m}(d-1)}\bigr)$ and even $o\bigl(\tau^{\frac{1}{m}(d-1)}\bigr)$ under the standard condition non-periodicity condition to the Hamiltonian trajectories of a_0 on the energy level $\tau=1$. Here $\kappa_{\frac{1}{2}}=0$ provided $a_{\frac{1}{2}}=0$.

Therefore we should treat $\mathsf{N}_1^{\mp}(\tau)$ in the same way we treated $\mathsf{N}^{\pm}_{\mathsf{cusp}}(\tau)$ before. First of all we need the following assertion

Proposition 12.5.9. *Let $A\in\boldsymbol{\Psi}^{m,\mu}(X,\Lambda,\mathbb{K})$ be a self-adjoint operator and $\psi_1\in\boldsymbol{\Psi}^0(X)$ satisfy (12.5.15).*

(i) Let $\phi\in\mathscr{C}_0^{K'}(\mathbb{R}^{d''})$ be supported in $B(0,\epsilon)$ with a small enough constant $\epsilon>0$. Then estimate

$$\bigl|\bigl(\phi(\tau^{-\frac{1}{m}}D')\psi_1 e\bigr)(x,y,\tau',\tau)\bigr|\le C\tau^{-s} \tag{12.5.17}$$

holds with arbitrarily large s for $\tau>0$, $\tau'\in[\epsilon_0\tau,\tau]$.

(ii) Let A be maximally hypoelliptic and let $(1-\phi)\in\mathscr{C}_0^{K'}(\mathbb{R}^{d'})$ and ϕ be supported in $\{|\xi'|\ge C_0\}$ with a large enough constant C_0. Then estimate

$$\bigl|\bigl(\phi(\tau^{-\frac{1}{l}}D')\psi_1 e\bigr)(x,y,\tau',\tau)\bigr|\le C\tau^{-s} \tag{12.5.18}$$

holds with arbitrarily large s for $\tau>0$, $\tau'\in[\epsilon_0\tau,\tau]$.

(iii) Let $\phi\in\mathscr{C}_0^{K'}(\mathbb{R}^{d'})$, $0\notin\operatorname{supp}(\phi)$ and $\varphi\in\mathscr{C}_0^{K'}(\mathbb{R}^{d''})$. Let $r\ge\tau^{\delta}$, $\epsilon\ge\sigma\ge r^{-\frac{1}{2}}$ and $\rho^{\mu}r^m\le\varepsilon\tau$ with an arbitrarily small exponent $\delta>0$ and with a small enough constant $\varepsilon>0$. Then estimate

$$\bigl|\bigl(\varphi(\rho^{-1}x'',\rho^{-1}r^{-1}D'')\phi(r^{-1}D')\psi_1 e\bigr)(x,y,\tau',\tau)\bigr|\le C\tau^{-s} \tag{12.5.19}$$

holds with arbitrarily large s for $\tau>0$, $\tau'\in[\epsilon_0\tau,\tau]$.

Proof. Statement (i) follows from the ellipticity of operator $(\tau-A)$ in the zone $\bigl\{|\xi|\le\varepsilon\tau^{\frac{1}{m}}\bigr\}$ with a small enough constant $\varepsilon>0$.

Further, Statement (iii) follows from the ellipticity of operator $(\tau-A)$ in the zone $\bigl\{\epsilon r\le|\xi'|\le cr, |x''|+|\xi''|r^{-1}\le c\rho\bigr\}$ with the indicated r and ρ.

On the other hand, Statement (ii) follows from the maximal hypoellipticity of operator $(\tau - A)$ in the zone $\{|\xi'| \ge C_0\tau^{\frac{1}{l}}, \sigma(x,\xi)\le\epsilon\}$. More precisely, estimate

$$\|Au\|_s + \|u\|_s \ge \epsilon_0 \|A'u\|_s \tag{12.5.20}$$

holds where as usual $\|.\|$ means the norm in the Sobolev space $\mathscr{H}^s(X,\mathbb{K})$ and A' means any operator belonging to $\boldsymbol{\Psi}^{m,\mu}(X,\Lambda,\mathbb{K})$ (Boutet de Monvel [2]). In particular,

$$\|Au\|_s + \|u\|_s \ge \epsilon_0 \|u\|_{s+l} \tag{12.5.21}$$

and estimate (12.5.20) yields Statement (ii). □

Therefore, in what follows we need to treat zone

$$\begin{multline}\mathcal{U} = \{\sigma(x,\xi)\le\epsilon, \epsilon_0\tau^{\frac{1}{m}} := r_*(\tau) \le |\xi'| \le r^*(\tau) := C_0\tau^{\frac{1}{l}},\\ \epsilon_0 \le \sigma(x,\xi)|\xi'|^{-\frac{m}{\mu}}\tau^{\frac{1}{\mu}}\}.\end{multline}\tag{12.5.22}$$

In this zone let us make a $(\epsilon_1, \epsilon_1|\xi'|; \epsilon_1\sigma(x,\xi), \epsilon_1\sigma(x,\xi)|\xi'|)$-admissible partition of unity on $(x',\xi';x'',\xi'')$ with a small enough constant $\epsilon_1>0$ and let us treat some element $\bar{q}$ of this partition; let $\bar{r}$ and $\bar{\sigma}$ be the values of $|\xi'|$ and σ calculated at $\bar{z}\in\bar{q}$; then

$$\epsilon_0\tau \le \bar{r}^m\bar{\sigma}\mu \le c_0\tau, \qquad r_*(\tau)\le\bar{r}\le r^*(\tau). \tag{12.5.23}$$

12.5.3 Weyl Asymptotics

Let us start with the completely Weyl theory. For this let us consider an $(\epsilon_1, \epsilon_1\sigma^2 r; \epsilon_1\sigma r, \epsilon_1\sigma)$-admissible partition of unity on $(x',\xi';x'',\xi'')$. Then rescaling $x''_{\mathsf{new}} = x''\bar{\sigma}^{-1}$ we transform A to an h-pseudodifferential operator $\bar{r}^m\bar{\sigma}^\mu\tilde{A}_h$ provided

$$h := \bar{r}^{-1}\bar{\sigma}^{-2} \le h_0 \tag{12.5.24}$$

with a small constant h_0.

Then the contribution of this element to the principal part of the asymptotics is 0 and its contribution to the remainder estimate is $O(h^s)$ with an arbitrarily large exponent s provided one of the following two conditions is fulfilled:

(a) a_0 is negative definite on $T^*X\setminus 0$;

(b) $\bar{r}^m\bar{\sigma}^\mu \ge C_0\tau$.

Then after summation over such elements we conclude that

(12.5.25) The total contribution of all such elements of the partition lying in $\mathcal{U}$ to the principal part of the Weyl asymptotics is 0 and to the remainder estimate is $O(\tau^{d'l^{-1}})$.

Really, it does not exceed

$$\iint_{\mathcal{U}} \sigma^{-2d'-1}r^{-1}\,d\sigma dr \asymp \int_{r\le \tau^{l^{-1}}} r^{d'-1}\,dr \asymp \tau^{d'l^{-1}}.$$

In particular, applying Proposition 12.5.9 we arrive to the following two statements

(12.5.26) If A is maximally hypoelliptic and a_0 is negative definite then the contribution of $\complement\mathcal{V}$ to the asymptotics is $O(\tau^{d'l^{-1}})$ where

$$\mathcal{V} = \bigl\{|\xi|\ge c_0, \sigma(x,\xi)\le\epsilon, \sigma^2(x,\xi)|\xi'|\le c_0\bigr\}.$$

and

(12.5.27) In the general case the contribution of zone

$$\mathcal{W} = \bigl\{|\xi|\ge c_0, |\xi'|^m\sigma(x,\xi)^\mu \ge C_0\tau\bigr\}$$

to the asymptotics is $O(\tau^{d'l^{-1}})$ as well.

On the other hand, let us consider zone $\mathcal{V}$. Again due to Proposition 12.5.9 we should consider only elements with $r_*(\tau)\le|\xi'|\le r^*(\tau)$. Applying the previous arguments with σ replaced by $\rho(\xi') = |\xi'|^{-\frac{1}{2}}$ and $h=1$ we see that

(12.5.28) The total contribution of $\mathcal{V}$ to the asymptotics is $O(\tau^{d'l^{-1}})$ as well.

So we obtain as a corollary the following rather well known estimate

$$\mathsf{N}^+(\tau) = O\bigl(\tau^{\frac{1}{l}d'}\bigr) \qquad \text{as } \tau \to +\infty \tag{12.5.29}$$

for maximally hypoelliptic operators with negative definite symbols.

Let us consider the general case and let us treat the elements of the above partition with $r\sigma^2 \ge C_0$, $r \ge c_0$, $r^m\sigma^\mu \asymp \tau$. Let us note that

(12.5.30) After the coordinate transform $x''_{\mathsf{new}} = x''\bar{\sigma}^{-1}$ and multiplication by τ^{-1} operator A is microhyperbolic in the direction $\langle x, \partial_\xi\rangle$ provided C_0 is large enough.

Then the contribution of each element to the principal part is given by Weyl expression

$$(2\pi)^{-d}\iint \bar{q}_1^{\mathsf{w}}(x,\xi) n^+(x,\xi,\tau)\, dxd\xi$$

where $n^+(x,\xi,\tau)$ is the number of eigenvalues of the symbol $a_0(x,\xi) + a_{\frac{1}{2}}(x,\xi)$ in the interval $[0,\tau)$; one can introduce $n^-(x,\xi,\tau)$ and $n(x,\xi,\tau)$ in the same way.

On the other hand, the contribution of this element to the remainder is $O(h^{1-d}) = O\bigl(\bar{r}^{d-1}\bar{\sigma}^{2(d-1)}\bigr)$. Then the contribution of the element of a $(\epsilon_1, \epsilon_1|\xi'|; \epsilon_1\sigma, \epsilon_1\sigma|\xi'|)$-admissible partition to the remainder is

$$O\bigl(\bar{r}^{d-1}\bar{\sigma}^{2(d''-1)}\bigr) = O\bigl(\bar{r}^{d-1-2(d''-1)\frac{m}{\mu}}\tau^{\frac{2}{\mu}(d''-1)}\bigr)$$

and after summation on this partition of unity we obtain the remainder estimate

$$R^{\mathsf{W}}(\tau) = \begin{cases} O\bigl(\epsilon'\tau^{\frac{1}{m}(d-1)}\bigr) & \text{for } (d-1)l > d'm\\ O\bigl(\tau^{\frac{1}{m}(d-1)}\log\tau\bigr) & \text{for } (d-1)l = d'm\\ O\bigl(\tau^{\frac{1}{l}d'}\bigr) & \text{for } (d-1)l < d'm \end{cases} \tag{12.5.31}$$

where $\epsilon' = \epsilon'(\epsilon) \to +0$ as $\epsilon \to +0$.

So (also taking into account estimates for other zones) we arrive to the following theorem

Theorem 12.5.10. *Let $A \in \Psi^{m,\mu}(X, \Lambda, \mathbb{K})$ be a maximally hypoelliptic (self-adjoint) operator, $l > 0$ and suppose a_0 is not a negative definite symbol. Then*

(i) For $(d-1)l > d'm$ asymptotics

$$\mathsf{N}^+(\tau) = \mathsf{N}^{\mathsf{W}+}(\tau) + O(\tau^{\frac{d-1}{m}}) \tag{12.5.32}$$

holds with

$$\mathsf{N}^{\mathsf{W}+}(\tau) = (2\pi)^{-d}\iint n^+(x,\xi,\tau)\,dxd\xi. \tag{12.5.33}$$

(ii) Moreover, if $(d-1)l > d'm$ and if the symbol a_0 satisfies the standard non-periodicity condition on the energy level $\tau = 1$ then asymptotics

$$\mathsf{N}^+(\tau) = \mathsf{N}^{\mathsf{W}+}(\tau) + \kappa'_1\tau^{\frac{d-1}{m}} + o(\tau^{\frac{d-1}{m}}) \tag{12.5.34}$$

hold with the standard coefficient κ'_1.

(iii) Further, for $(d-1)l = d'm$ asymptotics

$$\mathsf{N}^+(\tau) = \mathsf{N}^{\mathsf{W}+}(\tau) + O(\tau^{\frac{d-1}{m}}\log\tau) \tag{12.5.35}$$

holds.

(iv) Furthermore, for $(d-1)l < d'm \le dl$ asymptotics

$$\mathsf{N}^+(\tau) = \mathsf{N}^{\mathsf{W}+}(\tau) + O(\tau^{\frac{d'}{l}}) \tag{12.5.36}$$

holds where for $d'm = dl$ $\mathsf{N}^{\mathsf{W}+}(\tau)$ is given by cut-off Weyl formula (12.5.33) with integration only in the domain $\{|\xi'| \ge \tau^{\frac{1}{l}}, \sigma(x,\xi) \ge |\xi'|^{-\frac{1}{2}}\}$ and $\mathsf{N}^{\mathsf{W}+}(\tau) \asymp \tau^{\frac{d'}{l}}\log\tau$.

(v) Finally, for $dl < d'm$ estimate

$$\mathsf{N}^+(\tau) = O(\tau^{\frac{d'}{l}}) \tag{12.5.37}$$

holds. Moreover, the same estimate holds if $a_0(x,\xi)$ is negative definite.

(vi) Moreover, in all these asymptotics one can replace $\mathsf{N}^{\mathsf{W}+}(\tau)$ by

(12.5.38) $\mathsf{N}_1^{\mathsf{W}+}(\tau) =$

$$\begin{cases} \kappa_0\tau^{\frac{1}{m}d} + \kappa_{\frac{1}{2}}\tau^{\frac{1}{m}(d-\frac{1}{2})} + \kappa''_1\tau^{\frac{1}{m}(d-1)} & \textit{for } (d-1)l > d'm,\\ \kappa_0\tau^{\frac{1}{m}d} + \kappa_{\frac{1}{2}}\tau^{\frac{1}{m}(d-\frac{1}{2})} & \textit{for } (d-\frac{1}{2})l > d'm \ge (d-1)m,\\ \kappa_0\tau^{\frac{1}{m}d} + \kappa'_{\frac{1}{2}}\tau^{\frac{1}{m}(d-\frac{1}{2})}\log\tau & \textit{for } (d-\frac{1}{2})l = d'm,\\ \kappa_0\tau^{\frac{1}{m}d} & \textit{for } (d-\frac{1}{2})l < d'm < (d-1)m < dl,\\ \kappa'_0\tau^{\frac{1}{m}d}\log\tau & \textit{for } d'm < (d-1)m = dl \end{cases}$$

with positive constants κ_0 *and* κ'_0.

Moreover, $\kappa_{\frac{1}{2}} = \kappa'_{\frac{1}{2}} = 0$ *provided* $a_{\frac{1}{2}} \equiv 0$.

We leave to the reader the following easy problem

Problem 12.5.11. (i) Similarly prove that for Riesz means the remainder estimate is

$$R^{\mathsf{W}}_{(\vartheta)}(\tau) = \begin{cases} O\bigl(\epsilon'\tau^{\frac{1}{m}(d-1-\vartheta)+\vartheta}\bigr) & \text{for } (d-1-\vartheta)l > d'm,\\ O\bigl(\tau^{\frac{1}{m}(d-1-\vartheta)+\vartheta}\log\tau\bigr) & \text{for } (d-1-\vartheta)l = d'm,\\ O\bigl(\tau^{\frac{1}{l}d'+\vartheta}\bigr) & \text{for } (d-1-\vartheta)l < d'm \end{cases} \tag{12.5.39}$$

while the principal part contains terms $\kappa_j\tau^{\frac{1}{m}(d-j)+\vartheta}$ with $j \in \frac{1}{2}\mathbb{Z}$; for $(d-j)l = d'm$ one should replace this term by $\kappa'_j\tau^{\frac{1}{m}(d-j)+\vartheta}\log\tau$ while the remainder estimate is $O\bigl(\tau^{\frac{1}{m}(d-j)+\vartheta}\bigr)$ (if $j < 1+\vartheta$).

(ii) Moreover, let $k \notin \mathbb{Z}$ and $a_j \equiv 0$ for all $j \notin \mathbb{Z} : j \le k$. Prove that then $\kappa_k = 0$.

So, if a_0 is not negatively definite in what follows we need only to treat the case $(d-1)l \le d'm$ (or $(d-1-\vartheta)l \le d'm$ if we are interested in Riesz means).

12.5.4 Sharp Asymptotics

Let us treat x'' as an additional variable; however, we do not now introduce the special Hilbert space $\mathbb{H}$. First of all, let us note that standard ellipticity arguments yield the following

Proposition 12.5.12. *Let $q' \in \mathscr{C}_0^{K'}(\mathbb{R}^{2d'})$ and $q'' \in \mathscr{C}_0^{K'}(\mathbb{R}^{2d''})$ be fixed functions vanishing in neighborhoods of $\{\xi' = 0\}$ and $\{x'' = \xi'' = 0\}$ respectively. Moreover, let $\sigma > 0$ be a small enough constant and let $\bar{r} \geq c_\sigma \tau^{\frac{1}{m}}$ where σ and c_σ depend on q', q'' and c_σ is large enough.*

Then for the operators $Q' = q'(x', \bar{r}^{-1}D')$ and $Q'' = q''(\sigma^{-1}x'', \sigma^{-1}\bar{r}^{-1}D'')$ estimate

$$\text{(12.5.40)}\quad \bigl|F_{t\to\lambda}\chi_T(t)Q'_x\psi_1 Q''_x U\bigr| \leq C\hbar^s \qquad \forall x, y \in X \quad \forall\lambda : |\lambda - \tau| \leq \epsilon\tau \quad \forall T \in [\bar{r}^\delta\tau^{-1}, T_0\bar{r}\tau^{-1}]$$

holds with an arbitrarily small exponent $\delta > 0$, an arbitrarily large exponent s and with small enough constants $T_0 > 0$ and $\epsilon > 0$; here and below $U(x, y, t)$ is the Schwartz kernel of the operator $\exp(itA)$ and $\hbar = \bar{r}^{-1}$.

This proposition permits us to include x''-mollifying operator $q(x'', \bar{r}^{-1}D'')$ with compactly supported q in our expressions whenever necessary.

Let us note that if $\phi = \phi(x', \hbar D')$ then in the ϵ-neighborhood of the point $(\bar{x}', \bar{\xi}')$ with $|\bar{\xi}'| \asymp \bar{r}$ operator $\bar{r}[A, \phi]$ belongs to $\boldsymbol{\Psi}^{m,\mu}(X, \Lambda, \mathbb{K})$ uniformly in $(\bar{x}', \bar{\xi}')$.

Proposition 12.5.13. *Let $q'_1, q'_2 \in \mathscr{C}_0^{K'}(\mathbb{R}^{2d'})$ be fixed functions and let q' vanish in a neighborhood of $\{\xi' = 0\}$, $q'_2 = 1$ in a neighborhood of $\operatorname{supp}(q'_1)$. Then for the operators $Q'_j = q'_j(x', \bar{r}^{-1}D')$ estimate*

$$\text{(12.5.41)}\quad \bigl|F_{t\to\lambda}\psi_1\chi_T(t)Q'_{1x}(I - Q'_{2x})U\bigr| \leq C\hbar^s \qquad \forall x, y \in X \quad \forall\lambda : |\lambda - \tau| \leq \epsilon\tau \quad \forall T \in [\bar{r}^\delta\tau^{-1}, T_0\bar{r}\tau^{-1}]$$

holds with an arbitrarily small exponent $\delta > 0$, an arbitrarily large exponent s and with small enough constants $T_0 > 0$ and $\epsilon > 0$.

Proof. Proof uses the standard energy estimate technique of Chapter 2 (see proof of Theorem 2.1.2 and takes into account estimate (12.5.20) and Proposition 12.5.12.

Easy details are left to the reader. □

This proposition shows us that singularities with respect to (x', ξ') of U propagate with the speed not exceeding $(\bar{r}^{-1}\tau, \tau)$ along (x', ξ') (without rescaling on ξ') while in the zone which is semiclassical with respect to all the variables singularities propagate along (x'', ξ'') with the speed not exceeding $\bar{\sigma}^{-1}(\bar{r}^{-1}\tau, \tau)$.

Let us consider the $(\epsilon, \epsilon\bar{r})$-neighborhood of $(\bar{x}', \bar{\xi}')$ with $|\bar{\xi}'| = \bar{r}$. Let $a(x', \xi') = A(x', x'', \xi', D'')$ be an operator acting on x'' only. We claim that

Proposition 12.5.14. *For the operator $a(x', \xi)$ the microhyperbolicity estimate*

$$\operatorname{Re}\langle((\bar{\xi}', \partial_{\xi'}\rangle a)v, v\rangle \geq \epsilon_0|\tau| \cdot \|v\|^2 - c_2|\tau|^{-1}\|(\tau - a)v\|^2 - C\bar{r}^{-s}\|w\|^2 \tag{12.5.42}$$

holds for $v = \psi_1 w$ where norms and inner products are here considered only on x'', (x', ξ') belongs to this neighborhood, and $\tau > 0$ is such that $r_*(\tau) \leq \bar{r} \leq r^*(\tau)$; here $\epsilon_0 > 0$ is a small enough constant and c_2 and C are large enough constants; none of the constants depends on $(\bar{x}', \bar{\xi}')$ and τ; s is an arbitrarily large exponent.

Proof. It is sufficient to prove this estimate for $\xi' = \bar{\xi}'$. Let us apply Taylor's formula to A; we obtain that

$$a(x', \xi'; x'', D'') \equiv \sum_{|\alpha|+|\beta|+n=\mu} a^{(\alpha)}_{\frac{n}{2}(\beta)}(x', \xi')\frac{1}{\alpha!\beta!}x''^{\beta}D''^{\alpha} \qquad \mod \Psi^{m,\mu+1}$$

where here and below we take $x'' = \xi'' = 0$ in $a^{(\alpha)}_{\frac{n}{2}(\beta)}$. Then for $\xi' = \bar{\xi}'$ we obtain the equality

$$\langle\xi', \partial_{\xi'}\rangle a(x', \xi'; x'', D'') \equiv \sum_{|\alpha|+|\beta|+n=\mu} (m - n - |\alpha|)a^{(\alpha)}_{\frac{n}{2}(\beta)}(x', \xi')\frac{1}{\alpha!\beta!}x''^{\beta}D''\alpha \equiv la - i[L, a] \qquad \mod \Psi^{m,\mu+1}$$

with $L = \frac{1}{2}\langle x'', D''\rangle - \frac{1}{4}id''$. Then estimates (12.5.40)–(12.5.41) together with the assumption that $l > 0$ yield estimate (12.5.42). □

Again applying the analysis of Chapter 2 (see proof of Proposition 2.1.16 and Corollary 2.1.17) we conclude that

Proposition 12.5.15. *In the framework of Proposition 12.5.12 estimate*

$$|F_{t\to\lambda}\Gamma'_x\psi_1\chi_T(t)Q'_{1x}U| \le C\bar{r}^{-s} \tag{12.5.43}$$

holds for x', y', τ and T indicated in this proposition provided χ vanishes in a neighborhood of 0; Γ' means that we set $y'=x'$ and then integrate over x'.

Then we want to apply the successive approximation method but we lack auxiliary spaces $\mathbb{H}$ and $\mathbb{H}_j$. In order to overcome this difficulty we treat two cases:

(a) $\bar{r}\le r^*(\tau)\tau^{-\delta}$,

(b) $\bar{r}\ge r_*(\tau)\tau^{\delta}$

with $\delta>0$; these cases overlap and together cover the case $\bar{r}\in[r_*(\tau),r^*(\tau)]$.

In the former Case (a) we can apply our standard technique with an unperturbed operator $\bar{A}=a(y,D_x)$ and we obtain estimate

$$|F_{t\to\lambda}\bar{\chi}_T(t)\operatorname{tr}(\Gamma Q_xU)-\sum_{0\le j\le J-1}\varkappa'_{\frac{j}{2}}(\tau)|\le C\tau^{-s} \tag{12.5.44}$$
$$\forall\lambda:|\lambda-\tau|\le\epsilon\tau$$

with arbitrary $T\in[\tau^{-1}h^{-\delta'},T_0\tau^{-1}h^{-1}]$ where $T_0>0$ is a small enough constant, $\delta'>0$ and s are arbitrary and $h=\bar{r}^{-1}\bar{\sigma}^{-2}$; in contrast to the previous $\bar{\sigma}$ is not a small constant but a parameter such that $h\le 1$.

Here $J=J(d,m,\mu,s,\delta,\delta')$ is sufficiently large and $\varkappa'_{\frac{j}{2}}$ are Weyl terms (rather than coefficients) depending on Q and satisfying estimate

$$|\varkappa'_{\frac{j}{2}}|\le C(\bar{r}\bar{\sigma}^2)^{d-\frac{j}{2}}\tau^{-1} \tag{12.5.45}$$

provided the symbol of Q is supported in the $\epsilon(1,\bar{\sigma}^2\bar{r},\bar{\sigma},\bar{\sigma}\bar{r})$-neighborhood of $(\bar{x}',\bar{\xi}',\bar{x}'',\bar{\xi}'')$, $|\bar{\xi}'|\asymp\bar{r}$; here and below we refer to the standard 1-symbol of Q instead of the $\hbar$-symbol.

These estimates (12.5.44) and (12.5.45) are due to the standard microhyperbolicity of a_0 in the direction $\langle\xi,\partial_\xi\rangle$ after the coordinate transform $x'_{\mathsf{new}}=x'_{\mathsf{old}}$, $\xi'_{\mathsf{new}}=\xi'_{\mathsf{old}}\bar{\sigma}^{-2}$, $x''_{\mathsf{new}}=x''_{\mathsf{old}}\bar{\sigma}^{-1}$, $\xi''_{\mathsf{new}}=\xi''_{\mathsf{old}}\bar{\sigma}^{-1}$.

It should be noted that in Case (a) in virtue of standard ellipticity arguments we automatically obtain that only points with $h\le\tau^{-\delta''}$ with sufficiently small exponent $\delta''>0$ should be treated. Then for operator Q with symbol supported in the $\epsilon(1,\bar{r},\bar{\sigma},\bar{\sigma}\bar{r})$-neighborhood of $(\bar{x}',\bar{\xi}',\bar{x}'',\bar{\xi}'')$ we obtain the same result with another estimate of these terms:

$$|\varkappa'_{\frac{j}{2}}|\le C\bar{r}^{d-\frac{j}{2}}\bar{\sigma}^{2(d''-\frac{j}{2})}\tau^{-1}. \tag{12.5.46}$$

Moreover, (12.5.40) yields that the same estimates hold with Q replaced by $\psi_1 Q'$ where $Q'=q'(x',\bar{r}^{-1}D')$ and q' is supported in the ϵ-neighborhood of $(\bar{x}',\bar{\xi}'\bar{r}^{-1})$. Furthermore, estimate (12.5.43) then yields that estimate (12.5.44) holds for arbitrary $T\in[\tau^{-1}h^{-\delta'},T_0\tau^{-1}\hbar^{-1}]$.

Applying the standard Tauberian technique we obtain that

$$\begin{aligned}\Bigl|\Bigl(\int \operatorname{tr}(Q'_x\psi_1 e)(x,x,\tau)\,dx-\sum_{0\le j\le J-1}\varkappa_{\frac{j}{2}}(\tau)\Bigr)_{\tau=\tau'}^{\tau=\tau}\Bigr|\le\\ C\bar{r}^{d-1}\bar{\sigma}^{2d''}\asymp C\bar{r}^{d-1-2d''\frac{m}{\mu}}\tau^{\frac{2d''}{\mu}}\end{aligned} \tag{12.5.47}$$

with $\tau'\in[\epsilon_0\tau,\tau]$, $\tau>0$ where $\varkappa_{\frac{j}{2}}(\tau)$ are standard Weyl terms equal to the integrals of $\varkappa'_{j/2}(\tau)$ with respect to τ and therefore

$$|\varkappa_{\frac{j}{2}}(\tau)|\le C\bar{r}^{d-\frac{j}{2}}\bar{\sigma}^{2(d''-\frac{j}{2})}. \tag{12.5.48}$$

Let us discuss the terms $\varkappa_{\frac{j}{2}}(\tau)$. Recall that

$$\sum_{0\le j\le J-1}\varkappa'_{\frac{j}{2}}\sim(2\pi)^{-d-1}i^{-1}\operatorname{Res}_{\mathbb{R}}\int\bigl((\tau-A)^{-1}Q'\psi_1\bigr)_{\mathsf{w}}(x,\xi)\,dxd\xi \tag{12.5.49}$$

where the subscript "$_{\mathsf{w}}$" means that we take the Weyl symbol of the corresponding operator and $(\tau-A)^{-1}$ is a parametrix of $(\tau-A)$ in the framework of the formal calculus of pseudodifferential operators; we cut this symbol after we have taken a large enough number of $(m,1)$-quasihomogeneous terms (on (τ,ξ)) of the symbol of $(\tau-A)^{-1}$ and derivatives of these terms on (x,ξ) and therefore derivatives of the symbol of $Q'\psi_1$ on (ξ,x).

Let us note that any differentiation of ψ_1 gives us a term which does not exceed the right-hand expression in (12.5.47); the same is true for any differentiation with respect to x' and therefore with respect to ξ' (and *vice versa*) made in the calculation of $(\tau-A)^{-1}$ and of its product with Q'.

Therefore estimate (12.5.47) remains true if, in the definition of terms coefficients, we replace the right-hand expression of (12.5.49) by

$$(2\pi)^{-d-1}i^{-1}\operatorname{Res}_{\mathbb{R}}\int \bigl((\tau-a)^{-1}\bigr)_{\mathrm{w}}(x,\xi)Q'(x',\xi')\psi_1(x,\xi)\,dxd\xi \tag{12.5.50}$$

where $(\tau-a)^{-1}$ is the parametrix of $(\tau-a)$ in the framework of the formal calculus of pseudodifferential operators with respect to x' and the subscript "w" means the Weyl symbol of the corresponding operator; here $a(x',\xi')=A(x',\xi';x'',D'')$ but later we will change this definition.

Recall that $\bigl((\tau-a)^{-1}\bigr)_{\mathrm{w}}$ is a distribution. However, if we use spherical coordinates $\xi=(r,\varphi)$ with $\varphi\in\mathbb{S}^{d-1}$ and integrate on r (multiplying by a $\mathscr{C}_0^\infty$ function beforehand) we obtain a regular function of (τ,x,φ) provided $(x'',\varphi'')\ne 0$.

Let us introduce the spherical coordinate system $(\zeta,z)\in\mathbb{R}^+\times\mathbb{S}^{2d''-1}$ in $\mathbb{R}^{2d''}\ni(x'',\varphi'')$; then one can rewrite the term of $\bigl((\tau-a)^{-1}\bigr)_{\mathrm{w}}$ which is $(m,1)$-quasihomogeneous of order $(-m-\frac{j}{2})$ in the form

$$\sum_{0\le k\le J-1}\tau^{-1}b_{jk}(\tau,r,z)\zeta^{k-j}+b'_j(\tau,r,z,\zeta)$$

where b'_j are regular functions. Then a similar representation holds for $\operatorname{Res}_{\mathbb{R}}\bigl((\tau-a)^{-1}\bigr)_{\mathrm{w}}$. Let us integrate it with respect to the measure $dx'dzd\varphi'\times\zeta^{2d''-1}d\zeta\times r^{d-1}dr$. Then after integration over τ from τ' to τ and summation over a partition of unity in the zone $\{r_*(\tau)\le r\le\eta\}$ we obtain the following

Proposition 12.5.16. *Let conditions* (12.5.1)–(12.5.3), (12.5.6)–(12.5.9) *and* (12.5.11) *be fulfilled and let*

$$Q'=q_2(x',\varepsilon\tau^{-\frac{1}{m}}D')-q_1(x',\eta^{-1}D') \tag{12.5.51}$$

with fixed smooth functions q_j supported in $\{|\xi'|\le c\}$ and equal to 1 for $|\xi'|\le\epsilon_0$. Then estimate

$$\Bigl|\Bigl(\int \operatorname{tr}(Q'_x\psi_1 e)(x,x,\tau)dx-\mathcal{N}(\tau,\eta)\Bigr)\Bigr|_{\tau=\tau'}^{\tau=\tau}\Bigr|\le CR^*_1(\tau) \tag{12.5.52}$$

holds with

$$\mathcal{N}(\tau,\eta)=\mathcal{N}_0(\tau)+\mathcal{N}_1(\tau,\eta), \tag{12.5.53}$$

$$(12.5.54)\quad \mathcal{N}_0(\tau)=\sum_{j=0,1,2}\kappa_{\frac{j}{2}}\tau^{\frac{1}{m}(d-\frac{j}{2})}+\sum_{j,k=0,1,2:(d-\frac{j}{2})l=(d'-\frac{k}{2})m}\kappa'_{\frac{j}{2}}\tau^{\frac{1}{m}(d-\frac{j}{2})}\log\tau,$$

$$(12.5.55)\quad \mathcal{N}_1(\tau,\eta)=\sum_{j,k\le J:(d-\frac{j}{2})l\ne(d'-\frac{k}{2})m}\kappa_{jk}\tau^{\frac{1}{\mu}(2d''+k-j)}\eta^{\frac{2}{\mu}((d'-\frac{k}{2})m-(d-\frac{j}{2})l)}-\\ \sum_{j,k=0,1,2:(d-\frac{j}{2})l=(d'-\frac{k}{2})m}\kappa'_{\frac{j}{2}}m\tau^{\frac{1}{m}(d-\frac{j}{2})}\log\eta,$$

$$(12.5.56)\qquad R_1^*(\tau)=\begin{cases}\varepsilon'\tau^{\frac{1}{m}(d-1)} & \textit{for}\ \ (d-1)l>(d'-1)m\\ C\tau^{\frac{1}{m}(d-1)}\log\tau & \textit{for}\ \ (d-1)l=(d'-1)m\\ C\tau^{\frac{1}{l}(d'-1)-\delta''} & \textit{for}\ \ (d-1)l<(d'-1)m\end{cases}$$

where the coefficients κ_j, κ'_j *and* κ_{jk} *do not depend on* τ *and* η, $J=J(d,m,\delta)$ *is large enough,* $\delta''>0$ *and* $\varepsilon'\to+0$ *as* $\varepsilon\to+0$.

Let us now treat Case (b) $\bar{r}\ge r_*(\tau)\tau^\delta$. In this case in order to obtain the unperturbed operator $\bar{A}$ we freeze $x'=y'$ in the operator $A_{(x',D')}(x'',D'')$ where $A_{(x',\xi')}(x'',D'')$ is the microlocalization (12.5.4) of $A(x,D)$ at the point (x',ξ') and we take Weyl quantization on x'.

We now treat this operator $\bar{A}$, and A itself, as pseudodifferential operators on x' with operator-valued symbols in $\mathbb{H}=\mathscr{L}^2(\mathbb{R}^{d''},\mathbb{K})$. Then all the constructions of the propagation of singularities of Chapter 2 are applicable including the proof of microhyperbolicity (we explained this proof above) as well as the successive approximation method.

Moreover, no sandwich procedure is necessary now because we apply a pseudodifferential operator ψ_1 mollifying on x''. So, we obtain that if $Q'=q'(x',\bar{r}^{-1}D')$ with q' supported in the ϵ-neighborhood of $(\bar{x}',\bar{\xi}'\bar{r}^{-1})$ with $\bar{r}\asymp|\bar{\xi}'|$ and $r^*(\tau)\ge\bar{r}\ge r_*(\tau)\tau^\delta$ then for $T\in[\tau^{-1}\bar{r}^{\delta'},T_0\tau^{-1}\bar{r}]$ with a sufficiently small constant $T_0>0$ asymptotics

$$(12.5.57)\quad \bigl|F_{t\to\lambda}\chi_T(t)\Gamma\psi_1Q'_x(U-\sum_{0\le n\le M}U_n)\bigr|\le C\tau^{-s}\qquad\forall\lambda:|\lambda-\tau|\le\epsilon\tau$$

holds where U_n are the terms of the successive approximation method, the number $M=M(d,m,\mu,\delta,s)$ is large enough and

$$(12.5.58)\qquad \bigl|F_{t\to\lambda}\chi_T(t)\Gamma\psi_1Q'_xU_n\bigr|\le C\bar{r}^{d-2d''m\mu^{-1}-\nu(n)}\tau^{2d''\mu^{-1}-1}$$

for the same values of T,τ with $\nu(n)\to+\infty$ as $n\to+\infty$. Therefore

$$|F_{t\to\lambda}\chi_T(t)\Gamma\psi_1 Q'_x U|\le C\bar{r}^{d-2d''m\mu^{-1}}\tau^{2d''\mu^{-1}-1} \tag{12.5.59}$$

for the same values of T,τ. We leave all the easy details to the reader.

Then applying the standard Tauberian technique we obtain estimate

$$\begin{aligned}|\int \operatorname{tr}(Q'_x\psi_1 e)(x,x,\tau',\tau)\,dx-\\ \sum_{0\le n\le M}\int_{\tau'}^{\tau}(F_{t\to\lambda}\bar{\chi}_T(t)\Gamma\psi_1 Q'_x U_{n(q)})\,d\lambda|\le C\bar{r}^{d-1-2d''m\mu^{-1}}\tau^{2d'\mu^{-1}}\end{aligned} \tag{12.5.60}$$

for $\tau'\in[\epsilon_0\tau,\tau]$ where the conditions on $\bar{r}$ and T were given above, the subscript "(q)" has no meaning at this time and $\bar{\chi}\in\mathscr{C}_0^K([-1,1])$ is equal to 1 on $[-\frac{1}{2},\frac{1}{2}]$.

Let us note that if we sum the right-hand expression of estimate (12.5.60) on a partition of unity on the interval for the $\bar{r}$ indicated above we obtain that the remainder does not exceed $CR_2^*(\tau)$ with

$$R_2^*(\tau)=\begin{cases}\tau^{\frac{1}{m}(d-1)-\delta''} & \text{for } (d-1)l>(d'-1)m,\\ \tau^{\frac{1}{m}(d-1)-\delta''}\log\tau & \text{for } (d-1)l=(d'-1)m,\\ \varepsilon'\tau^{\frac{1}{l}(d'-1)} & \text{for } (d-1)l<(d'-1)m\end{cases} \tag{12.5.61}$$

where $\varepsilon'=1$ at this moment.

Further, let $(d-1)l<(d'-1)m$ and let us sum only over the partition in the zone $\{\varepsilon r^*(\tau)\ge\bar{r}\ge r_*\tau^\delta\}$ then $\varepsilon'=\varepsilon'(\varepsilon)\to+0$ as $\varepsilon\to+0$ and C does not depend on ε.

On the other hand, for $\bar{r}\in[\varepsilon r^*,r^*(\tau)]$ operator $\mathbf{a}(x',\xi')=A_{(x',\xi')}(x'',D'')$ has no more than $L(\varepsilon)$ eigenvalues in $[-2\tau,2\tau]$ and operator $A(x',\xi';x'',D'')$ is close to this operator in the natural sense.

Therefore for $\bar{r}$ from this interval and under the standard non-periodicity condition for the generalized Hamiltonian flow generated by operator-valued symbol $\mathbf{a}(x',\xi')$ on the energy level $\tau=1$ (see (12.2.51)) one can replace the right-hand expression in estimate (12.5.60) by $o(\tau^{\frac{1}{l}(d'-1)})$; this and the arbitrary choice of $\varepsilon'>0$ yield that

(12.5.62) For $(d-1)l<(d'-1)m$ under the standard non-periodicity condition for $\mathbf{a}(x',\xi')$ on the energy level $\tau=1$ the remainder estimate is $o(\tau^{\frac{1}{l}(d'-1)})$ even for the sum over the partition of unity in the zone $\{r^*(\tau)\ge\bar{r}\ge r_*\tau^\delta\}$.

12.5.5 Calculations

Let us consider the terms $F_{t\to\lambda}\bar{\chi}_T(t)\Gamma\psi_1 Q'_x U_{n(q)}$ with $\lambda\in[\epsilon_0\tau,\tau]$ in (12.5.60). Then the standard arguments of Section 4.4 associated with the microhyperbolicity yield that one can replace $\bar{\chi}_T$ with 1 in these terms; the error caused by this operation is $O(\tau^{-s})$. Moreover, explicit calculation yields that one can replace ψ_1 by 1 in these terms.

On the other hand, one can easily see that for $Q'=q(x',r^*(\tau)^{-1}D')$ with $q'(x',\xi')$ supported in $\{|\xi'|\ge\frac{1}{2}\}$ and equal to 1 for $|\xi'|\ge 1$ we obtain from Proposition 12.5.9(ii) that the first term in the left-hand expression of (12.5.60) is $O(\tau^{-s})$ and all the other terms modified above (i.e., with $\psi_1=I,\bar{\chi}=1$) simply vanish (we can write them explicitly and note that the operator $a(x',\xi')$ has no eigenvalue in $[-2\tau,2\tau]$. So for this operator Q' the modified formula (12.5.60) also holds. Therefore we obtain the following

Proposition 12.5.17. *(i) Consider* $Q'=q'(x',\eta^{-1}D')$ *with* q' *supported in* $\{|\xi'|\ge\frac{1}{2}\}$ *and equal to* 1 *for* $|\xi'|\ge 1$.

Then for $\eta\in[r_*(\tau)\tau^\delta, r^*(\tau)]$, $\tau'\in[\epsilon_0\tau,\tau]$ *asymptotics*

$$\int \operatorname{tr}(Q'_x\psi_1 e)(x,x,\tau',\tau)dx = \sum_{0\le n\le M}\int_{\tau'}^{\tau}\bigl(F_{t\to\lambda}\Gamma Q'_x U_{n(q)}\bigr)d\lambda + O\bigl(R_2^*(\tau)\bigr) \tag{12.5.63}$$

holds.

(ii) Moreover, let $(d-1)l<(d'-1)m$ *and let the standard non-periodicity condition* (12.2.51) *for the generalized Hamiltonian flow generated by operator-valued symbol* $a(x',\xi)$ *on the energy level* $\tau=1$ *be fulfilled.*

Then asymptotics (12.5.61) *holds with* $O\bigl(R_2^*(\tau)\bigr)$ *replaced by* $o\bigl(R_2^*(\tau)\bigr)$.

This proposition is not our final goal. Let us consider all the terms in the expansion in the right-hand expression of (12.5.61). Let us first note that any differentiation on x' (which is accompanied by differentiation on ξ' and *vice versa*) decreases the magnitude of the term in (12.5.60) by an additional factor $\bar{r}^{-1}$ and therefore

(12.5.64) Asymptotics (12.5.60) remain true even if we skip all the terms containing differentiation on (x',ξ').

Moreover,

(12.5.65) The sharper asymptotics with remainder estimate $o\bigl(R_2^*(\tau)\bigr)$ remains true if we preserve terms with only one differentiation on x' (and one on ξ'). Furthermore, even in this case one can skip all terms with differentiation of the symbol of Q' provided $\eta = o\bigl(r^*(\tau)\bigr)$.

In fact, such a term is supported in the zone $\bar{r} \asymp \eta$ and its order coincides with the order of the right-hand expression of (12.5.60), i.e., it is "$o\bigl(R_2^*(\tau)\bigr)$" provided $(d-1)l < (d'-1)m$ and $\eta = o\bigl(r^*(\tau)\bigr)$.

Therefore one should consider only terms of the form

$$U_{n(q)} = F_{\lambda\to t}^{-1} \operatorname{Res}_{\mathbb{R}} F_n(x',\xi',\lambda)Q(x',\xi') \tag{12.5.66}$$

where as $\lambda \in \mathbb{C}\setminus\mathbb{R}$ $F_n(x',\xi',\lambda)$ is a product of $(k+1)$ copies of the operator $\bigl(\lambda - A_{(x',\xi')}(x'',D'')\bigr)^{-1}$ separated by k operators $B_j(x',\xi',x'',D'')$ of the form

$$B_j = \sum_{\alpha,\beta:|\alpha|+|\beta|\le\mu+q} b_{j\alpha\beta}(x',\xi')x''^{\alpha}D''^{\beta}$$

and $b_{j\alpha\beta}$ are positively homogeneous of degrees $m + \frac{1}{2}(|\alpha| - |\beta| - \mu - q)$ on ξ' and $k \ge n$, $q = 1, 2, \dots$.

Therefore, assigning the index q to the operator B_j we also assign the index q to F_n as the sum of all the indices of factors of this type in the product. Then (12.5.58) holds with $\nu(n) = \delta'' q$.

Further, let us introduce spherical coordinates $\xi' = (r,\varphi)$. Then after the coordinate transform $x''_{\mathsf{new}} = x''r^{\frac{1}{2}}$, $D''_{\mathsf{new}} = D''r^{-\frac{1}{2}}$ we rewrite the corresponding term in the decomposition of $F_{t\to\lambda}U_{n(q)}$ in the form

$$\int_0^\infty \operatorname{Res}_{\mathbb{R}} F_n(\lambda r^{-l})Q(r)r^{d'-1-l-\frac{q}{2}}\,dr \tag{12.5.67}$$

where here and below F_n is now calculated for operator $A_{(x',\varphi)}(x'',D'')$ with $\varphi \in \mathbb{S}^{d'-1}$ and we do not indicate the dependence on (x',φ) and integration on these variables. Moreover, one can assume that $Q(r) = \phi(r^{-l}\eta^l)$ where $\phi \in \mathscr{C}_0^K([-1,1])$ is a fixed function equal to 1 in a neighborhood of 0. Then after substituting $\lambda = zr^l$ we rewrite (12.5.67) in the form

$$\frac{1}{l}\lambda^{\frac{1}{l}(d'-\frac{q}{2})-1}\int_0^\infty \operatorname{Res}_{\mathbb{R}} F_n(z)\phi(z\eta^l\lambda^{-1})z^{-\frac{1}{l}(d'-\frac{q}{2})}\,dz. \tag{12.5.68}$$

For $\omega = \lambda\eta^{-l}$ let us consider expression

$$\Phi_{n(q)}(\omega) = \int_0^\infty \operatorname{Res}_{\mathbb{R}} F_n(z)\phi(\frac{z}{\omega}) z^{\frac{1}{l}(\frac{q}{2}-d')}\, dz. \tag{12.5.69}$$

Proposition 12.5.18. *For $\omega \to +\infty$ asymptotics*

$$\operatorname{Tr}\Phi_{n(q)} = \kappa_q + \kappa'_q \log\omega + \sum_{p:\frac{1}{l}(d'-\frac{q}{2})\neq\frac{2}{p}(d''-\frac{p}{2})} \kappa'_{qp}\omega^{\frac{2}{\mu}(d''-\frac{p}{2})-\frac{1}{l}(d'-\frac{q}{2})} + O(\omega^{-s}) \tag{12.5.70}$$

holds with arbitrarily large s. Here we sum over $p\in\mathbb{Z}^+$ such that $p\le J(d,m,\mu,s)$ and the term $\kappa'_q\log\omega$ arises if and only if there exists $p\in\mathbb{Z}^+$ with $\frac{1}{l}(d'-\frac{q}{2}) = \frac{2}{\mu}(d''-\frac{p}{2})$; here we do not indicate the dependence on n.

Proof. Let us differentiate $\Phi_{(q)}(\omega)$:

$$\operatorname{Tr}\Phi'_{(q)} = \omega^{-\frac{1}{l}(d'-\frac{q}{2})}\int \operatorname{Tr}\operatorname{Res}_{\mathbb{R}} F(z)\psi(\frac{z}{\omega})\,\frac{dz}{\omega}$$

where $\psi = z_+^{1-\frac{1}{l}(d'-\frac{q}{2})}\phi'(z)$ obviously belongs to $\mathscr{C}_0^K([\epsilon_0,c_0])$; the prime here means the derivative.

One can easily see that modulo $O(\omega^{-s})$ the integral in the right-hand expression is equal to

$$\operatorname{tr}\bigl(\psi(\omega^{-1}D_t)\bar\chi_T(t)\Gamma' V_k\bigr)|_{t=0} \tag{12.5.71}$$

with arbitrary $T\ge\omega^{-\delta-1}$ and an arbitrarily small exponent $\delta>0$; here $\Gamma' v = \int v(x'',x'',.)dx''$ and one defines the functions V_k from the recurrence relations

$$(D_t - a)V_0 = 0, \qquad V_0|_{t=0} = \delta(x''-y'')I, \tag{12.5.72$_0$}$$

$$(D_t - a)V_k = B_k V_{k-1}, \qquad V_k|_{t=0} = 0 \quad (k\ge 1), \tag{12.5.72$_1$}$$

$a(x',\xi)$ here and below means microlocalization, $(k+1)$ is the number of factors $(\omega-a)^{-1}$ in the definition of F and B_j are the other factors in this definition; further, operators $a=a(x'',D'')$ and $B_j = B_j(x'',D'')$ are calculated at (x',ξ').

However, one can apply the standard Weyl successive approximation method of Chapter 4 on x'' with the semiclassical parameter $\omega^{-\mu^{-1}}$; therefore the reader can easily prove (there are no Tauberian arguments!) that expression (12.5.71) is equal to

$$\sum_{p\le J}\kappa''_{qp}\omega^{\frac{1}{\mu}(2d'-p)-1}$$

modulo $O(\omega^{-s})$. Finally, multiplying by $\omega^{-\frac{1}{\mu}(d'-\frac{q}{2})}$ and integrating we obtain (12.5.70). □

Let us set $\eta\le\tau^{\frac{1}{l}(1-\delta)}$ (in which case $\omega\ge\tau^{\delta}$). Then (12.5.71) yields that modulo $O(\lambda^{-s})$ the trace of (12.5.68) is equal to

$$\big(\kappa_q+\kappa'_q\log\lambda-\kappa'_q l\log\eta\big)\lambda^{\frac{1}{l}(d'-\frac{q}{2})-1}+\\ \sum_{p:\frac{1}{l}(d'-\frac{q}{2})\ne\frac{2}{\mu}(d''-\frac{p}{2})}\kappa'_{qp}\lambda^{\frac{1}{\mu}(2d''-p)-1}\eta^{-(2d''-p)\frac{l}{\mu}+(d'-\frac{q}{2})}. \tag{12.5.73}$$

After integration over λ from τ' to τ we obtain terms of expansion (12.5.60). Therefore we have proven the following

Proposition 12.5.19. *(i) For $\eta\in[\tau^{\frac{1}{m}+\delta},\tau^{\frac{1}{l}-\delta}]$ asymptotics*

$$\int \operatorname{tr}(Q'_x\psi_1 e)(x,x,\tau',\tau)\,dx=\bar{\mathcal{N}}(\tau,\eta)\big|_{\tau=\tau'}^{\tau=\tau}+O\big(R^*_2(\tau)\big) \tag{12.5.74}$$

holds with

$$\bar{\mathcal{N}}(\tau,\eta)=\bar{\mathcal{N}}_0(\tau)+\bar{\mathcal{N}}_1(\tau,\eta), \tag{12.5.75}$$

$$\bar{\mathcal{N}}_0(\tau)=\sum_q\bar\kappa_q\tau^{(d'-\frac{q}{2})}+\bar\kappa_0\log\tau, \tag{12.5.76}$$

and

$$\bar{\mathcal{N}}_1(\tau,\eta)=\bar\kappa'_0\big(\log^2\tau-2l\log\tau\log\eta\big)+\\ \sum_{q,p:\frac{1}{l}(d'-\frac{q}{2})=\frac{1}{m}(2d''-p)}\bar\kappa''_q\tau^{\frac{1}{l}(d'-\frac{q}{2})}(\log\tau-l\log\eta)+\sum_{q\ne 2d'}\bar\kappa'''_q\eta^{(d'-\frac{q}{2})}\log\tau+\\ \sum_{q,p:\frac{1}{l}(d'-\frac{q}{2})\ne\frac{1}{\mu}(2d''-p)}\bar\kappa'_{qp}\tau^{\frac{1}{\mu}(2d''-p)}\eta^{(d'-\frac{q}{2})-(2d''-p)\frac{l}{\mu}} \tag{12.5.77}$$

and $R_2^*(\tau)$ *given by* (12.5.61)*. Here all the coefficients* $\bar{\kappa}^*_{**}$ *do not depend on* τ, τ', η *and we sum over* $p, q \le J = J(d, m, \mu, \delta)$.

(ii) Moreover, for $(d-1)l < (d'-1)m$ *and the standard non-periodicity condition (see* (12.2.51)*) on the Hamiltonian flow for operator-valued symbol* $\mathsf{a}(x', \xi')$ *on the energy level* $\tau = 1$ *the sharper asymptotics*

$$\int \operatorname{tr}(Q'_x \psi_1 e)(x, x, \tau', \tau)\, dx = \bar{\mathcal{N}}(\tau, \eta)\big|_{\tau=\tau'}^{\tau=\tau} + o\bigl(\tau^{\frac{1}{l}(d'-1)}\bigr) \tag{12.5.78}$$

holds.

Let us combine propositions 12.5.16 and 12.5.19 and note that the final answer *should not* depend on the choice of η (lying in the given interval). Moreover, we know the Weyl asymptotics for $\mathsf{N}_0^+(\tau)$ in (12.5.14). Then we obtain the following statement for $\mathsf{N}(\tau', \tau)$ with $\tau' \in [\epsilon_0 \tau, \tau]$ instead of $\mathsf{N}^+(\tau)$ and with $\mathcal{N}(\tau', \tau) = \mathcal{N}^+(\tau)|_{\tau=\tau'}^{\tau=\tau}$ instead of $\mathcal{N}^+(\tau)$:

Theorem 12.5.20. *Let* $A \in \boldsymbol{\Psi}^{m,\mu}(X, \Lambda, \mathbb{K})$ *be a self-adjoint operator satisfying conditions* (12.5.1)–(12.5.9) *and* (12.5.11)*. Then*

(i) Asymptotics

$$\mathsf{N}^+(\tau) = \mathcal{N}^+(\tau) + O\bigl(R^*(\tau)\bigr) \tag{12.5.79}$$

holds with

$$R^*(\tau) = \begin{cases} \tau^{\frac{1}{m}(d-1)} & \text{for } (d-1)l > (d'-1)m \\ \tau^{\frac{1}{m}(d-1)} \log \tau & \text{for } (d-1)l = (d'-1)m \\ \tau^{\frac{1}{l}(d'-1)} & \text{for } (d-1)l > (d'-1)m \end{cases} \tag{12.5.80}$$

where

$$\mathsf{N}^+(\tau) = \sum_{p=0,1,2} \kappa_{\frac{p}{2}} \tau^{\frac{1}{m}(d-\frac{p}{2})} + \sum_{q=0,1,2} \bar{\kappa}_{\frac{q}{2}} \tau^{\frac{1}{l}(d'-\frac{q}{2})} \tag{12.5.81}$$

provided $(d-\frac{p}{2})l \ne (d'-\frac{q}{2})m$ *for all* p, q *indicated above.*

On the other hand, if $(d-\frac{p}{2})l = (d'-\frac{q}{2})m$ *for some* p, q *then one should replace term* $\kappa_{\frac{p}{2}} \tau^{\frac{1}{m}(d-\frac{p}{2})}$ *by* $\kappa_{\frac{p}{2}} \tau^{\frac{1}{l}(d-\frac{p}{2})} \log \tau$.[52]

[52] One can easily see that there is usually at most one logarithmic term; there are two logarithmic terms if $m = 2l$, $d = 2d'$.

(ii) Moreover, let $(d-1)l > (d'-1)m$ *and let the standard non-periodicity condition on the Hamiltonian flow for* $a_0(x,\xi)$ *be fulfilled on the energy level* $\tau = 1$. *Then asymptotics*

$$\mathsf{N}^+(\tau) = \mathcal{N}^+(\tau) + o\bigl(\tau^{\frac{1}{m}(d-1)}\bigr) \tag{12.5.82}$$

holds.

(iii) On the other hand, let $(d-1)l < (d'-1)m$ *and let non-periodicity condition (see* (12.2.51)*) to the Hamiltonian flow for operator-valued symbol* $a(x',\xi')$ *on the energy level* $\tau = 1$ *be fulfilled. Then asymptotics*

$$\mathsf{N}^+(\tau) = \mathcal{N}^+(\tau) + o\bigl(\tau^{\frac{1}{l}(d'-1)}\bigr) \tag{12.5.83}$$

holds.

Finally, let us replace τ and τ' by $2^{-n}\tau$ and $2^{-n-1}\tau$ and sum over $n = 1,\dots,n_0 \le [\log\tau]+1$. Then we obtain this theorem in the final form (i.e., with $\tau' = 0$).

Remark 12.5.21. (i) Moreover, one obtains similar statements for Riesz means with

(12.5.84)

$$R^*_{(\vartheta)}(\tau) \quad = \quad \begin{cases} \tau^{\frac{1}{m}(d-1-\vartheta)+\vartheta} & \text{for } (d-1-\vartheta)l > (d'-1-\vartheta)m \\ \tau^{\frac{1}{m}(d-1-\vartheta)+\vartheta}\log\tau & \text{for } (d-1-\vartheta)l = (d'-1-\vartheta)m \\ \tau^{\frac{1}{l}(d'-1-\vartheta)+\vartheta} & \text{for } (d-1-\vartheta)l > (d'-1-\vartheta)m \end{cases}$$

and with principal part

$$\mathcal{N}^+_{(\vartheta)}(\tau) = \sum_{p\le 2+2\vartheta} \kappa_{\frac{p}{2}}\tau^{\frac{1}{m}(d-\frac{p}{2})+\vartheta} + \sum_{q\le 2+2\vartheta} \bar\kappa_{\frac{q}{2}}\tau^{\frac{1}{l}(d'-\frac{q}{2})+\vartheta} \tag{12.5.85}$$

provided $(d-\frac{p}{2})l \ne (d'-\frac{q}{2})m$ for all p,q indicated above.

On the other hand, if $(d-\frac{p}{2})l = (d'-\frac{q}{2})m$ for some p,q then one should replace the term $\kappa_{\frac{p}{2}}\tau^{\frac{1}{m}(d-\frac{p}{2})}$ by $\kappa_{\frac{p}{2}}\tau^{\frac{1}{m}(d-\frac{p}{2})}\log\tau$ [53]. Coefficients κ_*, $\bar\kappa_*$ depend on $\vartheta \ge 0$.

Surely, some terms are smaller than the remainder estimate and should be removed. One can see that for $\vartheta > \frac{2}{\mu}(ld-md')-1$ we obtain a Weyl remainder estimate (but the asymptotics contain non-Weyl terms).

[53] So as two terms have equal powers of τ one of them has an extra logarithmic factor.

(ii) Moreover, let $p \notin \mathbb{Z}$ and $a_j \equiv 0$ for all $j \notin \mathbb{Z} : j \le k$. Then $\kappa_p = 0$ which is also true in the case $(d-p)l = (d-q)m$ for some $q \in \frac{1}{2}\mathbb{Z}$: so the logarithmic term does not arise.

12.5.6 Operators with Negative Principal Symbols

Let us consider the case when $a_0(x,\xi)$ is a negative definite symbol. However, we assume that $a(x',\xi') = A_{(x',\xi')}(x'',D'')$ is not a negative definite operator (otherwise A is semibounded from above). However in this case $a(x',\xi')$ has only finite number of positive eigenvalues and thus its positive part is essentially a matrix.

Our goal is to prove the following

Theorem 12.5.22. *Let the conditions of Theorem 12.5.20 be fulfilled and let the condition*

(12.5.86) $a_0(x,\xi)$ *is a negative definite symbol on* $T^*X\setminus 0\setminus\Lambda$

also be fulfilled.

Then

(i) Asymptotics

$$\mathsf{N}^+(\tau) = \mathcal{N}^+(\tau) + O\bigl(\tau^{\frac{1}{l}(d'-1)}\bigr) \tag{12.5.87}$$

holds with

$$\mathcal{N}^+(\tau) = \bar{\kappa}_0\tau^{\frac{1}{m}d'} + \bar{\kappa}_{\frac{1}{2}}\tau^{\frac{1}{l}(d'-\frac{1}{2})} + \bar{\kappa}_1\tau^{\frac{1}{l}(d'-1)}. \tag{12.5.88}$$

(ii) Moreover, let the non-periodicity condition (see (12.2.51)*) for Hamiltonian flow for operator-valued symbol* $a(x',\xi')$ *on the energy level* $\tau = 1$ *be fulfilled. Then asymptotics*

$$\mathsf{N}^+(\tau) = \mathcal{N}^+(\tau) + o\bigl(\tau^{\frac{1}{l}(d'-1)}\bigr) \tag{12.5.89}$$

holds.

Proof. First of all, (12.5.57) and the explicit formula for U_n yield that for $\chi \in \mathscr{C}_0^K(\mathbb{R})$ vanishing in a neighborhood of 0 one can replace (12.5.59) by estimate

$$|F_{t\to\lambda}\chi_T(t)\Gamma\psi_1 Q'_x U| \le C\bar{r}^{d-\frac{2}{\mu}d''m}\tau^{\frac{2}{\mu}d''-1-s}T^{-sl} \tag{12.5.90}$$

for $T\in[T_0\tau^{-1}, T_0\tau^{-1}\bar{r}]$, $\lambda\in[\epsilon_0\tau, c\tau]$. Therefore the same estimate holds for

$$|F_{t\to\lambda}(\bar{\chi}_{T'}(t)-\bar{\chi}_T(t))\Gamma\psi_1 Q'_x U|$$

for $T, T'\in[T_0\tau^{-1}, T_0\tau^{-1}\bar{r}]$, $T<T'$ and $\bar{\chi}\in\mathscr{C}_0^K(\mathbb{R})$ equal to 1 in a neighborhood of 0.

On the other hand, for $T\in[T_0, \tau^{-1}h^{-\delta}, T_0\tau^{-1}h^{-1}]$ with $h=\bar{r}^{-\frac{2}{\mu}l}\tau^{\frac{2}{\mu}}$, the standard Weyl formula (which is derived by setting $\bar{A}=a_0(y,D)$) holds modulo terms which do not exceed the right-hand expression in (12.5.90).

Condition (12.5.74) also yields that all the coefficients in this formula vanish. Taking $T=T_0\tau^{-1}\bar{h}'^{-1}$ we replace (12.5.59) by the estimate

$$|F_{t\to\lambda}\bar{\chi}_T(t)\Gamma\psi_1 Q'_x U| \le C\bar{r}^{d-\frac{2}{\mu}d''m+s}\tau^{\frac{2}{\mu}d''-1-s}T^{-sl} \tag{12.5.91}$$

with $T\in[T_0, \tau^{-1}\tau^{-1}\bar{h}'^{-\delta}, T_0\tau^{-1}\bar{r}]$, $\lambda\in[\epsilon_0\tau, c\tau]$ and with another still arbitrarily large s. Here we have omitted the prime on T.

Therefore one can replace $\bar{\chi}$ with a Hörmander function χ and apply the routine Tauberian technique to obtain that the left hand expression in (12.5.60) now does not exceed

$$C\bar{r}^{d-1-\frac{2}{\mu}d''m+sl}\tau^{\frac{2}{\mu}d''-s}.$$

Then after summation over a partition of unity in the zone $\{r_*(\tau)\le|\xi'|\le\varepsilon r^*(\tau)\}$ we obtain remainder estimate $\varepsilon'\tau^{\frac{1}{l}(d'-1)}$ with $\varepsilon'=\varepsilon'(\varepsilon)\to+0$ as $\varepsilon\to+0$. Therefore under condition (12.5.86) Proposition 12.5.17 holds with the remainder estimate $O(\tau^{\frac{1}{l}(d'-1)})$; moreover it holds with the remainder estimate $o(\tau^{\frac{1}{l}(d'-1)})$ under standard non-periodicity condition to the Hamiltonian flow for matrix-valued symbol $a(x',\xi')$ on the energy level $\tau=1$.

On the other hand, Proposition 12.5.9 yields that the remainder estimate in Proposition 12.5.16 is now $O(\tau^{-s})$. Therefore for $\mathsf{N}(\tau',\tau)$ with $\tau'\in[\epsilon_0,\tau]$ the remainder estimate is $O(\tau^{\frac{1}{l}(d'-1)})$ [54]. This estimate yields the same remainder estimate for $\mathsf{N}^+(\tau)$ for $d'\ge 2$.

For $d'=1$ standard arguments with τ, τ' replaced by $2^{-n}\tau$, $2^{-n-1}\tau$, etc., yield the remainder estimate $O(\log\tau)$ but one can also easily obtain the

[54] And even $o(\tau^{\frac{1}{l}(d'-1)})$ under the standard non-periodicity condition.

remainder estimate $O(1)$. In fact, one has the remainder estimate $O(1)$ for $\mathsf{N}(\tau',\tau)$ and the remainder estimate $O(\tau^{-s})$ for

$$\int \phi(\frac{\lambda}{\tau})\, d_\lambda \mathsf{N}^+(\lambda) d\lambda$$

with $\phi \in \mathscr{C}_0^K([\epsilon_0, c])$ (this estimate is the routine estimate with mollification, the deduction of which requires neither the Tauberian technique nor condition (12.5.86)). Replacing $\phi(\lambda)$ by $\big(\bar\phi(\tau) - \bar\phi(2\tau)\big)$ with $\bar\phi \in \mathscr{C}^K(\mathbb{R})$ supported in $[-\infty, 1]$ and equal to 1 on $[-\infty, \frac{1}{2}]$ and replacing τ by $2^{-n}\tau$, etc., we obtain the remainder estimate $O(1)$ for

$$\int \bar\phi(\frac{\lambda}{\tau})\, d_\lambda \mathsf{N}^+(\lambda) d\lambda,$$

and this remainder estimate and the remainder estimate for $\mathsf{N}(\tau',\tau)$ obviously yield the remainder estimate $O(1)$ for $\mathsf{N}^+(\tau)$. So, both remainder estimates in Theorem 12.5.22 have been proven.

Moreover, in Proposition 12.5.16 we obtain $\mathcal{N}(\tau,\eta) = 0$ because of condition (12.5.86) (recall that standard Weyl asymptotics were applied here) and in Proposition 12.5.18 one should replace (12.5.70) by the asymptotics

$$\operatorname{Tr} \Phi_{n(q)}(\omega) = \kappa_q + O(\omega^{-s}). \tag{12.5.92}$$

Therefore expression (12.5.73) is reduced to

$$\kappa_q \tau^{\frac{1}{l}(d'-\frac{q}{2})-1}$$

and $\bar{\mathcal{N}}(\tau,\eta) = 0$.

Thus for $d' \ge 2$ we obtain expression (12.5.88) exactly for $\mathcal{N}^+(\tau)$, but for $d' = 1$ an additional term $\bar\kappa_1' \log \tau$ arises (this term is less than the remainder estimate for $d' \ge 2$). Then we apply the Remark 12.5.23 below. □

Remark 12.5.23. By explicit calculation prove that $\bar\kappa_1' = 0$ for $d' = 1$. Note that the same calculation was done in the analysis of non-semibounded boundary value problems for operators with positive principal symbols in Subsection 7.3.3 (one can apply these arguments without any modification).

Remark 12.5.24. The remainder estimate for Riesz means is $O\big(\tau^{\frac{1}{l}(d-1-\vartheta)+\vartheta}\big)$ and even $o\big(\tau^{\frac{1}{l}(d-1-\vartheta)+\vartheta}\big)$ under the standard condition to the Hamiltonian flow.

12.5.7 Coefficients in Spectral Asymptotics

Let us discuss the coefficients κ_p and $\bar{\kappa}_q$ in the asymptotics of Theorems 12.5.20, 12.5.22 and Remarks 12.5.21, 12.5.24. First of all the following obvious remark holds (we discuss coefficients for $\vartheta = 0$; the coefficients of $\tau^{k+\vartheta}$ for $\vartheta > 0$ have the extra factor $k(k+\vartheta)^{-1}$ compared with the coefficient of τ^k for $\vartheta = 0$).

Remark 12.5.25. (i) If $p\,(=0, \frac{1}{2}, 1, \dots)$ is such that $l(d-p) > d'm$ then for κ_p the standard Weyl formula holds, the integral in which surely converges; in particular,

$$\kappa_0 = (2\pi)^{-d} \int n_0^+(x,\xi,1)\, dxd\xi \tag{12.5.93}$$

where $n_0^+(x,\xi,1)$ is the number of eigenvalues of $a_0(x,\xi)$ lying in $[0,1]$. Therefore $\kappa_0 > 0$ provided $a_0(x,\xi)$ is not negative definite.

(ii) On the other hand, for $p\,(=0, \frac{1}{2}, 1, \dots)$ let equality $(d-p)l = d'm$ hold. Then the integral providing the standard Weyl coefficient of $\tau^{\frac{1}{m}(d-p)}$ diverges logarithmically; namely, modulo convergent integrals we have

$$\int \omega(x',\xi';\varphi)\, dx'd\xi'd\varphi \frac{d\zeta}{\zeta}$$

with $\varphi = \varrho^{-1}\cdot(x'',\xi''|\xi'|^{-1}) \in \mathbb{S}^{2d''-1}$, $\varrho = \bigl(|x''|^2 + |\xi''|^2|\xi'|^{-2}\bigr)^{\frac{1}{2}}$; then the coefficient $\bar{\kappa}_p$ at $\tau^{(\frac{1}{m}d-p)}\log\tau$ is given by formula

$$\bar{\kappa}_p = \frac{1}{2l}\int \omega(x',\xi',\varphi)\, dx'd\xi'd\varphi.$$

In particular, for $dl = d'm$ the coefficient $\bar{\kappa}_0$ at $\tau^{\frac{d}{m}}\log\tau$ is

$$\bar{\kappa}_0 = \frac{1}{2l(2\pi)^d}\int_{\Lambda\times\mathbb{S}^{2d''-1}} \tilde{n}^+(x',\xi',\varphi,1)\, dx'd\xi'd\varphi \tag{12.5.94}$$

where $\tilde{n}^+(x',\xi',\varphi,1)$ is the number of eigenvalues of the symbol

$$\tilde{a}_0(x',\xi',\varphi) = \sum_{|\alpha|+|\beta|=\mu} \frac{1}{\alpha!\beta!} a^{(\alpha)}_{0(\beta)}(x',\xi')x''^{\beta}\xi''^{\alpha} \tag{12.5.95}$$

again with $\varphi = \varrho^{-1}\cdot(x'',\xi''|\xi''|^{-1})(x,\xi) \in \mathbb{S}^{2d''-1}$.

(iii) Either let $q = (0, \frac{1}{2}, 1, \dots)$ be such that $(d' - q)m > dl$ or let $a_0(x, \xi)$ be a negative definite symbol. Then the coefficient $\bar{\kappa}_q$ is given by a convergent integral

$$\frac{1}{2d' - q}\int_0^\infty \operatorname{Tr}\operatorname{Res}_{\mathbb{R}} F_{(q)}(z) z^{-\frac{1}{l}(d'-q)}\, dz$$

where $F_{(q)}$ is the sum of all the symbols with the given index q. In particular, if either $d'm > dl$ or $a_0(x, \xi)$ is a negative definite symbol then

$$\bar{\kappa}_0 = (2\pi)^{-d'} \int_\Lambda n^+(x', \xi', 1)\, dx' d\xi' = \frac{1}{(2\pi)^d d}\int_{\Lambda\cap S^*X} \omega(z)\, dz \tag{12.5.96}$$

where $\omega(z) = \sum_j \eta_j^{-\frac{d}{l}}$, η_j are all the positive eigenvalues of $A_{(x',\xi')}$ and dz is the natural density on $\Lambda \cap S^*X$ provided a Riemannian metric is introduced on X.

(iv) $a_{\frac{1}{2}} = 0$ yields that $\kappa_{\frac{1}{2}} = 0$. See also problem 12.5.11(ii).

(v) Moreover, let μ be even. Let us again treat the zone $\{|\xi'| \le r^*(\tau)\tau^{-\delta}\}$. Let us consider a term in the expansion of $\varkappa_{\frac{j}{2}}(x, \xi)$ in terms which are homogeneous with respect to $(x'', \xi''|\xi'|^{-1})$. Obviously, if this term is odd with respect to $(x'', \xi''|\xi'|^{-1})$ then its integral vanishes. Therefore a term with a logarithmic factor appears only if the equality $\bigl(d - \frac{j}{2}\bigr)l = \bigl(d' - \frac{k}{2}\bigr)m$ holds with $j \equiv k \mod 2$. So, if μ is even then a term $\kappa_{\frac{j}{2}}\tau^{\frac{1}{m}(d-\frac{j}{2})}\log\tau$ appears only if $\bigl(d - \frac{j}{2}\bigr)l = \bigl(d' - \frac{k}{2}\bigr)m$ and $j \equiv k \pmod 2$.

(vi) Let $a_{\frac{1}{2}} \equiv 0$. Then in the previous analysis one should only treat even j (and therefore only even k). On the other hand, analysis in the zone $\{|\xi'| \ge r_*(\tau)\tau^{\delta}\}$ proves that $\bar{\kappa}_{\frac{q}{2}} = 0$ for odd q. So, if $a_{\frac{1}{2}} \equiv 0$ and μ is even then in (12.5.81), (12.5.88) one should treat only $p = 0, 2$ and $q = 0, 2$ in the calculation of powers and in the criterion for terms with a logarithmic factor. Similarly, for $\vartheta > 0$ in this case ($a_{\frac{n}{2}} \equiv 0$ for odd n and μ is even) one should consider only even p, q in the criterion for terms with a logarithmic factor.

(vii) On the other hand, one can easily see that for an operator with the symbol $\begin{pmatrix} \lambda & \mathcal{L} \\ \mathcal{L}^* & \lambda \end{pmatrix}$ with $\mathcal{L} \in \mathbf{\Psi}^{1,1}$ and $\lambda \in \mathbb{C}$ the previous statement for $\bar{\kappa}_q$ is wrong.

(viii) Finally, if $a_{\frac{1}{2}} \equiv 0$, μ and m are even and a_0 and a_1 are homogeneous (not only positive homogeneous) of degrees m and $(m-1)$ in ξ respectively then in (12.5.81) one should treat only $p = 0$ in the calculation of powers and in the criterion for terms with a logarithmic factor.

We leave to the reader the following set of problems

Problem 12.5.26. (i) Consider the asymptotics of Riesz means.

(ii) Consider manifolds with boundary. In this case one should assume that every connected component of Λ either does not intersect $T^*X|_{\partial X}$ or is contained in this manifold. In the second case one should consider $A_{(x',\xi')}(x'', D'')$ as an operator in the half-space of $\mathbb{R}^{d''}$ and the problem of boundary conditions arises.

Moreover, there are more variants for the leading term of the asymptotics because we know that the condition that the operator is semi-bounded on a manifold with boundary consists of two conditions, and if the first condition is fulfilled and the second condition is violated then the first term on the asymptotics vanishes in the non-degenerate case.

(iii) The other problem left to the reader is to return to Sections 12.2–12.4 and obtain (under appropriate conditions) the non-Weyl term in the asymptotics in a simpler form (as we did here, and in the form used here).

12.6 Riesz Means Asymptotics for Operators Singular at a Point

12.6.1 Problem Set-up

In this section we consider h-differential operator $A = A_h$ with Weyl symbol

$$A(x,\xi) = \sum_{|\alpha|\le m} a_\alpha(x)\xi^\alpha, \tag{12.6.1}$$

Hermitian and such that in $B(0,1)$

$$|D^\beta a_\alpha| \le c|x|^{\mu(m-|\alpha|)-|\beta|} \qquad \forall \alpha : |\alpha| \le m \quad \forall \beta : |\beta| \le K \tag{12.6.2}$$

with

$$\mu > \max(-1, -\frac{d}{m}). \tag{12.6.3}$$

Moreover, we assume that

$$\langle a(x,\xi)v, v\rangle \ge \epsilon_0|\xi|^m\|v\|^2 \qquad \forall (x,\xi)\in T^*B(0,1)\quad \forall v\in \mathbb{K} \tag{12.6.4}$$

where $a(x,\xi)=\sum_{|\alpha|=m} a_\alpha(x)\xi^\alpha$ is the senior symbol of A.

Let us consider the operator A in the domain $X\subset\mathbb{R}^d$ assuming that

$$B(0,1)\subset X,\quad d\ge 2 \tag{12.6.5}$$

and

(12.6.6) A is a self-adjoint operator with $\mathfrak{D}(A)\supset \mathscr{C}_0^m(B(0,1),\mathbb{K})$.

We are interested in semiclassical asymptotics of the Riesz mean

$$\int \psi(x)\,\mathrm{tr}(e_{(\vartheta)}(x,x,0))\,dx \tag{12.6.7}$$

where $\psi\in\mathscr{C}_0^K(B(0,\frac{1}{2}))$ and $e_{(\vartheta)}(.,.,\tau)$ is the Riesz mean of order ϑ of the Schwartz kernel $e(.,.,\tau)$ of the spectral projector A, $\vartheta>0$:

$$e_{(\vartheta)}(.,.,\tau)=\frac{1}{\Gamma(\vartheta+1)}\int_0^\tau (\tau-\lambda)^\vartheta\, d_\lambda e(.,.,\lambda) \tag{12.6.8}$$

where $\Gamma(.)$ is an Euler' Γ-function.

In Section 11.1 we proved that under an appropriate condition associated with microhyperbolicity (namely, condition (12.6.34) below) asymptotics

$$\int \psi(x)\,\mathrm{tr}(e_{(\vartheta)}(x,x,0))\,dx = \sum_{0\le j\le J} h^{-d+j}\int \psi(x)\kappa_{(\vartheta)j}(x)\,dx+ O(R_{(\vartheta)}(\tau)) \tag{12.6.9}$$

holds with

$$R_{(\vartheta)}(\tau)=\begin{cases} h^{-d+\vartheta+1} & \text{for }\ \vartheta<\bar{\vartheta}_p,\\ h^{-d+\vartheta+1}\log h & \text{for }\ \vartheta=\bar{\vartheta}_p,\\ h^{l} & \text{for }\ \vartheta>\bar{\vartheta}_p\end{cases} \tag{12.6.10}$$

where here and below

$$\bar{\vartheta}_p = \begin{cases} ((d-1)(\mu+1)+p)(-m\mu+\mu+1)^{-1} & \text{for } m\mu < \mu+1, \\ +\infty & \text{for } m\mu \geq \mu+1, \end{cases} \tag{12.6.11}$$

and

$$l = l_0 = m\mu\vartheta(\mu+1)^{-1}, \qquad l_p = (m\mu\vartheta+p)(\mu+1)^{-1}, \tag{12.6.12}$$

$\kappa_{(\vartheta)j}$ are standard Weyl coefficients and

$$J = \begin{cases} \lfloor \vartheta \rfloor + 1 & \text{for } \vartheta \leq \bar{\vartheta}_p, \\ d + \lfloor l \rfloor & \text{for } \vartheta \geq \bar{\vartheta}_p \end{cases} \tag{12.6.13}$$

excluding the special case

$$J = \vartheta \qquad \text{for } \vartheta = \bar{\vartheta}_p \in \mathbb{N}. \tag{12.6.13$'$}$$

Further, for $\vartheta > \bar{\vartheta}_p$ and $l \in \mathbb{N}$ one should cut the corresponding term near the origin, i.e., replace $\psi(x)$ by $\psi(x)\phi(xr_*(h)^{-1})$ in the term with $j = d + l$ of the right-hand expression of (12.6.9) where here and below

$$r_*(h) = h^{1/(\mu+1)}, \tag{12.6.14}$$

$(1-\phi) \in \mathscr{C}_0^K(\mathbb{R}^d)$ and ϕ vanishes in a neighborhood of 0.

Furthermore, we proved that if $\vartheta < \bar{\vartheta}_p$ and the standard non-periodicity condition is fulfilled on the energy level $\tau = 0$ then one can replace $O(h^{-d+\vartheta+1})$ by $o(h^{-d+\vartheta+1})$ in asymptotics (12.6.9).

The principal part of the asymptotics is $O(h^{-d})$ for $m\mu\vartheta > -d(\mu+1)$ and $O(h^{-d}\log h)$ for $m\mu\vartheta+p = -d(\mu+1)$ and $O(h^{l})$ for $m\mu\vartheta+p < -d(\mu+1)$ [55]; so in the last case we have only an estimate instead of an asymptotics.

Remark 12.6.1. All these statements hold for $p = 0$ but also remain true with $l = l_p$ for $p \geq 0$ provided

$$\operatorname{supp}(\psi) \subset B(0, \frac{1}{2}), \tag{12.6.15}$$

$$|D^\alpha \psi| \leq c|x|^{p-|\alpha|} \qquad \forall \alpha : |\alpha| \leq K \tag{12.6.16}$$

instead of the previous assumptions. One can see easily that $\vartheta_p > p$.

[55] There are equivalence relations under obvious conditions.

We leave to the reader the following easy problem:

Problem 12.6.2. Consider case $-d < p < 0$, $m\mu < \mu+1$; consider separately $m\mu < \mu + 1$ and $m\mu \geq \mu + 1$.

12.6.2 The Scott Correction Term

Our goal is to treat the case $\vartheta \geq \bar{\vartheta}_p$ and to improve this remainder estimate. This better estimate, obtained under appropriate conditions, requires an additional non-Weyl term ωh^l in the asymptotics. Physicists call this term the *Scott correction term*. Later we will obtain more than one such term for large ϑ. In this analysis, without any loss of the generality we can treat

$$\int x^\alpha \psi(x)\,\mathrm{tr}(e_{(\vartheta)}(x,x,0))\,dx$$

with $|\alpha| = p$ and $\psi \in \mathscr{C}_0^K(\mathbb{R}^d)$ as

$$\bar{\vartheta}_p < +\infty \qquad (\iff m\mu < \mu + 1). \tag{12.6.17}$$

Instead we consider more general case

(12.6.18) $\psi \in \mathscr{C}^K(\mathbb{R}^d \setminus 0)$ is positively homogeneous of degree p in $B(0,\frac{1}{2})$ and $\mathsf{supp}(\psi) \subset B(0,1)$.

First of all we should treat the positive homogeneous case:

Proposition 12.6.3. *Let conditions* (12.6.2)–(12.6.5) *be fulfilled. Moreover, let* $X = \mathbb{R}^d$, A *be a self-adjoint operator with* $\mathfrak{D}(A) \supset \mathscr{C}_0^m(\mathbb{R}^d, \mathbb{K})$ *and let us assume that*

(12.6.19) *The coefficients* a_α *are positively homogeneous of degrees* $(m-|\alpha|)\mu$ *in* x.

Finally, let us assume that

(12.6.20) $A(x,\xi)$ *is microhyperbolic in* (x,ξ) *on the energy level* $\tau = 0$ *for* $x \neq 0$ [56].

[56] Surely, the constants in the microhyperbolicity condition depend on $|x|$.

Then for any $\beta \in \mathbb{N}^d$ *and for* $W \in \mathscr{C}^K(\mathbb{R}^d \setminus 0, \operatorname{Hom}(\mathbb{K},\mathbb{K}))$ *positively homogeneous of degree* $(p-\mu|\beta|)$ *in* x *equality*

$$\int W(x)\Bigl(\bigl((hD_x)^\beta e_{(\vartheta)}\bigr)(x,x,0)- \phi\bigl(\frac{x}{r_*}\bigr)\sum_{0\le j\le J} \varkappa_{(\vartheta)\beta,j}(x)h^{-d+j}\Bigr)\,dx = \omega h^l \tag{12.6.21}$$

holds with the coefficients $\kappa_{(\vartheta)\beta,j}(x)$ [57] *and a certain constant coefficient* $\omega \in \operatorname{Hom}(\mathbb{K},\mathbb{K})$ *which depends on* W, β, ϕ [58] *and* $J \ge [l]+d$ *and the integral in the left-hand expression converges at infinity* [59].

Moreover, in terms with $j < l+d$ *one can replace* ϕ *by* 1 [60]*; in particular, if either* $l \notin \mathbb{N}$ *or* $l \in \mathbb{N}$ *but* $\kappa_{(\vartheta)\beta,l+d} \equiv 0$ *then one can set* $J = \lfloor l \rfloor + d$ *and replace* ϕ *by* 1 *in all the terms.*

Proof. Let us first note that in the context of the proposition

(12.6.22) Coefficients $\kappa_{(\vartheta)\beta,j}(x)$ are positively homogeneous in x of degrees $(d\mu + m\mu\vartheta - j(\mu+1) - |\beta|)$.

Moreover, let us note that after rescaling $x_{\text{new}} = x_{\text{old}}/r_*(h)$ with $r_*(h) = h^{-1/(\mu+1)}$ operator $A^{\mathrm{w}}(x,hD)$ is transformed into operator $h^{m\mu(\mu+1)^{-1}}A^{\mathrm{w}}(x,D)$ and therefore

$$e_{(\vartheta)}(x,y,0;h) = h^{(m\mu\vartheta+d)/(\mu+1)}e_{(\vartheta)}(xh^{-1/(\mu+1)}, yh^{-1/(\mu+1)}, 0; 1) \tag{12.6.23}$$

due to the definition of $e_{(\vartheta)}(.,.,.,.)$; here the fourth variable is the semiclassical parameter h.

This equality and (12.6.22) yield that the left-hand expression in (12.6.21) is positively homogeneous of degree l on h provided it converges; therefore we should prove only that the integral in (12.6.21) converges at infinity.

However, Theorem 4.4.6 yields that if $\psi' \in \mathscr{C}_0^K(\mathbb{R}^d, \operatorname{Hom}(\mathbb{K},\mathbb{K}))$ vanishes in a neighborhood of 0 then the estimate

---

[57] Here we do not take the trace and therefore the coefficients κ_* and ω are matrix-valued in this statement.

[58] Recall that ϕ is a fixed function such that $(1-\phi) \in \mathscr{C}_0^K(\mathbb{R}^d)$, $\phi = 1$ near 0.

[59] It surely converges at 0 due to the cutting function ϕ.

[60] Surely, the constant ω changes.

$$|\int \psi'(\frac{x}{r})\Bigl(((hD_x)^\beta e_{(\vartheta)})(x,x,0) - \phi(\frac{x}{r_*})\sum_{0\le j\le J} \kappa_{(\vartheta)\beta,j}(x)h^{-d+j}\Bigr)\,dx| \le \\ Ch^{-d+J+1}r^{l(\mu+1)(l-J-1)}$$

holds uniformly on $r \ge r_* = h^{(\mu+1)^{-1}}$; this estimate immediately yields that the integral in the left-hand expression of (12.6.21) converges for $J \ge \lfloor l \rfloor + d$. □

12.6.3 Improved Asymptotics: the First Attempt

Our first goal is to improve the remainder estimate $O(h^l)$ for $\vartheta > \bar{\vartheta}_p$ in the general situation. To do this we will pass from one operator to another to reduce the problem.

Proposition 12.6.4. *Let the operators A and A' satisfy conditions* (12.6.2)–(12.6.6) *and coincide in $B(0,1)$. Let $\psi \in \mathscr{C}_0^K(B(0,\frac{1}{2}))$, $\psi \ge 0$. Moreover, let estimate*

$$|\int \psi(x)e(x,x,\tau+h,\tau)\,dx| \le Ch^{1-d} \qquad \forall \tau : |\tau| \le \epsilon_0 \tag{12.6.24}$$

hold. Then the same estimate holds for A' as well and, moreover, estimate

$$|\int \psi(x)\bigl(e_{(\vartheta)}(x,x,0) - e'_{(\vartheta)}(x,x,0)\bigr)\,dx| \le Ch^{1-d+\vartheta} \tag{12.6.25}$$

holds where here and below $e(.,.,.)$ with a superscript is associated with the operator A with the same superscript.

Proof. Proof follows from the same analysis as in Chapter 9. We leave the easy details to the reader. □

Therefore the remainder estimate which will not be better than $O(h^{1-d+\vartheta})$ is preserved if we change the operator A outside of $B(0,\epsilon_0)$ with some fixed $\epsilon_0 > 0$.

(12.6.26) Let us pick an operator A' coinciding with A in $B(0,\epsilon_0)$ and such that the negative spectrum of A' is discrete for every $h \in (0,h_0]$ (we can even pick an operator with negative spectrum, finite for every $h \in (0,h_0]$).

Let us now assume that $\vartheta \geq 1$. In fact the arguments below are really only good for $\vartheta = 1$ but we will later modify them to work in the case $\vartheta < 1$ and to provide a really good estimate for $\vartheta > 1$.

Proposition 12.6.5. *For $\theta \in [0, 1]$ let the operator $A^\theta = A^0 + \theta B$ satisfy conditions (12.6.2)–(12.6.6), let the negative spectrum of A be discrete and let B be supported in $B(0, \frac{1}{2})$. Then equality*

$$\int \operatorname{tr}\bigl(e^1_{(\vartheta)}(x, x, \tau) - e^0_{(\vartheta)}(x, x, \tau)\bigr)\, dx = \\ -\int_0^1 d\theta \int \operatorname{tr}\bigl(B_x e^\theta_{(\vartheta-1)}\bigr)(x, x, \tau)\, dx \tag{12.6.27}$$

holds for $\tau \leq 0$ and $\vartheta \geq 1$.

Proof. Let $E^\theta(\tau)$ be the spectral projector of A^θ. Let us differentiate the equality

$$\partial_\tau \operatorname{Tr}\bigl(E^\theta(\tau)\bigr) = \operatorname{Tr}\bigl(\operatorname{Res}_{\mathbb{R}}(\tau - A^\theta)^{-1}\bigr), \tag{12.6.28}$$

which holds for $\tau < 0$ in the distribution sense, with respect to θ. We obtain

$$\partial_\tau \partial_\theta \operatorname{Tr}\bigl(E^\theta(\tau)\bigr) = -\partial_\tau^2 \operatorname{Tr}\bigl(BE^\theta(\tau)\bigr).$$

This equality and the semiboundedness of A^θ from below yield that

$$\partial_\eta \operatorname{Tr}\bigl(E^\theta(\tau)\bigr) = -\partial_\tau \operatorname{Tr}\bigl(BE^\theta(\tau)\bigr). \tag{12.6.29}$$

Integrating over θ from 0 to 1 and convolving with $\tau_+^{(\vartheta-1)}/\Gamma(\theta)$ we obtain after integration by part (12.6.27) for $\tau < 0$. Semi-continuity from the left also yields (12.6.27) for $\tau = 0$. $\square$

Let us assume that the coefficients b_α of the operator $B = A - A^0$ satisfy estimates

$$|D^\beta b_\alpha| \leq c|x|^{\mu(m-|\alpha|)+\nu-|\beta|} \qquad \forall \alpha : |\alpha| \leq m \quad \forall \beta : |\beta| \leq K \tag{12.6.30}$$

with $\nu > 0$ and

$$\operatorname{supp}(b_\alpha) \subset B(0, r), \qquad 1 \geq r \geq r_*(h) = h^{1/(\mu+1)} \tag{12.6.31}$$

where r is a parameter as well as h.

Let us apply the Weyl approximation to the right-hand expression in (12.6.27); this means that we should replace the integrand by the convolution of

$$(2\pi)^{-d}\int \operatorname{Res}_{\mathbb{R}} \operatorname{tr}\bigl(B(\tau-A)^{-1}\bigr)_{\mathsf{w}}(x,\xi)\,d\xi \tag{12.6.32}$$

with $\tau_+^{\vartheta-1}/\Gamma(\vartheta)$; recall that the subscript w means the Weyl symbol and all calculations are done in the context of the formal calculus of pseudodifferential operators.

In this situation let us reverse the deduction of (12.6.27). We then obtain the routine answer

$$\sum_{0\le j\le \lfloor\vartheta\rfloor} h^{-d+j}\int \bigl(\kappa^1_{(\vartheta)j}(x)-\kappa^0_{(\vartheta)j}(x)\bigr)\,dx \tag{12.6.33}$$

where $\kappa^\theta_{(\vartheta)j}$ are the Weyl coefficients for the operator A^θ. However, we gain a remainder estimate: calculating the remainder estimate in our Weyl approximation we obtain

Proposition 12.6.6. *Let the operator $A^\theta = A+\theta B$ satisfy conditions (12.6.2)–(12.6.6), (12.6.30), (12.6.31) uniformly on $\theta\in[0,1]$.*

In addition, let the negative spectrum of A^θ be discrete and let us assume that

(12.6.34) *Symbol $A(xr,\xi r^\mu)r^{-m\mu}$ is microhyperbolic in (x,ξ) on the energy level 0 uniformly with respect to $x\in\mathbb{S}^{d-1}$, $\xi\in\mathbb{R}^d$, $r\in(0,1)$.*

Then estimate

$$\begin{aligned}|\int \operatorname{tr}\bigl(e^1_{(\vartheta)}(x,x,0)-e^0_{(\vartheta)}(x,x,0)\bigr)-\\ \sum_{0\le j\le\lfloor\vartheta\rfloor+1}\bigl(\varkappa^1_{(\vartheta)j}(x)-\varkappa^0_{(\vartheta)j}(x)\bigr)\phi\bigl(\frac{x}{r_*}\bigr)h^{-d+j}\,dx| \le CR^*_1(r,h)\end{aligned} \tag{12.6.35}$$

holds with

$$R^*_1(r,h)=\begin{cases} h^{-d+\vartheta}r^{m\mu\vartheta+(d-\vartheta)(\mu+1)+\nu} & \text{for } \vartheta<\bar\vartheta_\nu,\\ h^{-d+\vartheta}|\log r| & \text{for } \vartheta=\bar\vartheta_\nu,\\ h^{l'} & \text{for } \vartheta>\bar\vartheta_\nu\end{cases} \tag{12.6.36}$$

where according to (12.6.11) *and* (12.6.12)

$$\bar{\vartheta}_\nu = \bigl(d(\mu+1)+\nu\bigr)(\mu+1-m\mu)^{-1} > \bar{\vartheta}_0, \tag{12.6.37}$$

and

$$l' = l_\nu = (m\mu\vartheta+\nu)(\mu+1)^{-1} > l. \tag{12.6.38}$$

Recall that $r_*(h) = h^{1/(\mu+1)}$, ϕ *is a fixed function with* $(1-\phi)\in\mathscr{C}_0^K(\mathbb{R}^d)$, *and* ϕ *vanishes in a neighborhood of* 0.

Moreover, for $\vartheta<\bar{\vartheta}_\nu$ *one can replace* ϕ *by* 1 *in all the terms; one can also do this for* $\vartheta=\bar{\vartheta}_\nu$, *skipping terms with* $j=\vartheta$; *furthermore, for* $\vartheta>\bar{\vartheta}_\nu$ *one can skip terms with* $j>l'+d$ *and one can replace* ϕ *by* 1 *in terms with* $j<l'+d$.

Proof. Proof follows the scheme we indicated but with the following details:

We apply the Weyl approximation only in the semiclassical zone, i.e., for $r\ge r_*$. In the ball $B(0,r_*)$ we apply a routine estimate of Chapter 9 and obtain that the contribution of this ball to the asymptotics does not exceed $h^{l'}$ with l' given by (12.6.38).

Surely we only treated the case of a bounded function B in Chapter 9. However, it follows from (12.6.3) that on $\mathscr{C}_0^m\bigl(B(0,2r_*),\mathbb{K}\bigr)$ operator A satisfies inequalities

$$A\ge -C_0h^l,\qquad A+C_0h^l\ge h^{l-2mq/(m+1)}L^*L \tag{12.6.39}$$

for any operator L of the same type as A but with m replaced by $\frac{q}{2}$, $q\le m$; then the routine estimates of Chapter 9 yield the estimate

$$\operatorname{Tr}\bigl(ME(-\infty,0)\bigr)\le h^{mq/(m+1)} \tag{12.6.40}$$

for $M=L^*L$ with M supported in $B(0,\frac{3}{2}r_*)$ and therefore the same estimate (12.6.40) holds for every operator M supported in $B(0,r_*)$ of the same type as A but with m replaced by q. □

Furthermore, let $X=\mathbb{R}^d$ and let conditions (12.6.2), (12.6.4) be fulfilled in $\mathbb{R}^d$ instead of $B(0,1)$. Applying the arguments of Chapter 9 one can easily prove estimate

$$\Bigl|\int\bigl(1-\psi(\frac{x}{r_*})\bigr)\bigl(e^1_{(\vartheta)}(x,x,0)-e^0_{(\vartheta)}(x,x,0)\bigr)\,dx\Bigr|\le Ch^{-d+\vartheta+1} \tag{12.6.41}$$

provided $\psi = 1$ in $B(0, \frac{1}{3})$ [61] and $r \le \epsilon r^*$ where $r^* > 0$ is an arbitrarily small constant and $\epsilon > 0$ is sufficiently small.

Moreover, the local spectral asymptotics of Chapter 4 applied in the zone $\{1 \ge |x| \ge \frac{1}{4} r\}$ yield that

Proposition 12.6.7. *Let microhyperbolicity condition* (12.6.34) *be fulfilled for operator* $A = A^0$ [62]. *Then estimate*

$$\begin{multline}
|\int \operatorname{tr}\bigl(\psi(\frac{x}{r^*}) - \psi(\frac{x}{r})\bigr)\bigl(e^\theta_{(\vartheta)}(x,x,0) - \\
\sum_{0\le j\le \lfloor\vartheta\rfloor+1} \kappa^\theta_{(\vartheta)j} h^{-d+j}\bigr)\,dx| \le CR^*_2(r,h)
\end{multline} \tag{12.6.42}$$

holds with

$$R^*_2(r,h) = \begin{cases} h^{-d+\vartheta+1} r^{m\mu\vartheta + (d-\vartheta-1)(\mu+1)} & \text{for } \vartheta > \bar\vartheta_\nu, \\ h^{-d+\vartheta+1}\log r & \text{for } \vartheta = \bar\vartheta_\nu. \end{cases} \tag{12.6.43}$$

Then (12.6.35)–(12.6.36), (12.6.41) and (12.6.42)–(12.6.43) yield estimate

$$\begin{multline}
|\int \psi(\frac{x}{r^*}) \operatorname{tr}\Bigl(e^1_{(\vartheta)}(x,x,0) - e^0_{(\vartheta)}(x,x,0) - \\
\sum_{0\le j\le\lfloor\vartheta\rfloor+1} \bigl(\kappa^1_{(\vartheta)j}(x) - \kappa^0_{(\vartheta)j}(x)\bigr)\phi(\frac{x}{r_*}) h^{-d+j}\Bigr)\,dx| \le CR^*_1(r,h) + CR^*_2(r,h)
\end{multline} \tag{12.6.44}$$

with R^*_1, R^*_2 given by (12.6.36) and (12.6.43).

Further, Proposition 12.6.4 yields that one can skip the assumptions that $X = \mathbb{R}^d$, that condition (12.6.2) is fulfilled in X instead of $B(0,1)$ and that the negative spectrum of A is discrete. In fact, it is easy to construct an appropriate extension of A to $\mathbb{R}^d$.

Furthermore, let $\bar{A}^0$ coincide with A^0 in $B(0, \frac{1}{2}r)$ and be self-adjoint. Then the arguments of Chapter 9 easily yield that

$$|\int \psi(\frac{x}{r}) \operatorname{tr}\bigl(e^0_{(\vartheta)}(x,x,0) - \bar{e}^0_{(\vartheta)}(x,x,0)\bigr)\,dx| \le CR^*_1. \tag{12.6.45}$$

---

[61] This condition is also assumed to be fulfilled below.

[62] Then it is also fulfilled for $A = A^\eta$ for $r \le r^*$.

Finally, in this case $\kappa^0_{(\vartheta)j}(x)=\bar{\kappa}^0_{(\vartheta)j}(x)$ in a neighborhood of $\mathsf{supp}(\psi(x/r))$. Therefore, one can skip condition (12.6.31) on the support of B in the deduction of estimate (12.6.44).

Therefore we conclude that

(12.6.46) Estimate (12.6.44) remains true under the following weakened assumptions: A^0 is an operator in $\mathbb{R}^d$ satisfying conditions of Proposition 12.6.3, A^1 is an operator in the domain $X\supset B(0,1)$ satisfying conditions (12.6.6) and $b_\alpha=a_\alpha-a^0_\alpha$ satisfying (12.6.30).

Here and below we skip the superscript "1".

Again applying estimates (12.6.42)–(12.6.43) we obtain that

(12.6.47) Estimate (12.6.44) remains true under weakened assumptions if we replace r by r_* in its left-hand expression. Moreover, dividing the left-hand expression into two parts (associated with A, $\kappa_{(\vartheta)j}$ and with A^0, $\kappa^0_{(\vartheta)j}$) we see that in the second part one can replace r by arbitrary $r'>r$; moreover, one can even pick $r'=+\infty$, i.e., replace ψ by 1 here.

Minimizing the right-hand expression by $r\in[r_*,\epsilon r^*]$ (the left-hand expression surely does not depend on r and the optimal value is $r\asymp\bar{r}=h^{1/(\mu+\nu+1)}$) we obtain estimate

$$\begin{aligned}(12.6.48)\quad \Bigl|\int\psi\bigl(\frac{x}{r^*}\bigr)\,\mathrm{tr}\Bigl(e_{(\vartheta)}(x,x,0)-\sum_{0\le j\le\lfloor\vartheta\rfloor+1}\kappa_{(\vartheta)j}\phi\bigl(\frac{x}{r_*}\bigr)h^{-d+j}\Bigr)\,dx-\\ \int\psi\bigl(\frac{x}{r_2}\bigr)\,\mathrm{tr}\Bigl(e^0_{(\vartheta)}(x,x,0)-\sum_{0\le j\le\lfloor\vartheta\rfloor+1}\kappa^0_{(\vartheta)j}\phi\bigl(\frac{x}{r_*}\bigr)h^{-d+j}\Bigr)\,dx\Bigr|\le CR^*\end{aligned}$$

with

$$(12.6.49)\qquad R^*=R^*(h)=\begin{cases}h^{-d+\vartheta+1}|\log h| & \text{for }\ \vartheta=\bar{\vartheta}_0,\\ h^{l'''} & \text{for }\ \bar{\vartheta}_0<\vartheta<\bar{\vartheta}_\nu,\\ h^{l''}|\log h| & \text{for }\ \vartheta=\bar{\vartheta}_\nu,\\ h^{l''} & \text{for }\ \vartheta>\bar{\vartheta}_\nu\end{cases}$$

where

$$(12.6.50)\qquad l''=(m\mu\vartheta-(d-\vartheta-1)\nu)(\mu+\nu+1)^{-1}.$$

One can easily see that $l < l'' < l'$ for $\bar{\vartheta}_0 < \vartheta \le \bar{\vartheta}_\nu$, $l'' = l = -d + \vartheta + 1$ for $\vartheta = \bar{\vartheta}_0$, $l'' = l'$ for $\vartheta = \bar{\vartheta}_\nu$.

In virtue of Proposition 12.6.3 with $J = \lfloor \vartheta \rfloor + 1$ the second line in the left-hand expression of (12.6.48) is equal to $\omega h^{l''}$; picking up $r_2 = +\infty$ we arrive to asymptotics

$$\int \psi(x)\,\mathrm{tr}(e_{(\vartheta)}(x,x,0))\,dx = \sum_{0\le j\le \lfloor \vartheta \rfloor +1} h^{-d+j} \int \psi(x)\kappa_{(\vartheta)j}(x)\phi(\frac{x}{r_*})\,dx + \omega h^{l''} + R^* \tag{12.6.51}$$

with R^* given by (12.6.49) and $\psi \in \mathscr{C}_0^\infty$, $\psi = 1$ in the vicinity of 0.

Therefore, we have proven the following

Theorem 12.6.8. *Let the operator A satisfy conditions (12.6.2)–(12.6.6). Furthermore, let $b_\alpha = a_\alpha - a^0_\alpha$ satisfy (12.6.30) with (matrix-valued) coefficients a^0_α which are positively homogeneous of degrees $(m - |\alpha|)\mu$. Let $\vartheta \ge \max(\bar{\vartheta}_0, 1)$.*

Then asymptotics (12.6.51) holds with Weyl coefficients $\kappa_{(\vartheta)j}(x)$ and some (non-Weyl) coefficient ω and with the remainder estimate given by (12.6.49).

Recall that $r_* = h^{(\mu+1)^{-1}}$ and that in the terms with $j < \vartheta + 1$ one can replace ϕ by 1 (changing ω).

Let us note that here we gained nothing for $\vartheta = \bar{\vartheta}_0$.

12.6.4 Adjustment for $\vartheta < 1$

Let us consider the case $\vartheta < 1$ when one should modify the above arguments. Let us pick $\tau = 0$. Moreover, let us consider a partition unity on $\mathbb{R}$ corresponding to the scaling function $\frac{1}{2}(|\tau| + L)$ where we choose parameter L later. Then $\psi(x)e_\vartheta(x,x,0)$ is equal to a sum of terms of the form

$$\vartheta \iint \lambda_-^{\vartheta-1}\varphi(\frac{\lambda}{\ell})\psi(x)e(x,x,\lambda)\,d\lambda dx \tag{12.6.52}$$

with <u>either</u> $\varphi = \varphi_0 \in \mathscr{C}_0^K([-1,1])$ and $\ell = L$ <u>or</u> $\varphi = \varphi_1 \in \mathscr{C}_0^K([\frac{1}{2},1])$ and $\ell \ge L$.

We apply this decomposition (12.3.54) *only* with $\psi(x/r)$ instead of ψ where we will choose $r\in[r_*,1]$ later. Let us apply the arguments of the previous subsection which are not associated with Proposition 12.6.5 (and therefore do not require the assumption $\vartheta\geq 1$). In particular, let us apply estimates (12.6.41) and (12.6.42)[63]. We then obtain that one needs to evaluate

$$\frac{1}{\Gamma(\vartheta)}\iint \lambda_-^{\vartheta-1}\varphi_0(\frac{\lambda}{\ell})\psi(\frac{x}{r})\operatorname{tr}(e^\theta(x,x,\lambda))\,d\lambda dx \tag{12.6.53}$$

and

$$\frac{1}{\Gamma(\vartheta)}\iint \lambda_-^{\vartheta-1}\varphi_1(\frac{\lambda}{\ell})\operatorname{tr}\bigl(e(x,x,\lambda)-e^0(x,x,\lambda)\bigr)\,d\lambda dx \tag{12.6.54}$$

where $\theta=0,1$, the error $O(R_2^*)$ (which is $=O\bigl(h^{-d+1+\vartheta}r^{m\mu\vartheta+(d-\vartheta-1)(\mu+1)}\bigr)$ in this case) is already accounted for, R_2^* is given by (12.6.43) and without any loss of the generality one can assume that conditions (12.6.2), (12.6.4) are fulfilled in $\mathbb{R}^d$.

Applying standard Weyl asymptotics to (12.6.53) one gets the remainder estimate $O\bigl(L^\vartheta h_{\mathsf{eff}}^{1-d}\bigr)$. On the other hand, one can apply the arguments of the proof of Proposition 12.6.5 to (12.6.54) and represent it in the form

$$\int_0^1 dt\iint \bigl(\partial_\tau^2\lambda_-^\vartheta\varphi_1(\frac{\lambda}{\ell})\bigr)\operatorname{tr}(B_xe^\eta)(x,x,\lambda)\,d\lambda dx. \tag{12.6.55}$$

Let us apply Weyl asymptotics to this expression. We obtain then the remainder estimate $O\bigl(r^{m\mu+\nu}\ell^{\vartheta-1}h_{\mathsf{eff}}^{1-d}\bigr)$ provided

$$m\mu+\nu+(d-1)(\mu+1)>0. \tag{12.6.56}$$

After summation on the partition of unity we obtain that the total contribution of the terms of (12.6.54) type to the remainder is $O\bigl(r^{m\mu+\nu}L^{\vartheta-1}h_{\mathsf{eff}}^{1-d}\bigr)$ while the contribution of (12.6.53) is $O\bigl(L^\vartheta h_{\mathsf{eff}}^{1-d}\bigr)$. Minimizing the sum with respect to L (the optimal value is $L\asymp r^{m\mu+\nu}$) we obtain

$$O\bigl(h^{1-d}r^{(m\mu+1)\vartheta+(\mu+1)(d-1)}\bigr). \tag{12.6.57}$$

[63] In order to get these with the convolutions, including the cut-off function φ, one should apply Theorem 4.4.1.

Recall that there is also an error $O\bigl(h^{-d+1+\vartheta} r^{m\mu\vartheta+(d-\vartheta-1)(\mu+1)}\bigr)$. Minimizing their sum over r (the optimal value is again $\asymp \bar{r} = h^{1/(\mu+\nu+1)}$) we obtain the final remainder estimate $O\bigl(h^{l''}\bigr)$ where l'' is given by (12.6.50).

So, the following statement has been proven:

Theorem 12.6.9. *Let all the conditions of Theorem 12.6.8 be fulfilled excluding* $\vartheta \geq 1$. *Let* $\bar{\vartheta}_0 < \vartheta < 1$ *and condition* (12.6.56) *be fulfilled (which means exactly that* $\bar{\vartheta}_\nu > 1$*).*

Then asymptotics (12.6.51) *remains true with the Weyl coefficients* $\kappa_{\vartheta j}(x)$ *and some (non-Weyl) coefficient* ω *and with the remainder estimate given by* (12.6.49).

Remark 12.6.10. Surely, even if condition (12.6.56) is violated one can get semiclassical asymptotics for (12.6.55) but the remainder estimate will be worse and the final remainder estimate will be also worse. However, this condition is only technical, and we are going to get rid of it without deterioration of the remainder estimate in the general approach.

12.6.5 The General Case

Let us pick up $\bar{r} = h^{(1+\mu+\nu)^{-1}}$ and then apply the same arguments as before in the zone $\{\bar{r} \leq |x| \leq 1\}$.

Then

(12.6.58) The contribution of this zone $\{\bar{r} \leq |x| \leq 1\}$ to the remainder is $O\bigl(R_2^*(\bar{r})\bigr)$ as before.

Let us consider $B(0, \bar{r})$ and apply the same partition of unity here. Thus one should consider the same expression

$$\frac{1}{\Gamma(\vartheta+1)} \int \tau_-^\vartheta \operatorname{Tr}\bigl(\psi\, d_\tau E(\tau)\bigr) \tag{12.6.59}$$

as before. We skip all the subscripts at ψ. Let us define $r = \frac{1}{2}(|x| + r_*)$ at some point of $\mathsf{supp}(\psi)$ and let us pick up

$$L = L(r) = hr^{m\mu+\nu-\mu-1} \tag{12.6.60}$$

for this element. Let us consider the same partition on $\mathbb{R}$ as before and let us replace $\operatorname{Tr}\psi\, d_\tau E(\tau)$ with its Weyl expression (with cut-off for $|x| \leq r_*$).

Then the contribution of the element with $\ell \le 2L$ to the remainder does not exceed $CR(r,h)$ with

$$R(r,h) = h^{-d+1} r^{(d-1)(\mu+1)} L^{\vartheta} = h^{-d+1+\vartheta} r^{(m\mu+\nu)\vartheta+(d-1-\vartheta)(\mu+1)}. \tag{12.6.61}$$

Summation (12.6.61) over the partition of unity on x results in the same expression with $r = r_*$ or $r = \bar{r}$ as $(m\mu+\nu)\vartheta + (d-1-\vartheta)(\mu+1) \lesseqgtr 0$ with an extra factor $\log(\bar{r}/r_*)$ in the case of equality; so we get

$$R^*(h) = \begin{cases} h^{l''} & \text{for } (m\mu+\nu)\vartheta + (d-1)(\mu+1) > 0 \\ h^{l'} \log h & \text{for } (m\mu+\nu)\vartheta + (d-1)(\mu+1) = 0 \\ h^{l'} & \text{for } (m\mu+\nu)\vartheta + (d-1)(\mu+1) < 0 \end{cases} \tag{12.6.62}$$

where

$$l'' = \bigl((m\mu+\nu)\vartheta - (d-1)\nu\bigr)(\mu+\nu+1)^{-1}, \tag{12.6.63}$$

$$l' = (m\mu+\nu)\vartheta(\mu+1)^{-1}. \tag{12.6.64}$$

Moreover, the contribution of $B(0,3r_*)$ to the remainder does not exceed $Ch^{l''}$.

Let us consider elements with $\ell \ge L$. Then one can apply the same arguments as before with k differentiations instead of 1 and easily get the following generalization of (12.6.27):

$$\begin{aligned} &\int \varphi(\tau)\tau_-^{\vartheta}\, d_\tau E^1(\tau) = \\ &\sum_{0\le j\le k-1} \int \varphi_{\vartheta,j}(\lambda_1,\ldots,\lambda_{j+1}) E^0(\lambda_1) B E^0(\lambda_2)\cdots BE^0(\lambda_{j+1})\, d\lambda_1\cdots d\lambda_{j+1} + \\ &\int_0^1 (k+1)(1-\theta)^k\, d\theta \times \\ &\qquad \int \varphi_{\vartheta,k+1}(\lambda_1,\ldots,\lambda_{k+1}) E^\theta(\lambda_1) BE^\theta(\lambda_2)\cdots BE^\theta(\lambda_{k+1})\, d\lambda_1\cdots d\lambda_{k+1} \end{aligned} \tag{12.6.65}$$

where $E^\theta(\lambda)$ is the spectral projector of A^θ and

$$\varphi_{\vartheta j} = (-1)^{j+1} \int \delta(\tau-\lambda_1)\cdots\delta(\tau-\lambda_j)\Bigl(\partial_\tau^{j+1}\bigl(\tau_-^{\vartheta}\varphi(\tau)\bigr)\Bigr)\, d\tau, \tag{12.6.66}$$

understood as a distribution on $(\lambda_1, \dots, \lambda_j)$, is in fact a smooth function such that

$$\operatorname{supp}(\varphi_{j,\vartheta}) \subset [\frac{1}{2}\ell, 2\ell]^j, \qquad |D_\lambda^\beta \varphi_{\vartheta j}| \le c\ell^{\vartheta+1-2j-|\beta|}. \tag{12.6.67}$$

Let us multiply by ψ and take the trace. Our goal is to evaluate the result.

First of all we are going to evaluate the elements with $r \ge 2r_*$. For this purpose let us return to Chapters 4 and 7. Under the standard assumptions of those chapters let us consider the following problems:

$$(hD_t - A)U_0(t) = 0, \qquad U_0(0) = I \tag{12.6.68$_0$}$$

$$(hD_t - A)U_j(t) = B_jU_{j-1}(t), \qquad U_j(0) = 0, \qquad j \ge 1 \tag{12.6.68$_j$}$$

where the operators B_j are also regular h-pseudodifferential operators of order 0. Then the results of Chapters 2 and 3 concerning the oscillatory front set of $U = U_0$ obviously remain true for U_j as well.

In particular, we get[64]

(12.6.69) Under the microhyperbolicity condition for the operator A on the energy level 0 (with the boundary conditions, if there is a boundary), $\varphi(hD_t)\chi_T(t)(\Gamma\psi U_j)$ is negligible for $\varphi \in \mathscr{C}_0^K$ supported in a small neighborhood of 0, $\chi \in \mathscr{C}_0^K([-1, -\frac{1}{2}] \cup [\frac{1}{2}, 1])$ and a sufficiently small constant $T > 0$.

Applying rescaling-decomposition arguments one can prove easily that[65]

(12.6.70) Estimate (2.1.58) remains true for $U = U_j$ as well

On the other hand, one can apply the successive approximation method in the interval $|t| \le h^{\frac{1}{2}+\delta}$. Let us note that for $B_1 = B_2 = \dots = B_j = B$,

$$\varphi(hD_t)\Gamma \operatorname{tr}(\psi U_j(.,t))\bigr|_{t=0}$$

is exactly the term to be evaluated (for $j = k + 1$ one should integrate over the parameter θ, and A depends on this parameter.

Thus for $r \asymp 1$ the remainder estimate (after we replace the whole thing by its Weyl expression) is $Ch^{1-d}\ell^\vartheta(1 + \ell/h)^{-s}$ where we have taken into account condition (12.6.67) on φ.

[64] Cf. Corollary 2.1.17 and Theorem 3.3.5.

[65] Cf. Theorems 2.1.19 and 3.3.6.

Remark 12.6.11. The procedure to get the Weyl approximation is obvious: one replaces $d_\tau E(\tau)$ by the formal parametrix $(\tau - A)^{-1}$ of the operator A (with the corresponding boundary conditions), then takes $A = A^\theta = A_0 + \theta B$ and calculates the j-th derivative with respect to θ. Afterward one applies $\mathsf{Res}_\mathbb{R}$ and formally calculates the trace of this operator.

Rescaling, we obtain that for $r \ge r_*$ the contribution to the remainder estimate is

$$C\ell^{\vartheta-j} r^{(m\mu+\nu)j} \bigl(\frac{h}{r^{\mu+1}}\bigr)^{1-d+\vartheta-j} \bigl(1 + \frac{\ell r^{\mu+1-m\mu}}{h}\bigr)^{-s}. \tag{12.6.71}$$

For $j > \vartheta$ we obtain that the sum with respect to $\ell \ge L(r)$ does not exceed this expression with $\ell = L(r)$ which is equal to $CR(r,h)$ with $R(r,h)$ given by (12.6.61) and after summation over the partition on r we get $CR^*(h)$ with with $R^*(h)$ given by (12.6.62). *Let us pick $k > \vartheta$.*

In order to treat $j \le \vartheta$ let us replace $\sum_{r\le \bar r}\sum_{\ell \ge L(r)}$ by $\sum_{\ell\ge L^*}\sum_{r\le r(L)}$ where $L^* = L(r_*) = h^{m\mu(\mu+1)^{-1}}$ and

$$r(\ell) = \bigl(\ell/h\bigr)^{1/(m\mu+\nu-\mu-1)} \tag{12.6.72}$$

inverse function to $L(r)$ defined by (12.6.60). Let us now consider the sum on r only.

(i) For

$$f(j) := m\mu\vartheta + (\mu+1)(d-1-\vartheta) + (\mu+\nu+1)j > 0 \tag{12.6.73}$$

we replace the corresponding term in the partition of the right-hand expressions of (12.6.65) by its Weyl approximations. Then the error will not exceed $CR'(\ell)$ with $R'(\ell) = R\bigl(r(\ell)\bigr)$. After summation on $\ell \ge L^*$ one again obtains $CR^*(h)$.

(ii) Observe that for $f(j) = 0$ one gets the same answer with the additional factor $|\log h|$ in the right-hand expression. Thus, we need to replace $R^*(h)$ by

(12.6.74) $R^{**}(h) =$

$$\begin{cases} h^{l'''} & \text{for } (m\mu+\nu)\vartheta+(d-1)(\mu+1)>0, \\ & \qquad \vartheta-(\mu+1)(d-1)(\mu+\nu+1)^{-1}\notin\mathbb{Z}^+, \\ h^{l'''}|\log h| & \text{for } (m\mu+\nu)\vartheta+(d-1)(\mu+1)>0, \\ & \qquad \vartheta-(\mu+1)(d-1)(\mu+\nu+1)^{-1}\in\mathbb{Z}^+, \\ h^{l''}|\log h| & \text{for } (m\mu+\nu)\vartheta+(d-1)(\mu+1)=0, \\ h^{l''} & \text{for } (m\mu+\nu)\vartheta+(d-1)(\mu+1)<0. \end{cases}$$

(iii) Let us consider the terms with $f(j)<0$. We replace them by their Weyl expressions as well but treat these terms in a different way. First, we consider the difference between the terms and their Weyl approximations on each element of the partition of unity, but we *do not estimate it by* (12.6.71).

Instead we sum this difference with respect to $r\le r(\ell)$, then we replace $\sum_{r\le r(\ell)}$ by $\sum_{r<\infty}-\sum_{r>r(\ell)}$ and only then for $r>r(\ell)$ let us apply the above remainder estimate (12.6.71) and get the same answer as before. However we also get $\sum_{r<\infty}$.

One can easily see that the contribution of $B(0,r_*)\times[0,3L^*]$ to the original expression does not exceed $Ch^{l''}$.

Later we will prove that

(12.6.75) The contribution of $B(0,2r_*)$ to terms with $f(j)>0$ does not exceed $Ch^{l'''}$ as well.

Then modulo $O\bigl(R^{**}(h)\bigr)$ our original expression is equal to

$$\text{(12.6.76)}\quad \mathsf{Weyl}+\sum_{j:f(j)<0} c_j\iint \tau_-^{\vartheta-1}\operatorname{tr}\bigl(\mathcal{K}_{(B(\tau-A^0)^{-1})^j}(x,x)-\mathsf{Weyl}_j(x,\tau)\bigr)\,dxd\tau$$

where the first Weyl is some combination of the Weyl expressions and Weyl_j is the Weyl approximation for $\mathcal{K}_{(B(\tau-A^0)^{-1})^j}(x,x)$ where $\mathcal{K}_S$ denotes the Schwartz kernel of operator S and c_j are some numerical coefficients; these terms are coming from $\sum_{r<\infty}$.

Recall that all these Weyl expressions contain the factor $\bigl(1-\phi(x/r_*)\bigr)$ cutting them in $B(0,r_*)$.

Now, let us assume for the sake of simplicity that

(12.6.77) Operator B is quasihomogeneous; i.e., $B(\zeta x, \zeta^\mu \xi) = \zeta^{m\mu+\nu} B(x,\xi)$ $\forall x, \xi, \zeta > 0$.

Later we will get rid of this condition with the possible deterioration of the remainder estimate, and it is not necessary from the very beginning if $f(1) \geq 0$.

Under this condition (12.6.77) let us consider terms with $f(j) < 0$ and with integration on $x \in \mathbb{R}^d$. Then same homogeneity arguments as in Proposition 12.6.3 imply that

(12.6.78) The terms with $f(j) < 0$ in (12.6.76) are equal to $\omega_{(\vartheta)j} h^{(m\mu\vartheta+\nu j)(\mu+1)^{-1}}$ with unknown numerical coefficients $\omega_{(\vartheta)j}$.

We will call them *Scott correction terms*. Note that

$$f(j) < 0 \iff (m\mu\vartheta + \nu j)(\mu+1)^{-1} < \min(l', l''). \tag{12.6.79}$$

Further, one can easily see that modulo $O\bigl(R^{**}(h)\bigr)$ the first Weyl is equal to

$$\sum_{p<d+\min(l',l'')} \varkappa_{(\vartheta)p} h^{-d+p} + \sum_{p<d+\min(l',l''),j:-d+p=(m\mu\vartheta+\nu j)/(\mu+1)} \varkappa'_{(\vartheta)p} h^{-d+p} |\log h| \tag{12.6.80}$$

plus the sum of the "superficial" terms $c_{(\vartheta)p,j} h^{-d+p+\nu j} \bar{r}^{\alpha_{p,j}}$ with exponents $\alpha_{p,j} := (\mu+1)(d-p-\nu j) \neq 0$ and $c_{(\vartheta)p,j} h^{-d+p+\nu j} \log \bar{r}$ with $\alpha_{p,j} = 0$.

Here $\varkappa_{(\vartheta)p}$ are standard Weyl coefficients (one can easily see that the integrals defining them converge (at 0) for $p < d + m\mu\vartheta 1/(\mu+1)$. Taking into account the fact that one can pick any $\bar{r} \asymp h^{(\mu+\nu+1)^{-1}}$ and the final answer should not depend on $\bar{r}$ we see that all these coefficients $c_{(\vartheta)p,j}$ should vanish and only (12.6.80) survives.

Therefore (12.6.76) modulo $O\bigl(R^{**}(h)\bigr)$ is equal to (12.6.80) plus sum of the terms from (12.6.78):

$$\sum_{p<d+\min(l',l'')} \varkappa_{(\vartheta)p} h^{-d+p} + \sum_{p<d+\min(l',l''),j:-d+p=(m\mu\vartheta+\nu j)/(\mu+1)} \varkappa'_{(\vartheta)p} h^{-d+p} |\log h| + \sum_{j:(m\mu\vartheta+\nu j)/(\mu+1)<\min(l',l'')} \omega_{(\vartheta)j} h^{(m\mu\vartheta+\nu j)/(\mu+1)} \tag{12.6.81}$$

where $\omega_j h^{(m\mu\vartheta+\nu j)/(\mu+1)}$ are (multiple) Scott correction terms. More precisely, defining

$$\bar{\vartheta}_J := \max\Bigl(\bigl((\mu+1)(d-1)+(\mu+\nu+1)J\bigr)(\mu+1-m\mu)^{-1}, J\Bigr) \tag{12.6.82}$$

we see that

(12.6.83) For $\bar{\vartheta}_{J-1} < \vartheta \le \bar{\vartheta}_J$ there are exactly J Scott correction terms $(j=1,\dots.)$

To prove pending claim (12.6.75) let us now consider an element of the partition supported in $B(0,2r_*)$. Let us replace every factor B by $B_1 = \bar{\psi}B$ where the cut-off function $\bar{\psi}$ is supported in $B(0,4r_*)$ and equal to 1 in $B(0,r_*)$. We will explain how to get rid of this cut-off function later.

Let us note that

(12.6.84) For $\tau \le 0$ the operator norm of $\bar{\psi}BE^\theta(\tau)$ does not exceed $r_*^{m\mu+\nu}$ while the trace norm of $\bar{\psi}E^\theta(\tau)$ does not exceed C [66].

One can prove this easily for $h=r_*=1$ and then extend to the general case by rescaling (recall that $r_*^{\mu+1}=h$).

Then the trace norm of the operator

$$\int \varphi_{\vartheta,k+1}(\lambda_1,\dots,\lambda_j)\psi E^\theta(\lambda_1)B_1E^\theta(\lambda_2)\cdots B_1E^\theta(\lambda_{k+1})\,d\lambda_1\cdots d\lambda_{k+1}$$

does not exceed $\ell^{\vartheta-k}r_*^{(m\mu+\nu)k}$ and after summation on the partition we get the same expression with $\ell=L=r_*^{m\mu+\nu}$ provided $k>\vartheta$. Namely, we get $r_*^{(m\mu+\nu)\vartheta} = h^{(m\mu+\nu)\vartheta/(\mu+1)}$.

Finally, we should consider similar expressions with some copies of B replaced by $\bar{B}$ and other copies replaced by $\psi'B$ where ψ' is supported outside of $B(0,3r_*)$. Taking partitions one can replace the i-th copy of ψ' by ψ_i supported in $\{\frac{1}{2}r_i \le |x| \le 2r_i\}$. Combining the above arguments one can estimate this term through $r_*^{(m\mu+\nu)\vartheta}\prod_i (r_*/r_i)^s$ and the total contribution will not exceed $r_*^{(m\mu+\nu)\vartheta}$ again.

Thus, all the estimates are done. Therefore the first statement of the following theorem has been proven:

[66] One can add the factor $\bigl(|\tau|r_*^{-m\mu}+1\bigr)^{-s}$ with arbitrarily large s to the right side of these estimates but it is not important here.

Theorem 12.6.12. *Let the conditions of Theorem 12.6.8 be fulfilled.*

(i) Moreover, let condition (12.6.77) *be fulfilled. Then asymptotics*

$$\int \psi(x)e_{(\vartheta)}(x,x,\tau)\,dx = \sum_{p:p<d+\min(l',l'')} \varkappa_{(\vartheta)p}h^{-d+p} + \sum_{p<d+\min(l',l''),j:-d+p=(m\mu\vartheta+\nu j)/(\mu+1)} \varkappa'_{(\vartheta)p}h^{-d+p}|\log h| + \sum_{j:(m\mu\vartheta+\nu j)/(\mu+1)<\min(l',l'')} \omega_{(\vartheta)j}h^{(m\mu\vartheta+\nu j)/(\mu+1)} + O\bigl(R^{**}(h)\bigr) \tag{12.6.85}$$

holds with $R^{**}(h)$ *given by* (12.6.74).

(ii) Moreover, for $\vartheta\le\bar\vartheta_1$ *one can drop condition* (12.6.77).

(iii) Let us assume that

$$\bigl|D^\beta(a_\alpha - a^0_\alpha - b^0_\alpha)\bigr| \le c|x|^{(m-|\alpha|)\mu+\nu'-|\beta|} \tag{12.6.86}$$

where a^0_α, b^0_α *are positively homogeneous of degrees* $(m-|\alpha|)\mu$, $(m-|\alpha|)\mu+\nu$ *respectively,* $\nu'>\nu$.

Then asymptotics (12.6.85) *holds with* $R^{**}(h)$ *replaced by* $R^{**}(h)+R^{*\prime}(h)$ *where* $R^{*\prime}(h)$ *is given by* ((12.6.49) *with* ν' *instead of* ν.

The proof of Statement (iii) is simple: we first make the transition from A to A^0+B^0 (then the remainder estimate $O\bigl(R^*(h)\bigr)$ arises) and then apply Statement (i).

We believe that it would not be difficult to

Problem 12.6.13. Generalize this theorem to the case when

$$D^\beta\bigl(b_\alpha - b^1_\alpha - \ldots - b^s_\alpha\bigr) = O\bigl(|x|^{(m-\alpha)\mu+\nu_{s+1}-|\beta|}\bigr)$$

and b^k_α are positively homogeneous of degrees $(m-|\alpha|)\mu+\nu_k$.

12.6.6 Propagation of Singularities

Now our goal is to improve the remainder estimate, which requires some condition of a global nature on the symbol $A^0(x,\xi)$. To formulate the condition the reader must recall the definitions of generalized bicharacteristics and generalized δ-bicharacteristics (see Definition 2.2.8 or similar notion in Chapter 3); otherwise we should assume that the eigenvalues of $A^0(x,\xi)$ which are close to 0 have constant multiplicities. Moreover, the reader should remember quasiconical sets and the quasiconical hull of a set; these notions are associated with the rescaling family $\Phi_\zeta : (x,\xi) \to (\zeta x, \zeta^\mu \xi)$ with $\zeta > 0$.

Definition 12.6.14. We say that the *escape condition* is fulfilled if there exist quasiconical sets Λ^+ and Λ^- closed in $(\mathbb{R}^d \setminus 0) \times \mathbb{R}^d$ and numbers $T_0 > 0$, $\epsilon_1 > 0$ such that

$$\Lambda^+ \cup \Lambda^- = \Sigma^0 = \bigl\{(x,\xi) \in (\mathbb{R}^d \setminus 0) \times \mathbb{R}^d : \det A^0(x,\xi) = 0\bigr\} \tag{12.6.87}$$

and for every point $(\bar{x},\bar{\xi})$ belonging to the ϵ_1-neighborhood of $\Lambda^\pm$ intersected with $\bigl\{(x,\xi) : |x| = 1\bigr\} \cap \Sigma^0$ all the generalized bicharacteristics of the symbol $-A^0(x,\xi)$ passing through $(\bar{x},\bar{\xi})$ in the direction of $\pm t > 0$ are infinitely long with respect to t, and along them

$$(x(t),\xi(t)) \in \Lambda^\pm, \quad |x(t)| \geq 2|\bar{x}| \qquad \forall t : \pm t \geq T_0; \tag{12.6.88}$$

these conditions should be fulfilled for each sign "+" and "−" separately.

We discuss this condition below; here we make certain corollaries and remarks concerning it.

Remark 12.6.15. (i) Condition (12.6.4) yields that $\Sigma^0 \cap \{|x| = 1\}$ is compact.

(ii) Under condition (12.6.88) for each generalized bicharacteristics $(x(t),\xi(t))$ with $(x(0),\xi(0)) \in \Lambda^\pm_{\epsilon_1}$ for $\pm t > 0$ inequalities

$$\epsilon_0 \leq |x(t)| \cdot \bigl(|x(0)| + |t|^\sigma\bigr)^{-1} \leq C_0 \tag{12.6.89}$$

hold with

$$\sigma^{-1} := \sigma' := 1 + \mu - m\mu > 0 \tag{12.6.90}$$

(otherwise $\bar{\vartheta}_0 = +\infty$) and with appropriate constants ϵ_0, C_0 which do not depend on t and $(x(0),\xi(0))$; here and below $\Lambda^\pm_\varepsilon$ denotes the intersection of the quasiconical hull of ε-neighborhood of $\Lambda^\pm \cap \{|x| + 1\}$ with Σ^0.

In fact, let us pick t_n: $t_0 = 0$, $t_{n+1} = t_n + T_0 r_n^{\sigma'}$ with $r_n = |x(t_n)|$. Then condition (12.6.88) and the quasihomogeneity of A^0 and the quasiconic nature of Λ^+ yield that $2r_n \le r_{n+1} \le c_0 r_n$ (we have taken $(x(0), \xi(0)) \in \Lambda^+$ for notational simplicity).

Therefore $r_n \to +\infty$ as $n \to +\infty$. Then these inequalities and the definition of t_{n+1} obviously yield the inequality $r_{n+1} \le c_1 t_{n+1}^{\sigma}$ (and hence $t_n \to +\infty$ as $n \to +\infty$) and the inequality $t_{n+1} r_{n+1}^{-\sigma'} \le c_2 + (1-\epsilon) t_n r_n^{\sigma'}$ with $\epsilon > 0$. This inequality yields by induction that $t_n r_n^{-\sigma'} \le c_3$ with a sufficiently large constant c_3. So, both inequalities (12.6.89) are fulfilled for $t = t_n$ with $n = 0, 1, \dots$ and therefore these inequalities are fulfilled for $t \in (t_n, t_{n+1}]$ as well.

(iii) Let us replace the symbol $A^0(x, \xi)$ with another symbol $A^1(x, \xi)$ also of the form (12.6.1) such that

$$|D^\beta (a^1_\alpha - a^0_\alpha)| \le \varepsilon |x|^{\mu(m-|\alpha|)-|\beta|} \qquad \forall \alpha : |\alpha| \le m \quad \forall \beta : |\beta| \le K \tag{12.6.91}$$

with a small enough constant $\varepsilon > 0$. We *do not* assume that $A^1(x, \xi)$ is a quasihomogeneous symbol. Then

$$|\xi| \le c_3 |x|^\mu \qquad \text{on } \Sigma^j \qquad j = 0, 1 \tag{12.6.92}$$

and the microhyperbolicity of A^0 on the energy level 0 yields that

$$\forall (x, \xi) \in \Sigma^0 \quad \exists (x', \xi') \in \Sigma^1 : \\ |x - x'| \le c_3 \varepsilon |x| \text{ and } |\xi - \xi'| \le c_4 \varepsilon |x|^\mu \tag{12.6.93}$$

and *vice versa* (one can permute Σ^0 and Σ^1 in this assertion), and c_4 does not depend on ε.

Let us introduce the sets

$$\Lambda_1^\pm = \{(x, \xi) \in \Sigma^1 : \\ \exists (x', \xi') \in \Lambda^\pm : |x - x'| \le \epsilon |x| \text{ and } |\xi - \xi'| \le \epsilon |x|^\mu\} \tag{12.6.94}$$

with sufficiently small $\epsilon > 0$. Moreover, let us introduce the sets $\Lambda_2^\pm$ in the same way but with ϵ replaced 2ϵ.

Then the properties of the generalized bicharacteristics (see Lemma 2.2.9) and condition (12.6.88) yield that if $\epsilon > 0$ is a small enough constant and if $\varepsilon = \varepsilon(\epsilon) > 0$ is also small enough then every generalized bicharacteristics $(x(t), (\xi(t))$ of the symbol $A^1(x,\xi)$ passing through the point $(\bar{x},\bar{\xi}) \in \Lambda_2^\pm$ in the direction of positive $\pm t$ can be extended to the time interval $\pm t \in [0, T_1|\bar{x}|^{\sigma'}]$, and for $\pm t \in [2T_0|\bar{x}|^{\sigma'}, T_1|\bar{x}|^{\sigma'}]$ along this bicharacteristics

$$(x(t),\xi(t)) \in \Lambda_1^\pm, \qquad |x(t)| \ge \frac{3}{2}|\bar{x}|; \tag{12.6.88$'$}$$

$T_1 > 2T_0$ is arbitrarily large and ϵ and ε depend on T_1.

Then the same arguments as in Statement (i) yield that along these trajectories inequalities (12.6.91) are fulfilled with constants c_0, ϵ_0 which do not depend on T_1. Then the arbitrary choice of T_1 yields that this statement remains true even for $T_1 = +\infty$.

(iv) The same analysis remains true if we assume that (12.6.91) is fulfilled in $B(0,r)$ instead of $\mathbb{R}^d$; surely, we should treat the trajectory only as long as it remains in $B(0,r)$. In this case (12.6.91) yields that all the statements remain true in the time interval $\pm t \in [0, \epsilon_2 r^{\sigma'}]$ with a small enough constant $\epsilon_2 > 0$ and $|\bar{x}| \le \frac{1}{2}r$.

(v) Therefore in the analysis of the trajectories on the energy level τ instead of $\tau = 0$ all the Statements (i)–(iii) remain true with the following modifications:

(a) For $\mu < 0$ one should replace r with $\min\bigl(r, \epsilon_2|\tau|^{(m\mu)^{-1}}\bigr)$, i.e., one should treat the starting points $(\bar{x},\bar{\xi})$ with $|\bar{x}| \le \min\bigl(r, \epsilon_2|\tau|^{(m\mu)^{-1}}\bigr)$ and the time interval $\pm t \in \bigl[0, \epsilon_2 \min(r^{\sigma'}, |\tau|^{(m\mu\sigma)^{-1}})\bigr]$.

(b) For $\mu > 0$ one should leave both r and the time interval $\pm t \in [0, \epsilon_2 r^{\sigma'}]$ unchanged and one should treat the starting points $(\bar{x},\bar{\xi})$ with $c_2|\tau|^{(m\mu)^{-1}} \le |\bar{x}| \le r$.

(c) rem-12-5-15-vc For $\mu = 0$ one should leave r and the time interval $\pm t \in [0, \epsilon_2 r^{\sigma'}]$ and the condition $|\bar{x}| \le r$ unchanged and one should only assume that $|\tau| \le \epsilon_2$ where $\epsilon_2 > 0$ is a small enough constant.

Remark 12.6.16. Further, let us consider a more complicated flow. Namely, let us assume that at the moments $\pm t_n = T_1^n|\bar{x}|^{\sigma'}$ with $n = 1, 2, \dots$ the point $(x(t),\xi(t))$ makes random "jumps"; namely it jumps to an arbitrary point in its εr_n-neighborhood intersected with the set Ω; $\Omega = \Pi_{\tau_0}$ for $\mu \le 0$ and

$\Omega = \Pi_{\tau_n}$ with $\tau_n = (r_n/r_*)^{m\mu}\tau_0$ for $\mu > 0$ where $\Pi_\varsigma = \cup_{\tau:|\tau|\le\varsigma}\Sigma^1_\tau$ and we assume that $\tau_0 > 0$.

So, we assume that the "jump" and the flow itself preserve the energy levels only approximately. This exotic flow will be useful later. We claim that for sufficiently small $\varepsilon > 0$ and for sufficiently large $T_1 > 0$, Statements (i)–(v) of Remark 12.6.15 remain true.

In fact, let $r_n = |x(\pm t_n \pm 0)|$. Then Remark 12.6.15 and the conditions on the jump at the end of a segment of the trajectory yield that if

$$0 < \epsilon_0 \le r_n t_n^{-\sigma} \le c_0 \tag{12.6.95$_n$}$$

then for sufficiently large $T_1 > 0$ depending on c_0, ϵ_0 estimates

$$r_n + \epsilon_1 t^\sigma_{n+1} \le r_{n+1} \le r_n + c_1 t^\sigma_{n+1}$$

hold with c_1, ϵ_1 which do not depend on c_0, ϵ_0. Then under an appropriate choice of the constants c_0, $\epsilon_0 > 0$ condition $(12.6.95)_n$ yields $(12.6.95)_{n+1}$ with the same constants c_0, ϵ_0; then induction on n yields (12.6.91).

Moreover, the energy level after a jump remains admissible.

Let us prove some assertions concerning the long-time propagation of singularities:

Proposition 12.6.17. *Let the operator A satisfy* (12.6.2)–(12.6.6). *Let $b_\alpha = a_\alpha - a^0_\alpha$ satisfy* (12.6.30) *and let a^0_α be positively homogeneous functions of degrees $(m-|\alpha|)\mu$. Let symbol A^0 satisfy conditions* (12.6.20) *and escape condition* (12.6.87)–(12.6.88).

Further, let $r \in [r_* h^{-\delta}, \varepsilon_0]$ with $r_* = h^{(\mu+1)^{-1}}$, an arbitrarily small exponent $\delta > 0$ and a sufficiently small constant $\varepsilon_0 > 0$. Let $h' = hr^{-\mu-1}$, $\gamma = h'^\rho$ with an exponent $\rho \in (0, \frac{1}{2})$. Let $\tau_0 = \epsilon_3 r^{m\mu}\gamma$ for $\mu < 0$ and $\tau_0 = \epsilon_3 r^{m\mu}$ for $\mu \ge 0$.

Furthermore, let $Q \in \boldsymbol{\Psi}_{h,\gamma r^\mu,\gamma r}$ be an operator with the symbol supported in the intersection of the $\varepsilon(r, r^\mu)$-neighborhood (in (x,ξ)) of the set $\Lambda^\pm \cap \{\epsilon_0 r \le |x| \le c_0 r\}$ and Π_{τ_0}.

Finally, let

$$T'' = \begin{cases} \epsilon_3 \min(1, r^{\sigma'}\gamma^{(m\mu\sigma)^{-1}}) & \text{for } \mu < 0,\\ \epsilon_3 & \text{for } 0 \le \mu < (m-1)^{-1} \end{cases} \tag{12.6.96}$$

and $T' = c_3 r^{\sigma'}$ *with small enough constants* $\epsilon_3 > 0$ *and* $\varepsilon > 0$ *and a large enough constant* c_3*. Let* $\bar{T} \in [T', T'']$.

Then there exists an operator $\bar{Q} = \bar{Q}_{\bar{T}} \in \boldsymbol{\Psi}_{h,\bar{\gamma}\bar{r}^{\mu},\bar{\gamma}\bar{r}}$ *with the symbol supported in the* $\varepsilon(\bar{r},\bar{r}^{\mu})$*-neighborhood (in* (x,ξ)*) of the set* $\Lambda^{\pm} \cap \{\bar{\epsilon}_0 r \le |x| \le \bar{c}_0 r\}$ *with* $\bar{r} = \bar{T}^{\sigma'}$, $\bar{h}' = h\bar{r}^{-\mu-1}$, $\bar{\gamma} = \bar{h}'^{\bar{\rho}}$ *such that the equality*

$$U(\pm\bar{T})Q \equiv \bar{Q}U(\pm\bar{T})Q \tag{12.6.97}$$

holds for the operator $U(t) = \exp(ih^{-1}tA)$ *modulo operators with operator norm less than* Ch^{s} *with arbitrarily large* s.

Moreover, constants $\bar{\epsilon}_0$, $\bar{c}_0$ *do not depend on* $\varepsilon > 0$ *(but the constant* c_3 *does) and one can choose the exponent* $\bar{\rho} > \rho$ *arbitrarily close to* ρ*, and the statement that the operator belongs to the operator class is uniform with respect to the parameters* h, r, $\bar{T}$.

Proof. Let us consider the sign "+" for definiteness. Let us pick $Q_0 = Q$ and define moments t_n according to Remark 12.6.16 with T_1 depending on ε. Let $r_n = t_n^{\sigma}$. Let us introduce γ_n, h'_n with r replaced by r_n and with ρ replaced by $\bar{\rho}$ as was done above. We want to construct operators $Q_n \in \boldsymbol{\Psi}_{h,\gamma_n r_n^{\mu},\gamma_n r_n}$ with symbols supported in the $\varepsilon(r_n, r_n^{\mu})$-neighborhoods of the sets $\Lambda \cap \{\bar{\epsilon}_0 r_n \le |x| \le \bar{c}_0 r_n\}$ intersected with Π_{τ_n} (where we define τ_n later) such that for all n with $t_n \le T''$ equalities

$$U(t_{n+1}-t_n)Q_n \equiv Q_{n+1}U(t_{n+1}-t_n)Q_n \tag{12.6.98}$$

are fulfilled, and all statements that operators belong to operator classes and all equivalence relations are uniform in h, r, n; here $t_0 = 0$.

Let us rescale $x_{\mathsf{new}} = x_{\mathsf{old}} r_n^{-1}$, $\xi_{\mathsf{new}} = \xi_{\mathsf{old}} r_n^{-\mu}$, $t_{\mathsf{new}} = t_{\mathsf{old}} r_n^{-\sigma'}$ and set $h_n = h r_n^{-\mu-1}$. Then we get a routine problem of propagation of singularities in bounded time in the domain $\{\epsilon \le |x_{\mathsf{new}}| \le c, |\xi_{\mathsf{new}}| \le c\}$.

This problem was treated in Sections 2.1, 2.2 and therefore Theorem 2.2.10 and the Remark 12.6.16 yield the following procedure to construct Q_n: let Q_n be constructed and let M_n be the support of its symbol. Let us consider all the generalized bicharacteristics passing through M_n at the moment $t = 0$; let M'_{n+1} be the set of all the ends of these trajectories at the moment $t = (t_{n+1} - t_n)$.

Then we can choose Q_n from the class indicated above, with symbol equal to 1 in the $\frac{1}{2}\varepsilon(r_{n+1}, r_{n+1}^{\mu})$-neighborhood of M'_{n+1} intersected with $\Pi_{\tau_{n+\frac{1}{2}}}$ and supported in the $\varepsilon(r_{n+1}, r_{n+1}^{\mu})$-neighborhood of M'_{n+1} intersected with

$\Pi_{\tau_{n+1}}$ where $\tau_k = \tau_0(1+kh^{\delta'})$ with a small enough exponent $\delta'>0$ for $\mu\le 0$ and $\tau_k=\epsilon' r_k^{m\mu}$ with a small enough constant $\epsilon'>0$ for $\mu>0$ [67]. Obviously, $t_k\le T''$ yields $k\le C_0|\log h|$ and therefore $\tau_k\le 2\tau_0$ for $\mu\le 0$.

Then (12.6.98) yields that for n with $t_{n+1}\le T''$

$$U(t_{n+1})Q_0 \equiv Q_{n+1}U(t_{n+1}-t_n)Q_n(t_n-t_{n-1})Q_{n-1}\cdots Q_1U(t_1)Q_0$$

and we immediately obtain (12.6.97) for $\bar{T}=t_{n+1}$. However, if $t_n<\bar{T}<t_{n+1}$ then one can always redefine t_n and take $t_{n+1}=\bar{T}$ in our analysis. □

So, in our analysis for $\mu\ge 0$ we can take small enough constants $\tau_0>0$ and $T''>0$. For $\mu<0$ the situation is more subtle: we took the smallest $\tau_0=r^{m\mu}\gamma$ admissible by the uncertainty principle (for coordinates and momenta) and the condition $T''\le \epsilon_3 r^{\sigma'}\gamma^{(m\mu\sigma)^{-1}}$ means that $\tau_0\le |x|^{m\mu}$ along the trajectory. In order to avoid this restriction we must use some subtle arguments.

Now we want to extend the time interval:

Proposition 12.6.18. *Let the conditions of Proposition 12.5.17 be fulfilled and let $\varphi\in\mathscr{C}_0^K(\mathbb{R}^d)$ be supported in $\{1\le|x|\le 2\}$, $\chi\in\mathscr{C}_0^K([\frac{1}{4},1])$ [68]. Then for the Schwartz kernel $u(x,y,t)$ of the operator $U(t)=\exp(ith^{-1}A)$ estimate*

$$|F_{t\to h^{-1}\tau}\operatorname{tr}\Bigl(\chi_T(t)\int\varphi(\frac{x}{r})u(x,x,t)\,dx\Bigr)|\le Ch^s \tag{12.6.99}$$
$$\forall\tau:|\tau|\le\frac{1}{2}\tau_0\quad\forall T\in[3T',T^*]$$

holds where $\tau^*=\min(\frac{1}{2}\tau_0,\epsilon')$, $T^*>0$ and $\epsilon'>0$ are small enough constants, $T'=c_1r^{\sigma'}$, c_1 is a large enough constant and $\chi\in\mathscr{C}_0^K([-1,1])$ is arbitrary. Recall that $r\in(r^*,\varepsilon_0)$.

Remark 12.6.19. (i) Therefore one can pick $\tau^*=\epsilon'$ for $\mu\le 0$ and $\tau^*=\epsilon' r^{m\mu}$ for $\mu>0$.

(ii) This proposition and Corollary 12.6.20 (see below) remain true if one replaces $u(x,y,t)$ by $B_{j+1}u_j(x,y,t)$ where u_j is a solution of the recurrence problem $(12.6.68)_0$–$(12.6.68)_j$.

[67] A neighborhood of $\Pi_{\tau_{n+\frac{1}{2}}}$ may be cut out by standard elliptic arguments.

[68] Then this proposition remains true for $\chi\in\mathscr{C}_0^K([-1,-\frac{1}{4}]\cup[\frac{1}{4},1])$.

Proof of Proposition 12.6.18. Let us make a partition of unity in a neighborhood of $\mathsf{supp}(\varphi_r)$ where $\varphi_r(x) = \varphi(x/r)$: $\varphi_r = Q^+ + Q^- + Q'$ where $Q^+, Q^-, Q' \in \boldsymbol{\Psi}_{h,\gamma r^\mu,\gamma r}$ with symbols of $Q^\pm$ supported in the $\varepsilon(r, r^\mu)$-neighborhoods of $\Lambda^\pm \cap \{\epsilon_0 r \le |x| \le c_0 r\}$ intersected with Π_{τ_0} and with symbol of Q' supported in $\{\epsilon_0 r \le |x| \le c_0 r\} \setminus \Pi_{\frac{2}{3}\tau_0}$. Then

$$U(t)\varphi_r = U(t)Q^+ + U(t)Q^- + U(t)Q' \tag{12.6.100}$$

and standard elliptic arguments yield that for $V(t) = U(t)Q'$ estimate

$$\|F_{t\to h^{-1}\tau}\chi_T(t)V(t)\| \le Ch^s \quad \forall T \in [3T', T^*] \quad \tau : \forall \tau : |\tau| \le \tau^* \tag{12.6.101}$$

holds where the norm in the left-hand expression is the operator norm and we use the definition of Q', τ_0, T' and the inequalities $r \ge h^{(\mu+1)^{-1}-\delta}$ and $\tau_0 T \ge h^{1-\delta'}$ due to $\mu < (m-1)^{-1}$.

We denote equivalence modulo operator-valued functions satisfying (12.6.101) by "$\cong$." So,

$$U(\pm t)\varphi_r \cong U(\pm t)Q^+ + U(\pm t)Q^-. \tag{12.6.102}$$

Then applying $F_{t\to h^{-1}\tau}\chi_T(t)\,\mathsf{Tr}$ we obtain that for $T \in [3T', T'']$, both terms in the right-hand expression of (12.6.102) will result in $O(h^s)$ in the final answer; this follows from Proposition 12.6.17, and for the second term one should pass to the adjoint operator and then permute Q^{-*} and $U(-t)$ because of the presence of "Tr".

Thus, for $\mu \ge 0$ the proposition is proven and for $\mu < 0$ it is proven only under the additional restriction $T < T''$.

Let $\mu < 0$ and $T > T''$; then $T'' < T^*$. Let us take $T_1 = \epsilon T''$ with sufficiently small $\epsilon > 0$ and let us construct the operators $\bar{Q}^+$ and $\bar{Q}^-$ associated with $Q = Q_0^+ = Q^+$ and with $Q = Q_0^- = Q^-$ respectively according to Proposition 12.6.17; $\bar{T} = T_1$. Then (12.6.97) yields that

$$U(\pm t)\,Q_0^\pm \cong U(\pm(t - T_1))\,\bar{Q}^\pm U(\pm T_1)\,Q_0^\pm. \tag*{(12.6.103)$_1$}$$

Let us represent $\bar{Q}^\pm$ in the form $\bar{Q}^\pm = Q_1^\pm + Q_1^{\pm\prime}$ where $Q_\bullet \in \boldsymbol{\Psi}_{h,\gamma_1 r^{*\mu},\gamma_1 r^*}$ and the symbols of $Q_1^\pm$ and $Q_1^{\pm\prime}$ are supported in the $\varepsilon(r^*, r^{*\mu})$-neighborhoods of $\Lambda^\pm \cap \{\bar{\epsilon}_1 r^* \le |x| \le \bar{c}_1 r^*\}$ intersected with Π_{τ_1} and in $\{\bar{\epsilon}_1 r^* \le |x| \le \bar{c}_1 r^*\} \setminus \Pi_{\frac{2}{3}\tau_1}$ respectively where $r_n = T_n^\sigma$, $\hbar_n = hr_n^{-\mu-1}$, $\gamma_n = \hbar_n{}^{\rho_n}$, $\rho < \rho_1 < \dots < \rho_n$ now and one can choose ρ_n arbitrarily close to ρ; now the subscript "1" replaces the "bar."

Elliptic arguments again yield that $U(t-T_1)Q^{\pm\prime}\cong 0$ and the presence of the factor $U(T_1)Q^{\pm}$ is not essential because here we do not apply "Tr". Therefore $(12.6.96)_1$ remains true with $\bar{Q}_1^{\pm}$ replaced by $Q_1^{\pm}$ where we replace τ_0 by τ_1 in the definition of "$\cong$." Repeating this process we obtain that

$$(12.6.103)_n \quad U(\pm t)\,Q_0^{\pm}\cong \\ U(\pm(t-T_n))\,Q_n^{\pm}U(\pm(T_n-T_{n-1}))\,Q_{n-1}^{\pm}\cdots Q_1^{\pm}U(\pm T_1)\,Q_0^{\pm}$$

where we replace τ_0 by τ_n in the definition of "$\cong$."

Therefore again applying Proposition 12.6.17 we obtain that

$$U(\pm t)Q_0^{\pm}\cong\bar{\varphi}_{r_n}U(\pm t)Q_0^{\pm}$$

provided $T\in[\bar{\epsilon}_n,T_{n+1}]$, $\bar{\varphi}=1$ in $\{|x|\ge\epsilon_4\}$ with a sufficiently small constant $\epsilon_4>0$ and χ is supported in $[\frac{1}{4},1]$. Hence, applying the operator $F_{t\to h^{-1}\tau}\chi_T(t)\operatorname{Tr}$ we obtain relations

$$F_{t\to h^{-1}\tau}\chi_T(t)\operatorname{Tr}U(t)Q^{\pm}=O(h^s)$$

where in the "$-$" case one should pass to the adjoint operator and move Q_0^{-*} to the right (and replace Q_0^{-*} by Q_0^- in the analysis).

Therefore, it remains to prove that in the case $\mu<0$ we can pass from $r_*\ge h^{1/(\mu+1)-\delta}$ to $r_n>\varepsilon_0$ with a small enough constant ε_0 in a finite number of steps depending only on the exponents. But this assertion obviously follows from the equality

$$hr_n^{-\mu-1}=\mathsf{const}_n\bigl(hr_{n-1}^{-\mu-1}\bigr)^{1-\rho_{n-1}/(m\mu)}$$

which is due to (12.6.96) for $r_n<\epsilon$; here $1-\rho_{n-1}(m\mu)^{-1}>1-\rho(m\mu)^{-1}>1$ and $hr_*^{-\mu-1}\le h^{\delta''}$ with $\delta''>0$ and therefore for appropriate n the inequality $hr_n^{-\mu-1}\le Ch$ holds. □

12.6.7 Improved Remainder Estimates

Applying Proposition 12.6.18 we obtain Statement (i) of the following assertion:

Corollary 12.6.20. *(i) In the framework of Proposition 12.6.18 estimate*

$$(12.6.104)\quad \bigl|F_{t\to h^{-1}\tau}\bigl(\bar{\chi}_{T_2}(t)-\bar{\chi}_{T_1}(t)\bigr)\int\varphi\bigl(\frac{x}{r}\bigr)\Gamma u(x,x,)\,dx\bigr|\le Ch^s \\ \forall\tau:|\tau|\le\tau^*$$

holds with a sufficiently small constant $T_2 > 0$ *and* $T_1 = C_0 r^{\theta'}$. *Recall that* $\bar{\chi} \in \mathscr{C}_0^K([-1,1])$ *and* $\bar{\chi} = 1$ *on* $[-\frac{1}{2}, \frac{1}{2}]$.

(ii) Moreover, $T_1 = \varepsilon r^{\sigma'}$ *works with an arbitrarily small constant* $\varepsilon > 0$.

Proof. To prove Statement (ii) note that in the time interval $|t| \le T_1$ Theorem 2.2.10 is applicable with no modification and escape condition (12.6.87)–(12.6.88) obviously forbids periodic generalized trajectories for the symbol A^0 on the energy level 0. So these trajectories are also forbidden for A under the restrictions $|x| = r$, $T \le Lr^{\sigma'}$, $r \le r_*$ with arbitrarily large L and sufficiently small $r_* = r_*(L) > 0$; moreover, one can consider energy levels τ with $|\tau| \le \tau^*$. ☐

Now we can get very precise remainder estimates:

Theorem 12.6.21. *Let conditions of Theorem 12.6.8 and escape condition* (12.6.87)–(12.6.88) *be fulfilled.*

(i) Moreover, let condition (12.6.77) *be fulfilled. Then asymptotics*

$$\begin{aligned}\text{R} := \Bigl| \int \psi(x) e_{(\vartheta)}(x,x,\tau)\,dx - \sum_p \kappa_{(\vartheta)p} h^{-d+p} - \\ \sum_{p,j:-d+p=(m\mu\vartheta+\nu j)/(\mu+1)} \kappa'_{(\vartheta)p} h^{-d+p} |\log h| - \\ \sum_j \omega_{(\vartheta)j} h^{(m\mu\vartheta+\nu j)/(\mu+1)} \Bigr| = O\bigl(R^{\#}(h)\bigr)\end{aligned} \tag{12.6.105}$$

holds with

$$R^{\#}(h) = h^{\min(-d+1+\vartheta, l')}; \tag{12.6.106}$$

recall that $l' = (m\mu + \nu)\vartheta(\mu+1)^{-1}$.

Moreover, for $\vartheta \le 1$ *one can skip condition* (12.6.77) *unless* $\vartheta = 1$ *and* $(m\mu + \nu) = (-d+p)(\mu+1)$ *with* $p = 0, 1$.

(ii) Furthermore, let conditions of Statement (i) be fulfilled and

$$\vartheta(m\mu + \nu - \mu - 1) + (d-1)(\mu+1) > 0 \tag{12.6.107}$$

(then $l' > -d+1+\vartheta$ and Statement 12.6.21 yields the remainder estimate $O(h^{-d+1+\vartheta})$) and let the standard non-periodicity condition for the operator A be fulfilled.

Then one can replace the remainder estimate $O\bigl(h^{-d+\vartheta+1}\bigr)$ with $o\bigl(h^{-d+\vartheta+1}\bigr)$.

(iii) Replacing condition (12.6.77) *with* (12.6.86) *one should replace $R^\#(h)$ by $R^\#(h)+R^{*\prime}(h)$ with $R^{*\prime}(h)$ defined by* (12.6.62) *with ν' instead of ν.*

Proof. Let us first consider the zone $\bigl\{|x|\ge r'_*=r_*h^{-\delta}\bigr\}$. Let us change the coordinate system $x'=xr^{-1}$ and set $h'=hr^{-\mu-1}$, $t'=tr^{-\sigma}$. Then

$$F_{t\to h^{-1}\tau}\chi_T(t)\int \varphi(\frac{x}{r})u(x,x,t)\,dx \tag{12.6.108}$$

transforms to

$$r^\sigma F_{t'\to h'^{-1}\tau'}\chi_{T'}(t)\int \varphi(x')u'(x',x',t')\,dx'$$

where $T'=Tr^{-\sigma}$, $\tau'=\tau r^{-m\mu}$ (then $|\tau|\le\tau_0 \implies |\tau'|\le\epsilon'$), $u'(.,.,t')$ is the Schwartz kernel of the operator $\exp(ith'^{-1}A')$, $A'=r^{-m\mu}A$ in the new coordinate system and then for $T=T_1=\epsilon_0 r^{\sigma'}$ with small enough constants ϵ_0, ϵ_1 the results of Chapter 4 yield the standard Weyl formula for expression (12.6.108) with $T=T_1$ modulo $O(h^s)$. Therefore this formula also holds for expression (12.6.108) with $T=T_2$.

Recall that in Proposition 12.6.17 $\gamma=h'^\rho$ with an exponent $\rho\in(0,\frac12)$ and τ_0 is defined there. Then, for sufficiently small $\rho>0$ we obtain that $\tau^*\ge h^{1-\delta''}$ with $\delta''=(\mu+1-m\mu)(\mu+1)^{-1}>0$ and therefore taking $T=T_2$ and applying standard Tauberian arguments we obtain estimate

$$\begin{gathered}\bigl|\int \varphi(\frac{x}{r})\bigl(e_{(\vartheta)}(x,x,0)-\sum_{0\le p\le\bar p}\kappa_{(\vartheta)p}(x)h^{-d+p}\bigr)\,dx\bigr|\le\\ Ch^{-d+\vartheta+1}r^{m\mu\vartheta+(d-\vartheta-1)(\mu+1)}\bigl(\frac{T_1}{T_2}\bigr)^{\vartheta+1}=Ch^{-d+\vartheta+1}r^{d(\mu+1)-m\mu}.\end{gathered}\tag{12.6.109}$$

Here the right-hand expression has an additional factor

$$\bigl(\frac{T_1}{T_2}\bigr)^{\vartheta+1}\asymp r^{(\vartheta+1)\theta'}=r^{(\vartheta+1)(\mu+1-m\mu)}$$

if we compare it with the right-hand expression obtained in Subsection 12.6.1, $\bar p=\bar p(d,m,\mu,\vartheta,s,\delta)$ is large enough.

Since $d(\mu+1)-m\mu>0$ we immediately obtain that

(12.6.110) The contribution of the zone $\{r_3 \le |x| \le \varepsilon\}$ to the remainder does not exceed $\varepsilon' h^{1-d+\vartheta}$ with $\varepsilon' \to 0$ as $\varepsilon \to 0$.

Let us apply the arguments of Subsection 12.6.3 with $r'_* = h^{(\mu+1)^{-1}-\delta}$, $L = L(r^*) = h^{(m\mu+\nu)(\mu+1)^{-1}}$ and sufficiently small $\delta > 0$ depending on $d, m, \mu, \vartheta < \bar{\vartheta}_1, \nu$.

(a) Let us first assume that condition (12.6.107) is fulfilled. Then the terms with $m\mu\vartheta+\nu j > (-d+1+\vartheta)(\mu+1)$ are not essential (their total contribution is less than $\varepsilon' h^{-d+1+\vartheta}$).

So, we only need to consider terms with

$$m\mu\vartheta + \nu j \le (-d+1+\vartheta)(\mu+1) \implies j < \vartheta.$$

In order to handle these terms by the same method as before we should estimate

$$\iint \tau_-^{\vartheta-1} \operatorname{tr}\bigl(\mathcal{K}_{(B(\tau-A^0)^{-1})^j} - \mathsf{Weyl}_j\bigr)\, dx d\tau. \tag{12.6.111}$$

We do this exactly as before but due to escape condition (12.6.87)–(12.6.88) an extra factor $T^{-1-\vartheta+j}$ appears in the right-hand expression where one can take $T(x) \asymp h_{\mathsf{eff}}^{-s}$ with arbitrarily large s; recall that $h_{\mathsf{eff}} = hr^{-\mu-1}$. One can easily see that for an appropriate s the corresponding integrals converge and are negligible.

Thus, under condition (12.6.107) Statement (i) has been proven. Then Statement (ii) follows by standard arguments (recall that we have also proven that the contribution of $B(0,\varepsilon)$ to the remainder estimate does not exceed $\varepsilon' h^{1-d+\vartheta}$). The proof of Statement (iii) is obvious.

(b) Let us now assume that condition (12.6.107) is violated. Then applying the same arguments as before we obtain the required estimates for the contributions of $B(0,2r_*)$ and $B(0,1)\setminus B(0,r'_*)$. Further, in $B(0,2r'_*)\setminus B(0,r_*)$ the only source of trouble is the presence of Weyl approximation errors of the form $h^{l'}\bigl(r/r_*\bigr)^{\varsigma}$ with exponents $\varsigma \ge 0$.

To handle them let us apply the above propagation results. In order to do this let us rescale $x' = \gamma^{-1}x$, then one obtains $r' = \gamma^{-1}$ and $h' = h\gamma^{-1-\mu}$ instead of r and h. To fulfill conditions $1 \ge r' \ge h'^{1/(\mu+1)-\delta'}$ let us pick $\gamma = r\bigl(r/r_*\bigr)^s$ with large s. Then for arbitrarily large s and sufficiently small

δ and δ' depending on s one can set $T = \bigl(r/r_*\bigr)^s$ which means that we should multiply the right-hand expression of the remainder estimate by T^{-1} and obtain $h''\bigl(r/r_*\bigr)^\varsigma$ with $\varsigma < 0$. Then the contribution of such terms does not exceed Ch'. ☐

12.6.8 Final Remarks

Let us discuss escape condition (12.6.87)–(12.6.88) and some generalizations of our results.

Remark 12.6.22. (i) Escape condition (12.6.87)–(12.6.88) is fulfilled for $A^0 = \bigl(|\xi|^2 - V(x)\bigr)^{\frac{m}{2}}$ with $V(x)$ positively homogeneous of degree 2μ, $\mu > -1$ and $V(x) \asymp |x|^{2\mu}$. Really, one needs to consider only $m = 2$. Then for $\varphi = \langle x, \xi\rangle$, on the energy level 0

$$\{A^0, \varphi\} = 2\bigl(|\xi|^2 + \mu V(x)\bigr) = 2(\mu + 1)|\xi|^2 \asymp |x|^{2\mu}$$

due to positive homogeneity of $V(x)$ and therefore one can apply Theorem 12.6.21.

(ii) For $\mu = -\frac{1}{2}$, $d = 3$, $m = 2$, $V(x) = |x|^{-\frac{1}{2}}$ and $\vartheta = 1$ the numerical value of the coefficient ω_0 is well known (see V. Ivrii and I. M. Sigal [1], for example).

Remark 12.6.23. (i) One can easily generalize Theorems 12.6.8, 12.6.9 and 12.6.12 to the case when X is a conical domain (in a neighborhood of 0) and Dirichlet boundary condition are given on ∂X.

(ii) In particular, consider domain X with conical points, operator with smooth coefficients in the vicinity of $\bar{X}$, $h = 1$ and asymptotics of $\mathsf{N}^-_{(\vartheta)}(\tau)$ as $\tau \to +\infty$. Then if the domain was smooth the remainder estimate would be $O(\tau^{\vartheta + (d-1-\vartheta)/m})$.

Applying the generalization we mentioned (with $\mu = 0$, $\nu = 1$ and with h replaced by $\tau^{-1/m}$) we arrive to the same remainder estimate and with Scott correction terms $\omega_{(\vartheta)j}\tau^{\vartheta - j/m}$ and the main (with $j = 0$) is above remainder estimate as $\vartheta > (d-1)$; as $\vartheta = (d-1)$ singularity manifests itself through cut Weyl term $\kappa'_{(\vartheta)\vartheta}\tau^\vartheta |\log \tau|$ and if the remainder estimate is "o" rather than "O" through Scott correction term as well.

Remark 12.6.24. (i) Moreover, one could extend Theorem 12.6.21 under nice propagation of singularities results. Namely, escape condition has the following important corollaries:

(12.6.112) For $t = T_0$ and $(x,\xi) \in \Lambda^{\pm}_{\epsilon_1} \cap \{|x| = 1\}$ we conclude that $K^{\pm}_t(x,\xi) \subset \Lambda^{\pm} \cap \{|x| \geq 2\}$ where $K^{\pm}_t$ mean propagation of singularities "cones" on the energy level 0 for A^0 (see Sections 2.2 and 3.3). Moreover, this property is stable with respect to small perturbations of A^0

and

(12.6.113) $(x,\xi) \notin K^{\pm}_t(x,\xi)$ for $t > 0$. Moreover, this property is also stable.

Exactly these conditions along with a stability condition apply to boundary-value problems. By the geometric interpretation of Section 3.3 one can replace $K_t(x,\xi)$ by a union of generalized bicharacteristics (in which case stability is achieved automatically).

(ii) Let us consider $\mu = 0$ and $A^0(x,\xi)$ which is $(1,0)$-homogeneous of degree m in (x,ξ), and the energy level $\tau = 1$ instead of $\tau = 0$. We obtain this problem when we treat the asymptotics of Riesz means when the spectral parameter τ tends to $+\infty$.

In order to prove escape condition note that one can take the function $\phi = \langle x,\xi\rangle$ in the analysis of propagation of singularities in Section 3.1. For $x \in \partial X$ this function is constant on the layers $\{(x, \xi + n\lambda), \lambda \in \mathbb{R}\}$ where n is a normal to ∂X at x and we assume that X is a conical set; moreover, the microhyperbolicity conditions are fulfilled with $\ell = H_\phi$. Therefore, for $(x,\xi) \in \Sigma_{1,f}$, $(y,\eta) \in K^{\pm}(x,\xi)$ (see definitions of $\Sigma_{\tau,b}$, $\Sigma_{\tau,f}$ in Chapter 3) the inequality $\pm\langle y,\eta\rangle \geq \pm\langle x,\xi\rangle + \epsilon_0 t$ holds.

Therefore condition (12.6.113) is fulfilled. Moreover, ellipticity yields that ξ is bounded on Σ_1 and $\langle x,\xi\rangle |x|^{-1}$ is bounded on $\Sigma_{1,b}$. Therefore picking $\Lambda^{\pm} = \{\pm\langle x,\xi\rangle \geq -\epsilon_1 |x|\}$ with sufficiently small ϵ_1 we obtain $K^{\pm}_t(x,\xi) \subset \{(y,\eta), |y| \geq \epsilon(|x| + t)\}$ for $t \geq 0$ and therefore condition (12.6.112) is also fulfilled.

Note that condition (12.6.107) is now "$(\nu - 1)\vartheta > 1 - d$". Moreover, the non-Weyl terms are $\omega_{(\vartheta)j}\tau^{\vartheta - \frac{1}{m}\nu j}$ while the Weyl terms are $\kappa_{(\vartheta)p}\tau^{\vartheta + \frac{1}{m}(d-p)}$

and a logarithmic factor appears as soon as these exponents coincide or one of them vanishes. For $\nu = 1$ condition (12.6.107) is fulfilled for all ϑ and there are many terms with logarithms.

(iii) All the arguments in Statements (i) and (ii) apply to other boundary conditions provided that for fixed h the problem is elliptic in the classical sense and the operator is semibounded from below.

Remark 12.6.25. Let us assume that $X = \mathbb{R}^d$ (or X is a cone). Further, let stabilization condition (12.6.30) be fulfilled as $|x| \to \infty$ as well but with the operator A^∞ and exponents $\mu' \in (-1, 0), \nu' < 0$ instead of A^0 and μ, ν. Furthermore, let us assume that escape condition is fulfilled for this operator A^∞ as well with the sets $\Lambda^\pm_\infty$ instead of $\Lambda^\pm$.

Let us define the sets $\Lambda^\mp_1$ according to (12.6.94); we define these sets for $|x| \le \epsilon$ and $|x| \ge R$ with $\epsilon > 0$ very small and R very large. Finally, let us assume that there exists t_0 such that if $(x(0), \xi(0))$ lies in the (small) set $\Lambda^\mp_1 \cap \{|x| \le \epsilon\}$ then $(x(\pm t_0), \xi(\pm t_0)) \in \Lambda^\mp_1 \cap \{|x| \ge R\}$.

Under these assumptions we control the propagation of singularities for very large T ($\asymp h_{\mathsf{eff}}^{-s}$ with an arbitrarily large s) and therefore

(12.6.114) In the framework of Theorem 12.6.21 and the assumptions of this remark the remainder estimate is $O(h^{l'})$ with $l' = (m\mu + \nu)\vartheta(\mu + 1)^{-1}$ even if $l' > -d + 1 + \vartheta$. Moreover, for $(\mu + 1)d + m\mu\vartheta < 0$ one can take $\psi = 1$.

As an example one can consider $X = \mathbb{R}^d$ and the Schrödinger operator with potential $W(|x|)$ where $W(r)$ is positively homogeneous of degrees 2μ and $2\mu'$ for $|x| \le \epsilon$ and $|x| \ge R$ respectively, $-1 < \mu' \le \mu < 0$, and $W(r) < 0$, $\frac{d}{dr} r^2 W(r) < 0\ \forall r$. Moreover, one can perturb this Schrödinger operator slightly and get rid of its spherical symmetry and homogeneity.

The following problem seems to be very interesting and challenging:

Problem 12.6.26. (i) Construct the similar theory as singularity is on manifolds Z of dimension d' rather than 0.

One may suspect that Scott correction terms should be in the form

$$\omega_{(\vartheta)d',j,k} h^{-d'+k+(m\mu\vartheta+\nu j)(\mu+1)^{-1}} \qquad j, k \in \mathbb{Z}^+. \tag{12.6.115}$$

(ii) In particular, consider the case of the regular operator in domain with $(d-2)$-dimensional edges (i.e. when ∂X consists of smooth manifolds which intersect only pairwise along submanifolds of dimension $(d-2)$).

(iii) Allow edges of any codimension. Then as $h=1$ and we consider $\mathsf{N}^-_{(\vartheta)}(\tau)$ as $\tau\to+\infty$, in accordance to (12.6.115) we should expect Scott terms associated with edges of dimension d' to be $\omega'_{(\vartheta),d',k}\tau^{\vartheta+(d'-k)/m}$ and the final answer should contain their sum with respect to d', k.

In Part (i) propagation along Z seems to be a difficult obstacle and in Part (iii) additional difficulties appear from "collision" of different manifolds Z_k.

12.7 Sharp Spectral Asymptotics for Neumann Laplacian in Thin Cusps

12.7.1 Problem Set-up

In this section we consider the Neumann Laplacian and similar operators in domains with ultrathin cusps (with $\exp(-|x|^{\mu+1})$-like width $(\mu>0)$) and recover eigenvalue asymptotics with accurate remainder estimate.

It is known that in some classes of domains the spectral theories of the Neumann and Dirichlet Laplacians could be very different. In particular, in domains with cusps the spectrum of Dirichlet Laplacian is always discrete and, moreover, Weyl formula holds as cusp is not very thick while for Neumann Laplacian the spectrum is essential unless cusp is very thin and the Weyl formula holds even for thinner cusps.

As in Section 12.2 we considered this operator in the domain $X\subset\mathbb{R}^d$ such of (12.2.7)-type i.e.

$$X=X_0\cup X_{\mathsf{cusp}},\ \partial X\in\mathscr{C}^K,\ X_0\Subset\mathbb{R}^d,\qquad X_{\mathsf{cusp}}=\bigl\{x=(x',x''): x'\in Z,\ x''\in\sigma(x')Y\bigr\},\quad Y\Subset\mathbb{R}^{d''},\ \partial Y\in\mathscr{C}^K \tag{12.7.1}$$

We consider here only second-order $\mathsf{D}\times\mathsf{D}$-matrix operators

$$A:=\sum_{j,k}D_ja_{jk}D_k+\sum_j(a_jD_j+D_ja_j^\dagger)+a_0 \tag{12.7.2}$$

which are defined by a quadratic form

$$Q(u) = \int \Bigl(\sum_{j,k} \langle a_{jk} D_k u, D_j u\rangle + 2\operatorname{Re}\langle a_k D_k u, u\rangle + \langle a_0 u, u\rangle\Bigr)\, dx \tag{12.7.3}$$

where we assume that this operator is self-adjoint

$$a_{kj} = a_{jk}^{\dagger}, \qquad a_0 = a_0^{\dagger}, \qquad a_b = a_b^{\dagger} \tag{12.7.4}$$

in $\mathscr{L}^2(X, \mathbb{C}^{\mathsf{D}})$ and (uniformly) elliptic

$$C_0|\xi|^2 \ge a(x,\xi) := \sum a_{jk}\xi_j\xi_k \ge \epsilon_0|\xi|^2 \qquad \forall x, \xi. \tag{12.7.5}$$

We supply quadratic form with no boundary conditions which leads to the *Neumann boundary conditions*

$$Bu := \Bigl(\sum_j n_j\bigl(\sum_k a_{jk} D_k + a_j^{\dagger}\bigr) + i a_b\Bigr) u|_{\partial X} = 0 \tag{12.7.6}$$

with $a_b = 0$ where (n_j) is an interior unit normal to ∂X.

More generally we consider quadratic form

$$Q(u) = \int_X \Bigl(\sum_{j,k} \langle a_{jk} D_k u; D_j u\rangle + 2\operatorname{Re}\langle a_k D_k u, u\rangle + \langle a_0 u, u\rangle\Bigr)\, dx + \int_{\partial X} \langle a_b u, u\rangle\, dS \tag{12.7.7}$$

where dS is a Lebesgue measure on ∂X. Then a_b in (12.7.6) does not necessary vanish.

Let us recall analysis of Section 12.3.1. Let us introduce new variables $x''_{\mathsf{new}} = x''_{\mathsf{old}}\sigma(x')^{-1}$ transforming a cusp X_{cusp} into a cylinder $Z \times Y$ as on Figure 12.1; then $dx_{\mathsf{old}} = \sigma(x')^{d} dx''_{\mathsf{new}}$ and to preserve $\mathscr{L}^2(X, \mathbb{C}^{\mathsf{D}})$ we also plug $u_{\mathsf{old}} = \sigma(x')^{-\frac{d''}{2}} u_{\mathsf{new}}$ resulting in the quadratic form

$$\begin{gathered} Q_{\mathsf{new}}(u) := \\ \int_{Z\times Y} \Bigl(\sum_{j,k} \langle a_{jk} P_k u, P_j u\rangle + 2\operatorname{Re}\sum_k \langle a_k P_k u, u\rangle + \langle a_0 u, u\rangle\Bigr)\, dx'dx'' + \\ \int_{Z\times\partial Y} \sigma^{-1}(x')\langle a_b u, u\rangle\, dx'dS'' \end{gathered} \tag{12.7.8}$$

where

$$P_j = \begin{cases} D_j - b_j(x')L & \text{as } j = 1, \dots, d', \\ \sigma^{-1}(x')D_j & \text{as } j = d'+1, \dots, d \end{cases} \tag{12.7.9}$$

with

$$b_j(x') = \partial_{x_j} \log \sigma(x') \tag{12.7.10}$$

and

$$L = \langle x'', D''\rangle - \frac{1}{2}d''i = \frac{1}{2}\bigl(\langle x'', D''\rangle + \langle D'', Dx''\rangle\bigr). \tag{12.7.11}$$

This is another version of (12.2.23). Sure, we changed coordinates in all coefficients.

Our first question is *if all the spectrum of operator discrete?* The problem now is that condition (12.3.14) fails for D lowest eigenvalues of $\boldsymbol{a}$ corresponding to constant $w(x'')$. Therefore, at least heuristically, to answer this question we need to restrict our quadratic form to the space of such functions (which coincides with $\mathcal{L}^2(Z, \mathbb{C}^{\mathsf{D}})$)

$$Q'(v) := \int_Z \Bigl(\sum_{j,k} \langle a'_{jk}P'_k v, P'_j v\rangle + 2\operatorname{Re}\sum_k \langle a'_k P'_k v, v\rangle + \langle a'_0 v, v\rangle\Bigr)\, dx' \tag{12.7.12}$$

where summation is taken over $j, k = 1, \dots, d'$,

$$a'_{jk}(x') = \int_Y a_{jk}\, dx'', \qquad a'_k(x') = \int_Y a_k\, dx'',$$
$$a'_0(x') = \int_Y a_0\, dx'' + \int_{\partial Y} a_b(x)\, dS'', \tag{12.7.13$_{1-3}$}$$

$$P'_j = D_j + \frac{i}{2}d''b_j, \qquad P'^*_j = D_j - \frac{i}{2}d''b_j \tag{12.7.14}$$

and corresponding operator is

$$A' = \sum_{j,k} P'^*_j a'_{jk} P'_k + \sum_k \bigl(P'^*_k a'^{\dagger}_k + a'_k P'_k\bigr) + a'_0 = \sum_{j,k} D_j a'_{jk} D_k + \sum_k \bigl(D_k a''^{\dagger}_k + a''_k D_k\bigr) + a''_0 \tag{12.7.15}$$

with

$$a''_0 = a'_0 + \frac{1}{4} d''^2 \sum_{jk} a_{jk} b_j b_k + \frac{i}{2} d'' \sum_k b_k (a'_j - a'^{\dagger}_j), \tag{12.7.16}$$

$$a''_k = a'_k - \frac{i}{2} d'' \sum_j b_j a_{jk}. \tag{12.7.17}$$

Assuming that the lower order coefficients a_j, a_0 and a_b are not important (we will ensure this imposing conditions) we conclude that in order to ensure that the spectrum of A' and thus of A is discrete we should assume that a''_0 grows at infinity i.e.

$$|\nabla \log \sigma| \to \infty \qquad \text{as } |x'| \to \infty. \tag{12.7.18}$$

This is definitely not the case for cusps described in Section 12.7.1.

The spectral asymptotics for operator A' may play an important role in this section. To fall into conditions of Section 11.5 let us assume that

$$-\log \sigma \asymp |x'|^{\mu+1}, \tag{12.7.19}$$

$$|\nabla \log \sigma| \asymp |x'|^{\mu} \tag{12.7.20}$$

and

$$|\nabla^\alpha \log \sigma| \le c|x|^{\mu+1-|\alpha|} \qquad \forall \alpha : |\alpha| \le K \tag{12.7.21}$$

with $\mu > 0$. Further, let us assume that

$$|\nabla^\alpha a''_{jk}| \le c|x'|^{-|\alpha'|}, \qquad |\nabla^\alpha a''_j| \le c|x'|^{-|\alpha'|-1}, \qquad |\nabla^\alpha a''_0| \le c|x'|^{-|\alpha'|-2} \tag{12.7.22$_{1-3}$}$$

and

$$a_0(x', \xi') \ge \epsilon(|\xi'|^2 + |x'|^{2\mu}) \qquad \text{as } |x'| \ge c \tag{12.7.23}$$

with

$$a_0(x', \xi') = \sum_{j,k} a''_{jk} \xi'_j \xi'_k + \sum_j (a''_j + a''^{\dagger}) \xi'_j + a''_0. \tag{12.7.24}$$

Let us also assume that

$$|\nabla^\alpha a_{jk}| \le c\gamma^{-|\alpha'|} \sigma^{-|\alpha''|}, \qquad |\nabla^\alpha a_j| \le c\gamma^{-|\alpha'|-1} \sigma^{-|\alpha''|}, \tag{12.7.25$_{1-2}$}$$

$$|\nabla^\alpha a_0| \le c\gamma^{-|\alpha'|-2} \sigma^{-|\alpha''|} \qquad |\nabla^\alpha a_b| \le c\gamma^{-|\alpha'|-1} \sigma^{-|\alpha''|} \tag{12.7.25$_{3-4}$}$$

with $\gamma(x') = |x'|^{-\mu}$; these and all the other inequalities are supposed to hold as $|\alpha| \le K = K(d, m)$, $|x'| \ge c$.

12.7.2 Weyl Part of Asymptotics

Let us consider the "core domain" and the "near cusp". Let us introduce the scaling function $\epsilon\sigma(x')$. Let $\psi(x)$ be a ($\sigma(x')$-admissible) function supported in

$$\mathcal{X}(\tau)=\{x:\ \sigma(x)\geq \epsilon\tau^{-\frac{1}{2}}\} \tag{12.7.26}$$

and equal 1 in $\mathcal{X}(\tau/2)$. Then the theory of Section 11.2 and and 11.3 implies that

$$|\int \psi(x)\bigl(e(x,x,\lambda)-\mathsf{Weyl}(x)\bigr)\,dx|\leq C\tau^{\frac{d-1}{2}}\int_{\mathcal{X}(\tau)}\sigma^{-1}\,dx \tag{12.7.27}$$

where $\mathsf{Weyl}(x)$ is the standard Weyl expression for $A(x,D)$ and we should assume that a_j, a_0 and a_b do not grow very fast as $|x'|\to\infty$.

In particular, if

$$\int_X \sigma(x')^{-1}\,dx \asymp \int_Y \sigma(x')^{d''-1}\,dx' \tag{12.7.28}$$

converges which is the case as $d''\geq 2$ due to superfast decay of $\sigma(x')$, the right-hand expression of (12.7.27) is $O\bigl(\tau^{\frac{1}{2}(d-1)}\bigr)$.

Furthermore, the same arguments imply that then one can replace $\int\psi(x)\mathsf{Weyl}(x)\,dx$ by $\int \mathsf{Weyl}(x)\,dx$ and by $\varkappa_0\tau^{\frac{d}{2}}$ in the very end.

Moreover, in this case under standard non-periodicity assumption to Hamiltonian billiard trajectories generated by $a(x,\xi)$ the remainder estimate is $o\bigl(\tau^{\frac{1}{2}(d-1)}\bigr)$ but one should include in the left-hand expression lower order and boundary terms so the final expression is $\varkappa_0\tau^{\frac{1}{2}d}+\varkappa_1\tau^{\frac{1}{2}(d-1)}$.

On the other hand, as $d''=1$ no matter how fast $\sigma(x')$ decays, integral (12.7.28) does not converge and the right-hand expression of (12.7.27) is of magnitude $\tau^{\frac{1}{2}(d-1)}r_*^{d'}(\tau)$ where

$$r_*(\tau)\asymp (\log\tau)^{1/(\mu+1)} \tag{12.7.29}$$

is defined from $|x'|\leq r_*(\tau)\implies \sigma(x')\leq\tau^{-\frac{1}{2}}$.

In Subsection 12.7.4 we will improve it by the long-range propagation arguments. So far we proved

Proposition 12.7.1. *Let operator A satisfy assumptions of Theorem 11.2.7 with $m=2$ and σ satisfy* (12.7.19), (12.7.21).

Then[69)]

(i) Asymptotics

$$\int \psi_\tau(x')e(x,x,\tau)\,dx = \varkappa_0\tau^{\frac{1}{2}d} + O\bigl(R(\tau)\bigr) \tag{12.7.30}$$

holds where

$$R(\tau)=\begin{cases}\tau^{\frac{1}{2}(d-1)} & \text{as } d''\geq 2,\\ \tau^{\frac{1}{2}(d-1)}|\log\tau|^{(d-1)/(\mu+1)} & \text{as } d''=1,\end{cases} \tag{12.7.31}$$

coefficients $\varkappa_0$ *and* $\varkappa_1$ *are standard Weyl coefficients.*

(ii) Let $d''\geq 2$ *and the standard non-periodicity condition be fulfilled to Hamiltonian billiard trajectories generated by* $a(x,\xi)$*. Then*

$$\int \psi_\tau(x')e(x,x,\tau)\,dx = \varkappa_0\tau^{\frac{1}{2}d} + \varkappa_1\tau^{\frac{1}{2}(d-1)} + o\bigl(\tau^{\frac{1}{2}(d-1)}\bigr). \tag{12.7.32}$$

12.7.3 Cusp Part of Asymptotics

Now we need to calculate the contribution of the remaining part of the cusp

$$\mathcal{X}'(\tau) := \{x, \sigma(x')\leq 2\epsilon\tau^{-\frac{1}{2}}\}. \tag{12.7.33}$$

We would like to transform our cusp to a cylinder and apply the theory of operators with the operator-valued symbols. According to Subsection 12.7.1 after change of coordinates $x'_{\mathsf{new}}=x'_{\mathsf{old}}$, $x''_{\mathsf{new}}=\sigma(x')^{-1}x''_{\mathsf{old}}$ we arrive to operator A_{new} corresponding to quadratic form Q_{new} defined by (12.7.8)

$$A_{\mathsf{new}}(x,D)=\sum_{j,k}P_j^*a_{jk}P_k+\sum_k\bigl(a_kP_k+P_k^*a_k^\dagger\bigr)+a_0 \tag{12.7.34}$$

with P_j defined by (12.7.9). Then "no boundary condition" to quadratic form translates to boundary condition

$$B_{\mathsf{new}}u := \Bigl(\sum_j \nu_j\bigl(\sum_k a_{jk}P_k + a_j^\dagger\bigr)+i\sigma^{-1}a_b\Bigr)u|_{\partial X}=0 \tag{12.7.35}$$

with

[69)] No matter if we consider Dirichlet or Neumann boundary conditions.

(12.7.36) $\nu_j = n_j$ for $j = d'+1, \dots, d$, $\nu_j = -(\partial_{x_j}\sigma_j)\langle x'', n''\rangle$ for $j = 1, \dots, d'$ as $n'' = (n_{d'+1}, \dots, n_d)$ is a unit normal to ∂Y.

However in contrast to the Dirichlet boundary problem now the boundary operator contains differentiation with respect to x' and we cannot apply directly theory of Chapter 4. Instead we decompose $\mathbb{H} = \mathscr{L}^2(Y, \mathbb{C}^D)$ into $\mathbb{H}_0 \oplus \mathbb{H}_1$ where $\mathbb{H}_0$ consists of constants and $\mathbb{H}_1 = \mathbb{H}_0^\perp$. Then

(12.7.37)
$$a(x', \xi') = \begin{pmatrix} a_{00}(x', \xi') & a_{01}(x', \xi') \\ a_{10}(x', \xi') & a_{11}(x', \xi') \end{pmatrix}, \qquad b(x', \xi') = \begin{pmatrix} b_0(x', \xi') \\ a_1(x', \xi') \end{pmatrix}$$

and the problem $(hD_t - A)U = 0$, $BU = 0$ (we skip subscript "new") becomes a system

$$(hD_t - a_{00})U_{0j} - a_{01}U_{1j} = 0, \tag{12.7.38$_1$}$$

$$-a_{10}U_{0j} + (hD_t - a_{11})U_{1j} = 0, \tag{12.7.38$_2$}$$

$$(b_0 u_0 + b_1 u_1)\big|_{\partial Y} = 0. \tag{12.7.39}$$

We will use $(12.7.38)_2$–(12.7.39) to express U_{1j} through U_{0j} and to plug it into $(12.7.38)_1$.

One can see easily that the first step is possible in the zone in question provided

$$\langle a(x', \xi'; x'', D'')v, v\rangle \ge \epsilon_0\bigl(|\xi'|^2 + \sigma(x')^{-2}\bigr)\|v\|^2 + \|D''v\|^2 \qquad \forall v \in \mathbb{H}_1 \tag{12.7.40}$$

which is ensured by

$$|a_j(x)| \le c\rho(x'), \qquad |a_0(x)| \le c\rho(x')^2, \qquad |a_b(x)| \le c\sigma(x')\rho(x')^2 \tag{12.7.41$_{2-4}$}$$

with $\rho(x') = |x'|^\mu$ and we get that $a_{01}U_{1j} = r(x', hD_t, hD')U_{0j}$ with h-pseudodifferential operator $r(x', hD_t, hD')$ with $h = \tau^{-\frac{1}{2}}$.

Further, one can see easily that $r(x', hD_t, hD')$ decays as $\tau \to +\infty$ provided

$$|(a_{jk}(x) - \bar{a}_{jk}(x')| \le c\sigma(x'), \tag{12.7.41$_1$}$$

Then one arrives to the following proposition:

Proposition 12.7.2. *Assume that* $(12.7.41)_{1-4}$, (12.7.19), (12.7.21) *and*

$$c^{-1}\bigl(|\xi'|^2+\rho(x')^2\bigr)\le \mathsf{a}_0(x',\xi')\le c\bigl(|\xi'|^2+\rho(x')^2\bigr) \tag{12.7.42}$$

are fulfilled where $\mathsf{a}_0 = A'$ *is defined by* (12.7.15).

Then contribution of zone (12.7.33) *to the asymptotics is given by the Weyl approximation for* d'*-dimensional operator* $\mathsf{a}_0^{\mathsf{w}}(x',D')$

$$\mathsf{N}_{\mathsf{cusp}}(\lambda)=(2\pi)^{-\mathsf{d}'}\int \boldsymbol{n}(x',\xi')\,dx'd\xi', \tag{12.7.43}$$

$\boldsymbol{n}(x',\xi')$ *is an eigenvalue counting function for symbol* $\mathsf{a}_0(x',\xi)$ *and the contribution of this zone to the remainder estimate is* $O\bigl(R'(\tau)+\tau^{\mathsf{d}'/2}\bigr)$ *where* $R'(\tau)$ *is the corresponding remainder estimate*

(i) $R'(\tau)=O\bigl(\tau^{\mathsf{d}'(\mu+1)/2\mu}\bigr)$ *in the general case.*

(ii) $R'(\tau)=O\bigl(\tau^{(\mathsf{d}'-1)(\mu+1)/2\mu}\bigr)$ *provided* <u>*either*</u> a_0 *is* (x',ξ')*-microhyperbolic on the energy level* τ <u>*or*</u> $\mathsf{D}=1$ *and* $\mathsf{d}'\ge 2$.

(iii) $R'(\tau)=O\bigl(\tau^{\delta}\bigr)$ *provided* $\mathsf{D}=1$ *and* $\mathsf{d}'=1$.

Proof. One needs to prove only that extra term $\boldsymbol{r}$ does not contribute much to the principal part of the asymptotics. One can see easily that its norm does not exceed $C\tau\sigma$ and then one can see easily that its contribution to the principal part is $O(\tau^{\mathsf{d}'/2})$. □

Further, we arrive to

Proposition 12.7.3. *Let assumptions of Proposition 12.7.2(ii) be fulfilled. Furthermore, assume that* $\mathsf{d}'\ge 2$ *and* $\mathsf{a}_0(x',x')$ *stabilizes to quasihomogeneous symbol* $\mathsf{a}_{0,\infty}(x',\xi')$ *and the standard non-periodicity condition is fulfilled for the Hamiltonian flow generated by* $\mathsf{a}_{0,\infty}(x',\xi)$.

Then the contribution of zone (12.7.33) *to the asymptotics is given by improved two-term Weyl approximation* $\mathsf{N}^*_{\mathsf{cusp}}(\tau)$ *for* a_0 *and the contribution of this zone to the remainder estimate is* $o\bigl(R'(\tau)\bigr)+O\bigl(\tau^{\mathsf{d}'/2}\bigr)$.

12.7.4 Main Results. I

Combining Propositions 12.7.1–12.7.3 we arrive to the following theorem:

Theorem 12.7.4. *Let conditions* (12.7.1)–(12.7.5) *be fulfilled and let* σ *satisfy* (12.7.19), (12.7.21).

Further, let conditions $(12.7.41)_{1-4}$ *and* (12.7.42) *be fulfilled where* $\mathsf{a}_0 = A'$ *is defined by* (12.7.15).

Then

(i) For $d'' \geq 2$ *asymptotics*

$$\mathsf{N}(\tau) = \varkappa_0 \tau^{\frac{d}{2}} + \mathsf{N}_{\mathsf{cusp}}(\tau) + O(\tau^{\frac{d-1}{2}}) + O(R'(\tau)) \tag{12.7.44}$$

holds where $\varkappa_0$ *and* $\varkappa_1$ *(see below) are standard Weyl coefficients for operator* A *and* $\mathsf{N}_{\mathsf{cusp}}(\tau)$ *is the Weyl approximation for* d'*-dimensional operator* $\mathsf{a}_0^{\mathsf{w}}(x', D')$ *given by* (12.7.43) *and* $R'(\tau)$ *is defined in Proposition 12.7.2.*

(ii) For $d'' = 1$ *the following estimate holds*

$$\mathsf{N}(\tau) = \varkappa_0 \tau^{\frac{d}{2}} + \mathsf{N}_{\mathsf{cusp}}(\tau) + O(\tau^{\frac{d-1}{2}} (\log \tau)^{\frac{d-1}{\mu+1}}) + O(R'(\tau)). \tag{12.7.45}$$

(iii) Moreover, if $d'' \geq 2$ *and the Hamiltonian billiard flow generated by* $a(x, \xi)$ *on* T^*X *satisfies the standard non-periodicity condition then one can replace* $O(\tau^{(d-1)/2})$ *by* $(\varkappa_1 + o(1))\tau^{\frac{1}{2}(d-1)}$.

(iv) On the other hand, if $d' \geq 2$ *if* $\mathsf{a}_0(x', \xi')$ *stabilizes to quasihomogeneous symbol* $\mathsf{a}_{0,\infty}(x', \xi')$ *and the Hamiltonian flow generated by* $\mathsf{a}_{0,\infty}(x', \xi')$ *on* T^*Z *satisfies the standard non-periodicity condition then one can replace* $\mathsf{N}_{\mathsf{cusp}}(\tau)$ *by improved two-term Weyl approximation* $\mathsf{N}^*_{\mathsf{cusp}}(\tau)$ *and* $O(\tau^{(d-1)(\mu+1)/2\mu})$ *by* $o(\tau^{(d-1)(\mu+1)/2\mu})$.

(v) Skipping conditions ensuring that the Weyl zone contribution to the remainder estimate is $O(\tau^{(d-1)/2})$ *we should replace* $\mathsf{N}^{\mathsf{W}}(\tau)$ *by* $O(\tau^{d/2})$.

12.7.5 Improved Estimate as $d'' = 1$

In this case $d'' = 1$ the semiclassical part of the cusp (also called *near cusp*, cf. (12.7.26) where we use the same notation)

$$\mathcal{X}(\tau) = \{x : |x'| \geq R, \sigma(x') \geq \epsilon \tau^{-\frac{1}{2}}\} \tag{12.7.46}$$

contributes a bit more (by the logarithmic factor) then usual to the remainder estimate as it contributes $\mathsf{N}^{\mathsf{W}}(\tau)$ to the main part. This is important only as $(d-1)/2 \ge (d-2)(\mu+1)/(2\mu)$ i.e.

$$\mu \ge d-2. \tag{12.7.47}$$

So we need to consider

$$\int \operatorname{tr}(e(x,x,\tau))\psi_1(x')\,dx \tag{12.7.48}$$

where $\psi_1(x')$ is supported in $\{x' : |x'| \ge R, \sigma(x') \ge \epsilon\tau^{-\frac{1}{2}}\}$ and equals 1 in $\{x' : |x'| \ge 2R, \sigma(x') \ge 2\epsilon\tau^{-\frac{1}{2}}\}$ with constant R.

To improve this remainder estimate we need (x',ξ')-microhyperbolicity for operator $\boldsymbol{a}(x',\xi') = A(x',\xi'; x'', D_{x''})$ under boundary condition $\boldsymbol{b}(x',\xi)v = 0$ with $\boldsymbol{b} = B(x',\xi'; x'', D_{x''})$. This condition is obviously defined provided "mixed" coefficients a_{jk} (i.e. with $j = 1,\dots,d'$, $k = d'+1,\dots,d$) decay as $|x'| \to \infty$ because according to $(12.7.41)_{2-4}$ lower order terms are not large enough to influence it. In the general case we do not have the notion of (x',ξ')-microhyperbolicity in Chapter 2 but we essentially have it in Chapter 3 (see Definition 3.1.4.

Let us assume that condition (12.7.20) is fulfilled.

Then the error does not exceed

$$C\tau^{(d-1)/2}\int_{\mathcal{X}(\tau)} \gamma(x')^{-1}\,dx \asymp C\tau^{(d-1)/2}\int_{\mathcal{X}(\tau)} \gamma(x')^{-1}\sigma(x')\,dx' \tag{12.7.49}$$

where $\gamma(x') \asymp |x'|^{-\mu}$ is the scale with respect to x'. This integral does not exceed $\varepsilon(R)$ (with $\varepsilon(R) = o(1)$ as $r \to \infty$) if taken over $\{|x'| \ge R\}$.

Further, if we consider

$$\tau^{(d-j)/2}\int_{\mathcal{X}(\tau)} \sigma(x')^{-j}\,dx \asymp \tau^{(d-j)/2}\int_{\mathcal{X}(\tau)} \sigma(x')^{1-j}\,dx' \tag{12.7.50}$$

we conclude that it is $o(\tau^{(d-1)/2})$ as $\mu > d-2$ and is $O(\tau^{(d-1)/2})$ as $\mu = d-2$. Therefore we need to consider only terms with $j = 0,1$ unless $\mu = d-2$ and we are interested in the remainder estimate $o(\tau^{(d-1)/2})$ in which case we need to consider also $j \ge 2$.

Furthermore,

$$\int_{\mathcal{X}'(\tau)} \tau^{d/2}\,dx \asymp \int_{\mathcal{X}'(\tau)} \tau^{d/2}\sigma(x')^1\,dx' \tag{12.7.51}$$

is $o(\tau^{(d-1)/2})$ as $\mu > d-2$ and $O(\tau^{(d-2)/2})$ as $\mu = d-2$; here and below $\mathcal{X}'(\tau) = \{x : \sigma(x') \le \epsilon\tau^{-\frac{1}{2}}$ is a *far cusp*.

To provide (x', ξ')-microhyperbolicity we need the following proposition:

Proposition 12.7.5. *Let conditions of Theorem 12.7.4 be fulfilled. Further, let condition* (12.7.20) *be fulfilled and*

$$|\nabla^\alpha (a_{jk}(x) - a_{jk,\infty}(x'))| \le c|x'|^{-\varsigma-|\alpha'|}\sigma(x')^{-|\alpha''|} \tag{12.7.52}$$

with $\varsigma > 0$, $ajk,\infty(x')$ *positively homogeneous of degree* 0. *Then operator* $A(x', \xi'; x'', D_{x''})$ *with* $B(x', \xi'; x'', D_{x''})v = 0$ *is uniformly* (x', ξ')*-microhyperbolic.*

Proof. Proof is trivial as one needs to consider operator with variables (ξ', σ) where σ is now independent variable and to prove that it is (ξ', σ)-microhyperbolic. □

Now we arrive to Statements (i) and (ii) of the following theorem:

Theorem 12.7.6. *Let* $d'' = 1$ *and conditions* (12.7.47), (12.7.1)–(12.7.5) *be fulfilled and let* σ *satisfy* (12.7.19)–(12.7.21); *let* (12.7.52) *be fulfilled.*

Further, let conditions $(12.7.41)_{1-4}$ *and* (12.7.42) *be fulfilled where* $\mathbf{a}_0 = A'$ *is defined by* (12.7.15).

Then

(i) Asymptotics

$$\mathsf{N}(\tau) = \varkappa_0\tau^{\frac{d}{2}} + \nu(\tau) + \mathsf{N}^{\mathsf{W}}_{\mathsf{cusp}}(\tau) + O(\tau^{\frac{d-1}{2}}) \tag{12.7.53}$$

holds with

$$\nu(\tau) = \tau^{(d-1)/2} \int (1 - \psi'_\tau(x))\kappa_1(x)\, dx \tag{12.7.54}$$

with $\operatorname{supp}(\psi'_\tau) \subset \mathcal{X}'(\tau)$.

(ii) Moreover, if $\mu > d-2$ *and the standard non-periodicity condition is fulfilled for Hamiltonian billiard flow generated by symbol* $a(x, \xi)$ *one can replace* $O(\tau^{\frac{d-1}{2}})$ *by* $o(\tau^{\frac{d-1}{2}})$.

(iii) Let $\mu = \mathsf{d} - 2$ *and the standard non-periodicity condition be fulfilled. Further, let* $\mathsf{d} \geq 3$, $\mathbf{a}_0(x', \xi')$ *stabilize to quasihomogeneous symbol* $\mathbf{a}_{0,\infty}(x', \xi')$ *and the Hamiltonian flow generated by* $\mathbf{a}_{0,\infty}(x', \xi')$ *on* T^*Z *satisfy the standard non-periodicity condition.*

Then asymptotics

$$\mathsf{N}(\tau) = \varkappa_0 \tau^{\frac{d}{2}} + \nu_1(\tau) + \mathsf{N}^{\mathsf{W}}_{\mathsf{cusp}}(\tau) + \varkappa_1 \tau^{\frac{d-1}{2}} + o(\tau^{\frac{d-1}{2}}) \tag{12.7.55}$$

holds with

$$\nu_1(\tau) = \nu(\tau) - \tau^{\frac{d}{2}} \int \kappa_0(x)\psi'_\tau(x)\, dx. \tag{12.7.56}$$

(iv) Under stabilization condition alone one can replace $\nu(\tau)$ *and* $\nu_1(\tau)$ *by*

$$\varkappa'_1 \tau^{\frac{d-1}{2}} |\log \tau|^{(d-\mu-2)/(\mu+1)} \tag{12.7.57}$$

and

$$\tau^{\frac{d-1}{2}} \bigl(\varkappa'_1 |\log \tau|^{(d-\mu-2)/(\mu+1)} + \varkappa''_1\bigr) \tag{12.7.58}$$

respectively.

Proof. To prove Statement (iii) one needs just to notice that $|\kappa_j| \ll c\sigma^{-j}$ as $j \geq 2$ because only terms which do not satisfy this inequality could come from the analysis of 1-dimensional operator $A(x', \xi'; D'')$ with boundary condition $B(x', \xi'; x'', D_{x''})v = 0$ on each side of the segment Y and one can prove easily by formal calculations that in this case $\kappa_j = 0$ as $j \geq 2$.

Proof of Statement (iv) is trivial. □

Remark 12.7.7. All the above results hold for manifolds with the compact boundary and with cusps, like $Z \times Y$ with the metrics $dx'^2 + \sigma(x')^2 dx''^2$ where now Y is a compact closed manifold. In particular we can consider surfaces of revolutions.

Remark 12.7.8. Let $\mathsf{D} = 1$. Then we can get rid of derivatives with respect to x' in the boundary conditions. One can see easily that one can introduce variables $x'_{\mathsf{new}} = x'_{\mathsf{old}} - F(x)$ with

$$|\nabla^\alpha F| \leq c|x'| \cdot \sigma(x')\gamma(x')^{-1-|\alpha'|} \tag{12.7.59}$$

such that in the new coordinates the quadratic form is of the same type with the same inequalities and $\tilde{a}_{jk} = 0$ as $j \leq d', k \geq d'' + 1$ and $x'' \in \partial Y$ and we replaced $\sigma(x')$ by $\sigma(x'_{\mathsf{new}})$ which leads to the negligible perturbations.

12.7.6 Maxwell System

Remainder

We are interested in eigenvalue asymptotics for Maxwell operator A in $X \subset \mathbb{R}^d$. Namely, let $\mathfrak{H} = \mathscr{L}^2(X, \Lambda^k) \oplus \mathscr{L}^2(X, \Lambda^{k+1})$ where Λ^k is a space of exterior forms of a degree $k = 0, \dots, d$ [70], and let for

$$\phi = \sum_I \phi_I dx_I, \qquad dx_I = dx_{i_1} \wedge \dots dx_{i_k}: \qquad I = (i_1, \dots, i_k), 1 \le i_1 < \dots < i_k \le d \tag{12.7.60}$$

we define

$$\|\phi\|^2 = \sum_I \|\phi_I\|^2_{\mathscr{L}^2(X)} \qquad \text{and} \qquad \| \begin{pmatrix} \phi \\ \psi \end{pmatrix} \|^2 = \|\phi\|^2 + \|\psi\|^2. \tag{12.7.61}$$

Let us consider an operator

$$A = \begin{pmatrix} 0 & \beta^\dagger \mathsf{d}^* \alpha^\dagger \\ \alpha \mathsf{d} \beta & 0 \end{pmatrix} \tag{12.7.62}$$

with domain

$$\mathfrak{D}(A) = \{ \begin{pmatrix} \phi \\ \psi \end{pmatrix} \in \mathfrak{H}, Au \in \mathfrak{H}, \iota_{\partial X} \phi = 0\} \tag{12.7.63}$$

where $\mathsf{d} = i\langle dx, D\rangle \wedge : \mathscr{C}^\infty(X, \Lambda^k) \to \mathscr{C}^\infty(X, \Lambda^{k+1})$ is the operator of the exterior differentiation and $\mathsf{d}^* : \mathscr{C}^\infty(X, \Lambda^{k+1}) \to \mathscr{C}^\infty(X, \Lambda^k)$ is the formally adjoint operator[71]; for ϕ in form (12.7.60) we have

$$\mathsf{d}\phi = \sum_j (iD_j \phi_I) dx_j \wedge dx_I, \qquad \mathsf{d}^* \phi = \sum_{1 \le p \le k} (iD_{i_p} \phi_I)(-1)^p dx_{I \setminus i_p}, \tag{12.7.64}$$

(12.7.65) $\alpha(x) \in \mathcal{L}(\Lambda^{k+1}, \Lambda^{k+1})$; and $\beta(x) \in \mathcal{L}(\Lambda^k, \Lambda^k)$ are nondegenerate matrices smoothly depending on x and constant close to infinity (or quickly stabilizing to constant),

---

[70] At a given point, $\dim \Lambda^k = \frac{d!}{k!(d-k)!}$ and we consider complex rather than real spaces.

[71] We can define action of $\mathcal{D}$, $\mathcal{D}^*$ and $\mathcal{A}$ at distributions as well.

$\iota_Y : \mathscr{C}^\infty(X, \Lambda^k(X)) \to \mathscr{C}^\infty(X, \Lambda^k(Y))$ is the restriction of the exterior form to submanifold Y[72].

There is no problem with self-adjoint expansion of such an operator defined on functions with compact support first provided ∂X, α and β are smooth enough. Furthermore, we assume that

(12.7.66) α and β are Hermitian positive matrices

(otherwise one can reach it by means of the unitary transformation $\begin{pmatrix} T_1 & 0 \\ 0 & T_2 \end{pmatrix}$ with unitary matrices $T_1(x)$ and $T_2(x)$).

Remark 12.7.9. (i) Let $\begin{pmatrix} \phi \\ \psi \end{pmatrix}$ be an eigenfunction with eigenvalue $\lambda \ne 0$. Then automatically

$$d\alpha^{-1}\psi = d^*\beta^{-1}\phi = 0 \tag{12.7.67}$$

and further,

$$\iota_{\partial X}\alpha^{-1}\psi = 0. \tag{12.7.68}$$

(ii) Further, $\begin{pmatrix} \phi \\ -\psi \end{pmatrix}$ is an eigenfunction with an eigenvalue $(-\tau)$.

Then we conclude that

Remark 12.7.10. (i) The number of eigenvalues of operator A belonging to $(0, \tau)$ is

$$N_k(\tau) = N^-_{\Delta_{k,0}}(\tau^2) - N^-_{\Delta_{k,0}}(0 + 0) \tag{12.7.69}$$

where $N^-_{\Delta_{k,0}}(\lambda)$ is the number of the eigenvalues of operator $\Delta_{k,0}$ generated by a quadratic form

$$Q(\phi) = \|\alpha d\beta\phi\|^2 + \|\alpha_1^{-1} d^*\beta^{-1}\phi\|^2 \tag{12.7.70}$$

on the space

$$\mathbb{H}_{k,0} = \{\phi \in \mathscr{L}^2(X, \Lambda^k), d^*\beta^{-1}\phi = 0\} \tag{12.7.71}$$

[72] One can see easily that if Y is smooth of codimension 1 and $\phi \in \mathscr{L}^2(X, \Lambda^K)$, $d\phi \in \mathscr{L}^2$ then $\iota_Y\phi$ is well-defined; if $Y = \{x_1 = 0\}$ in appropriate coordinates and ϕ is of the form (12.7.60) then $\iota_Y\phi = \sum_{I \not\ni 1} \phi_I|_Y dx_I$.

with domain

$$\mathfrak{D}(Q) = \{\phi \in \mathbb{H}_{k,0},\, Q(\phi) < \infty,\, \iota_{\partial X}\beta\phi = 0\} \tag{12.7.72}$$

which are less than λ; later in our conditions it will be a finite number; $\alpha_1 \in \mathcal{L}(\Lambda^{k-1}, \Lambda^{k-1})$ is a nondegenerate matrix (usually constant close to infinity).

(ii) Furthermore, considering the same form $Q(u)$ on the space

$$\mathbb{H}_k = \{\phi \in \mathscr{L}^2(X, \Lambda^k)\} \tag{12.7.73}$$

with domain

$$\mathfrak{D}(Q) = \{\phi \in \mathbb{H}_k\colon\, Q(\phi) < \infty,\;\; \iota_{\partial X}\beta\phi = 0\} \tag{12.7.74}$$

we get an operator Δ_k; one can check easily that for $\tau > 0$ eigenspaces $\mathbb{H}_k(\tau)$ and $\mathbb{H}_{k,0}(\tau)$ of $\Delta_{k,0}$ and Δ_k satisfy

$$\mathbb{H}_k(\tau) \subset \mathbb{H}_{k,0}(\tau), \tag{12.7.75}$$

$$\mathbb{H}_k(\tau) \ominus \mathbb{H}_{k,0}(\tau) = \beta^{-1}\mathsf{d}\alpha_1^{-1}\bigl(\alpha_1^{-1}\mathsf{d}^*\beta^{-1}\mathbb{H}_k(\tau)\bigr) \tag{12.7.76}$$

and that $\alpha_1^{-1}\mathsf{d}^*\beta^{-1}\mathbb{H}_k(\tau)$ is an eigenspace of the operator $\Delta_{k-1,0}$ generated by a quadratic form $\|\beta^{-1}\mathsf{d}\alpha_1^{-1}\phi'\|^2 + \ldots$ on the space

$$\mathbb{H}_{k-1,0} = \{\phi \in \mathscr{L}^2(X, \Lambda^{k-1}),\, \mathsf{d}^*\alpha_1\phi = 0,\, \iota_{\partial X}\alpha_1^{-1}\phi = 0\}. \tag{12.7.77}$$

(iii) Therefore

$$\mathsf{N}(\tau) = \mathsf{N}^-_{\Delta_k}(\tau^2) - \mathsf{N}^-_{\Delta_k}(0+0) - \mathsf{N}^-_{\Delta_{k-1,0}}(\tau^2) + \mathsf{N}^-_{\Delta_{k-1,0}}(0+0) \tag{12.7.78}$$

(iv) This reduction is not necessarily correct for non-smooth problems unless we are able to prove that $\mathfrak{D}(\Delta_k) \subset \mathscr{H}^2_{\mathrm{loc}}(X)$ which is not true even for $\alpha = \beta = I$ and $X = W \oplus \mathbb{R}^{d-2}$ with a sector W with an angle between π and 2π.

(v) Note that for $k = d-1$, $\alpha = I$, $\beta = I$ and $\psi = u dx_1 \wedge \cdots \wedge dx_d$ we obtain exactly eigenvalue problem for the Neumann Laplacian.

Remark 12.7.11. (i) These formulae lead us to standard (known) asymptotics for compact domains with smooth boundaries:

$$\mathsf{N}(\tau^2) = \kappa_0\tau^d + O(\tau^{d-1}) \tag{12.7.79}$$

(here and below we omit indices k and may be 0) and even

$$\mathsf{N}(\tau^2) = \kappa_0\tau^d + \kappa_1\tau^{d-1} + o(\tau^{d-1}) \tag{12.7.80}$$

provided

$$\det\Bigl(\tau^2 - \beta\mathsf{d}(\xi)^\dagger\alpha^2\mathsf{d}(\xi)\beta\Bigr) = \tau^r\bigl(\tau^2 - g(x,\xi)\bigr)^s \tag{12.7.81}$$

where $\mathsf{d}(\xi) = i\langle dx,\xi\rangle\wedge$ is a principal symbol of d, g is a metrics on X and standard billiard condition holds.

(ii) The same results hold for other types of compact domains: as long as formula (12.7.78) holds, the irregularity of the boundary should be of the type described in Sections 11.2 and 11.3 and inner cone condition should be fulfilled.

Thin Cusps. Heuristics

Let us consider domains with cusps; we consider one cusp for simplicity. So we consider domains of the type described in Section 12.2.

Look first for operators Δ_k. As we already know, boundary conditions are very important for the Laplace operator: if we have the Dirichlet boundary condition then the spectrum of operator is discrete (provided cusp shrinks at infinity).

On the other hand for operator with a Neumann boundary condition spectrum is discrete only if cusp is very thin ($\log\sigma \asymp |x'|^{\mu+1}$ with $\mu>0$) and for such operators asymptotics with sharp remainder estimate are derived in Subsections 12.7.1–12.7.5.

So basically we should determine first if for operator Δ_k condition is "Dirichlet"-like or "Neumann"-like at infinity; then for "Neumann"-like cusp assume that it is ultra-thin and write the cusp contribution; for "Dirichlet"-like cusp we need to describe non-Weyl contribution.

Let us consider the space

$$\begin{aligned}\Phi_k = \bigl\{\phi \in \mathscr{C}^\infty(Y, \Lambda^k(X)),\\ \mathsf{d}''\beta\phi = \mathsf{d}''^*\beta^{-1}\phi = 0, \iota_{Z\times\partial Y}\beta\phi = 0\bigr\}\end{aligned} \tag{12.7.82}$$

(so we consider full forms with coefficients depending on x'' only); one can see easily that $\dim\Phi_k$ does not depend on the choice of matrix β and

$$\dim\Phi_k = \sum_j \frac{d'!}{j!(d'-j)!}\dim\Phi''_{k-j} \tag{12.7.83}$$

where

$$\Phi''_j = \bigl\{\phi\in\mathscr{C}^\infty(Y,\Lambda(Y)^j),\, \mathsf{d}''\phi = \mathsf{d}''^*\phi = 0,\, \iota_{\partial Y}\phi = 0\bigr\}. \tag{12.7.84}$$

From the point of view of operator Δ_k the cusp is "Dirichlet"-like iff $\dim\Phi_k = 0$; we will discuss this case later and assume that $\dim\Phi_k\ge 1$. Let us consider operator

$$\begin{aligned}\ell_k(x',D') = \boldsymbol{P}_k\beta M_k(x',D')^*\alpha^2 M_k(x',D')\beta +\\ \beta^{-1}M_{k-1}(x',D')\alpha_1^{-2}M_{k-1}(x',D')^*\beta^{-1}\phi\end{aligned} \tag{12.7.85}$$

acting from $\mathscr{C}^\infty(Z,\Phi_k)$ to $\mathscr{C}^\infty(Z,\Phi_k)$ with

$$M(x',D') = \mathsf{d}' + \frac{d''}{2}(D'\log\sigma)\wedge \tag{12.7.86}$$

acting from : $\mathscr{C}^\infty(Z,\Phi_k)$ to $\mathscr{C}^\infty(Z,\Lambda(Y)^{k+1})$, $M(x',D')^*$ a formally adjoint operator and $\boldsymbol{P}_k$ orthogonal projector from $\mathscr{L}^2(Y,\Lambda^k)$ to Φ_k.

Then the cusp term for operator Δ_k will be

$$\mathsf{N}_{k,\mathsf{cusp}}(\tau) = (2\pi)^{-d'}\iint \boldsymbol{n}_k(x',\xi',\tau^2)\,dx'd\xi' \tag{12.7.87}$$

where $\boldsymbol{n}_k(x',\xi',\tau)$ is the eigenvalue counting function for finite-dimensional[73] symbol $\ell_k(x',\xi')$ of an operator $\ell_k(x',D')$.

Applying then formula (12.7.78) we conclude that the cusp contribution for operator $\Delta_{k,0}$ is given by the formula (12.7.87) with $\boldsymbol{n}_k(x',\xi',\tau^2)$ replaced by $\boldsymbol{n}_{k,0}(x',\xi',\tau^2)$ which is the eigenvalue counting function for $\mathcal{L}(\Phi_{k,0},\Phi_{k,0})$-valued symbol $\ell_{k,0}(x',\xi') = \boldsymbol{P}_{k,0}(x',\xi')\ell_{k,0}(x',\xi')$; here

$$\Phi_{k,0}(x',\xi') = \bigl\{\phi\in\Phi_k,\, M_{k-1}(x',\xi')^*\beta^{-1}\phi = 0\bigr\} \tag{12.7.88}$$

and $\boldsymbol{P}_{k,0}(x',\xi')$ is the orthogonal projection onto $\Phi_{k,0}(x',\xi')$.

[73] $\mathcal{L}(\Phi_k,\Phi_k)$-valued.

One can see easily that

$$\dim \Phi_{k,0} = \dim \Phi_k - \dim \Phi_{k-1,0}, \qquad \dim \Phi_{0,0} = 0 \tag{12.7.89}$$

and (12.7.83), (12.7.89) yield that

$$\dim \Phi_{k,0} = \sum_j \frac{(d'-1)!}{j!(d'-1-j)!} \dim \Phi''_{k-j}. \tag{12.7.90}$$

Furthermore, for $\alpha = I$, $\beta = I$ near infinity

$$\mathsf{N}_{k,\mathsf{cusp}}(\tau^2) = (\dim \Phi_k) N_{\mathsf{cusp}}(\tau^2), \tag{12.7.91}$$

$$\mathsf{N}_{k,0,\mathsf{cusp}}(\tau^2) = (\dim \Phi_{k,0}) N_{\mathsf{cusp}}(\tau^2) \tag{12.7.92}$$

where $\mathsf{N}_{\mathsf{cusp}}(\tau^2)$ is a cusp contribution to the asymptotics for Neumann Laplacian on $\mathscr{L}^2(X, \mathbb{C})$.

One can expect that Δ_k or $\Delta_{k,0}$ are "Neumann-like" iff $\dim \Phi_k \ne 0$ or $\dim \Phi_{k,0} \ne 0$ respectively.

Thin Cusps. Results

Theorem 12.7.12 [74]. *Let* <u>*either*</u> $\Delta := \Delta_k$ *and* $\dim \Phi_k \ge 1$ <u>*or*</u> $\Delta = \Delta_{k,0}$ *and* $\dim \Phi_{k,0} \ge 1$. *Let*

$$\alpha = I, \ \beta = I \qquad as \ \ |x| \ge C \tag{12.7.93}$$

and let conditions (12.7.19)–(12.7.21) *be fulfilled with* $\mu > 0$. *Then*

(i) For $d'' \ge 2$ *the following asymptotics holds*

$$\mathsf{N}(\tau) = \varkappa_0 \tau^d + \mathsf{N}_{\mathsf{cusp}}(\tau^2) + O(\tau^{d-1}) + O(\tau^q) \tag{12.7.94}$$

where $\varkappa_0$ *and* $\varkappa_1$ *(see below) are standard Weyl coefficients,* $\mathsf{N}_{\mathsf{cusp}}(\tau) \asymp \tau^q$ *is defined by* (12.7.87) *for Neumann Laplacian,* $p = d'(\mu+1)/\mu$, $q = (d'-1)(\mu+1)/\mu$ *and we skip operator-related indices* k *and may be* 0.

(ii) For $d'' = 1$ *the following asymptotics holds*

$$\mathsf{N}(\tau) = \varkappa_0 \tau^d + \mathsf{N}_{\mathsf{cusp}}(\tau^2) + O\bigl(\tau^{d-1}(\log \tau)^{(d-1)/(\mu+1)}\bigr) + O(\tau^q). \tag{12.7.95}$$

[74] Cf. Theorem 12.7.4.

(iii) Moreover, if $\mathsf{d}''\geq 2$, (12.7.81) *holds,* $\Delta=\Delta_{k,0}$ *(and similar condition for* $\Delta=\Delta_k$*) and standard Hamiltonian condition is fulfilled, then one can replace* $O(\tau^{d-1})$ *in asymptotics* (12.7.94) *by* $(\varkappa_1+o(1))\tau^{d-1})$.

(iv) Moreover, as $\mathsf{d}''\geq 2$ and stabilization and non-periodicity condition of Theorem 12.7.4(iv) are fulfilled then one can replace $O(\tau^q)$ *by* $\bigl(\varkappa_1^*+o(1)\bigr)\tau^q$.

Theorem 12.7.13[75]**.** *Let* $\mathsf{d}''=1$, $\mu\geq \mathsf{d}-2$ *and conditions* (12.7.93) *and* (12.7.19)–(12.7.21) *be fulfilled and let* $\log\sigma$ *be positively homogeneous of degree* $(\mu+1)$. *Then*

(i) Asymptotics

$$\mathsf{N}(\tau)=c_0\tau^d+\nu(\tau^2)+N_{\mathsf{cusp}}(\tau^2)+O\bigl(\lambda^{fracd-12}\bigr) \tag{12.7.96}$$

holds with

$$\nu(\tau^2)=\varkappa_1'\tau^{d-1}(\log\tau)^{(d-1)/(\mu+1)}+O(\tau^{d-1}) \tag{12.7.97}$$

the Weyl expression for second order term in domain $\{\sigma(x')\tau\leq 1\}$.

(ii) Moreover, if $\mu>\mathsf{d}-2$, (12.7.81) *holds,* $\Delta=\Delta_{k,0}$ *(and similar condition for* $\Delta=\Delta_k$*) and standard Hamiltonian condition is fulfilled, then one can replace* $O(\tau^{d-1})$ *in asymptotics* (12.7.96) *by* $\bigl(\kappa_1+o(1)\bigr)\tau^{d-1})$.

(iii) Further, as $\mathsf{d}''=1$, $\mu=\mathsf{d}-2$ *and stabilization and non-periodicity condition of Theorem 12.7.6(iii) are fulfilled then one can replace* $\nu(\tau)+O(\tau^q)$ *by* $\nu_1(\tau)+o(\tau^q)$.

Remark 12.7.14. One can weaken condition (12.7.93). Moreover, same asymptotics holds for manifolds and for manifolds with compact boundary one can skip condition (12.7.93).

Thick Cusps. Sketch

In the case of the "Dirichlet"-like' cusp condition of being ultra-thin is no longer necessary and in this case we leave to the reader rather interesting and challenging

Problem 12.7.15. Derive asymptotics with the sharp remainder estimates exactly in the same type as in Section 12.2.

[75] Cf. Theorem 12.7.6.

The only one real difficulty is in the case of operator $\Delta_{k,0}$ as $\dim \Phi_{k,0} = 0$ and $\dim \Phi_k \geq 1$ but one can overcome it by taking α_1 fast growing rather than constant at infinity.

Another difficulty is that we need to study operators with the operator-valued symbols with domain depending on (x', ξ') but it already was discussed in Problem 12.2.21.

12.7.7 Generalizations

Higher Order Operators–Discussion

Remark 12.7.16. As we mentioned in Problem 12.2.21 results of Section 12.2 should hold for appropriate boundary value problem such that all eigenvalues $\lambda_k(x', \xi')$ of $a(x', \xi')$ satisfy (12.3.14) (actually this is condition only for few lowest eigenvalues).

Let us recall that $a(x', \xi')$ is an operator obtained by replacing in $A(x, D)$ $D' \mapsto (x', \xi')$ and $(x'', D'') \mapsto (\sigma(x')x'', \sigma(x')^{-1}D'')$ and considered on Y where Y is a fixed domain in $\mathbb{R}^{d''}$ defining the shape of the cusp cross-section.

Now we want to consider the case when condition (12.3.14) fails for few lowest eigenvalues. This failure can be of very different nature: it may fail just for some x' or fail mildly in the sense that $\lambda_1(x', \xi') \geq (|\xi'| + \sigma(x')^{-\theta})^{2m}$; and the failure is basically very unstable.

Consider operator

$$\sum_{\alpha,\beta:|\alpha|=m,|\beta|=m} D^\alpha a_{\alpha,\beta}(x) D^\beta \tag{12.7.98}$$

defined by the corresponding quadratic form

$$\int \sum_{\alpha,\beta:|\alpha=m,|\beta|=m} \langle a_{\alpha\beta}(x) D^\beta u, D^\alpha u\rangle \, dx \tag{12.7.99}$$

with the boundary condition

$$D^\alpha u|_{\partial X} = 0 \qquad \forall \alpha : |\alpha| \leq n - 1. \tag{12.7.100}$$

Remark 12.7.17. Note that different quadratic forms may lead to the same differential expression but to different boundary conditions f.e. consider

$$Q_1(u) = \|D_1^2 u\|^2 + \|D_2^2 u\|^2 + 2\|D_1 D_2 u\|^2 \tag{12.7.101}$$

and

$$Q_2(u) = \|D_1^2 u\|^2 + \|D_2^2 u\|^2 + 2\,\mathrm{Re}(D_1^2 u, D_2^2 u) \tag{12.7.102}$$

leading to operator Δ^2 but with the different boundary conditions (as $n = 0$ in (12.7.100)) and only (12.7.101) leads to elliptic problem. However as $n = 1$ these forms coincide and lead to elliptic problem.

Assume that quadratic form

$$q(x)(v) = \sum_{\alpha,\beta:|\alpha=m,|\beta|=m} \langle a_{\alpha\beta} v_\beta, v_\alpha\rangle \tag{12.7.103}$$

satisfies

$$q(x)(v) \asymp \|v\|^2 \text{ uniformly with respect to } x' \text{ and } v = (v_\alpha)_{|\alpha|=m}. \tag{12.7.104}$$

Only form (12.7.101) satisfies this condition.

Then we need to consider $u(x'; x'')$ which are polynomial of degree $(m-1)$ with respect to x'' and check if such polynomials satisfying (12.7.100) must be identically 0. The existence of non-zero such polynomials is rather unstable. So if originally Y depended mildly on x' and we arrive to constant Y by a change of variables x'' then the resulting operator will not be of form (12.7.98) (lower order term come into play) and violation of assumption (12.3.14) for the lowest λ_k could be mild.

Let us consider quadratic forms after rescaling (which means that originally these forms contained lower order terms)

$$Q_1(u) = \|D_1^2 u\|^2 + \|\sigma^{-2} D_2^2 u\|^2 + 2\|\sigma^{-1} D_1 D_2 u\|^2 + (Wu, u) \tag{12.7.105}$$

and

$$Q_2(u) = \|D_1^2 u\|^2 + \|\sigma^{-2} D_2^2 u\|^2 + (Wu, u) \tag{12.7.106}$$

with $\sigma = \sigma(x_1)$ rapidly decaying at infinity and $W = W(x_1)$ polynomially growing at infinity.

Obviously only form (12.7.105) satisfies (12.7.104) after rescaling back; however it does not prevent problem associated with (12.7.106) from being uniformly elliptic.

Consider quadratic forms with D_1 replaced by ξ_1 and integration over x_2 only. Both forms behave the same way on $v=\mathsf{const}$; so for both forms $\lambda_1(x_1,\xi_1)=\xi_1^4+W(x_1)$. However on $v=\mathsf{const}x_2$ we get $\lambda_2(x_1,\xi_1)=\xi_1^4+c^2\xi_1^2\sigma^{-2}+W(x_1)$ and $\lambda_2(x_1,\xi_1)=\lambda_1(x_1,\xi_1)=\xi_1^4+W(x_1)$ for forms (12.7.105) and (12.7.106) respectively.

We leave to the reader the following series of the problems

Problem 12.7.18. (i) Develop theory covering (12.7.105); in particular explore if $\lambda_2(x',\xi')$ produces any noticeable contribution to $\mathsf{N}_{\mathsf{cusp}}(\tau)$.

For this purpose one needs to construct the theory similar to one of Section 11.5 but for matrix operators with eigenvalues of the symbol having different growth as (x',ξ') escaping to infinity.

(ii) Develop theory covering (12.7.106). This theory seems to be rather straightforward and easy.

More General Cusps

One can consider more general cusps. We believe that the first problem is rather challenging:

Problem 12.7.19. Consider cusps as in Problem 12.2.23(i) and (ii) albeit with "superexponential" as in this section rather than power decay as in Section 12.2 and with the Neumann-like rather than the Dirichlet-like boundary conditions.

The next problem seems to be rather easy:

Problem 12.7.20. Analyze the case when $\mathbf{a}_0$ is a Schrödinger operator with a potential having valleys as in Section 12.3.

It will be the case as $|\nabla\log\sigma|\asymp|x'|^{\nu}|x'''|^{\varsigma}$ where $x'''=(x_{d'''+1},\dots,x_{d'})$ and $1\le d'''\le d'$, $\nu>0$, $\varsigma>0$; then $|\log\sigma|\asymp|x'|^{\nu}|x'''|^{\varsigma+1}$ and thus $\sigma\cong\exp(-|x'|^{\nu}|x'''|^{\varsigma+1})$.

Domains with Spikes

We consider domains with the *spikes* which are "cusps of the finite length" (see Figure 11.2). Now Z is a compact domain with a smooth boundary and we can assume locally that $Z = \{x_1 > 0\}$ and $X = \{x : x' \in Z, x'' \in \sigma(x')Y\}$ with compact domain Y with the smooth boundary ∂Y and $\sigma(x_1) \le cx_1$,

$$|D^\alpha \nu(x')| \le c\sigma(x')|x_1|^{-|\nu|}. \tag{12.7.107}$$

We also assume that

$$|D^\nu a_{\alpha\beta}(x')| \le c|x_1|^{-|\nu'|+\mu(2m-|\alpha|-|\beta|)}\sigma(x')^{-|\nu''|} \tag{12.7.108}$$

with $\mu > \max(-1, -d/2m)$.

We consider quadratic form with the boundary terms satisfying conditions

$$|D^\alpha a'_{\alpha\beta}(x')| \le c|x_1|^{-|\nu'|+\mu(2m-2-|\alpha|-|\beta|)}\sigma(x')^{-|\nu''|}. \tag{12.7.109}$$

The following problem seems to be rather easy:

Problem 12.7.21. Using arguments of this section prove that for such domains under Neumann (or reduced order Dirichlet leading to mixed boundary conditions) standard asymptotics as in the domains with the smooth boundaries hold (despite cone condition does not necessarily hold).

More precisely, after coordinates $x''_{\mathsf{new}} = x''_{\mathsf{old}}/\sigma(x')$ transforming spike into finite cylinder $Z \times Y$ singular at $x_1 = 0$ but with singularities not violating positivity. Moreover the possible degeneration of the lowest eigenvalues of $A(x'', \xi''; x'', D'')$ is compensated by lower order term $W(x')u$ with $W \ge \epsilon|x_1|^{-2m}$.

12.7.8 Operators with Degenerate Potentials

While not covered by this section, the following topic is related:

Example 12.7.22. Consider Schrödinger operator

$$A = D_1^2 + |D''|^2 + V(x), \qquad V(x) = e^{2|x_1|^{\mu+1}}|x''|^2 - e^{|x_1|^{\mu+1}}k \tag{12.7.110}$$

with $k = d''$. Without degeneration such operators were considered in Section 11.5; one can see easily that even with this degeneration but with $k < d''$ such operators would be still covered by the general theory of that section.

However now the lower eigenvalue of $|D''|^2 + V(x)$ is 0. Then changing variables $x''_{\mathsf{new}} = x''_{\mathsf{old}} \exp(|x_1|^{\mu+1})$ we arrive to the operator not much different from

$$A = D_1^2 + cx_1^{2\mu} + e^{x_1^{\mu+1}} B, \qquad \text{with } B = |D''|^2 + |x''|^2 - d''. \tag{12.7.111}$$

Problem 12.7.23. Develop the theory, covering Example 12.7.22.

12.8 Periodic Operators

In this section we consider number $\mathsf{N}(\tau)$ of eigenvalues of operator $A_t(x, D) = A(x, D) - tW(x, x)$ crossing level E as t runs from 0 to τ, $\tau \to \infty$. Here A is periodic matrix operator, matrix W is positive, periodic with respect to the first copy of x and decaying as second copy of x goes to infinity, E either belongs to a spectral gap of A or is one its ends.

12.8.1 Periodic Operators

Let us consider in $\mathbb{R}^d$ $\mathsf{D} \times \mathsf{D}$-matrix operator $A(x, x, D) = A^{\mathrm{w}}(x, x, D)$ with the Weyl symbol

$$A(z, y, \xi) = \sum_{|\sigma| \le m} a_\sigma(z, y)\xi^\sigma \tag{12.8.1}$$

where

(12.8.2) Operator $A(z, y, D)$ is z-periodic with period lattice

$$\Gamma = \mathbb{Z}e_1 \oplus \mathbb{Z}e_2 \oplus \ldots \mathbb{Z}e_d \tag{12.8.3}$$

and transformation matrices $\{T_1, \ldots, T_d\}$, which means that

$$A(z + e_j, y, D) = T_j^\dagger A(z, y, D) T_j \qquad \forall j = 1, \ldots, d \tag{12.8.4}$$

with unitary commuting matrices T_j.

Let $T^n = T_1^{n_1} \cdots T_d^{n_d}$ for $n = n_1 e_1 + \cdots + n_d e_d \in \Gamma$.

Further, let us assume that

$$|\nabla_z^\alpha \nabla_y^\beta (a_\sigma(z, y) - a_\sigma^\infty(z))| \le C\langle y\rangle^{-\delta-|\beta|} \qquad \forall \alpha : |\alpha| \le K \ \forall |\beta| : |\beta| \le K \tag{12.8.5}$$

$$|\nabla_z^\alpha a_\sigma^\infty(z)| \le C \qquad \forall \alpha : |\alpha| \le K \tag{12.8.6}$$

with $\delta > 0$ and large enough $K = K(m, d, \mu)$ where

(12.8.7) Operator $A^\infty(z, D_z)$ is z-periodic with period lattice (12.8.3) and transformation matrices $\{T_1, \dots, T_d\}$.

Furthermore, as usual we assume that A is elliptic:

$$|A(x,\xi)v| \ge \epsilon_0\bigl(|\xi|^m - C\bigr)|v| \qquad \forall x,\xi \in \mathbb{R}^d\ \forall v \in \mathbb{C}^D \tag{12.8.8}$$

and that

$$A(x,\xi)^\dagger = A(x,\xi) \qquad \forall (x,\xi). \tag{12.8.9}$$

To describe the spectrum of operator A^∞ (and the essential spectrum of A) we need the following definition:

Definition 12.8.1. Function $v(x)$ is *quasiperiodic with quasimomentum* ξ *and transformation matrices* $\{T\} = \{T_1, \dots, T_d\}$ if

$$u(x+n) = T^{-n}u(x)e^{i\langle n,\xi\rangle} \qquad \forall n \in \Gamma. \tag{12.8.10}$$

We denote by $\mathbb{H}_{\xi,\{T\}}$ the space of such functions with an inner product as in $\mathscr{L}^2(\mathcal{Q}, \mathbb{C}^d)$ but multiplied by $(2\pi)^{-d}\operatorname{vol}(\mathcal{Q}')$ where $\mathcal{Q} = [0,1]e_1 \oplus \dots [0,1]e_d$ is an *elementary cell*, $\mathcal{Q}' = [0,1]e'_1 \oplus \dots \oplus [0,1]e'_d$ is a *dual elementary cell* and $\langle e'_j, e_k\rangle = 2\pi\delta_{jk}\ \ \forall j,k$.

Then we arrive to the following proposition:

Proposition 12.8.2. *Both operators* A *and* A^∞ *are self-adjoint in* $\mathscr{L}^2(\mathbb{R}^d, \mathbb{C}^D)$ *and both essential spectra of* A *and* A^∞ *and spectrum of* A^∞ *coincide with*

$$\mathcal{E} := \bigcup_k \mathcal{E}_k, \qquad \mathcal{E}_k := \bigcup_{\xi\in\mathcal{Q}'} \lambda_k(\xi) \tag{12.8.11}$$

where $\lambda_k(\xi)$ *are eigenvalues of operator* A^∞ *restricted to* $\mathbb{H}_{\xi,\{T\}}$.

Remark 12.8.3. (i) Proposition 12.8.2 as well as Statements (i)–(iv) below are well-known and their proofs are easy.

(ii) k runs $\mathcal{K}$ where $\mathcal{K} = \{1, 2, 3, \dots\}$ if and only if A^∞ is semi-bounded from below, $\mathcal{K} = \{0, -1, -2, \dots\}$ if and only if A^∞ is semi-bounded from above and $\mathcal{K} = \mathbb{Z}$ otherwise.

(iii) Spectrum of A^∞ restricted to $\mathbb{H}_{\xi,\{T\}}$ is discrete and consists of eigenvalues $\lambda_k(\xi)$.

(iv) Functions $\lambda_k(\xi)$ are periodic with respect to the *dual lattice* $\Gamma' = \mathbb{Z}e'_1 \oplus \mathbb{Z}e_2 \oplus ... \mathbb{Z}e'_d$ and thus we can consider them as defined on $\mathcal{Q}'$.

(v) Without any loss of the generality we assume that

$$\lambda_k(\xi) \geq \lambda_{k-1}(\xi) \qquad \forall k. \tag{12.8.12}$$

So, $\mathsf{Spec}(A^\infty)$ has is *zone spectrum* with possible overlapping of *spectral zones* (*spectral bands*)

$$\mathcal{E}_k = [\min_\xi \lambda_k(\xi), \max_\xi \lambda_k(\xi)] \tag{12.8.13}$$

separated by *spectral gaps*

$$\mathcal{G}_k := (\max_\xi \lambda_k(\xi), \min_\xi \lambda_{k+1}(\xi)) \qquad \text{provided} \quad \max_\xi \lambda_k(\xi) < \min_\xi \lambda_{k+1}(\xi). \tag{12.8.14}$$

If operator A^∞ is semibounded from below (above) we introduce also the spectral gap $\mathcal{G}_0 := (-\infty, \min_\xi \lambda_1(\xi))$ or $\mathcal{G}_0 := (\max_\xi \lambda_0(\xi), \infty)$ respectively.

It follows from Proposition 12.8.2 that

Corollary 12.8.4. $\mathsf{Spec}(A)$ *intersected with the spectra gap* $\mathcal{G}_k$ *is discrete and may be accumulating to one or both ends of it.*

12.8.2 Statement of the Problem

Let us pick some energy level which is in the spectral gap: $E \in \mathcal{G}_k$.

Later we consider also cases when E is either lower end or upper end of the spectral gap: $[E, E+\epsilon) \cap \mathsf{Spec}(A^\infty) = \{E\}$ or $(E-\epsilon, E] \cap \mathsf{Spec}(A^\infty) = \{E\}$.

Let $W(x) = W(x,x)$ where

(12.8.15) Matrices $W(z,y)$ and $W^\infty(z,y)$ are Hermitian and z-periodic with the period lattice Γ and transformation matrices $\{T\} = \{T_1, \dots, T_d\}$ with respect to x,

and

$$|\nabla_z^\alpha \nabla_y^\beta W(z,y)| \le C\langle y\rangle^{-m\mu-|\beta|} \qquad \forall\alpha:|\alpha|\le K\ \forall|\beta|:|\beta|\le K, \qquad \mu>0 \tag{12.8.16}$$

$$|\nabla_z^\alpha \nabla_y^\beta (W(z,y)-W^\infty(z,y))| = o(|y|^{-m\mu-|\beta|}) \qquad \text{as } |y|\to\infty \quad \forall\beta:|\beta|=1 \tag{12.8.17}$$

$$\langle W(z,y)v,v\rangle \ge C|v|^2\langle y\rangle^{-m\mu} \qquad \forall x,y\in\mathbb{R}^d\ \forall v\in\mathbb{C}^d \tag{12.8.18}$$

where

(12.8.19) $W^\infty(z,y)$ is positive homogeneous with respect to y of degree $-m\mu$ and satisfies condition (12.8.18) as well.

In the the framework of the previous assumptions let us consider operator $A_t = A - tW$ with $t>0$. Then

Remark 12.8.5. (i) Let E reside within spectral gap; then for each t only finite number of eigenvalues of A_t belong to $(E-\epsilon, E+\epsilon)$;

(ii) Further, if E is the lower end of the spectral gap then under appropriate conditions for each t only finite number of eigenvalues of A_t belong to $(E, E+\epsilon)$;

(iii) Similarly, if E is the upper end of the spectral gap then under appropriate conditions for each t only finite number of eigenvalues of A_t belong to $(E-\epsilon, E)$;

(iv) All the eigenvalues are monotone decreasing functions of t.

Let $\mathsf{N}_E(\tau)$ be a number of the eigenvalues of $A-tW$ (counting multiplicities) passing through the *observation point* E in in the framework of (i) (reaching E from above, leaving E to below respectively in the frameworks of (ii) and (iii)) as t changes from $0+$ to $\tau-$.

We are interested in asymptotics of $\mathsf{N}_E(\tau)$ as $\tau\to+\infty$. There is also a lite version of the problem when there is no periodic variable at all; then formally $\mathcal{Q}=\{0\}$ and $\mathcal{Q}'=\mathbb{R}^d$.

12.8.3 Reformulation of the Problem

We start from the following proposition:

Proposition 12.8.6. *In the framework of our assumptions*

(i) Equality

$$\mathsf{N}_E(t) = \mathsf{N}(L_E; [0, t)) \tag{12.8.20}$$

holds where $\mathsf{N}(L; [0, t))$ *is a number of eigenvalues of the operator*

$$L = L_E = W^{-\frac{1}{2}}(A - E)W^{-\frac{1}{2}} \tag{12.8.21}$$

belonging to $[0, t)$.

(ii) L_E *is a self-adjoint operator in* $\mathscr{L}^2(\mathbb{R}^d, \mathbb{C}^D)$.

Proof. Both statement are obvious for $A = A^\infty$ and

$$L_E^\infty = W^{-\frac{1}{2}}(A^\infty - E)W^{-\frac{1}{2}}$$

because $(A^\infty - E)$ is invertible if $E \notin \mathsf{Spec}(A^\infty)$ and this is true for perturbed operator L_E since L_E^∞-bound of the operator

$$L'_E = W^{-\frac{1}{2}}(A - A^\infty)W^{-\frac{1}{2}}$$

is 0 which means that

$$\|L'_E u\| \le \epsilon \|L_E^\infty u\| + C_\epsilon \|u\| \qquad \forall u \ \forall \epsilon > 0.$$

Furthermore,

$$\mathsf{Spec}_{\mathsf{ess}}(L_E) = \emptyset. \tag{12.8.22}$$

Remark 12.8.7. Moreover, even if E is the extreme of a spectral gap, conditions of our theorems will assure that L_E^∞ is self-adjoint unbounded operator with compact inverse and therefore all above statements (save that $(A^\infty - E)$ is invertible) remain true.

12.8.4 Reduction to Operators with Operator-Valued Symbols

One can try to apply arguments of Chapters 9 and 10 directly but they lead to sharp remainder estimates only if $\rho^{d-1}\in\mathscr{L}^1$ even in the framework of more relaxed Theorem 12.8.12 below. We need to involve operator-valued theory of Chapter 4) and we introduce Gelfand' transformation:

Definition 12.8.8. Let Γ and $\{T\}$ be introduced above. Then

$$\mathcal{F}u(\xi,x)=(2\pi)^{\frac{d}{2}}(\operatorname{vol}(\mathcal{Q}'))^{-1}\sum_{n\in\Gamma}T^n e^{-i\langle x-n,\xi\rangle}u(x-n). \tag{12.8.23}$$

is a *Gelfand' transformation* of u.

Proposition 12.8.9. *(i) As defined for x in $\mathbb{R}^d$ $\mathcal{F}(\xi,x)$ is x-quasiperiodic with quasimomentum ξ and transformation matrices $\{T\}$.*

(ii) $\mathcal{F}$ is an unitary operator from $\mathscr{L}^2(\mathbb{R}^d,\mathbb{C}^\mathsf{D})$ to $\mathscr{L}^2(\mathcal{Q}',\mathbb{H}_{\xi,\{T\}})$.

(iii) Inverse Gelfand's transform is defined by

$$u(x)=(2\pi)^{-\frac{d}{2}}\int_{\mathcal{Q}'}e^{i\langle x,\xi\rangle}(\mathcal{F}u)(\xi,x)\,d\xi. \tag{12.8.24}$$

Proof. Statement (i) is obvious.

Plugging (12.8.23) into the right-hand expression of (12.8.24) and using the first of equalities

$$\int_{\mathcal{Q}'}e^{i\langle n,\xi\rangle}\,d\xi=c\delta_{n0},\qquad \sum_{n\in\Gamma}e^{i\langle n,\xi\rangle}=c\sum_{k\in\Gamma'}\delta(\xi-k),\qquad \text{with}\quad c=\operatorname{vol}(\mathcal{Q}') \tag{12.8.25$_{1,2}$}$$

we get $u(x)$; similarly, plugging (12.8.24) into the right-hand expression of (12.8.23) with $\mathcal{F}u(\xi,x)$ replaced by $v(\xi,x)$ which is x-quasiperiodic with quasimomentum ξ and transformation matrices $\{T\}$ and using the $(12.8.25)_2$ we get $v(\xi,x)$. This proves Statement (iii).

Finally, calculating $\int_{\mathcal{Q}'}\|\mathcal{F}(\xi,x)\|^2_{\mathbb{H}_{\xi,\{T\}}}\,d\xi$ and using $(12.8.25)_1$ again we get $\sum_{n\in\Gamma}\int_{\mathcal{Q}}\|u(x+n)\|^2\,dx=\|u\|^2_{\mathscr{L}^2(\mathbb{R}^d,\mathbb{C}^\mathsf{D})}$ which proves Statement (ii). $\square$

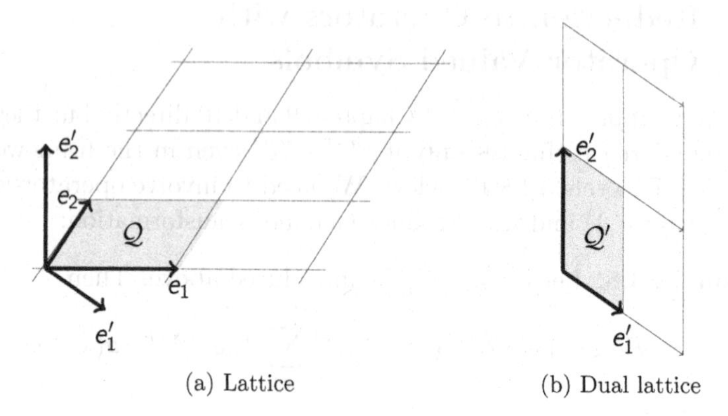

(a) Lattice (b) Dual lattice

Figure 12.7: Lattice, elementary cell $\mathcal{Q}$, dual lattice and dual elementary cell $\mathcal{Q}'$.

Remark 12.8.10. (i) Operator

$$u \to \mathcal{F}' u(x,\xi) := (2\pi)^{\frac{d}{2}} (\mathsf{vol}(\mathcal{Q}'))^{-1} \sum_{n\in\Gamma} T^n e^{i\langle n,\xi\rangle} u(x-n) \tag{12.8.26}$$

with

$$u(x) = (2\pi)^{-\frac{d}{2}} \int_{\mathcal{Q}'} (\mathcal{F}' u)(\xi, x)\, d\xi \tag{12.8.27}$$

has the similar properties albeit transforms $\mathscr{L}^2(\mathbb{R}^d, \mathbb{C}^{\mathsf{D}})$ into $\mathscr{L}^2(\mathcal{Q}', \mathbb{H}_{0,\{T\}})$.

(ii) Consider $T = I$, replace Γ by $\varepsilon\Gamma$ and calculate limits as $\varepsilon \to +0$. Then $\mathcal{F}(\xi, x) \to (2\pi)^{d/2} F_{x\to\xi} u$.

Proposition 12.8.11. *Operator $\mathcal{F}$ transforms our operators in the following way:*

$$\mathcal{F} A^{\mathsf{w}} \mathcal{F}^* = A^{\mathsf{w}}(x - D_\xi, x, \xi + D_x), \tag{12.8.28}$$

$$\mathcal{F} A^{\infty\,\mathsf{w}} \mathcal{F}^* = A^{\infty\,\mathsf{w}}(x, \xi + D_x), \tag{12.8.29}$$

$$\mathcal{F} W \mathcal{F}^* = W^{\mathsf{w}}(x - D_\xi, x). \tag{12.8.30}$$

Proof. Proof is easy by direct calculations from (12.8.24). The only thing to notice is that we can treat the fast copy of x in $\mathbb{R}^d$ in different ways than the slow one using periodicity with respect to fast x. □

12.8.5 Main Theorem: E Inside Spectral Gap

Theorem 12.8.12. *Let conditions* (12.8.2), (12.8.15), (12.8.5)–(12.8.9) *and* (12.8.16)–(12.8.18) *be fulfilled and let*

$$E \notin \operatorname{Spec}(A^\infty). \tag{12.8.31}$$

(i) Further, let us assume that there is an infinite number of $\lambda_k(\xi)$ *exceeding* E *(or equivalently* A *is not semibounded from above). Then the following asymptotics holds as* $\tau \to +\infty$*:*

$$|\mathsf{N}(\tau) - \mathcal{N}(\tau)| \le R(\tau) = \begin{cases} \tau^{(d-1)/m} & \text{for } \mu > 1,\\ \tau^{(d-1)/m}\log\tau & \text{for } \mu = 1,\\ \tau^{(d-1)/(m\mu)} & \text{for } \mu < 1,\end{cases} \tag{12.8.32}$$

with

$$\mathcal{N}(\tau) = \int_{\mathbb{R}^d}\int_{Q'} \mathsf{n}(y,\xi,\tau)\,d\xi dy \tag{12.8.33}$$

where $\mathsf{n}(y,\xi,\tau)$ *is the number of eigenvalues of operator* $A(x,y,D_x) - tW(x,y)$ *crossing* E *as* t *runs from* $0+$ *to* $\tau - 0$*; this operator depends on* y *and is restricted to* $\mathbb{H}_{\xi,\{T\}}$.

Moreover,

$$\mathcal{N}(\tau) \asymp \begin{cases} \tau^{d/m} & \text{for } \mu > 1,\\ \tau^{d/m}\log\tau & \text{for } \mu = 1,\\ \tau^{d/(m\mu)}, & \text{for } \mu < 1.\end{cases} \tag{12.8.34}$$

(ii) Further, if $\mu > 1$ *and the standard non-periodicity condition for symbol*

$$L_m(x,\xi) = A_m W^{-1}(x), \qquad A_m(x,\xi) = \sum_{\sigma:|\sigma|=m} a_\sigma(x)\xi^\sigma \tag{12.8.35}$$

is fulfilled then asymptotics

$$\mathsf{N}(\tau) = \mathcal{N}(\tau) + \bigl(\varkappa_1 + o(1)\bigr)\tau^{(d-1)/m} \tag{12.8.36}$$

holds.

(iii) Finally, let us assume that there is finite number of λ_k exceeding E (or equivalently A is semibounded from above). Then

$$|\mathsf{N}(\tau) - \mathcal{N}(\tau)| \le C\tau^{(d-1)/(m\mu)} \tag{12.8.37}$$

and

$$\mathsf{N}(\tau) \asymp \tau^{d(m\mu)}. \tag{12.8.38}$$

To prove Theorem 12.8.12 let us consider operator $L = L_E(x, D)$ defined by (12.8.21) and the Schwartz kernel $e_L(x, x, \tau)$ of its spectral projector. Then Theorem 12.8.12 follows from Proposition 12.8.13 below as we pick up $\rho(y) = \langle y \rangle^{-\mu}$ and $\gamma_*(\tau) = \tau^{1/(m\mu)}$ and consider γ-admissible partition of unity with $\gamma = \frac{1}{2}\langle x \rangle$.

Proposition 12.8.13. *Let conditions* (12.8.2), (12.8.15), (12.8.4)–(12.8.9) *and* (12.8.16)–(12.8.18) *be fulfilled.*

(i) Let ψ be γ-admissible function supported in $B(\bar{x}, \frac{1}{2}\gamma)$ with $\gamma \le C\gamma_*(\tau)$. Then

$$\mathsf{R}_1 := |\int \psi(x)\Bigl(e_L(x, x, \tau) - e_L(x, x, \frac{1}{2}\tau) - (2\pi)^{-d}\int \bigl(\nu_{L_m}(x, \xi, \tau) - \nu_{L_m}(x, \xi, \frac{1}{2}\tau)\bigr)\, d\xi\Bigr)\, dx| \le C\tau^{(d-1)/m}\rho^{d-1}\gamma^d \tag{12.8.39}$$

where $\nu_{L_m}(x, \xi, \tau)$ is the eigenvalue counting function for symbol $L_m(x, \xi)$ defined by (12.8.35).

(ii) Further, for fixed r under non-periodicity condition for symbol $L_m(x, \xi)$

$$\mathsf{R}_1 := |\int \psi(x)\Bigl(e_L(x, x, \tau) - e_L(x, x, \frac{1}{2}\tau) - (2\pi)^{-d}\int \bigl(\nu_{L_m}(x, \xi, \tau) - \nu_{L_m}(x, \xi, \frac{1}{2}\tau)\bigr)\, d\xi - \kappa_1(x)\tau^{(d-1)/m}\Bigr)\, dx| = o\bigl(\tau^{(d-1)/m}\rho^{d-1}\gamma^d\bigr). \tag{12.8.40}$$

(iii) Furthermore, for $2 < \gamma \le C\gamma_*(\tau)$

$$|\int \psi(x)\Bigl(e_L(x, x, \tau) - e_L(x, x, \frac{1}{2}\tau) - (2\pi)^{-d}\int \bigl(n_M(x, \xi, \tau) - n_M(x, \xi, \frac{1}{2}\tau)\bigr)\, d\xi\Bigr)\, dx| \le C\tau^{(d-1)m}r^{(1-\mu)(d-1)}\rho^{d-1}\gamma^{d-1} \tag{12.8.41}$$

where

(12.8.42) *$n_M(y,\xi,\tau)$ is the number of eigenvalues between 0 and τ of operator*

$$M(y,\xi;z,D_z)=W^{-\frac{1}{2}}(z,y)\bigl(A(z,y,\xi+D_z)-E\bigr)W^{-\frac{1}{2}}(z,y) \tag{12.8.43}$$

restricted to the space $\mathbb{K}_{\xi,\{T\}}$ for fixed "slow" argument y.

(iv) Let $E\notin\operatorname{Spec}(A^\infty)$. Then for $\gamma\ge C_0\gamma_*(\tau)$ with large enough constant C_0

$$|\int \psi(x)\bigl(e_L(x,x,\tau)-e_L(x,x,\frac{1}{2}\tau)\bigr)\,dx|\le C\tau^{-s}\gamma^{-s} \tag{12.8.44}$$

with arbitrarily large s.

(v) Finally, let A be semibounded from above. Then estimate (12.8.44) also holds for $\gamma\le\epsilon_0\gamma_*(\tau)$ with small enough constant ϵ_0.

Proof. (a) Statements (i) and (ii) follow from the standard local spectral asymptotics.

(b) Statement (iii) follows from the spectral asymptotics for operator $M(D_\xi,\xi;x,D_x)$ with the operator-valued symbol $M(y,\xi;z,D_z)$.

Let us apply Theorems 4.4.5, 4.4.6, 4.4.8 and 4.4.9) with $B=(I+\Delta_z)^{m/2}$ in $\mathbb{H}$; due to Remark 12.8.10(i) we can pass from $\mathbb{H}_\xi$ to $\mathbb{H}_0$.

Remark 12.8.14. Here and during all remaining part of the section ξ plays the role of the spatial variable, $z\in\mathcal{Q}$ is an "inner variable" defining $\mathcal{K}=\mathscr{L}^2(\mathcal{Q},\mathbb{C}^D)$ and $y\to D_\xi$ is a dual to ξ variable, so $(y,\xi)\in T^*\mathcal{Q}'$.

To prove Statement (iii) we need to establish that

(12.8.45) $M(y,\xi;z,D_z)$ is (y,ξ)-microhyperbolic on the energy level $\tau\gg 1$

which is due to (12.8.5) and (12.8.17).

(c) Finally, Statements (iv) and (v) and follow from the fact that zones in question are classically forbidden for operators with operator-valued symbols introduced in Part (b) of this proof and uses Theorem 4.4.2. □

Corollary 12.8.15. *(i) In the framework of Theorem 12.8.12(i) with* $\mu > d(d-1)^{-1}$ *the standard Weyl asymptotics holds with the remainder estimate* $O(\tau^{(d-1)/m})$ *and even* $o(\tau^{(d-1)/m})$ *under standard non-periodicity condition as in Theorem 12.8.12(ii).*

(ii) In the framework of Theorem 12.8.12(i) with $\mu > 1$ *and*

$$|\nabla_y^\beta (W(z,y) - W^\infty(z,y))| = o(|y|^{-\delta - m\mu - |\beta|}) \qquad \text{as } |y|\to\infty\ \forall \beta : |\beta| = 1 \tag{12.8.46}$$

with $\delta > \delta_0 = 1-(d-1)(\mu-1)$ *asymptotics* (12.8.32) *holds with*

$$\mathcal{N}(\tau) = \int_{\mathbb{R}^d}\int_{Q'} \mathbf{n}^\infty(y,\xi,\tau)\,d\xi dy + (2\pi)^{-d}\int_{\mathbb{R}^{2d}} \Bigl(\nu(x,\xi,\tau) - \nu(x,\xi,0) - \nu^\infty(x,\xi,\tau) + \nu^\infty(x,\xi,0)\Bigr)\,dxd\xi \tag{12.8.47}$$

where $\nu^\infty(x,\xi,\tau)$ *is defined in the same way as* $\nu(x,\xi,\tau)$ *but for* $A^\infty(x,\xi)$ *and* $W^\infty(x,x)$ *and* $\mathbf{n}^\infty(y,\xi,\lambda)$ *is defined in the same way as* $\mathbf{n}(y,\xi,\tau)$ *but for operators* A^∞ *and* W^∞.

(iii) In the framework of Theorem 12.8.12(i) with $\mu < (d-1)d^{-1}$ *and* (12.8.46), (12.8.17) *with* $\delta > (1-\mu)(d-1)(d-d\mu-1)^{-1}$ *asymptotics* (12.8.32) *holds with*

$$\mathcal{N}(\tau) = \int_{\mathbb{R}^d}\int_{Q'} \mathbf{n}^\infty(y,\xi,\tau)\,d\xi dy. \tag{12.8.48}$$

(iv) In the framework of Theorem 12.8.12(ii) and (12.8.46), (12.8.17) *with* $\delta > 1$ *asymptotics* (12.8.32) *holds with* $\mathcal{N}$ *defined by* (12.8.48).

Proof. Statement (iii) is due to Proposition 12.8.13(i) and (ii).

Note that if $\gamma^{-\delta} < h = \tau^{-1/m}\gamma^{\mu-1}$ then one can replace in Proposition 12.8.13(iii) $\mathbf{n}$ by $\mathbf{n}^\infty$. Combining with second part of Statement (iv) of Proposition 12.8.13 we get Statement (iv) and with Proposition 12.8.13(i) we get Statement (iii).

To prove Statement (ii) note that if $\tau^{d/m}\gamma^{-\delta+d(1-\mu)} < \tau^{(d-1)/m}$ then one can replace in Proposition 12.8.13(iii) $\mathbf{n}$ by $\mathbf{n}^\infty$. On the hand, one can employ the theory of operators with operator-valued symbols for propagation of singularities only. This yields standard Weyl asymptotics as in Proposition 12.8.13(i) but with the remainder estimate

$$C\bigl(\tau^{(d-2)/m}\gamma^{(1-\mu)(d-2)+2}+\tau^{(d-1)/m}\gamma^{(1-\mu)(d-1)}\bigr)$$

where the first part comes from the third term in complete Weyl asymptotics. One can find γ satisfying first condition and making last expression less than $O(\tau^{(d-1)/m})$ iff $\delta > \delta_0$. □

Now let us assume that <u>either</u> $\mu < 1$ <u>or</u> A is semi-bounded from above. Then the main contribution to the remainder is delivered by zone

$$\{\epsilon\gamma_*(\tau) \le |x| \le C_0\gamma_*(\tau)\}. \tag{12.8.49}$$

To improve remainder estimate $O\bigl(\tau^{(d-1)/(m\mu)}\bigr)$ we need to consider Hamiltonian flow generated by operator-valued symbol

$$\begin{aligned} M^\infty(D_\xi,\xi;z,D_z) :=\\ W^{\infty\,-\frac{1}{2}}(-D_\xi,z)\bigl(A^{\infty,\mathrm{w}}(x,\xi+D_z)-E\bigr)W^{\infty\,-\frac{1}{2}}(-D_\xi,z)\end{aligned} \tag{12.8.50}$$

which leads immediately to the following theorem:

Theorem 12.8.16. *Let in the framework of Theorem 12.8.12 <u>either</u> $\mu < 1$ <u>or</u> A be semi-bounded from above. Further let the standard non-periodicity condition be fulfilled for operator-valued symbol* (12.8.50).

Then asymptotics

$$\mathsf{N}(\tau) = \mathcal{N}(\tau) + \bigl(\varkappa_1^* + o(1)\bigr)\tau^{(d-1)/(m\mu)} \tag{12.8.51}$$

holds with an appropriate coefficient $\varkappa_1^*$.

Now let us consider the case $\bar{\mu} = 1$ but W contains extra-logarithmic factor.

Theorem 12.8.17[76]**.** *(i) Let all the conditions of Theorem 12.8.12 be fulfilled excluding* (12.8.16)–(12.8.18) *which are replaced by*

$$|\nabla_z^\alpha\nabla_y^\beta W(z,y)| \le C\rho(y)^m\langle y\rangle^{-|\beta|}\qquad \forall\alpha:|\alpha|\le K\ \forall\beta:|\beta|\le K, \tag{12.8.16$'$}$$

[76] Cf. Theorem 12.8.12.

$$|\nabla_y^\beta (W(z,y) - W^\infty(z,y))| = o(\rho(y)^m \langle y\rangle^{-|\beta|}) \qquad \textit{as}\ |y|\to\infty\quad \forall\beta: |\beta|=1, \tag{12.8.17$'$}$$

$$\langle W(z,y)v, v\rangle \ge C\rho(y)^m |v|^2 \qquad \forall x,y\in\mathbb{R}^d\ \forall v\in\mathbb{C}^d \tag{12.8.18$'$}$$

with

$$\rho(y) = |y|^{-\mu}(\log|y|)^{-\mu_1}(\log\log|y|)^{-\mu_2}\cdots(\underbrace{\log\ldots\log}_{l\ \textit{times}}|y|)^{-\mu_l} \qquad (|y|>C) \tag{12.8.52}$$

with

$$\mu > 0. \tag{12.8.53}$$

Then Statements (i) and (ii) of Theorem 12.8.12 remain true as soon as

$$\int \rho^{d-1}\gamma^{d-2}\,d\gamma < \infty. \tag{12.8.54}$$

On the other hand, if this assumption is violated, Statement (i) of Theorem 12.8.12 modified in the obvious way remains true (modification affects remainder estimate which is now

$$R(\tau) = \tau^{(d-1)/m}\int_{\gamma<\gamma_*(\tau)} \rho^{d-1}\gamma^{d-2}\,d\gamma \tag{12.8.55}$$

with $\gamma_*(\tau)$ defined from

$$\tau^{\frac{1}{m}}\rho(\gamma_*(\tau)) = 1. \tag{12.8.56}$$

Moreover, as soon as

$$\int \rho^{d}\gamma^{d-1}\,d\gamma < \infty \tag{12.8.57}$$

magnitude of the principal part does not change; if this assumption is violated, the principal part will be

$$\mathcal{N}(\tau) \asymp \int_{\gamma<\gamma_*(\tau)} \rho^{d}\gamma^{d-1}\,d\gamma. \tag{12.8.58}$$

(ii) Further, Statement (iii) of Theorem 12.8.12 remains true after modification of the remainder estimate and the the principal part (which are now

$$R_0(\tau) = \gamma_*(\tau)^{d-1} \tag{12.8.59}$$

and

$$\mathcal{N}(\tau) \asymp \gamma_*(\tau)^d \tag{12.8.60}$$

respectively).

12.8.6 E on the Bottom of Spectral Gap

Now let us consider the case when E is a boundary of the spectral gap: either that either

$$[E, E+\epsilon) \cap \mathsf{Spec}(A^\infty) = \{E\} \tag*{(12.8.61)$^+$}$$

or

$$(E-\epsilon, E] \cap \mathsf{Spec}(A^\infty) = \{E\}. \tag*{(12.8.61)$^-$}$$

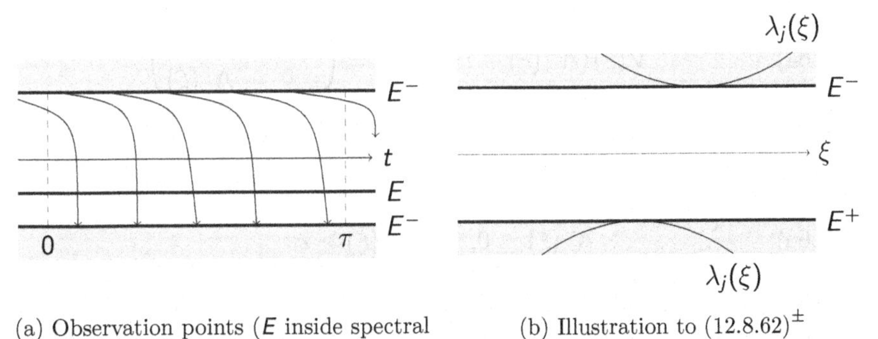

(a) Observation points (E inside spectral gap, E^+ on the bottom and E^- on the top) and eigenvalues of $A - tW$;

(b) Illustration to (12.8.62)$^\pm$

Figure 12.8: Spectral gap, observation points and behavior of $\lambda_j(\xi)$.

In the former case

$$\exists j : \lambda_k(\xi) \le E \quad \forall k \le j, \qquad \lambda_k(\xi) \ge E+\epsilon \quad \forall k > j \qquad \forall \xi, \tag*{(12.8.62)$^+$}$$

while in the latter

$$\exists j : \lambda_k(\xi) \le E \quad \forall k \le j, \qquad \lambda_k(\xi) \ge E + \epsilon \quad \forall k > j \qquad \forall \xi \tag{12.8.62$^-$}$$

In virtue of Proposition 12.8.13 we need to consider only an *external zone*

$$X_{\mathsf{ext}} := \{x :\ |x| \ge \gamma_*(\tau)\} \tag{12.8.63}$$

as an *inner zone* $X_{\mathsf{inn}} := \{x :\ |x| \ge \gamma_*(\tau)\}$ is already covered.

However situation is really different under assumptions $(12.8.61)^+$ and $(12.8.61)^-$. In this subsection we consider only the much easier former case.

Theorem 12.8.18. *Let assumptions of one of Theorems 12.8.12, 12.8.16 or 12.8.17 be fulfilled, except assumption* (12.8.31) $E \notin \mathsf{Spec}(A^\infty)$, *which is replaced by* $(12.8.61)^+$. *Also let us assume that* $A = A^\infty$.

Then all assertions of the corresponding theorem remain true.

Proof. The Gelfand's transform transforms A^∞ into $A^\infty(\xi)$ which in turn due to $(12.8.62)^+$ one transforms by unitary operator $V(\xi)$ in $\mathbb{H}_{\xi,\{T\}}$ into the block-diagonal form

$$V(\xi)\bigl(A^\infty(\xi) - E\bigr)V^*(\xi) = \begin{pmatrix} \Lambda^-(\xi) & 0 \\ 0 & \Lambda^+(\xi)\end{pmatrix}, \tag{12.8.64}$$

$$V(\xi)WV^*(\xi) = \begin{pmatrix} W^{--} & W^{-+} \\ W^{+-} & W^{++}\end{pmatrix} \tag{12.8.65}$$

with

$$\Lambda^-(\xi) \le 0, \qquad \Lambda^+(\xi) \ge \epsilon. \tag{12.8.66}$$

and $W^{\cdot\cdot}$ ξ-pseudodifferential operators.

As $|y| \ge \gamma_*(\tau)$ [77] we can invert $\bigl(\Lambda^+(\xi) - E - tW^{++}\bigr)$ which is elliptic operator as $t \ge 0$, plug into the second equation and derive that the system is elliptic and negative.

Then under condition $(12.8.61)^+$ contribution of external zone to the remainder is negligible. We do not need the uncertainty principle assumption here because we have a define sign. □

[77] With a large enough constant in its definition, so $\rho^m\tau \le \epsilon$ there.

12.8.7 E on the Top of Spectral Gap

Now let us consider the most complicated case when condition (12.8.61)$^-$ is fulfilled.

Let us assume that

(12.8.67) $\lambda_k(\xi)=E$ implies that $\lambda_k(\xi)$ is the simple eigenvalue of $A^\infty(x,D_x)$ on $\mathbb{H}_{\xi,\{T\}}$ [78])

and

(12.8.69) $\lambda_j(\xi)=E$ implies that $\operatorname{Hess}\lambda_j(\xi)$ is non-degenerate.

Then

(12.8.70) $\Lambda=\{\xi:\ \lambda_j(\xi)=E\}=\{\bar\xi_0^{(i)}, i=1,\dots \bar{i}\}$ with $\bar{i}<\infty$ and

$$\epsilon\omega(\xi)\le \mp\bigl(\lambda_j(\xi)-E\bigr)\le C\omega^2(\xi) \tag{12.8.71}$$

with $\omega(\xi):=\min_i|\xi-\bar\xi^{(i)}|$.

Considering operator $A-E-tW$ near $\bar\xi^{(i)}\in\Sigma$ and assuming for simplicity that $\bar\xi^{(i)}=0$ we observe that it is similar to $\Delta-tW(y)$ "projected" to the null-space of operator $(A-E)$ and unless W decays faster than $|y|^{-2}$ it may happen that this operator has a sequence of eigenvalues accumulating to -0. To avoid this we assume that

(12.8.72) Either $\mu':=m\mu/2>1$, or $\mu'=1$ and $\mu_1=\dots=\mu_{l-1}=0$ and $\mu_l>0$ $(k=1,\dots,n-1)$.

So, we need to consider X_{ext} defined by (12.8.63) and only as $\omega(\xi)\le\epsilon$. In this zone we can make reduction similar to (12.8.64)–(12.8.66)

$$V(\xi)\bigl(A^\infty(\xi)-E\bigr)V^*(\xi)=\begin{pmatrix}\lambda(\xi) & 0\\ 0 & \Lambda'(\xi)\end{pmatrix}, \tag{12.8.73}$$

$$V(\xi)WV^*(\xi)=\begin{pmatrix}W^{00} & W^{0\prime}\\ W^{\prime 0} & W^{\prime\prime}\end{pmatrix} \tag{12.8.74}$$

---

[78]) Then in virtue of (12.8.62)$^+$ there exists j such that only for $k=j$

$$\Sigma_k:=\{\xi:\ \lambda_k(\xi)=E\}\ne\emptyset. \tag{12.8.68}$$

In what follows $\Sigma:=\Sigma_j$.

with

$$|\lambda(\xi)| \le \epsilon, \qquad |(\Lambda'(\xi) - E)^{-1}| \le c_0 \tag{12.8.75}$$

where $\lambda(\xi) = \lambda_j(\xi) - E$.

Due to our methods we can consider in the zone X_{ext} a different operator, namely,

$$\lambda(D) - \tau V \tag{12.8.76}$$

with

$$V = W^{00} - \tau W^{0'}(\Lambda'(D) - \tau - W'')^{-1} W^{0'} \tag{12.8.77}$$

and problem is to find the number of negative eigenvalues of operator (12.8.76); this is basically the very special case of our original problem (one we mentioned as the "lite" problem: when A is not a periodic but a constant coefficients operator.

Due to (12.8.68) this is a semiclassical problem with an effective semiclassical parameter

$$\hbar = (\tau\rho^m)^{\frac{1}{2}}\gamma \tag{12.8.78}$$

and then $\hbar \ll 1 \iff |x| \le \gamma^*(\tau)$ with $\gamma^*\tau$ defined from equation

$$\tau\rho^m\gamma^2 = 1. \tag{12.8.79}$$

In the normal Weyl theory contribution of the zone

$$X'_{\mathsf{ext}} := \{x : \gamma_*(\tau) \le |x| \le \gamma^*(\tau)\} \tag{12.8.80}$$

to the principal part of asymptotics is

$$\mathcal{N}'_{\mathsf{ext}} := (2\pi)^{-d} \iint_{\{y \in X'_{\mathsf{ext}}, \lambda(y,\xi) - \tau V(y,\xi,\tau) < 0\}} \psi_{\mathsf{ext}}(y)\, dy d\xi \tag{12.8.81}$$

and has a magnitude

$$\tau^{\frac{d}{2}} \int_{\{\gamma_*(\tau) \le |x| \le \gamma^*(\tau)\}} (\rho^m\gamma^2)^{d/2} \gamma^{-d}\, dx \tag{12.8.82}$$

while contribution to the remainder does not exceed

$$R'_{\mathsf{ext}} = C\tau^{\frac{1}{2}(d-1)} \int_{\{\gamma_*(\tau) \le |x| \le \gamma^*(\tau)\}} (\rho^m\gamma^2)^{(d-1)/2} \gamma^{-d}\, dx. \tag{12.8.83}$$

In this subsection we assume that $\mu'>1$ i.e. $\mu>m/2$. Then contribution of the zone in question to the remainder does not exceed $C(\tau\rho^m\gamma^2)^{(d-1)/2}$ calculated as $\gamma=\gamma_*(\tau)$ and in turn it is equal to $C\gamma_*(\tau)^{d-1}$ which is no more than the contribution of X_{inn}.

It clears remainder estimate "O" obtained before and even "o" as $\mu\geq 1$; however as $\mu<1$ we get here "o" as well due to increasing old $\gamma_*(\tau)$ to $C_0(\tau)\gamma_*(\tau)$ with slowly growing $C_0(\tau)$.

Further, to estimate contribution of the zone $X''_{\mathsf{ext}}:=\{x:|x|\geq a=\gamma^*(\tau)\}$ to the remainder one needs to consider the quadratic form $\|Du\|^2-\tau(Vu,u)$ under the Dirichlet boundary condition on $\partial X''_{\mathsf{ext}}$ and estimate properly the dimension of its negative subspace $\mathsf{N}''_{\mathsf{ext}}(\tau)$, applying then arguments of Chapter 9.

As $d\geq 3$

$$\mathsf{N}''_{\mathsf{ext}}(\tau)\leq\tau^{\frac{d}{2}}\int_{X''_{\mathsf{ext}}}\rho^d\,dx \tag{12.8.84}$$

due to Rozenblioum estimate (9.A.11). Note that the right hand expression does not exceed C as $\mu'>1$.

As $d=2$ however we should be a bit more subtle and refer to Rozenblioum–Solomyak estimate (see Theorem 12.A.1) which says that in our assumptions Rozenblioum' estimate (9.A.11) with $q=1$ still is applicable. Then we arrive to the following assertion:

(12.8.85) As $\mu'>1$ a contribution of the zone

$$X''_{\mathsf{ext}}=\{x:|x|\geq\gamma^*(\tau)\} \tag{12.8.86}$$

to both main part of the asymptotics and the remainder does not exceed C.

Therefore as $\mu'=\mu m/2>1$ we arrive to the same remainder estimate as before but we need also include into asymptotics term (12.8.81) of magnitude $O(\tau^{-d/m\mu})$ [79] Obviously as the result here is given by Weyl formula the final answer is also given by Weyl formula and we arrive to the following theorem:

Theorem 12.8.19 [80]**.** *Let assumptions of one of Theorems 12.8.12, 12.8.16 or 12.8.17 be fulfilled, except assumption* (12.8.31) $E\notin\mathsf{Spec}(A^\infty)$, *which is*

[79] As $\rho=\gamma^{-\mu}$; otherwise extra logarithmic factors come into game.

[80] Cf. Theorem 12.8.18.

replaced by assumption (12.8.61)$^-$ *and conditions* (12.8.67)–(12.8.79). *Also let us assume that* $A = A^\infty$ *and* $m\mu > 2$.

Then all assertions of the corresponding theorem remain true.

Situation may drastically change as $m\mu = 2$. In this case to ensure that $\rho^{dm/2}$ and $\rho^{(d-1)m/2}\gamma^{-1}$ belong to $\mathscr{L}^1$ we need to assume that either $\mu_1' = m\mu_1/2 > 1$ or $\mu_1' = 1$ and then the same condition recurrently goes to $\mu_2' = m\mu_2/2$.

This implies the following theorem:

Theorem 12.8.20. *Let assumptions of one of Theorems 12.8.12, 12.8.16 or 12.8.17, except assumption* (12.8.31) $E \notin \operatorname{Spec}(A^\infty)$ *which is replaced by assumption* (12.8.61)$^-$, *and conditions* (12.8.67)–(12.8.79). *Also let us assume that* $A = A^\infty$ *and* $m\mu = 2$.

(i) Let

$$m\mu_1(d-1) = 2,\ \ldots,\ m\mu_{l-1}(d-1) = 2,\ m\mu_l(d-1) > 2. \tag{12.8.87}$$

Then all assertions of the corresponding theorem in question.

(ii) Let

$$m\mu_1(d-1) = 2,\ \ldots,\ m\mu_{k-1}(d-1) = 2,\ m\mu_k(d-1) \le 2. \tag{12.8.88}$$

Then all assertions of the corresponding theorem remain true with the remainder estimate

$$R(\tau) = \tag{12.8.89}$$
$$C\tau^{1/m\mu_1} \times \begin{cases} 1 & \text{as } k = 1, m\mu_1(d-1) < 2,\\ |\log \tau| & \text{as } k = 1, m\mu_1(d-1) = 2,\\ |\log_{k-1}\tau|^{1-m\mu_k(d-1)/2} & \text{as } k > 1, m\mu_k(d-1) < 2,\\ |\log_k \tau| & \text{as } k > 1, m\mu_k(d-1) = 2, \end{cases}$$

where $\log_l \tau = \underbrace{\log \ldots \log}_{l \text{ times}} \tau$.

(iii) Meanwhile the main part of asymptotics has the magnitude as in the theorem in question provided $m\mu_1 d > 2$.

Proof. Observe that as $m\mu = -2$ we have $\rho^{m(d-1)/2}\gamma^{-1} \in \mathscr{L}^1$ if and only if (12.8.87) holds; otherwise $\tau^{(d-1)/2}\int_{X'_{\mathsf{ext}}} \rho^{m(d-1)/2}\gamma^{-1}\,dx$ is given by (12.8.89) and is larger than that what we got before. So, Statements (i) and (ii) are proven.

Note that $\rho^{md/2} \in \mathscr{L}^1$ as $m\mu_1 d > 2$. It implies Statement (iii). □

12.8.8 E on the Top of Spectral Gap. II. Improved Asymptotics

Now we want to consider case of $\rho^{m(d-1)/2}|x|^{-1} \notin \mathscr{L}^1$ and improve the remainder estimate (12.8.89) adding some correction; this correction will be even larger than the previous type of asymptotics as $\rho^{md/2} \notin \mathscr{L}^1$.

Let us consider first

$$A = \Delta - tW(x), \qquad W(x) = |x|^{-2}V(\log|x|) \tag{12.8.90}$$

where

$$|D^\alpha V(z)| \le C\varrho(z)^2|z|^{-|\alpha|} \qquad \forall\alpha : |\alpha| \le K, \tag{12.8.91}$$

$$|\partial_z|V(z) \asymp \varrho^2(z)|z|^{-1} \qquad \text{as}\ \ |z| \ge C \tag{12.8.92}$$

$$\varrho(z) = |z|^{-\nu_\pm} \qquad \text{as} \quad \pm z \ge C, \nu_\pm > 0. \tag{12.8.93}$$

As $V(z)$ decays fast enough as $z \to \pm\infty$ non-degeneracy condition (12.8.92) will not be needed.

Then introducing spherical coordinates $(r,\theta) \in \mathbb{R}^+ \times \mathbb{S}^{d-1}$, separating variables and then substituting $z = \log r$ we arrive to the equivalent problem for the operator

$$A' = D_z^2 + \lambda + \frac{1}{4}(d-2)^2 - tV(z) \tag{12.8.94}$$

where

(12.8.95) $\lambda \in \mathsf{Spec}(\Lambda)$ and Λ is Laplace-Beltrami operator on $\mathbb{S}^{d-1}$.

To this operator we can apply Weyl asymptotics:

$$\mathsf{N}^-_\lambda(\tau) = \mathsf{N}^{\mathsf{W}}_\lambda(\tau) + O(1) \tag{12.8.96}$$

with

$$\mathsf{N}^{\mathsf{W}}_{\lambda}(\tau)=\frac{1}{\pi}\int \Bigl(\tau V(z)-\lambda-\frac{1}{4}(d-2)^2\Bigr)_+^{\frac{1}{2}}\,dz, \tag{12.8.97}$$

as $\lambda+\frac{1}{4}(d-2)^2\ge \epsilon_0>0$ and after summation over eigenvalues of Λ we arrive to the asymptotics with the principal part

$$\begin{aligned}\mathcal{N}(\tau):=\frac{1}{\pi}\sum_{\lambda\in\mathsf{Spec}(\Lambda)}\int \bigl(\tau V(z)-\lambda-\frac{1}{4}(d-1)^2\bigr)_+^{\frac{1}{2}}\,dz=\\ \frac{1}{\pi}\sum_n P(n,d)\int \bigl(\tau V(z)-(n+\frac{1}{2}d-1)^2\bigr)_+^{\frac{1}{2}}\,dz\end{aligned} \tag{12.8.98}$$

where $P(n,d)$ is the multiplicity of the eigenvalue $n(n+d-2)$ of Λ and with the remainder estimate

$$\sum_{\lambda\in\mathsf{Spec}(\Lambda)\le\tau} 1\asymp \tau^{(d-1)/2}.$$

Obviously

$$\mathcal{N}(\tau)\asymp\begin{cases}\tau^{d/2} & \text{as } \ -d\nu>1,\\ \tau^{d/2}|\log\tau| & \text{as } \ -d\nu=1,\\ \tau^{1/2\nu} & \text{as } \ -d\nu<1,\end{cases} \tag{12.8.99}$$

where $\nu=\min(\nu_+,\nu_-)$.

Formula (12.8.99) holds as either $d\ge 3$ or $\lambda\in\mathsf{Spec}(\Lambda)$ is not 0.

So, as $d=2$ we need to investigate $\lambda=0$ separately. In this case we arrive to the operator $D_z^2-\tau V(z)$ and to avoid the infinite negative spectrum we must assume that $V(z)=o(|z|^{-2})$ as $|z|\to\infty$.

Recall that as $d=2$ the critical value of the previous subsection is exactly $\alpha=1$. Thus we now are interested in $V(z)=|z|^{-2}V_1(\log z)$ with $V_1(z_1)$ decaying at infinity as $|z_1|^{-2}$ or slower. We consider only $z\ge 0$; the second semi-axis is treated in the same way.

Then plugging into quadratic form $u=e^{-z_1/2}u_1$, $z=e^{z_1}$ we arrive to quadratic form

$$\begin{aligned}\|D_z u\|^2-t(z^{-2}V_1(\log z)u,u)=\\ \|D_{z_1}u_1\|^2+\frac{1}{4}\|u_1\|^2-t(V_1(z_1)u_1,u_1)\end{aligned} \tag{12.8.100}$$

and a $V_1(z_1) \to 0$ as $|z_1| \to \infty$ we guarantee that there is only finite number negative eigenvalues. Moreover, as $V(z_1) \asymp |z_1|^{-2\beta_1}$ with $\beta_1 \le 1$ the remainder estimate is $O(\log \tau)$ and the main part is given by the Weyl formula

$$\mathsf{N}_0^{\mathsf{W}}(\tau) = \frac{2}{\pi} \int \Bigl(\tau V_1(z_1) - \frac{1}{4}\Bigr)^{\frac{1}{2}}\, dz_1 \tag{12.8.101}$$

While remainder estimate is below one we received earlier, the main part is of magnitude $\tau^{(\beta_1+1)/(2\beta_1)}$ and it is above of what we got earlier.

So, we have proven the following theorem:

Theorem 12.8.21. *(i) Let $d \ge 3$, $W(x) = |x|^{-2}V(\log|x|)$ where $V(z)$ is a function of the type described above.*

Then asymptotics

$$\mathsf{N}(\tau) = \mathcal{N}(\tau) + O(\tau^{\frac{1}{2}(d-1)}) \qquad \text{as } \tau \to \infty \tag{12.8.102}$$

holds with the principal part (12.8.98) *of magnitude* (12.8.99).

(ii) Let $d = 2$, $W(x) = |x|^{-2}|\log|x||^{-2}V_1(\log|\log|x||, \pm)$ where $V(z)$ is a function of the type described above and $\pm$ is responsible for different behavior as $|x| \to \infty$ and $|x| \to 0$.

Then asymptotics

$$\mathsf{N}(\tau) = \mathcal{N}^{\#}(\tau) + O(\tau^{\frac{1}{2}(d-1)}) \qquad \text{as } \tau \to \infty \tag{12.8.103}$$

holds where superscript $^{\#}$ means that the term with $\lambda = 0$ is replaced by (12.8.101) *of magnitude $\tau^{(\beta+1)/(2\beta)}$.*

We leave to the reader the following series of interesting problems and probably not very difficult problems:

Problem 12.8.22. (i) Prove that the similar results hold as $W(x) = W(|x|)$ is replaced by $W(x) = W_r(|x|)w(\theta)$ with smooth disjoint from 0 function $w(\theta)$.

Investigate if we need condition $w(\theta) \ne 0$ or one can replace it by a condition $|w(\theta)| + |\nabla_\theta w(\theta)| \asymp 1$ or drop it completely (depending on dimension)?

(ii) Prove that the similar results hold as $W(x)$ fast enough stabilizes to $W^\infty(x) = W_r(|x|)w(\theta)$ described above; then there will be a mixture of Weyl terms associated with $(W - W^\infty)$ and non-Weyl (with respect to θ) terms depending only on W^∞.

(iii) Extend Part (ii) to the case when Δ is replaced by $\lambda^{\mathrm{w}}(x, D)$ with $\lambda(x,\xi) = g(x,\xi)$ where $g(x,\xi)$ is non-degenerate quadratic form with respect to ξ stabilizing to $g^\infty(\theta,\xi)$ as $|x| \to \infty$.

(iv) As $W(x)$ has no singularity at 0 extend Part (iii) to the case when $\lambda^{\mathrm{w}}(x, D)$ is a pseudodifferential operator with the symbol $\lambda(x,\xi) = g(x,\xi) + O(|\xi|^3)$ uniformly with respect to x with $g(x,\xi)$ described in Part (iii).

Problem 12.8.23. Using Problem 12.8.22 improve results of the previous Subsection 12.8.7 by using non-Weyl asymptotics of the type (12.8.102), (12.8.103) in the external zone.

12.8.9 Case of Positive $(A - E)$

Let us consider case when operator $(A - E)$ is positive definite albeit not necessarily disjoint from 0 but W is indefinite. So, E is either inside the spectral gap $\mathcal{G}_0$ or on its top.

Note that eigenvalues of $(A - tW)$ can cross E only in the negative direction. Let $\mathsf{N}_E([0,\tau])$ be a number of eigenvalues crossing 0 as t changes from 0 to τ. Then

$$\mathsf{N}_E([0,\tau]) = \mathsf{N}^-(A - E - \tau W). \tag{12.8.104}$$

Problem 12.8.24. (i) Prove that the corresponding results of the previous subsections remain true in these settings as well.

(ii) Generalize these results to the case when $W = W(x, D)$ is an operator of lower order than A.

12.8.10 Generalizations and Modifications

Let us consider the operators treated in M.S. Birman, A. Laptev and T. Suslina [1] and in M.S. Birman and T. Suslina [1–3] but without remainder estimates:

Problem 12.8.25. (i) Prove that the same asymptotics hold for operators not in $\mathbb{R}^d$ but in a domain X^∞ and with the boundary condition $Bu|_{\partial X} = 0$ satisfying the following periodicity conditions:

$$X^\infty + \Gamma = X^\infty \tag{12.8.105}$$

with the lattice Γ (see Figure 12.9(a)),

(12.8.106) There exist functions ϕ_j ($j = 1, \dots, d'$) such that

$$\phi_j(x + e_k) - \phi_j(x) = \delta_{jk} \qquad \forall j, k = 1, \dots, d' \tag{12.8.107}$$

and

$$B^\infty e^{i\langle\phi(x),\xi\rangle} u|_{\partial X^\infty} = B^\infty u|_{\partial X^\infty} \qquad \forall\xi \tag{12.8.108}$$

where $\langle\phi(x), \xi\rangle = \sum_j \phi_j(x)\xi_j$.

In this case $\mathcal{Q}$ should be replaced by $\mathcal{Q}\cap X$ with the previous definition of the elementary cell $\mathcal{Q}$ and $\mathcal{Q}'$ remains intact. In this case the auxiliary space $\mathbb{H}_{\xi,\{T\}}$ is subspace of $\mathscr{L}^2_{\mathsf{loc}}(\pi^{-1}X^\infty, \mathbb{C}^\mathsf{D})$ satisfying condition of quasiperiodicity with the quasimomentum ξ and transformation matrices $\{T\}$; $\mathbb{H}_{\xi,\{T\}}$ is identified with $\mathscr{L}^2(\pi^{-1}\mathcal{Q}\cap X^\infty, \mathbb{C}^\mathsf{D})$. Here $d' = d$ and $\pi = I$.

(ii) Generalize these results to the case when X and B coincide with X^∞ and B^∞ as $|x| \ge c$.

Problem 12.8.26. Prove that the similar asymptotics hold for operators in waveguides i.e. domains bounded in $x'' = (x_{d'+1}, \dots, x_d)$ and satisfying above conditions with $\Gamma \subset Z = \mathbb{R}^{d'} \ni x' = (x_1, \dots, x_{d'})$ (see Figure 12.9(b)) and matrices $T_1, \dots, T_{d'}$; here $d' < d$.

Now $\mathcal{Q} \subset Z$ is an elementary cell, $\pi : X \ni (x', x'') \to x' \in Z$ is a projection. Then one should expect $\mathsf{N}(\tau) \asymp \max(\tau^{d'/(m\mu)}, \tau^{d/m})$ and $R(\tau) \asymp \max(\tau^{(d'-1)/(m\mu)}, \tau^{(d-1)/m})$ as E is inside the spectral gap or on its bottom, whatever power is larger and the logarithmic factor appears as powers coincide; similar results hold as E is on the top of the spectral gap.

In this case $d' = 1$ should be also treated.

Note that in the previous two Problems 12.8.25 and 12.8.26 X is not necessarily a connected domain.

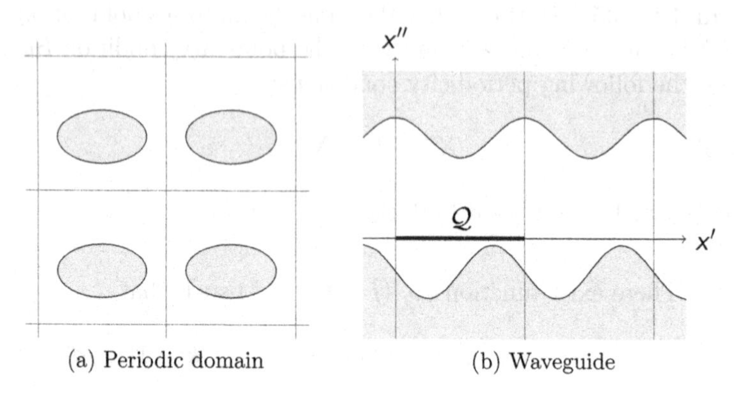

(a) Periodic domain (b) Waveguide

Figure 12.9: Periodic domain in full dimension ($d' = d$) and "waveguide" i.e. periodic domain in reduced dimension ($d' < d$)

Problem 12.8.27. Prove that the same asymptotics hold for operators with Weyl symbol $a(x, \xi - \alpha x)$ where α is a constant matrix such that

$$\alpha n \in \Gamma' \qquad \forall n \in \Gamma. \tag{12.8.109}$$

In this case $\mathbb{H}_{\xi,\{T\}}$ is the space of functions such that

$$u(x+n) = T^{-n}u(x)e^{i\langle n, \xi+\alpha x\rangle}u(x) \qquad \forall n \in \Gamma; \tag{12.8.110}$$

without condition (12.8.109) this is impossible equality.

Condition (12.8.109) needs to be satisfied only for skew-symmetric part of α as we can get rid of symmetric part by the gauge transformation. So, condition is

$$\frac{1}{2\pi}\int_{\mathcal{Q}_{jk}} d\omega \in \mathbb{Z} \qquad \forall j \ne k \tag{12.8.111}$$

where $\omega = \sum_j (\alpha x)_j dx_j$ and d denotes an exterior differential.

Therefore the problem is admissible as αx is replaced by non-linear vector-potential $V(x)$ both in operator and in condition (12.8.111): now $\omega = \sum_j V_j(x) dx_j$.

As an example one can consider Schrödinger operators with periodic magnetic field with "admissible" but not necessary vanishing flows over elementary cell.

Remark 12.8.28. As condition (12.8.111) is violated and some of the left-hand expressions are irrational operator seems to be beyond reach of the modern methods.

12.A Appendices

12.A.1 Variational Estimates as $m = 1$, $d = 1, 2$

In the $m = 1, d = 2$ case Rozenblioum estimate (9.A.11) is not often good enough as it contains $q > 1$ which often does not work (f.e. if potential is $|x|^{-2}|\log|x||^{-2\alpha}$). However there are other estimates albeit not of the same level of the generality and elegance, which work better.

First of all as $d = 1, 2$ replacing Δ by $\Delta + \epsilon|x|^{-2}$ with arbitrarily small constant ϵ saves the day restoring Rozenblioum estimate (9.A.11) with $q = 1$ (A. Laptev [2]). However consider unmodified Δ.

Let us start from $d = 1$ and half-line $\mathbb{R}^+ = \{x > 0\}$,

$$Q(u) = \int_0^\infty (|\nabla u|^2 - V(x)|u|^2)\, dx \tag{12.A.1}$$

with $V(x) \geq 0$.

In this case G. Rozenblioum-M.Z. Solomyak [2] paper provides estimate

$$\mathsf{N}^-(Q) \leq C \int_1^\infty \Bigl(t^{-1}\int_t^{2t} V(x)\, dx\Bigr)^{\frac{1}{2}}\, dt \tag{12.A.2}$$

which is the estimate following (5.3) in the mentioned paper modified in the obviously equivalent way. Then for potentials V such that

$$|x| \asymp |y| \implies V(x) \asymp V(y) \tag{12.A.3}$$

the right-hand expression of (12.A.2) is equivalent to the right-hand expression of

$$\mathsf{N}^-(Q) \leq C \int_{\mathbb{R}^+} V(x)^{\frac{1}{2}}\, dx \tag{12.A.4}$$

which is Rozenblioum estimate with formal $q = \frac{1}{2}$.

Consider now $d = 2$ and a potential satisfying (12.A.3). This is the same as considering radial potential $V(r)$ with $r = |x|$. Then we can go to polar coordinates (θ, r) and split u into two parts: one is radial $u_0(r)$ and another orthogonal $u_1(r,\theta) = \sum n \neq 0 u_n(r)e^{in\theta}$.

Then $\mathsf{N}^-(Q) = N_0^-(Q) + \mathsf{N}_2^-(Q)$ and due to A. Laptev [2] for $\mathsf{N}_1^-(Q)$ Rozenblioum estimate holds:

$$\mathsf{N}_1^-(Q) \le C\int_{\mathbb{R}^2} V(x)\,dx \asymp C\int_0^\infty V(r)r\,dr. \tag{12.A.5}$$

So, we need to consider only radial component. Then we arrive to quadratic form

$$\int_0^\infty r\bigl(|D_r|^2 - V(r)|u|^2\bigr)\,dr = \int_{-\infty}^\infty \bigl(|u|^2 - V(e^z)e^{2z}\bigr)\,dz \tag{12.A.6}$$

where we changed variables $r = e^z$ and skipped index "0". So we arrive to one-dimensional problem with "effective potential" $F(z) = e^{2z}V(e^z)$. This effective potential does not necessary satisfies condition (12.A.3) even if original potential V satisfied it. However assume that $F(z)$ satisfies this condition, or equivalently, that

(12.A.7) $V(r) = r^{-2}F(\log r)$ where $F(z)$ satisfies (12.A.3).

Then in virtue of (9.A.4)

$$\mathsf{N}_0^-(Q) \le C\int_{-\infty}^\infty e^{2z}V(e^z)\,dz = C\int_0^\infty rV(r)\,dr. \tag{12.A.8}$$

Thus we arrive to

Theorem 12.A.1. *(i) As $m = 1$, $d = 1$ and V satisfies* (12.A.3) *Rozenblioum' estimate* (9.A.11) *holds with $p = \infty$, $q = \frac{1}{2}$.*

(ii) As $m = 1$, $d = 2$ and V satisfies (12.A.7) *Rozenblioum' estimate* (9.A.11) *holds with $p = \infty$, $q = 1$.*

Comments

There are few papers devoted to the non-classical asymptotics in the domains with thick cusps (we are aware only about the existence of these papers); however, there are either no accurate remainder estimates or even no remainder estimate at all. The principal part of asymptotics in these papers, naturally, coincides with the one derived in Section 12.2.

The theory of operators with the strong degeneration of the potential at infinity also is not very well developed. I am aware of the paper M. Z. Solomyak [1] in which there is also no remainder estimate and again the principal part coincides with the one derived in Section 12.3.

There seem to be no papers[81] at all related to Section 12.4.

The list of papers devoted to the operators with the degenerations is much larger. We cannot list them all. The following papers are devoted to operators of the type we consider in Section 12.5: A. Menikoff and J. Sjöstrand [1–3], C. Iwasaki and N. Iwasaki [1], A. Mohamed [1, 2] etc. Remainder estimates were not obtained in them. A special case was treated in R. Melrose [5] and the remainder estimate was accurate in the case when it was Weyl and was not accurate otherwise.

There are many papers[81] dealing with the Scott correction term in mathematical physics. The reader can find a survey in V. Ivrii, V. and I.M. Sigal [1]. However, the method was completely different from the one we employ in Section 12.6 and only the spherically symmetric case was treated.

In V. Jakšić, S. Molčanov and B. Simon [1] the eigenvalue asymptotics for Neumann operators in domains with ultra-thin cusps was derived without remainder estimate. This paper inspired our research in Section 12.7. As cusp decay rate is exponential rather than superexponential the spectrum is continuous and the scattering theory was developed by many authors.

Finally, problems for periodic operators as in Section 12.8 were treated in papers of M. Sh. Birman [5, 8–10], M. Sh. Birman and -A. Laptev [2, 3], M.S. Birman, A. Laptev and T. Suslina [1], and M.S. Birman and T. Suslina [1–3]. These papers inspired our research.

Remark. We understand that these commentaries are completely inadequate and we are completely open and will be extremely grateful for suggestions.

[81] Prior to our analysis.

Bibliography

Agmon, S.

[1] *Lectures on Elliptic Boundary Value Problems*. Princeton, N.J.: Van Nostrand Mathematical Studies, 1965.

[2] *On kernels, eigenvalues, and eigenfunctions of operators related to elliptic problems*. Comm. Pure Appl. Math. 18, 627–663 (1965).

[3] *Asymptotic formulas with remainder estimates for eigenvalues of elliptic operators*. Arch. Rat. Mech. Anal., 28:165–183 (1968).

[4] *Spectral properties of Schrödinger operators and scattering theory*. Ann. Scuola Norm. Sup. Pisa Cl. Sci. (2) 4:151—218 (1975).

[5] *A perturbation theory of resonances*. Comm. Pure Appl. Math. 51(11-12):1255—1309 (1998).

Agmon, S.; Douglis, A.; Nirenberg, L.

[1] *Estimates near the boundary for solutions of elliptic partial differential equations satisfying general boundary conditions. I.*. Commun. Pure Appl. Math., 12(4):623–727 (1959).

[1] *Estimates near the boundary for solutions of elliptic partial differential equations satisfying general boundary conditions. II.*. Commun. Pure Appl. Math., 17(1):35–92 (1964).

Agmon, S.; Hörmander, L.

[1] *Asymptotic properties of solutions of differential equations with simple characteristics*. J. Analyse Math. 30: 1—38 (1976).

V. Ivrii, *Microlocal Analysis, Sharp Spectral Asymptotics and Applications II*, https://doi.org/10.1007/978-3-030-30541-3

Agmon, S.; Kannai Y.

[1] *On the asymptotic behavior of spectral functions and resolvent kernels of elliptic operators.* Israel J. Math. 5:1–30 (1967).

Agranovich, M. A.

[1] *Elliptic operators on closed manifolds.* Sovremennye problemy matematiki. Fundamental'nye napravleniya. VINITI (Moscow), 63:5–129 (1990). The English translation has been published by Springer-Verlag: Partial Differential Equations VI, EMS, vol. 63, (1994).

[2] *Elliptic boundary value problems.* Sovremennye problemy matematiki. Fundamental'nye napravleniya. VINITI (Moscow), 63 (1993). The English translation has been published by Springer-Verlag: Partial Differential Equations VIII, EMS, vol. 79 (1997).

Alekseev, A. B.; Birman, M. Sh.

[1] *A variational formulation of the problem of the oscillations of a resonator that is filled with a stratified anisotropic medium.* Vestn. Leningr. Univ., Math. 10:101–108 (1982).

Alekseev, A. B.; Birman, M. Sh.; Filonov, N

[1] *Spectrum asymptotics for one "nonsmooth" variational problem with solvable constraint.* St. Petersburg. Math. J. 18(5):681–697 (2007).

Ammari, K.; Dimassi, M.

[1] *Weyl formula with optimal remainder estimate of some elastic networks and applications.* Bull. Soc. Math. France 138(3):395–413 (2010).

Andreev, A. S.

[1] *On an estimate for the remainder in the spectral asymptotics of pseudodifferential operators of negative order.* Probl. Mat. Fiz., 11:31–46 (1986). In Russian.

Arendt W.; Nittka R., Peter W.; Steiner F.

[1] *Weyl's Law: Spectral Properties of the Laplacian in Mathematics and Physics*, Mathematical Analysis of Evolution, Information, and Complexity, by W. Arendt and W.P. Schleich, Wiley-VCH, pp. 1–71, 2009.

Arnold, V. I.

[1] *Geometrical Methods in the Theory of Ordinary Differential Equations*, Springer-Verlag, 1983.

[2] *Mathematical Methods of Classical Mechanics*. Springer-Verlag, 1990.

[3] *Singularities of caustics and wave fronts*. Kluwer, 1990.

Assal, M.; Dimassi, M.; Fujiié S. *Semiclassical trace formula and spectral shift function for systems via a stationary approach*, Int. Math. Res. Notices, rnx149 (2017).

Atiyah, M. F.; Bott, R. and Gårding, L.

[1] *Lacunas for hyperbolic differential operators with constant coefficients. I*. Acta Mathematica 124(1):109–189 (1970).

Avakumovič, V. G.

[1] *Über die eigenfunktionen auf geschlossen riemannschen mannigfaltigkeiten*. Math. Z., 65:324–344 (1956).

Avron, J.; Herbst, I.; Simon, B.

[1] *Schrödinger operators with magnetic fields, I: General interactions*. Duke Math. J., 45:847–883 (1978).

Babich, V. M.

[1] *Focusing problem and the asymptotics of the spectral function of the Laplace-Beltrami operator. II*. J. Soviet Math., 19(4):1288–1322 (1982).

Babich, V. M.; Buldyrev, V. S.

[1] *Short-wavelength Diffraction Theory*. Springer Series on Wave Phenomena, vol. 4. Springer-Verlag, New York,1991.

Babich, V. M.; Kirpicnikova, N. Y.

[1] *The Boundary-layer Method in Diffraction Theory*. Springer Electrophysics Series, vol. 3. Springer-Verlag, 1979.

Babich, V. M.; Levitan, B. M.

[1] *The focusing problem and the asymptotics of the spectral function of the Laplace-Beltrami operator*. Dokl. Akad. Nauk SSSR, 230(5):1017–1020 (1976).

[2] *The focusing problem and the asymptotics of the spectral function of the Laplace-Beltrami operator. I*. Zap. Nauchn. Sem. Leningr. Otd. Mat. Inst. 78:3–19 (1978).

Bach, V.

[1] *Error bound for the Hartree-Fock energy of atoms and molecules*. Commun. Math. Phys. 147:527–548 (1992).

Bañuelos R.; Kulczycki T.

[1] *Trace estimates for stable processes*. Probab. Theory Related Fields 142:313–338 (2008).

Bañuelos R.; Kulczycki T.; Siudeja B.

[1] *On the trace of symmetric stable processes on Lipschitz domains*. J. Funct. Anal. 257(10):3329–3352 (2009).

Barnett A.; Hassell A.

[1] *Boundary quasiorthogonality and sharp inclusion bounds for large Dirichlet eigenvalues*. SIAM J. Numer. Anal., 49(3):1046–1063 (2011).

Baumgartner, B.; Solovej, J. P.; Yngvason, J. *Atoms in Strong Magnetic Fields: The High Field Limit at Fixed Nuclear Charge*. Commun. Math. Phys.212:703–724 (2000).

Beals, R.

[1] *A general calculus of pseudo-differential operators* Duke Math. J., 42:1–42 (1975).

Beals, R.; Fefferman, C.

[1] *Spatially inhomogenous pseudo-differential operators*. Commun. Pure. Appl. Math., 27:1–24 (1974).

Ben-Artzi, M.; Umeda, T.

[1] *Spectral theory of first-order systems: from crystals to dirac operators*. (in preparation)

Benguria, R.

[1] *Dependence of the Thomas-Fermi energy on the nuclear coordinates*, Commun. Math. Phys., 81:419–428 (1981).

Benguria, R.; Lieb, E. H.

[1] *The positivity of the pressure in Thomas-Fermi theory*. Commun. Math. Phys., 63:193–218 (1978).

Bérard, P.

[1] *On the wave equation on a compact manifold without conjugate points*. Math. Zeit., 155(2):249–273 (1977).

[2] *Spectre et groupes cristallographiques. I: Domains Euclidiens*. Inv. Math., 58(2):179–199 (1980).

[3] *Spectre et groupes cristallographiques II: Domains sphériques*. Ann. Inst. Fourier, 30(3):237–248 (1980).

[4] *Spectral Geometry: Direct and Inverse Problems*. Lect. Notes Math. Springer-Verlag, 1207, 1985.

Berezin, F. A.; Shubin, M. A.

[1] *Symbols of operators and quantization*. In Hilbert Space Operators and Operator Algebras: Proc. Intern. Conf., Tihany, 1970, number 5, pages 21–52. Coloq. Math. Soc. Ja. Bolyai, North-Holland (1972).

[2] *The Schrödinger Equation*. Kluwer Ac. Publ., 1991. Also published in Russian by Moscow University, 1983.

Besse, A.

[1] *Manifolds all of whose Geodesics are Closed.* Springer-Verlag, 1978.

Birman, M. S.

[1] *The Maxwell operator for a resonator with inward edges*. Vestn. Leningr. Univ., Math., 19(3):1–8 (1986).

[2] *The Maxwell operator in domains with edges*. J. Sov. Math., 37:793–797 (1987).

[3] *The Maxwell operator for a periodic resonator with inward edges*. Proc. Steklov Inst. Math. 179: 21–34 (1989).

[4] *Discrete spectrum in the gaps of the continuous one in the large-coupling-constant limit*. Oper. Theory Adv. Appl., 46:17–25, Birkhauser, Basel, (1990).

[5] *Discrete spectrum in the gaps of a continuous one for perturbations with large coupling constant*. Adv. Soviet Math., 7:57–74 (1991).

[6] *Three problems in continuum theory in polyhedra*. Boundary value problems of mathematical physics and related questions in the theory of functions, 24. Zap. Nauchn. Sem. S.-Peterburg. Otdel. Mat. Inst. Steklov. (POMI) 200 (1992), 27–37.

[7] *Discrete spectrum of the periodic Schrödinger operator for non-negative perturbations*. Mathematical results in quantum mechanics (Blossin, 1993), 3-7. Oper. Theory Adv. Appl., 70, Birkhauser, Basel, (1994).

[8] *The discrete spectrum in gaps of the perturbed periodic Schrödinger operator. I. Regular perturbations.* Boundary value problems, Schrödinger operators, deformation quantization, Math. Top., Akademie Verlag, Berlin, (8):334–352 (1995).

[9] *The discrete spectrum of the periodic Schrödinger operator perturbed by a decreasing potential.* St. Petersburg Math. J., 8(1):1–14 (1997).

[10] Discrete spectrum in the gaps of the perturbed periodic Schrödinger operator. II. Non-regular perturbations. St. Petersburg Math. J., 9(6):1073–1095 (1998).

[11] *On the homogenization procedure for periodic operators in the neighbourhood of the edge of internal gap.* St. Petersburg Math. J. 15(4):507–513 (2004).

[12] *List of publications.* `http://www.pdmi.ras.ru/~birman/papers.html`.

Birman, M. Sh.; Borzov, V. V.

[1] *The asymptotic behavior of the discrete spectrum of certain singular differential operators.* Problems of mathematical physics, No. 5. Spectral theory., pp. 24–38. Izdat. Leningrad. Univ., Leningrad, 1971.

Birman, M. Sh.; Filonov, N

[1] *Weyl asymptotics of the spectrum of the Maxwell operator with non-smooth coefficients in Lipschitz domains.* Nonlinear equations and spectral theory, Amer. Math. Soc. Transl. Ser. 2, 220:27–44 (2007).

Birman, M. S.; Karadzhov, G. E.; Solomyak, M. Z.

[1] *Boundedness conditions and spectrum estimates for operators $b(X)a(D)$ and their analogs.* Adv. Soviet Math., 7:85–106 (1991).

Birman, M. Sh.; Koplienko, L. S.; Solomyak, M. Z. *Estimates of the spectrum of a difference of fractional powers of selfadjoint operators.* Izv. Vyssh. Uchebn. Zaved. Matematika, 3(154):3–10 (1975).

Birman, M. S.; Laptev A.

[1] *Discrete spectrum of the perturbed Dirac operator*. Ark. Mat. 32(1):13–32 (1994).

[2] *The negative discrete spectrum of a two-dimensional Schrödinger operator*.Commun. Pure Appl. Math., 49(9):967–997 (1996).

[3] *"Non-standard" spectral asymptotics for a two-dimensional Schrödinger operator*. Centre de Recherches Mathematiques, CRM Proceedings and Lecture Notes, 12:9–16 (1997).

Birman, M. Sh.; Laptev, A.; Solomyak, M.Z.

[1] *The negative discrete spectrum of the operator* $(-\Delta)^l - \alpha V$ *in* $L_2(R^d)$ *for* d *even and* $2l \geq d$. Ark. Mat. 35(1):87–126 (1997).

[2] *On the eigenvalue behaviour for a class of differential operators on semiaxis*. Math. Nachr. 195:17–46 (1998).

Birman M. S.; Laptev A.; Suslina T.

[1] *Discrete spectrum of the twodimensional periodic elliptic second order operator perturbed by a decreasing potential. I. Semiinfinite gap*. St. Petersbg. Math. J., 12(4):535–567 (2001).

Birman, M. Sh.; Pushnitski, A. B.

[1] *Discrete spectrum in the gaps of perturbed pseudorelativistic Hamiltonian*. Zap. Nauchn. Sem. POMI, 249:102–117 (1997).

[2] *Spectral shift function, amazing and multifaceted*. Integral Equations Operator Theory. 30(2):191–199 (1998).

Birman, M. S.; Raikov, G.

[1] Discrete spectrum in the gaps for perturbations of the magnetic Schrödinger operator. Adv. Soviet Math., 7:75–84 (1991).

Birman, M. S.; Sloushch, V. A. *Discrete spectrum of the periodic Schrodinger operator with a variable metric perturbed by a nonnegative potential.* Math. Model. Nat. Phenom. 5(4), 2010.

Birman, M. S.; Solomyak, M. Z.

[1] *The principal term of the spectral asymptotics for "non-smooth" elliptic problems.* Functional Analysis Appl. 4(4):265–275 (1970).

[2] *The asymptotics of the spectrum of "nonsmooth" elliptic equations.* Functional Analysis Appl. 5(1):56–57 (1971).

[3] *Spectral asymptotics of nonsmooth elliptic operators. I.* Trans. Moscow Math. Soc., 27:1–52 (1972).

[4] *Spectral asymptotics of nonsmooth elliptic operators. II.* Trans. Moscow Math. Soc., 28:1–32 (1973).

[5] *Asymptotic behaviour of spectrum of differential equations.* J. Soviet Math., 12:247–283 (1979).

[6] *Quantitative analysis in Sobolev imbedding theorems and application to the spectral theory.* AMS Trans. Ser. 2, 114 (1980).

[7] *A certain "model" nonelliptic spectral problem.* Vestn. Leningr. Univ., Math. 8:23–30 (1980).

[8] *Asymptotic behavior of the spectrum of pseudodifferential operators with anisotropically homogeneous symbols.* Vestn. Leningr. Univ., Math. 10:237–247 (1982).

[9] *Asymptotic behavior of the spectrum of pseudodifferential operators with anisotropically homogeneous symbols. II.* Vestn. Leningr. Univ., Math. 12:155–161 (1980).

[10] *Asymptotic behavior of the spectrum of variational problems on solutions of elliptic equations.* Sib. Math. J.20:1–15(1979), 1-15.

[11] *Asymptotic behavior of the spectrum of variational problems on solutions of elliptic equations in unbounded domains.* Funct. Anal. Appl. 14:267–274 (1981).

[11] *The asymptotic behavior of the spectrum of variational problems on solutions of elliptic systems*. J. Sov. Math. 28:633–644 (1985).

[12] *Asymptotic behavior of the spectrum of pseudodifferential variational problems with shifts*. Sel. Math. Sov. 5:245–256 (1986).

[13] *On subspaces that admit a pseudodifferential projector*. Vestn. Leningr. Univ., Math. 15:17–27 (1983).

[14] *Spectral Theory of Self-adjoint Operators in Hilbert Space*. D. Reidel (1987).

[15] *The Maxwell operator in domains with a nonsmooth boundary*. Sib. Math. J., 28:12–24 (1987).

[16] *Weyl asymptotics of the spectrum of the Maxwell operator for domains with a Lipschitz boundary*. Vestn. Leningr. Univ., Math., 20(3):15–21 (1987).

[17] *L_2-theory of the Maxwell operator in arbitrary domains*. Russ. Math. Surv., 42(6):75–96 (1987).

[18] *The self-adjoint Maxwell operator in arbitrary domains*. Leningr. Math. J., 1(1):99–115 (1990).

[19] *Interpolation estimates for the number of negative eigenvalues of a Schrodinger operator*. Schrodinger operators, standard and nonstandard (Dubna, 1988), 3–18, World Sci. Publishing, Teaneck, NJ, 1989.

[20] *Discrete negative spectrum under nonregular perturbations (polyharmonic operators, Schrodinger operators with a magnetic field, periodic operators)*. Rigorous results in quantum dynamics (Liblice, 1990), 25–36. World Sci. Publishing, River Edge, NJ, 1991.

[21] *The estimates for the number of negative bound states of the Schrödinger operator for large coupling constants*. In Proc. Conf. Inverse Problems, Varna, Sept. 1989, volume 235 of Pitman Res. Notes in Math. Sci., pages 49–57 (1991).

[22] *Negative Discrete Spectrum of the Schrödinger Operator with Large Coupling Constant: Qualitative Discussion*, volume 46 of Operator Theory: Advances and Applications. Birkhäuser (1990).

[23] *Schrodinger operator. Estimates for number of bound states as function-theoretical problem.* Spectral theory of operators (Novgorod, 1989), 1-54. Amer. Math. Soc. Transl. Ser. 2, 150, Amer. Math. Soc., Providence, RI, (1992).

[24] *Principal singularities of the electric component of an electromagnetic field in regions with screens.* St. Petersburg. Math. J. 5(1):125–139 (1994).

[25] *On the negative discrete spectrum of a periodic elliptic operator in a waveguide-type domain, perturbed by a decaying potential.* J. Anal. Math., 83:337–391 (2001).

Birman M. S., Suslina T.

[1] *The two-dimensional periodic magnetic Hamiltonian is absolutely continuous.* St. Petersburg Math. J. 9(1):21–32 (1998).

[2] *Absolute continuity of the two-dimensional periodic magnetic Hamiltonian with discontinuous vector-valued potential.* St. Petersburg Math. J. 10(4):1–26 (1999).

[3] *Two-dimensional periodic Pauli operator. The effective masses at the lower edge of the spectrum.* Mathematical results in quantum mechanics (Prague, 1998), Oper. Theory Adv. Appl., vol. 108, Birkhauser, Basel, pp. 13–31 (1999).

Birman, M. Sh.; Weidl, T.

[1] *The discrete spectrum in a gap of the continuous one for compact supported perturbations* Mathematical results in quantum mechanics (Blossin, 1993), 9–12. Oper. Theory Adv. Appl., 70, Birkhauser, Basel, (1994).

Blanshard, P.; Stubbe, J.; Reznde, J.

[1] *New estimates on the number of bound states of Schrödinger operator.* Lett. Math. Phys., 14:215–225 (1987).

Blumenthal, R. M.; Getoor, R. K.

[1] *Sample functions of stochastic processes with stationary independent increments*. J. Math. Mech., 10:493–516 (1961).

Boscain, U.; Prandi, D.; Seri, M.

[1] *ASpectral analysis and the Aharonov-Bohm effect on certain almost-Riemannian manifolds*. Comm. Part. Diff. Equats., 41(1):32–50 (2016).

Bolley, C.; Helffer, B.

[1] *An application of semi-classical analysis to the asymptotic study of the supercooling field of a superconducting material*. Annales de l'Ann. Inst. H. Poincaré, section Physique théorique, 58(2):189–233, (1993).

Bony, J. F..; Petkov, V.

[1] *Resolvent estimates and local energy decay for hyperbolic equations*. Annali dell'Universita' di Ferrara, 52(2):233-246 (2006).

Bony, J. M.; Lerner, N.

[1] *Quantification asymptotique et microlocalizations d'ordre supérier. I*. Ann. Sci. Ec. Norm. Sup., sér. IV, 22:377–433 (1989).

Borovikov V. A.

[1] *Diffraction by polygons and polyhedra*. Nauka (1966) (in Russian).

Borovikov V. A.; Kinber B. Ye.

[1] *Geometrical Theory of Diffraction* (IEEE Electromagnetic Waves Series). (1994).

Boutet de Monvel, L.

[1] *Boundary problems for pseudodifferential operators*. Acta Math., 126, 1–2:11–51 (1971).

[2] *Hypoelliptic operators with double characteristics and related pseudo-differential operators*. Comm. Pure Appl. Math., 27:585–639 (1974).

[3] *Selected Works*. Birkhäuser, 2017.

Boutet de Monvel L.; Grigis A.; Helffer, B.

[1] *Parametrixes d'opérateurs pseudo-différentiels à caractéristiques multiples*. Journées: Équations aux Dérivées Partielles de Rennes. Astérisque, 34-35:93–121 (1975).

Boutet de Monvel; L., Krée, P.

[1] *Pseudo-differential operators and Gevrey classes.* Annales de l'institut Fourier, 1:295–323 (1967).

Bouzouina, A.; Robert, D.

[1] *Uniform semiclassical estimates for the propagation of quantum observables*. Duke Mathematical Journal, 111:223–252 (2002).

Bovier A.; Eckhoff M.; Gayrard V.; Klein M.

[1] *Metastability in reversible diffusion processes. I. Sharp asymptotics for capacities and exit times*. J. Eur. Math. Soc., 6:399–424 (2004).

Brenner, A. V.; Shargorodsky, E. M.

[1] *Boundary value problems for elliptic pseudodifferential operators*. Partial Differential Equations IX, Encyclopaedia of Mathematical Sciences Volume 79:145–215 (1997).

Brezis, H.; Lieb E.

[1] *Long range potentials in Thomas-Fermi theory*. Commun. Math. Phys. 65:231–246 (1979).

Bronstein, M.; Ivrii V.

[1] Sharp Spectral Asymptotics for Operators with Irregular Coefficients. Pushing the Limits. Comm. Partial Differential Equations, 28, 1&2:99–123 (2003).

Brossard, J.; Carmona, R.

[1] *Can one hear the dimension of the fractal?* Commun. Math. Phys., 104:103–122 (1986).

Brummelhuis, R. G. M.

[1] *Sur les inégalités de Gårding pour les systèmes d'opérateurs pseudo-différentiels*. C. R. Acad. Sci. Paris, Sér. I, 315:149-152 (1992).

[2] *On Melins inequality for systems*. Comm. Partial Differential Equations, 26, 9&10:1559–1606 (2001).

pagebreak[2]

Brummelhuis, R. G. M.; Nourrigat, J.

[1] *A necessary and sufficient condition for Melin's inequality for a class of systems*. J. Anal. Math., 85:195–211 (2001).

Bruneau, V.; Petkov, V.

[1] *Representation of the scattering shift function and spectral asymptotics for trapping perturbations*. Commun. Partial Diff. Equations, 26:2081–2119 (2001).

[2] *Meromorphic continuation of the spectral shift function*. Duke Math. J., 116:389–430 (2003).

Brüning, J.

[1] *Zur Abschätzzung der Spectralfunktion elliptischer Operatoren*. Mat. Z., 137:75–87 (1974).

Bugliaro, L.; Fefferman, C.; Fröhlich; J., Graf, G. M.; Stubbe, J.

[1] *A Lieb-Thirring bound for a magnetic Pauli Hamiltonian*. Commun. Math. Phys. 187:567–582 (1997).

Burghelea D.; Friedlander L.; Kappeler T.

[1] *Meyer-Vietoris type formula for determinants of elliptic differential operators*, J. Funct. Anal. 107(1):34–65 (1992).

Buslaev, V. S.

[1] *On the asymptotic behaviour of spectral characteristics of exterior problems of the Schrödinger operator*. Math. USSR Izv., 39:139–223 (1975).

Calderón A. P.

[1] *On an inverse boundary value problem*. Seminar on Numerical Analysis and its Applications to Continuum Physics (Rio de Janeiro, 1980), pp. 65–73, Soc. Brasil. Mat., Rio de Janeiro, 1980.

Candelpergher, B.; Nosmas, J. C.

[1] *Propriétes spectrales d'opérateurs differentiels asymptotiques autoadjoints*. Commun. Part. Diff. Eq., 9:137–168 (1984).

Canzani, Y.; Hanin, B.

[1] *Scaling limit for the kernel of the spectral projector and remainder estimates in the pointwise Weyl law*. Analysis and PDE, 8(7):1707–1731 (2015).

[2] *C^∞ Scaling Asymptotics for the Spectral Projector of the Laplacian*. J of Geometric Analysis 1—12 (2017).

Carleman, T.

[1] *Propriétes asymptotiques des fonctions fondamentales des membranes vibrantes*. In C. R. 8-ème Congr. Math. Scand., Stockholm, 1934, pages 34–44, Lund (1935).

[2] *Über die asymptotische Verteilung der Eigenwerte partieller Differentialgleichungen*. Ber. Sachs. Acad. Wiss. Leipzig, 88:119–132 (1936).

Charbonnel, A. M.

[1] *Spectre conjoint d'opérateurs qui commutent*. Annales de Toulouse, 5(5):109–147 (1983).

[2] *Calcul fonctionnel à plusieurs variables pour des o.p.d. sur* $\mathbb{R}^n$. Israel J. Math., 45(1):69–89 (1983).

[3] *Localization et développement asymptotique des élements du spectre conjoint d'opérateurs pseudo-différentiels qui commutent*. Integral Eq. Op. Th., 9:502–536 (1986).

[4] *Comportement semi-classique du spectre conjoint d'opérateurs pseudodifférentiels qui commutent*. Asymp. Anal., 1:227–261 (1988).

Chazarain, J.

[1] *Formule de Poisson pour les variétés riemanniennes*. Inv. Math, 24:65–82 (1977).

[2] *Spectre d'un hamiltonien quantique et méchanique classique*. Commun. Part. Diff. Eq., 5(6):595–611 (1980).

[3] *Sur le Comportement Semi-classique du Spectre et de l'Amplitude de Diffusion d'un Hamiltonien Quantique*. Singularities in Boundary Value Problems, NATO. D. Reidel.

Cheeger, J.; Taylor, M. E.

[1] *On the diffraction of waves by conical singularities. I*. Comm. Pure Appl. Math. 35(3):275–331, 1982.

[2] *On the diffraction of waves by conical singularities. II*. Comm. Pure Appl. Math. 35(4):487–529, 1982.

Chen Hua

[1] *Irregular but nonfractal drums andn-dimensional weyl conjecture*. Acta Mathematica Sinica, 11(2):168—178 (1995).

Cheng, Q.-M.; Yang, H.

[1] *Estimates on eigenvalues of Laplacian*. Mathematische Annalen, 331(2):445—460 (2005).

Chervova O.; Downes R. J.; Vassiliev, D

[1] *The spectral function of a first order elliptic system*. Journal of Spectral Theory, 3(3):317–360 (2013).

[2] *Spectral theoretic characterization of the massless Dirac operator*. Journal London Mathematical Society-second series, 89 :301–320 (2014).

Colin de Verdiére, Y.

[1] *Spectre du laplacien et longeurs des géodésiques periodiques. I*. Comp. Math., 27(1):83–106 (1973).

[2] *Spectre du laplacien et longeurs des géodésiques periodiques. II*. Comp. Math., 27(2):159–184 (1973).

[3] *Spectre conjoint d'opérateurs qui commutent*. Duke Math. J., 46:169–182 (1979).

[4] *Spectre conjoint d'opérateurs qui commutent*. Mat. Z., 171:51–73 (1980).

[5] *Sur les spectres des opérateurs elliptiques à bicaractéristiques toutes périodiques*. Comment. Math. Helv., 54(3):508–522 (1979).

[6] *Sur les longuers des trajectories périodiques d'un billard*, pp. 122–139. in South Rhone seminar on geometry, III (Lyon, 1983) Travaux en Cours, Hermann, Paris, 1984, pp. 122-139.

[7] *Ergodicité et fonctions propres du laplacien*. Commun. Math. Phys., 102:497–502 (1985).

[8] *Comportement asymptotique du spectre de boitelles magnétiques*. In Travaux Conf. EDP Sant Jean de Monts, volume 2 (1985).

[9] *L'asymptotique du spectre des boitelles magnétiques*. Commun. Math. Phys., 105(2):327–335 (1986).

Colin De Verdière, Y.; Truc F.

[1] *Confining quantum particles with a purely magnetic field*. HAL-archives:00365828.

Colombini, F.; Petkov, V.; Rauch, J.

[1] *Spectral problems for non elliptic symmetric systems with dissipative boundary conditions*. J. Funct. Anal. 267:1637–1661 (2014).

Combescure, M.; Robert D.

[1] *Semiclassical spreading of quantum wavepackets an applications near unstable fixed points of the classical flow* Asymptotic Analysis, 14:377–404 (1997).

[2] *Quadratic quantum Hamiltonians revisited*. Cubo, A Mathematical Journal, 8, 1: 61–86 (2006).

Conlon, J.

[1] *A new proof of the Cwickel-Lieb-Rozenbljum bound*. Rocky Mountain J. Math., 15:117–122 (1985).

Cornean, H. D.; Fournais, S.; Frank, R. L.; Helffer, B.

[1] *Sharp trace asymptotics for a class of 2D-magnetic operators*. Annales de l'institut Fourier, 63(6):2457–2513 (2013).

Cornfeld, I. P.; Fomin, S. V.; Sinai, Y. G.

[1] *Ergodic Theory*, volume 245 of Grundlehren. Springer-Verlag (1982).

Courant, R.

[1] *Über die Eigenwerte bei den Differentialgleichungen der mathematischen Physik*. Mat. Z., 7:1–57 (1920).

[2] *Methods of Mathematical Physics, vol II, 2nd edition.* Wiley (1962) and reprints.

Courant, R.; Hilbert D.

[1] *Methods of Mathematical Physics, vol I,II*. Interscience (1953) and reprints.

Cwikel, M.

[1] *Weak type estimates for singular values and the number of bound states of Schrödinger operator*. Ann. Math, 106:93–100 (1977).

Cycon, H. L.; Froese, R. G.; Kirsh, W.; Simon, B.

[1] *Schrödinger Operators with Applications to Quantum Mechanics and Global Geometry*. Texts and Monographs in Physics. Springer-Verlag.

Danford, N.; Schwartz, J. T.

[1] *Linear Operators, Parts I–III*. Willey Classics Library.

Datchev K.; Gell-Redman J.; Hassell A.; Humphries P.

[1] *Approximation and equidistribution of phase shifts: spherical symmetry*. Comm. Math. Phys., 326(1):209–236 (2014).

Daubechies, I.

[1] *An uncertainty principle for fermions with generalized kinetic energy*. Commun. Math. Phys. 90(4):511–520 (1983).

Dauge, M.; Hellfer, B.

[1] *Eigenvalues variation. I, Neumann problem for Sturm-Liouville operators*. J. Differential Equations, 104(2):243–262 (1993).

Dauge, M.; Robert, D.

[1] *Formule de Weyl pour une classe d'opérateurs pseudodifférentiels d'ordre négatif sur* $L^2(\mathbb{R}^n)$. Lect. Notes Math., 1256:91–122 (1987).

Davies, E.B. *Heat Kernels and Spectral Theory*. Cambridge Univ. Press, Cambridge, (1989).

Davies, E.B.; Simon B. *Spectral properties of Neumann Laplacian of horns*. Geometric & Functional Analysis GAFA, 2(1):105–117 (1992).

De Bievre, S. ; Pulé, J. V.

[1] *Propagating edge states for a magnetic Hamiltonian*. Mathematical physics electronic journal, 5:1–17 (1999).

Dencker, N.

[1] *On the propagation of polarization sets for systems of real principal type*. J. Funct. Anal., 46(3):351–372 (1982).

[2] *The Weyl calculus with locally temperate metrics and weights*. Ark. Mat., 24(1):59–79 (1986).

[3] *On the propagation of polarization in conical refraction*. Duke Math. J., 57(1):85–134 (1988).

[4] *On the propagation of polarization in double refraction*. J. Func. Anal., 104:414–468 (1992).

Dimassi, M.

[1] *Spectre discret des opérateurs périodiques perturbés par un opérateur différentiel*. C.R..A.S., 326, 1181–1184, (1998).

[2] *Trace asymptotics formula and some applications.* Asymptotic Analysis, 18:1–32 (1998).

[3] *Semi-classical asymptotics for the Schrödinger operator with oscillating decaying potential.* Canad. Math. Bull. 59(4):734–747 (2016).

Dimassi, M.; Petkov, V.

[1] *Spectral shift function and resonances for non semi-bounded and Stark Hamiltonians.* J. Math. Math. Pures Appl. 82:1303–1342 (2003).

[1] *Spectral problems for operators with crossed magnetic and electric fields.* J. Phys. A: Math. Theor, 43(47): 474015 (2010).

[2] *Resonances for magnetic Stark Hamiltonians in two dimensional case.* Internat. Math. Res. Notices, 77:4147–4179 (2004).

Dimassi, M.; Sjöstrand, J.

[1] *Trace asymptotics via almost analytic extensions.* 21, Partial Progr. Nonlinear Differential Equation Appl., 126–142. Birkhäuser Boston, (1996).

[1] *Spectral Asymptotics in the semiclassical limit*, London Mathematical Society Lecture notes series 265, (1999), 227pp.

pagebreak[2]

Dimassi, M.; Tuan Duong, A.

[1] *Scattering and semi-classical asymptotics for periodic Schrödinger operators with oscillating decaying potential.* Math. J. Okayama Univ, 59:149–174 (2017).

[2] *Trace asymptotics formula for the Schrödinger operators with constant magnetic fields.* J. Math. Anal. Appl., 416(1):427–448 (2014).

Dolbeault, J.; Laptev, A.; Loss, M.

[1] *Lieb-Thirring inequalities with improved constants.* J. Eur. Math. Soc. 10(4):1121–1126 (2008).

Dolgopyat, D; Jakobson, D.

[1] *On small gaps in the length spectrum.* Journal of Modern Dynamics 10:339–352 (2016).

Dragoviĉ V.; Radnović, M.

[1] *Geometry of integrable billiards and pencils of quadrics.* J. Math. Pures Appl., 85:758–790 (2006).

Dubrovin, B. A.; Fomenko, A. T.; Novikov, S. P.

[1] *Modern Geometry—Methods and Applications 1*, volume 92 of Graduate Texts in Mathematics. Springer-Verlag (1984).

[2] *Modern Geometry—Methods and Applications 2*, volume 104 of Graduate Texts in Mathematics. Springer-Verlag (1985).

Duistermaat, J. J.

[1] *Fourier integral operators.* Birkhäuser (1996).

Duistermaat, J. J.; Guillemin, V.

[1] *The spectrum of positive elliptic operators and periodic bicharacteristics.* Invent. Math., 29(1):37–79 (1975).

Dunford N.; Schwartz J. T.

[1] *Linear Operators*, Parts 1–3.. Interscience, (1963).

Dyatlov S.; Zworski, M.

[1] *Mathematical theory of scattering resonances.* `http://math.mit.edu/~dyatlov/res/res_20170323.pdf`

Efremov, D. V.; Shubin, M. A.

[1] *Spectral asymptotics for elliptic operators of Schrödinger type on hyperbolic space.* Trudy sem. Petrovskogo, 15:3–32 (1991).

Egorov, Y. V.

[1] *Microlocal Analysis*. Partial Differential Equations IV, Encyclopaedia of Mathematical Sciences, 33:1–147 (1993).

Egorov, Y. V.; Kondrat'ev, V. A.

[1] *On the estimate of the negative spectrum of the Schrödinger operator*. Math. USSR Sbornik, 62(4):551–566 (1989).

[2] *On the negative spectrum of elliptic operator*. Math. USSR Sbornik, 69(4):155–177 (1991).

[3] *Estimates of the negative spectrum of an elliptic operator*. AMS Transl. Ser. 2, 150 (1992).

[4] *On Spectral Theory of Elliptic Operators*. Birkhäuser (1996), 317pp.

Egorov, Y. V.; Shubin, M. A.

[1] *Linear Partial Differential Equations. Elements of the Modern Theory*. Partial Differential Equations II, Encyclopaedia of Mathematical Sciences, 31:1–120 (1994).

Ekholm, T.; Frank, R.; Kovarik, H.

[1] *Eigenvalue estimates for Schrödinger operators on metric trees*. arXive: 0710.5500.

Erdös, L.

[1] *Magnetic Lieb-Thirring inequalities*. Commun. Math. Phys., 170:629–668 (1995).

Erdös, L.; Fournais, S.; Solovej, J. P.

[1] *Second order semiclassics with self-generated magnetic fields*. Ann. Henri Poincare, 13:671–730 (2012).

[2] *Relativistic Scott correction in self-generated magnetic fields.* Journal of Mathematical Physics 53, 095202 (2012), 27pp.

[3] *Scott correction for large atoms and molecules in a self-generated magnetic field.* Commun. Math. Phys., 25:847–882 (2012).

[4] *Stability and semiclassics in self-generated fields.* J. Eur. Math. Soc. (JEMS), 15(6):2093–2113 (2013).

Erdös, L.; Solovej, J. P.

[1] *Ground state energy of large atoms in a self-generated magnetic field.* Commun. Math. Phys., 294(1):229–249 (2010).

Eskin, G.

[1] *Parametrix and propagation of singularities for the interior mixed hyperbolic problem.* J. Anal. Math., 32:17–62 (1977).

[2] *General Initial-Boundary Problems for Second Order Hyperbolic Equations*, pages 19–54. D. Reidel Publ. Co., Dordrecht, Boston, London (1981).

[3] *Initial boundary value problem for second order hyperbolic equations with general boundary conditions, I.* J. Anal. Math., 40:43–89 (1981).

Fabio, N.

[1] *A lower bound for systems with double characteristics.* J. Anal. Math. 96:297–311 (2005).

Fang, Y.-L.; Vassiliev, D.

[1] *Analysis as a source of geometry: a non-geometric representation of the Dirac equation.* Journal of Physics A: Mathematical and Theoretical, 48, article number 165203, 19 pp. (2015).

[2] *Analysis of first order systems of partial differential equations.* Complex Analysis and Dynamical Systems VI: Part 1: PDE, Differential Geometry, Radon Transform. AMS Contemporary Mathematics Series, 653:163–176 (2015).

Fedoryuk, M. V.; Maslov, V. P.

[1] *Semi-classical Approximation in Quantum Mechanics*. D. Reidel (1981).

Fefferman, C. L.

[1] *The uncertainty principle*. Bull. Amer. Math. Soc., 9:129–206 (1983).

Fefferman, C. L.; Ivrii, V.; Seco, L. A.; Sigal, I. M.

[1] *The energy asymptotics of large Coulomb systems*. In Proc. Conf. "N-body Quantum Mechanics", Aarhus, Denmark 79–99 (1991).

Fefferman, C. L.; Seco, L.

[1] *On the Dirac and Schwinger corrections to the ground state energy of an atom* Adv. Math., 107(1): 1–185 (1994).

Feigin, V. I.

[1] *The asymptotic distribution of the eigenvalues of pseudodifferential operators in $\mathbb{R}^n$]*. Math. USSR Sbornik, 28:533–552 (1976).

[2] *The asymptotic distribution of eigenvalues and a formula of Bohr-Sommerfeld type*. Math. USSR Sbornik, 38(1):61–81 (1981).

Filonov, N.; Pushnitski, A.

[1] *Spectral asymptotics of Pauli operators and orthogonal polynomials in complex domains*. Comm. Math. Phys., 264:759–772 (2006).

Filonov; N., Safarov, Yu.

[1] *Asymptotic estimates of the difference between the Dirichlet and Neumann counting functions*. J. Funct. Anal. 260:2902–2932 (2011).

Fleckinger, J. and Lapidus, M.

[1] *Tambour fractal: vers une résolution de la conjecture de Weyl-Berry pour les valeurs propres du laplacien*. C. R. A. S. Paris, Sér. 1, 306:171–175 (1988).

Fleckinger, J.; Métivier, G.

[1] *Théorie spectrale des opérateurs uniforment elliptiques des quelques ouverts irréguliers*. C. R. A. S. Paris, Sér. A–B, 276:A913–A916 (1973).

Fleckinger, J.; Vassiliev, D. G.

[1] *Tambour fractal: Exemple d'une formule asymptotique à deux termes pour la "fonction de compatage"*. C. R. Acad. Sci. Paris, Ser. 1, 311:867–872 (1990).

Folland, G. B.

[1] *Real Analysis: Modern Techniques and Their Applications*. A Wiley-Interscience, 408pp (1999).

Frank, R.; Geisinger L.

[1] *Semi-classical analysis of the Laplace operator with Robin boundary conditions*. Bull. Math. Sci. 2 (2012), no. 2, 281–319.

[2] *Refined semiclassical asymptotics for fractional powers of the Laplace operator*. J. reine angew. Math. 712:1—37 (2016).

Frank, R.; Lieb, E.; Seiringer, R.

[1] *Hardy-Lieb-Thirring inequalities for fractional Schrödinger Operators*. J. Amer. Math. Soc. 21(4), 925—950 (2008).

Frank, R.; Siedentop, H.; Warzel, S.

[1] *The ground state energy of heavy atoms: relativistic lowering of the leading energy correction*. Commun. Math. Phys., 278(2):549–566 (2008).

Frank, R.; Siedentop, H.

[1] *The energy of heavy atoms according to Brown and Ravenhall: the Scott correction*. Documenta Mathematica, 14:463–516 (2009)

Friedlander, F. G.

[1] *Notes on closed billiard ball trajectories in polygonal domains. I*. Commun. Part. Diff. Eq., 12 (1990).

[2] *Notes on closed billiard ball trajectories in polygonal domains. II*. Commun. Part. Diff. Eq., 16(6):1687–1694 (1991).

Friedlander, L.

[1] *Some inequalities between Dirichlet and Neumann eigenvalues*. Arch. Rational Mech. Anal., 116(2):153–160 (1991).

Friedrichs, K.

[1] *Perturbations of Spectra in Hilbert Space*. AMS, Providence, RI (1965).

Fröhlich; J., Graf, G. M.; Walcher, J.

[1] *On the extended nature of edge states of Quantum Hall Hamiltonians*. Ann. H. Poincaré 1:405-442 (2000).

Fröhlich, J.; Lieb, E. H.; Loss, M.

[1] *Stability of Coulomb systems with magnetic fields. I. The one-electron atom*. Commun. Math. Phys., 104:251–270 (1986).

Fushiki, I.; Gudmundsson, E.; Pethick, C. J.; Yngvason, J.

[1] *Matter in a Magnetic Field in the Thomas-Fermi and Related Theories*. Annals of Physics, 216:29–72 (1992).

Geisinger, L.; Weidl T.

[1] *Sharp spectral estimates in domains of infinite volume.* Reviews in Math. Physics, 23(6):615–641 (2011).

Geisinger, L.; Laptev, A.; Weidl T.

[1] *Geometrical versions of improved Berezin–Li–Yau inequalities.* J. Spectral Theory, 1(1):87–109 (2011).

Gell-Redman J.; Hassell A.; Zelditch S.

[1] *Equidistribution of phase shifts in semiclassical potential scattering.* J. Lond. Math. Soc., (2) 91(1):159–179 (2015).

Gérard, P.; Lebeau, G.

[1] *Diffusion d'une onde par un coin.* J. Amer. Math. Soc. 6(2):341–424 (1993).

Gottwald, S.

[1] *Two-term spectral asymptotics for the Dirichlet pseudo-relativistic kinetic energy operator on a bounded domain.* arXiv:1706.08808

Graf, G. M.; Solovej, J. P.

[1] *A correlation estimate with applications to quantum systems with Coulomb interactions.* Rev. Math. Phys., 6(5a):977–997 (1994). Reprinted in The state of matter a volume dedicated to E. H. Lieb, Advanced series in mathematical physics, 20, M. Aizenman and H. Araki (Eds.), 142–166, World Scientific (1994).

Gravel, C.

[1] *Spectral geometry over the disk: Weyl's law and nodal sets.* arxiv.org:1208.5275

Grebenkov, D. S.; Nguyen, B.T.

[1] *Geometrical structure of Laplacian eigenfunctions*. SIAM Review, 55(4):601—667 (2013).

Grigoryan, A. *Heat kernels and function theory on metric measure spaces*. Heat Kernels and Analysis on Manifolds, Graphs, and Metric Spaces. (Paris, 2002), Am. Math. Soc., Providence, RI, 143–172 (2003).

Grubb, G.

[1] *Spectral sypmptotics for Douglis-Nirenberg elliptic systems and pseudodifferential boundary value problems*. Commun. Part. Diff. Eq., 2(9):1071–1150 (1977).

[2] *Remainder estimates for eigenvalues and kernels of pseudo-differential elliptic systems*. Math. Scand., 43:275–307 (1978).

[3] *Functional calculus of pseudo-differential boundary problems*. Birkhäuser (1986, 2012).

[4] *Local and nonlocal boundary conditions for μ-transmission and fractional elliptic pseudodifferential operators*. Analysis and Part. Diff. Equats., 7(71):649–1682 (2014).

[5] *Fractional Laplacians on domains, a development of Hörmander's theory of μ-transmission pseudodifferential operators*. Adv. Math. 268:478–528 (2015).

[6] *Spectral results for mixed problems and fractional elliptic operators*. J. Math. Anal. Appl., 421(2):1616–1634 (2015).

[7] *Regularity of spectral fractional Dirichlet and Neumann problems*. Math. Nachr., 289(7):831–844 (2016).

Grubb, G.; Hörmander, L.

[1] *The transmission property*. Math. Scand., 67:273-289 (1990).

Guillemin, V.

[1] *Lectures on spectral properties of elliptic operators.* Duke Math. J., 44(3):485–517 (1977).

[2] *Some spectral results for the Laplace operator with potential on the n-sphere.* Adv. Math., 27:273–286 (1978).

[3] *Some classical theorems in spectral theory revised.* Seminar on Singularities of solutions of partial differential equations, Princeton University Press, NJ, 219–259 (1979).

Guillemin, V.; Melrose, R. B.

[1] *The Poisson summation formula for manifolds with boundary.* Adv. Math., 32:204–232 (1979).

Guillemin, V.; Sternberg, S.

[1] *On the spectra of commuting pseudo-differential operators.* In Proc. Park City Conf., 1977, number 48 in Lect. Notes Pure Appl. Math., pages 149–165 (1979).

[2] *Geometric Asymptotics.* AMS Survey Publ., Providence, RI (1977).

[3] *Semi-classical Analysis.* International Press of Boston (2013).

Guo, J.; Wang, w.

[1] *An improved remainder estimate in the Weyl formula for the planar disk.* arXiv:math/0612039

Gureev, T.; Safarov, Y.

[1] *Accurate asymptotics of the spectrum of the Laplace operator on manifold with periodic geodesics.* Trudy Leningradskogo Otdeleniya Mat. Inst. AN SSSR, 179:36–53 (1988). English translation in Proceedings of the Steklov Institute of Math., 2, (1988).

Gutkin, E.

[1] *Billiard dynamics: a survey with the emphasis on open problems*. Regular and Chaos Dynamics, 8 (1): 1–14 (2003).

[2] *A few remarks on periodic orbits for planar billiard tables*. arXiv:math/0612039

Hainzl, C.

[1] *Gradient corrections for semiclassical theories of atoms in strong magnetic fields*. J. Math. Phys. 42:5596–5625 (2001).

Handrek, M.; Siedentop, H.

[1] *The ground state energy of heavy atoms: the leading correction*, Commun. Math. Phys., 339(2):589–617 (2015).

Harman, G.

[1] *Metric Number Theory* Clarendon Press, Oxford, xviii+297 (1998).

Hassel A. and Ivrii V.

[1] A. Hassel; V. Ivrii *Spectral asymptotics for the semiclassical Dirichlet to Neumann operator*. J. of Spectral Theory 7(3):881–905 (2017).

Havin, V.; Joricke B.

[1] *Uncertainty Principle in Harmonic Analysis*. Ergeb. Math. Grenzgeb. (3) 28, Springer-Verlag, Berlin (1994).

Helffer, B.

[1] *Théorie spectrale pour des opérateurs globalement elliptiques*. Astérisque, 112 (1984).

[2] *Introduction to the Semi-classical Analysis for the Schrödinger Operator and Applications*. Number 1336. Lect. Notes Math. (1984).

Helffer, B.; Klein M.; Nier, F.

[1] *Quantitative analysis of metastability in reversible diffusion process via a Witten complex approach.* Proceedings of the Symposium on Scattering and Spectral Theory. Matematica Contemporanea (Brazilian Mathematical Society), 26:41-86 (2004).

Helffer, B.; Knauf, A.; Siedentop, H.; Weikart, R.

[1] *On the absence of a first order correction for the number of bound states of a Schrödinger operator with Coulomb singularity.* Commun. Part. Diff. Eq., 17(3-4):615–639 (1992).

Helffer, B., Kordyukov, Y.

[1] *Accurate semiclassical spectral asymptotics for a two-dimensional magnetic Schrödinger operator.* Ann. Henri Poincaré, 16(7):1651—1688 (2015).

Helffer, B., Kordyukov, Y; Raymond, N.; Vu-Ngoc, S.

[1] *Magnetic wells in dimension three.* Analysis & PDE, Mathematical Sciences Publishers, 9(7):1575–1608 (2016).

Helffer, B.; Martinez, A.

[1] *Phase transition in the semiclassical regime.* Rev. Math. Phys., 12(11):1429–1450 (2000).

Helffer, B.; Martinez, A.; Robert, D.

[1] *Ergodicité et limite semi-classique.* Commun. Math. Phys., 109:313–326 (1987).

Helffer, B.; Mohamed, A. (Morame, A.)

[1] *Sur le spectre essentiel des opérateurs de Schrödinger avec champ magnétique.* Ann. Inst. Fourier, 38(2) (1988).

Helffer, B.; Nourrigat, J.; Wang, X. P.

[1] *Spectre essentiel pour l'équation de Dirac*. Ann. Sci. Ec. Norm. Sup., Sér. IV, 22:515–533 (1989).

Helffer, B.; Parisse, B.

[1] *Moyens de Riesz d'états bornés et limite semi-classique en liasion avec la conjecture de Lieb-Thirring. III*. Ec. Norm. Sup. Paris, Prep. LMENS-90–12 (1990).

Helffer, B.; Robert, D.

[1] *Comportement semi-classique du spectre des hamiltoniens quantiques elliptiques*. Ann. Inst. Fourier, 31(3):169–223 (1981).

[2] *Comportement semi-classique du spectre des hamiltoniens quantiques hypoelliptiques*. Ann. Ec. Norm. Sup. Pisa, Sér. IV, 9(3) (1982).

[3] *Propriétes asymptotiques du spectre d'opérateurs pseudo-differentiels sur* $\mathbb{R}^n$. Commun. Part. Diff. Eq., 7:795–882 (1982).

[4] *Etude du spectre pour un opérateur globalement elliptique dont le symbole présente des symétries. I: Action des groupes finis*. Amer. J. Math., 106(5):1199–1236 (1984).

[5] *Etude du spectre pour un opérateur globalement elliptique dont le symbole présente des symétries. II: Action des groupes compacts*. Amer. J. Math, 108:978–1000 (1986).

[6] *Asymptotique des niveaux d'énergie pour des hamiltoniens à un degré de liberté*. Duke Math. J., 49(4):853–868 (1982).

[7] *Puits de potentiel généralisés et asymptotique semi-classique*. Ann. de l'Inst. Henri Poincaré, sect. Phys. Theor., 41(3):291–331 (1984).

[8] *Riesz means of bound states and semi-classical limit connected with a Lieb-Thirring conjecture. I*. Asymp. Anal., 3(4):91–103 (1990).

[9] *Riesz means of bound states and semi-classical limit connected with a Lieb-Thirring conjecture. II*. Ann. de l'Inst. Henri Poincaré, sect. Phys. Theor., 53(2):139–147 (1990).

Helffer, B.; Sjöstrand, J.

[1] *Multiple wells in the semiclassic limit. I.* Commun. Part. Diff. Eq., 9(4):337–408 (1984).

[2] *On diamagnetism and Haas-van Alphen effect.* Ann. Inst. Henri Poincaré, 52(4):303–375 (1990).

Hempel R.; Seco L. A.; Simon B.

[1] *The essential spectrum of Neumann Laplacians on some bounded singular domains.* J. Func. Anal., 102(2):448–483 (1991).

Herbst, I. W

[1] *Spectral Theory of the Operator* $(p^+m^2)^{1/2} - Ze/r$, Commun. Math. Phys. 53(3):285–294 (1977).

Hörmander, L.

[1] *The Analysis of Linear Partial Differential Operators. I–IV*. Springer-Verlag (1983, 1985).

[2] Pseudo-differential operators. *Comm. Pure Appl. Math.*, 18, 501–517 (1965).

[3] *The spectral function of an elliptic operator*. Acta Math., 121:193–218 (1968).

[4] *On the Riesz means of spectral functions and eigenfunction expansions for elliptic differential operators*. In Yeshiva Univ. Conf., November 1966, volume 2 of Ann. Sci. Conf. Proc., pages 155–202. Belfer Graduate School of Sci. (1969).

[5] *The existence of wave operators in scattering theory*. Math. Z. 146(1):69–91 (1976).

[6] *The Cauchy problem for differential equations with double characteristics*. J. An. Math., 32:118–196 (1977).

[7] *On the asymptotic distribution of eigenvalues of p.d.o. in* $\mathbb{R}^n$. Ark. Math., 17(3):169–223 (1981).

Houakni, Z. E.; Helffer, B.

[1] *Comportement semi-classique en présence de symétries. action d'une groupe de Lie compact*. Asymp. Anal., 5(2):91–114 (1991).

Hughes, W.

[1] *An atomic energy lower bound that agrees with Scott's correction*. Adv. in Math., 79(2):213–270 (1990).

Hundertmark, D.; Laptev, A.; Weidl, T.

[1] *New bounds on the Lieb-Thirring constants*. Invent. Math. 140(3):693–704 (2000).

Ivrii, V.

[1] *Wave fronts of solutions of symmetric pseudo-differential systems*. Siberian Math. J., 20(3):390–405 (1979).

[2] *Wave fronts of solutions of boundary-value problems for symmetric hyperbolic systems. I: Basic theorems*. Siberian Math. J., 20(4):516–524 (1979).

[3] *Wave fronts of solutions of boundary value problems for symmetric hyperbolic systems. II. systems with characteristics of constant multiplicity*. Siberian Math. J., 20(5):722–734 (1979).

[4] *Wave Front Sets of Solutions of Certain Pseudodifferential Operators*. Trudy Moskov. Mat. Obshch. 39:49–86 (1979).

[5] *Wave Front Sets of Solutions of Certain Hyperbolic Pseudodifferential Operators*. Trudy Moskov. Mat. Obshch. 39:87–119 (1979).

[6] *Wave fronts of solutions of boundary value problems for symmetric hyperbolic systems. III. systems with characteristics of variable multiplicity*. Sibirsk. Mat. Zhurn., 21(1):54–60 (1980).

[7] *Wave fronts of solutions of boundary value problems for a class of symmetric hyperbolic systems*. Siberian Math. J., 21(4):527–534 (1980).

[8] *Second term of the spectral asymptotic expansion for the Laplace-Beltrami operator on manifold with boundary*. Funct. Anal. Appl., 14(2):98–106 (1980).

[9] *Propagation of singularities of solutions of nonclassical boundary value problems for second-order hyperbolic equations*. Trudy Moskov. Mat. Obshch. 43 (1981), 87–99.

[10] *Accurate spectral asymptotics for elliptic operators that act in vector bundles*. Funct. Anal. Appl., 16(2):101–108 (1982).

[11] *Precise Spectral Asymptotics for Elliptic Operators*. Lect. Notes Math. Springer-Verlag 1100 (1984).

[12] *Global and partially global operators. Propagation of singularities and spectral asymptotics*. Microlocal analysis (Boulder, Colo., 1983), 119–125, Contemp. Math., 27, Amer. Math. Soc., Providence, RI, (1984).

[13] *Weyl's asymptotic formula for the Laplace-Beltrami operator in Riemann polyhedra and in domains with conical singularities of the boundary*. Soviet Math. Dokl., 38(1):35–38 (1986).

[14] *Estimations pour le nombre de valeurs propres negatives de l'operateurs de Schrödinger avec potentiels singuliesrs*. C.R.A.S. Paris, Sér. 1, 302(13, 14, 15):467–470, 491–494, 535–538 (1986).

[15] *Estimates for the number of negative eigenvalues of the Schrödinger operator with singular potentials*. In Proc. Intern. Cong. Math., Berkeley, pages 1084–1093 (1986).

[16] *Precise spectral asymptotics for elliptic operators on manifolds with boundary*. Siberian Math. J., 28(1):80–86 (1987).

[17] *Estimates for the spectrum of Dirac operator*. Dokl. AN SSSR, 297(6):1298–1302 (1987).

[18] *Linear hyperbolic equations*. Sovremennye Problemy Matematiki. Fundamental'nye Napravleniya. VINITI (Moscow), 33:157–247 (1988).

[19] *Semiclassical spectral asymptotics*. (Proceedings of the Conference, Nantes, France, June 1991)

[20] *Asymptotics of the ground state energy of heavy molecules in the strong magnetic field. I*. Russian Journal of Mathematical Physics, 4(1):29–74 (1996).

[21] *Asymptotics of the ground state energy of heavy molecules in the strong magnetic field. II*. Russian Journal of Mathematical Physics, 5(3):321–354 (1997).

[22] *Heavy molecules in the strong magnetic field*. Russian Journal of Math. Phys., 4(1):29–74 (1996).

[23] *Microlocal Analysis and Precise Spectral Asymptotics.*, Springer-Verlag, (1998).

[24] *Heavy molecules in the strong magnetic field. Estimates for ionization energy and excessive charge*. Russian Journal of Math. Phys., 6(1):56–85 (1999).

[25] *Accurate spectral asymptotics for periodic operators*.Proceedings of the Conference, Saint-Jean-de-Monts, France, June (1999)

[26] *Sharp Spectral Asymptotics for Operators with Irregular Coefficients*. Internat. Math. Res. Notices (22):115–1166 (2000).

[27] *Sharp spectral asymptotics for operators with irregular coefficients. Pushing the limits. II*. Comm. Part. Diff. Equats., 28 (1&2):125–156, (2003).

[28] 100 years of Weyl's law. Bulletin of Mathematical Sciences, 6(3):379–452 (2016). Also in arXiv:1608.03963 and in this book.

[29] V. Ivrii. *Spectral asymptotics for fractional Laplacians*. In arXiv:1603.06364 and in this book.

[30] V. Ivrii. *Asymptotics of the ground state energy in the relativistic settings*. Algebra i Analiz (Saint Petersburg Math. J.), 29(3):76—92 (2018) and in this book.

[31] V. Ivrii. *Asymptotics of the ground state energy in the relativistic settings and with self-generated magnetic field.* In arXiv:1708.07737 and in this book.

[32] V. Ivrii. *Spectral asymptotics for Dirichlet to Neumann operator in the domains with edges.* In arXiv:1802.07524 and in this book.

[33] V. Ivrii. *Complete semiclassical spectral asymptotics for periodic and almost periodic perturbations of constant operator.* In arXiv:1808.01619 and in this book.

[34] V. Ivrii. *Complete differentiable semiclassical spectral asymptotics.* In arXiv:1809.07126 and in this book.

[34] V. Ivrii. *Bethe-Sommerfeld conjecture in semiclassical settings.* In arXiv:1902.00335 and in this book.

[35] V. Ivrii. *Upper estimates for electronic density in heavy atoms and molecules.* In arXiv:1906.00611.

Ivrii, V.; Fedorova, S.

[1] *Dilitations and the asymptotics of the eigenvalues of spectral problems with singularities.* Funct. Anal. Appl., 20(4):277–281 (1986).

Ivrii, V.; Petkov V.

[1] *Necessary conditions for the Cauchy problem for non-strictly hyperbolic equations to be well-posed.* Russian Math. Surveys, 29(5):1–70 (1974).

Ivrii, V.; Sigal, I. M.

[1] *Asymptotics of the ground state energies of large Coulomb systems.* Ann. of Math., 138:243–335 (1993).

Iwasaki, C.; Iwasaki, N.

[1] *Parametrix for a degenerate parabolic equation and its application to the asymptotic behavior of spectral functions for stationary problems.* Publ. Res. Inst. Math. Sci. 17(2):577–655 (1981).

Iwasaki, N.

[1] *Bicharacteristic curves and wellposedness for hyperbolic equations with non-involutive multiple characteristics*. J. Math. Kyoto Univ., 34(1):41–46 (1994).

Jakobson, D.; Polterovich, I.; Toth, J.

[1] *Lower bounds for the remainder in Weyl's law on negatively curved surfaces*. IMRN 2007, Article ID rnm142, 38 pages (2007).

Jakšić V.; Molčanov S.; Simon B.

[1] *Eigenvalue asymptotics of the Neumann Laplacian of regions and manifolds with cusps*. J. Func. Anal., 106:59–79 (1992).

Kac, M.

[1] *Can one hear the shape of a drum?* Amer. Math. Monthly, 73:1–23 (1966).

Kannai, Y.

[1] *On the asymptotic behavior of resolvent kernels, spectral functions and eigenvalues of semi-elliptic systems*. Ann. Scuola Norm. Sup. Pisa (3) 23:563—634 (1969).

Kapitanski L.; Safarov Yu.

[1] *A parametrix for the nonstationary Schrödinger equation*. Differential operators and spectral theory, Amer. Math. Soc. Transl. Ser. 2, 189:139–148, Amer. Math. Soc., Providence, RI (1999).

Kohn, J.J.; Nirenberg, L.

[1] *On the algebra of pseudo-differential operators*. Comm. Pure Appl. Math., 18, 269–305 (1965).

Kolmogorov;, A. N.; Fomin S. V.

[1] *Elements of the Theory of Functions and Functional Analysis.* Dover.

[2] *Introductory Real Analysis.* Dover.

Kovǎrik, H.; Vugalter, S.; Weidl, T.

[1] *Two-dimensional Berezin-Li-Yau inequalities with a correction term.* Comm. Math Physics, 287(3):959–881 (2009).

Kovǎrik, H.; Weidl, T.

[1] *Improved Berezin—Li—Yau inequalities with magnetic field.* Proc. Royal Soc. of Edinburgh Section A: Mathematics, 145(1)145–160 (2015).

Kozhevnikov, A. N.

[1] *Remainder estimates for eigenvalues and complex powers of the Douglis-Nirenberg elliptic systems.* Commun. Part. Diff. Eq., 6:1111–1136 (1981).

[2] *Spectral problems for pseudo-differential systems elliptic in the Douglis-Nirenberg sense and their applications.* Math. USSR Sbornik, 21:63–90 (1973).

Kozlov, V. A.

[1] *Asymptotic behavior of the spectrum of nonsemibounded elliptic systems.* Probl. Mat. Analiza, Leningrad Univ., 7:70–83 (1979).

[2] *Estimates of the remainder in formulas for the asymptotic behavior of the spectrum of linear operator bundles.* Probl. Mat. Analiza, Leningrad Univ., 9:34–56 (1984).

Kucherenko, V. V.

[1] *Asymptotic behavior of the solution of the system $A(x, ih\partial/\partial x)u = 0$ as $h \to 0$ in the case of characteristics with variable multiplicity.* (Russian) Izv. Akad. Nauk SSSR Ser. Mat. 38:625–662 (1974).

Kuznecov, N. V.

[1] *Asymptotic formulae for eigenvalues of an elliptic membrane*. Dokl. Akad. Nauk SSSR, 161:760–763 (1965).

[2] *Asymptotic distribution of eigenfrequencies of a plane membrane in the case of separable variables*. Differencial'nye Uravnenija, 2:1385–1402 (1966).

Kuznecov, N. V.; Fedosov, B. V.

[1] *An asymptotic formula for eigenvalues of a circular membrane*. Differencial'nye Uravnenija, 1:1682–1685 (1965).

Kwaśnicki, M.

[1] *Eigenvalues of the fractional laplace operator in the interval*. J. Funct. Anal., 262(5):2379–2402 (2012).

Lakshtanov, E.; Vainberg, B.

[1] *Remarks on interior transmission eigenvalues, Weyl formula and branching billiards*. Journal of Physics A: Mathematical and Theoretical, 45(12) (2012) 125202 (10pp).

Lapidus, M. L.

[1] *Fractal drum, inverse spectral problems for elliptic operators and a partial resolution of the Weyl-Berry conjecture*. Trans. Amer. Math. Soc., 325:465–529 (1991).

Lapointe H.; Polterovich I.; Safarov Yu.

[1] *Average growth of the spectral function on a Riemannian manifold*. Comm. Part. Diff. Equat., 34(4–6):581–615 (2009).

Laptev, A.

[1] *Asymptotics of the negative discrete spectrum of class of Schrödinger operators with large coupling constant*. Linköping University Preprint, (1991).

[2] *The negative spectrum of the class of two-dimensional Schrödinger operators with potentials that depend on the radius*. Funct. Anal. Appl., 34(4) 305–307 (2000).

[2] *Spectral inequalities for partial differential equations and their applications.*. AMS/IP Studies in Advanced Mathematics, 51:629–643 (2012).

Laptev A.; Netrusov, Yu.

[1] *On the negative eigenvalues of a class of Schrödinger operators*. Differential Operators and Spectral Theory, Am. Math. Soc., Providence, RI 173–186 (1999).

Laptev, A.; Geisinger, L.; Weidl, T.

[1] *Geometrical versions of improved Berezin-Li-Yau inequalities*. Journal of Spectral Theory 1:87–109 (2011).

Laptev A.; Robert D.; Safarov, Yu.

[1] *Remarks on the paper of V. Guillemin and K. Okikiolu: "Subprincipal terms in Szegö estimates"*. Math. Res. Lett. 4(1):173–179 (1997) and Math. Res. Lett. 5(1–2):57–61(1998).

Laptev A.; Safarov Yu.

[1] *Global parametrization of Lagrangian manifold and the Maslov factor*. Linköping University Preprint, (1989).

[2] *Szegö type theorems*. Partial differential equations and their applications (Toronto, ON, 1995), CRM Proc. Lecture Notes, Amer. Math. Soc., Providence, RI, 12:177–181 (1997).

Laptev, A.; Safarov, Y.; Vassiliev, D.

[1] *On global representation of Lagrangian distributions and solutions of hyperbolic equations*. Commun. Pure Applied Math., 47(11):1411-1456 (1994).

Laptev, A.; Weidl, T.

[1] *Sharp Lieb-Thirring inequalities in high dimensions*. Acta Math. 184(1):87–111 (2000).

[2] *Recent Results on Lieb-Thirring Inequalities*. J. Èquat. Deriv. Partielles" (La Chapelle sur Erdre, 2000), Exp. no. 20, Univ. Nantes, Nantes (2000).

Laptev, A.; Solomyak, M.

[1] *On spectral estimates for two-dimensional Schrödinger operators*. J. Spectral Theory 3(4):505–515 (2013).

Larson, S.

[1] *On the remainder term of the Berezin inequality on a convex domain*. Proc. AMS 145:2167–2181 (2017).

Lax, P.D.

[1] *Functional Analysis*. Wiley (2002).

Lazutkin, V. F.

[1] *Asymptotics of the eigenvalues of the Laplacian and quasimodes*. Math. USSR Sbornik, 7:439–466 (1973).

[2] *Convex billiards and eigenfunctions of the Laplace operators*. Leningrad Univ., (1981). In Russian.

[3] *Semiclassical asymptotics of eigenfunctions*. Sovremennye Problemy Matematiki. Fundamental'nye Napravleniya. VINITI (Moscow), 34:135–174 (1987)

Lebeau, G.

[1] *Régularité Gevrey* 3 *pour la diffraction*. Comm. Partial Differential Equations, 9(15):1437–1494 (1984).

[2] *Propagation des singularités Gevrey pour le problème de Dirichlet*. Advances in microlocal analysis (Lucca, 1985), NATO Adv. Sci. Inst. Ser. C Math. Phys. Sci., 168:203–223, Reidel (1986).

[3] *Propagation des ondes dans les dièdres*. Mém. Soc. Math. France (N.S.), 60, 124pp (1995).

[4] *Propagation des ondes dans les variétés à coins*. Ann. Sci. École Norm. Sup. (4), 30 (1):429–497 (1997).

Leray, J.

[1] *Analyse Lagrangienne et Mécanique Quantique*. Univ. Louis Pasteur, Strasbourg (1978).

Levendorskii, S. Z.

[1] *Asymptotic distribution of eigenvalues*. Math. USSR Izv., 21:119–160 (1983).

[2] *The approximate spectral projection method*. Acta Appl. Math., 7:137–197 (1986).

[3] *The approximate spectral projection method*. Math. USSR Izv., 27:451–502 (1986).

[4] *Non-classical spectral asymptotics*. Russian Math. Surveys, 43:149–192 (1988).

[5] *Asymptotics of the spectra of operators elliptic in the Douglis-Nirenberg sense*. Trans. Moscow Math. Soc., 52:533–587 (1989).

[6] *Asymptotic Distribution of Eigenvalues of Differential Operators*. Kluwer Acad. Publ. (1990).

Levin, D.; Solomyak, M.

[1] *The Rozenblum-Lieb-Cwikel inequality for Markov generators*. J. Anal. Math., 71:173–193 (1997).

Levitan, B. M.

[1] *On the asymptotic behaviour of the spectral function of the second order elliptic equation*. Izv. AN SSSR, Ser. Mat., 16(1):325–352 (1952). In Russian.

[2] *Asymptotic behaviour of the spectral function of elliptic operator*. Russian Math. Surveys, 26(6):165–232 (1971).

Levitin, M., Vassiliev D.

[1] *Spectral asymptotics, renewal Theorem, and the Berry conjecture for a class of fractals*. Proceedings of the London Math. Soc., (3) 72:188-–214 (1996)

Levy-Bruhl, P.

[1] *Spectre d'opérateurs avec potentiel et champ magnétique polynomiaux*. Preprint (1988?).

Li, P.; Yau, S.-T.

[1] *On the Schrödinger equation and the eigenvalue problem*. Comm. Math. Phys., 88:309–318 (1983).

Lieb, E. H.

[1] *The number of bound states of one-body Schrödinger operators and the Weyl problem*. Bull. of the AMS, 82:751–753 (1976).

[2] *Thomas-Fermi and related theories of atoms and molecules*. Rev. Mod. Phys. 65(4): 603–641 (1981).

[3] *Variational principle for many-fermion systems*. Phys. Rev. Lett., 46:457–459 (1981) and 47:69(E) (1981).

[4] *The stability of matter: from atoms to stars (Selecta)*. Springer-Verlag (2005).

Lieb, E. H.; Loss, M.

[1] *Analysis*. American Mathematical Society, 346pp (2001).

Lieb, E. H.; Loss, M.; Solovej, J. P.

[1] *Stability of matter in magnetic fields*. Phys. Rev. Lett., 75:985–989 (1995) arXiv:cond-mat/9506047

Lieb, E. H.; Oxford S.

[1] *Improved Lower Bound on the Indirect Coulomb Energy*. Int. J. Quant. Chem. 19:427–439, (1981)

Lieb, E. H.; Simon, B.

[1] *The Thomas-Fermi theory of atoms, molecules and solids*. Adv. Math. 23:22–116 (1977).

Lieb, E. H.; Solovej J. P.; Yngvason J.

[1] *Asymptotics of heavy atoms in high magnetic fields: I. Lowest Landau band regions*. Comm. Pure Appl. Math. 47:513–591 (1994).

[2] *Asymptotics of heavy atoms in high magnetic fields: II. Semiclassical regions*. Commun. Math. Phys., 161:77–124 (1994).

[3] *Asymptotics of Natural and Artificial Atoms in Strong Magnetic Fields*, in: D. Iagolnitzer (ed.), XIth International Congress of Mathematical Physics, pp. 185–205, International Press 1995.

Lieb, E. H.; Thirring, W.

[1] *Inequalities for the moments of the eigenvalues of the Schrödinger Hamiltonian and their relation to Sobolev inequalities*. Studies in Mathematical Physics, Princeton University Press, 269–303 (1976).

Lieb, E. H.; Yau, H.T.

[1] *The Stability and Instability of Relativistic Matter*. Commun. Math. Phys. 118(2): 177–213 (1988).

Lions, J. L.; Magenes, E.

[1] *Non-homogeneous Boundary Value Problems*, volume 181, 182, 183 of Die Grundlehren der mathematischen Wissenschaften in Einzeldarstellungen. Springer-Verlag, Berlin (1972–).

Liu, G.

[1] *Some inequalities and asymptotic formulas for eigenvalues on Riemannian manifolds*. J. Math. Anal.s and Applications 376(1):349–364 (2011).

[2] *The Weyl-type asymptotic formula for biharmonic Steklov eigenvalues on Riemannian manifolds*. Advances in Math., 228(4):2162–2217 (2011).

Lorentz, H.A.

[1] *Alte und neue Fragen der Physik*. Physikal. Zeitschr., 11, 1234–1257 (1910).

Malgrange, B.

[1] *Ideals of Differentiable Functions*. Tata Institute and Oxford University Press, Bombay (1966).

Markus, A. S.; Matsaev, V. I.

[1] *Comparison theorems for spectra of linear operators and spectral asymptotics*. Trans. Moscow Math. Soc., 45:139–187 (1982, 1984).

Martinet, J.

[1] *Sur les singularités des formes différentielles*. Ann. Inst. Fourier., 20(1):95–178 (1970).

Martinez, A.

[1] *An Introduction to semiclassical and microlocal analysis*. Springer-Verlag (2002).

Maslov, V. P.

[1] *Théorie des Perturbations et Méthodes Asymptotiques*. Dunod, Paris (1972).

Maz'ya, V. G.

[1] *Sobolev Spaces*. Springer-Verlag (1985).

Maz'ya, V. G.; Verbitsky, I. E.

[1] *Boundedness and compactness criteria for the one-dimensional Schrödinger operator*. Function spaces, interpolation theory and related topics (Lund, 2000), 369–382, de Gruyter, Berlin, 2002.

Melgaard, M.; Ouhabaz, E. M.; Rozenblum, G.

[1] *Negative discrete spectrum of perturbed multivortex Aharonov-Bohm Hamiltonians*. Annales Henri Poincaré, 5(5): 979–1012 (2004).

Melgaard, M.; Rozenblum, G.

[1] *Eigenvalue asymptotics for weakly perturbed Dirac and Schrödinger operators with constant magnetic fields of full rank*. Comm. Partial Differential Equations, 28(3-4):697–736 (2007).

Melrose, R.

[1] *Microlocal parametrices for diffractive boundary value problems*. Duke Math. J., 42:605–635 (1975).

[2] *Local Fourier-Airy operators*. Duke Math. J., 42:583–604 (1975).

[3] *Airy operators*. Commun. Part. Differ. Equat., 3(1):1–76 (1978).

[4] *Weyl's conjecture for manifolds with concave boundary*. In Proc. Symp. Pure Math., volume 36, Providence, RI. AMS (1980).

[5] *Hypoelliptic operators with characteristic variety of codimension two and the wave equation*. Séminaire Goulaouic-Schwartz, Exp. No. 11, 13 pp., École Polytech., Palaiseau, (1980).

[6] *Transformation of boundary problems*. Acta Math., 147:149–236 (1981).

[7] *The trace of the wave group*. Contemp. Math., 27:127–161 (1984).

[8] *Weyl asymptotics for the phase in obstacle scattering*. Commun. Part. Differ. Equat., 13(11):1441–1466 (1989).

Melrose, R. B.; Sjöstrand, J.

[1] *Singularities of boundary value problems. I*. Commun. Pure Appl. Math., 31:593–617 (1978).

[2] *Singularities of boundary value problems. II.*. Commun. Pure Appl. Math., 35:129–168 (1982).

Melrose, R. B.; Taylor, M. E.

[1] *Near peak scattering and the corrected Kirchhoff approximation for a convex obstacle*. Adv. in Math. 55(3):242–315 (1985).

Menikoff, A.; Sjöstrand, J.

[1] *On the eigenvalues of a class of hypoelliptic operators. I*. Math. Ann., 235:55–85 (1978).

[2] *On the Eigenvalues of a Class of Hypoelliptic Operators. II*. pages 201–247. Number 755 in Lect. Notes Math. Springer-Verlag.

[3] *On the eigenvalues of a class of hypoelliptic operators. III*. J. d'Anal. Math., 35:123–150 (1979).

Métivier, G. M.

[1] *Fonction spectrale et valeurs propres d'une classe d'opérateurs non-elliptiques*. Commun. Part. Differ. Equat., 1:467–519 (1976).

[2] *Valeurs propres de problèmes aux limites elliptiques irreguliers*. Bull. Soc. Math. France, Mém, 51–52:125–219 (1977).

[3] *Estimation du reste en théorie spectral*. In Rend. Semn. Mat. Univ. Torino, Facs. Speciale, pages 157–180 (1983).

Miyazaki, Y.

[1] *Asymptotic behavior of normal derivatives of eigenfunctions for the Dirichlet Laplacian*. J. Math. Anal and Applications, 388(1):205–218 (2012).

Mohamed, A. (Morame, A.)

[1] *Etude spectrale d'opérateurs hypoelliptiques à caractéristiques multiples. I*. Ann. Inst. Fourier, 32(3):39–90 (1982).

[2] *Etude spectrale d'opérateurs hypoelliptiques à caractéristiques multiples. II*. Commun. Part. Differ. Equat., 8:247–316 (1983).

[3] *Comportement asymptotique, avec estimation du reste, des valeurs propres d'une class d'opd sur* $\mathbb{R}^n$. Math. Nachr., 140:127–186 (1989).

[4] *Quelques remarques sur le spectre de l'opérateur de Schrödinger avec un champ magnétique*. Commun. Part. Differ. Equat., 13(11):1415–1430 (1989).

[5] *Estimations semi-classique pour l'operateur de Schrödinger à potentiel de type coulombien et avec champ magnétique*. Asymp. Anal., 4(3):235–255 (1991).

Mohamed, A. (Morame, A.); Nourrigat, J.

[1] *Encadrement du $N(\lambda)$ pour un opérateur de Schrödinger avec un champ magnétique et un potentiel électrique*. J. Math. Pures Appl., 70:87–99 (1991).

Morame, A.; Truc, F.

[1] *Eigenvalues of Laplacian with constant magnetic field on non-compact hyperbolic surfaces with finite area*. arXiv:1004.5291 (2010) 9pp.

ter Morsche, H.; Oonincx, P. J.

[1] *On the integral representations for metaplectic operators*. J. Fourier Analysis and Appls., 8(3):245–258 (2002).

Musina, R.; Nazarov, A. I.

[1] *On fractional Laplacians*. arXiv:1308.3606

[2] *On fractional Laplacians–2*. arXiv:1408.3568

[3] *OSobolev and Hardy-Sobolev inequalities for Neumann Laplacians on half spaces*. arXiv:708.01567

Naimark, K.; Solomyak, M.

[1] *Regular and pathological eigenvalue behavior for the equation $-\lambda u'' = Vu$ on the semiaxis*. J. Funct. Anal., 151(2):504–530 (1997).

Netrusov, Yu.; Safarov, Yu.

[1] *Weyl asymptotic formula for the Laplacian on domains with rough boundaries*. Commun. Math. Phys., 253:481–509 (2005).

[2] *Estimates for the counting function of the Laplace operator on domains with rough boundaries*. Around the research of Vladimir Maz'ya. III, Int. Math. Ser. (N. Y.), Springer, New York, 13:247–258 (2010).

Netrusov, Yu.; Weidl T.

[1] *On Lieb-Thirring inequalities for higher order operators with critical and subcritical powers*. Commun. Math. Phys., 182(2):355–370 (1996).

Nishitani, T.

[1] *Propagation of singularities for hyperbolic operators with transverse propagation cone*. Osaka J. Math., 27(1): 1–16 (1990).

[2] *Note on a paper of N. Iwasaki*. J. Math. Kyoto Univ., 38(3):415–418 (1998).

[3] *On the Cauchy problem for differential operators with double characteristics, transition from effective to non-effective characteristics*. arXiv:1601.07688, 1–31 (2016).

Nosmas, J. C.

[1] *Approximation semi-classique du spectre de systemes différentiels asymptotiques*. C.R.A.S. Paris, Sér. 1, 295(3):253–256 (1982).

pagebreak[2]

Novitskii, M.; Safarov, Yu.

[1] *Periodic points of quasianalytic Hamiltonian billiards*. Entire functions in modern analysis (Tel-Aviv, 1997), Israel Math. Conf. Proc., Bar-Ilan Univ., Ramat Gan, 15:269–287 (2001).

Otsuka, K.

[1] *The second term of the asymptotic distribution of eigenvalues of the laplacian in the polygonal domain*. Comm. Part. Diff. Equats., 8(15):1683–1716 (1983).

Pam The L.

[1] *Meilleurs estimations asymptotiques des restes de la fonction spectrale et des valeurs propres relatifs au Laplacien*. Math. Scand., 48:5–31 (1981).

Paneah, B.

[1] *Support-dependent weighted norm estimates for Fourier transforms*. J. Math. Anal. Appl. 189(2):552–574 (1995).

[2] *Support-dependent weighted norm estimates for Fourier transforms, II*. Duke Math. J., 92(1):335–353 (1998).

Parenti, C.; Parmeggiani, A.

[1] *Lower bounds for systems with double characteristics*. J. Anal. Math., 86: 49–91 (2002).

Parmeggiani, A.

[1] *A class of counterexamples to the Fefferman-Phong inequality for systems*. Comm. Partial Differential Equations 29(9-10):1281–1303 (2004).

[2] *On the Fefferman-Phong inequality for systems of PDEs*. in Phase space analysis of partial differential equations, 247–266, Progr. Nonlinear Differential Equations Appl., 69, Birkhäuser Boston, Boston, MA, (2006).

[3] *On positivity of certain systems of partial differential equations*. Proc. Natl. Acad. Sci. USA 104(3):723–726 (2007).

[4] *A remark on the Fefferman-Phong inequality for* 2×2 *systems*. Pure Appl. Math. Q. 6(4):1081-1103 (2010), Special Issue: In honor of Joseph J. Kohn. Part 2.

[5] *On the problem of positivity of pseudodifferential systems*. Studies in phase space analysis with applications to PDEs, 313–335, Progr. Nonlinear Differential Equations Appl., 84, Birkhäuser/Springer, New York, (2013).

[6] *Spectral theory of non-commutative harmonic oscillators: an introduction*. Lecture Notes in Mathematics, 1992. Springer-Verlag, Berlin, (2010). xii+254 pp.

Parnovski, L.; Shterenberg, R.

[1] *Complete asymptotic expansion of the integrated density of states of multidimensional almost-periodic Schrödinger operators*. arXiv:1004.2939v2, 1–54 (2010).

Paul, T.; Uribe, A.

[1] *Sur la formula semi-classique des traces*. C.R.A.S. Paris, Sér. 1, 313:217–222 (1991).

Petkov, V.; Popov, G.

[1] *Asymptotic behaviour of the scattering phase for non-trapping obstacles*. Ann. Inst. Fourier, 32:114–149 (1982).

[2] *On the Lebesgue measure of the periodic points of a contact manifold*. Math. Zeischrift, 218:91–102 (1995).

[3] *Semi-classical trace formula and clustering of eigenvalues for Schrödinger operators*. Ann. Inst. H. Poincare (Physique Theorique), 68:17–83 (1998).

Petkov, V.; Robert, D.

[1] *Asymptotique semi-classique du spectre d'hamiltoniens quantiques et trajectories classiques periodiques*. Commun. Part. Differ. Equat., 10(4):365–390 (1985).

Petkov, V.; Stoyanov, L.

[1] *Periodic geodesics of generic non-convex domains in $\mathbb{R}^2$ and the Poisson relation*. Bull. Amer. Math. Soc., 15:88–90 (1986).

[2] *Periods of multiple reflecting geodesics and inverse spectral results*. Amer. J. Math., 109:617–668 (1987).

[3] *Spectrum of the Poincare map for periodic reflecting rays in generic domains*. Math. Zeischrift, 194:505–518 (1987).

[4] *Geometry of reflecting rays and inverse spectral results*. John & Wiley and Sons, Chichester (1992).

[5] *Geometry of the generalized geodesic flow and inverse spectral problems*. John & Wiley and Sons, Chichester (2017).

Petkov, V.; Vodev, G. *Asymptotics of the number of the interior transmission eigenvalues*. Journal of Spectral Theory, 7(1):1–31 (2017).

Popov, G.

[1] *Spectral asymptotics for elliptic second order differential operators*. J. Math. Kyoto Univ., 25:659–681 (1985).

Popov, G.; Shubin, M. A.

[1] *Asymptotic expansion of the spectral function for second order elliptic operators on* $\mathbb{R}^n$. Funct. Anal. Appl., 17(3):193–199 (1983).

Pushnitski, A.; Rozenblum, G. V.

[1] *Eigenvalue clusters of the Landau Hamiltonian in the exterior of a compact domain*. Doc. Math. 12:569–586 (2007).

Raikov, G.

[1] *Spectral asymptotics for the Schrödinger operator with potential which steadies at infinity*. Commun. Math. Phys., 124:665–685 (1989).

[2] *Eigenvalue asymptotics for the Schrödinger operator with homogeneous magnetic potential and decreasing electric potential. I: Behavior near the essential spectrum tips*. Commun. Part. Differ. Equat., 15(3):407–434 (1990).

[3] *Eigenvalue asymptotics for the Schrödinger operator with homogeneous magnetic potential and decreasing electric potential. II: Strong electric field approximation*. In Proc. Int. Conf. Integral Equations and Inverse Problems, Varna 1989, pages 220–224. Longman (1990).

[4] *Strong electric field eigenvalue asymptotics for the Schrödinger operator with the electromagnetic potential*. Letters Math. Phys., 21:41–49 (1991).

[5] *Semi-classical and weak-magnetic-field eigenvalue asymptotics for the Schrödinger operator with electromagnetic potential*. C.R.A.S. Bulg. AN, 44(1) (1991).

[6] *Border-line eigenvalue asymptotics for the Schrödinger operator with electromagnetic potential*. Int. Eq. Oper. Theorem., 14:875–888 (1991).

[7] *Eigenvalue asymptotics for the Schrödinger operator in strong constant magnetic fields*. Commun. P.D.E. 23:1583–1620 (1998).

[8] *Eigenvalue asymptotics for the Dirac operator in strong constant magnetic fields*. Math. Phys. Electron. J. 5(2) (1999), 22 pp.

[9] *Eigenvalue asymptotics for the Pauli operator in strong non-constant magnetic fields*. Ann. Inst. Fourier, 49:1603–1636 (1999).

Raikov, G.; Warzel, S.

[1] *Quasiclassical versus non-classical spectral asymptotics for magnetic Schroödinger operators with decreasing electric potentials*. Rev. Math. Phys., 14 (2002), 1051–1072.

Randoll, B.

[1] *The Riemann hypothesis for Selberg's zeta-function and asymptotic behavior of eigenvalues of the Laplace operator*. Trans. AMS, 276:203–223 (1976).

Raymond, N.

[1] *Bound States of the Magnetic Schrödinger Operator*. Tract in Math., 27, European Math. Soc., xiv+280 pp (2015).

Raymond, N.; Vu-Ngoc, S.

[1] *Geometry and spectrum in 2D magnetic wells*. Annales de l'Institut Fourier, 2015, 65(1):137–169 (2015).

Reed, M.; Simon, B.

[1] *The Methods of Modern Mathematical Physics*. Volumes I–IV. Academic Press, New York (1972, 1975, 1979, 1978).

Robert, D.

[1] *Propriétés spectrales d'opérateurs pseudodifférentiels*. Commun. Part. Differ. Equat., 3(9):755–826 (1978).

[2] *Comportement asymptotique des valeurs propres d'opérateurs du type Schrödinger à potentiel dégénére*. J. Math. Pures Appl., 61(3):275–300 (1982).

[3] *Calcul fonctionnel sur les opérateurs admissibles et application*. J. Funct. Anal., 45(1):74–84 (1982).

[4] *Autour de l'Approximation Semi-classique*. Number 68 in Publ. Math. Birkhäuser (1987).

[5] *Propagation of coherent states in quantum mechanics and applications*. http://www.math.sciences.univ-nantes.fr/~robert/proc_cimpa.pdf.

Robert, D.; Tamura, H.

[1] *Asymptotic behavior of scattering amplitudes in semi-classical and low energy limits*. Ann. Inst. Fourier, 39(1):155–192 (1989).

Roussarie, R.

[1] *Modèles locaux de champs et de forms*. Astérisque, 30(1):3–179 (1975).

Rozenblioum, G. V.

[1] *The distribution of the discrete spectrum for singular differential operators*. Dokl. Akad. Nauk SSSR, 202:1012–1015 (1972). In Russian. Poor English translation in Soviet Math. Dokl. 13 245–249 (1972).

[2] *Distribution of the discrete spectrum for singular differential operators*. Soviet Math. Dokl., 13(1):245–249 (1972).

[3] *On the eigenvalues of the first boundary value problem in unbounded domains*. Math. USSR Sb., 89(2):234–247 (1972).

[4] *Asymptotics of the eigenvalues of the Schrödinger operator*. Math. USSR Sb., 22:349–371 (1974).

[5] *The distribution of the discrete spectrum of singular differential operators*. English transl.: Sov. Math., Izv. VUZ, 20(1):63–71 (1976).

[6] *An asymptotics of the negative discrete spectrum of the Schrödinger operator*. Math. Notes Acad. Sci. USSR, 21:222–227 (1977).

[7] *On the asymptotics of the eigenvalues of some 2-dimensional spectral Problems*. pages 183–203. Number 7 in Univ. Publ. Leningrad (1979).

Rozenblum, G. V.; Sobolev, A. V.

[1] *Discrete spectrum distribution of the Landau operator perturbed by an expanding electric potential*. Spectral theory of differential operators, 169–190, Amer. Math. Soc. Transl. Ser. 2, 225, Amer. Math. Soc., Providence, RI, 2008.

Rozenblioum, G. V.; Solomyak, M. Z.

[1] *The Cwikel-Lieb-Rozenblum estimates for generators of positive semigroups and semigroups dominated by positive semigroups*. St. Petersbg. Math. J., 9(6):1195–1211 (1998).

[2] *Counting Schrödinger boundstates: semiclassics and beyond*. Sobolev spaces in mathematics. II. Applications in Analysis and Partial Differential Equations, International Mathematical Series. Springer-Verlag, 9 (2008), 329–353.

Rozenblioum, G. V.; Solomyak, M. Z.; Shubin, M. A.

[1] *Spectral theory of differential operators*. Sovremennye problemy matematiki. Fundamental'nye napravleniya. VINITI, Moscow, 64 (1990). Translation to English has been published by Springer-Verlag: Partial Differential Equations VII, EMS volume 34 (1994).

Rozenblioum, G. V.; Tashchiyan, G.

[1] *On the spectral properties of the Landau Hamiltonian perturbed by a moderately decaying magnetic field. Spectral and scattering theory for quantum magnetic systems.* 169–186, Contemp. Math., 500, Amer. Math. Soc., Providence, RI, 2009.

[2] *On the spectral properties of the perturbed Landau Hamiltonian.* Comm. Partial Differential Equations 33(4–6), 1048–1081 (2008).

Ruskai, M. B. ; Solovej, J. P.

[1] *Asymptotic neutrality of polyatomic molecules.* In *Schrödinger Operators*, Springer Lecture Notes in Physics 403, E. Balslev (Ed.), 153–174, Springer Verlag (1992).

Safarov, Yu.

[1] *Asymptotic behavior of the spectrum of the Maxwell operator.* J. Soviet Math., 27(2):2655–2661 (1984).

[2] *The asymptotics of the spectrum of transmission problems.* J. Soviet Math., 32(5):519–525 (1986).

[3] *The asymptotics of the spectrum of boundary value problem with periodic billiard trajectories.* Funct. Anal. Appl., 21:337–339 (1987).

[4] *On the second term of the spectral asymptotics of the transmission problem.* Acta Appl. Math., 10:101–130 (1987).

[5] *On the Riesz means of the counting eigenvalues function of elliptic operator.* Zapiski Nauchnykh Seminarov Leningradskogo Otdeleniya Matematicheskogo Instituta im. V.A. Steklova AN SSSR, 163: 143–145 (1987). In Russian.

[6] *The asymptotics of the spectrum of pseudodifferential operator with periodic bicharacteristics.* In LOMI Proceedings, 152:94–104 (1986). English translation in J. Soviet Math., 40, no. 5 (1988).

[7] *The asymptotics of the spectral function of positive elliptic operator without non-trapping condition.* Funct. Anal. Appl., 22(3):213–223 (1988).

[8] *The precise asymptotics of the spectrum of boundary value problem and periodic billiards.* Izv. AN SSSR, 52(6):1230–1251 (1988).

[9] *Lower bounds for the generalized counting function.* The Maz'ya anniversary collection, Vol. 2 (Rostock, 1998), Oper. Theory Adv. Appl., Birkhäuser, Basel, 110:275–293 (1999).

[10] *Fourier Tauberian theorems and applications.* J. Funct. Anal. 185(1):111–128 (2001).

[11] *Berezin and Gårding inequalities.* Funct. Anal. Appl. 39(4):301–307 (2005).

[12] *Sharp remainder estimates in the Weyl formula for the Neumann Laplacian on a class of planar regions.* J. Functional Analysis, 250(1):21–41 (2007).

[13] *On the comparison of the Dirichlet and Neumann counting functions.* Spectral theory of differential operators, Amer. Math. Soc. Transl. Ser. 2, Amer. Math. Soc., Providence, RI, 225:191–204 (2008).

Safarov, Yu. G.; Vassiliev, D. G.

[1] *Branching Hamiltonian billiards.* Soviet Math. Dokl., 38(1) (1989).

[2] *The asymptotic distribution of eigenvalues of differential operators.* AMS Transl., Ser. 2, 150 (1992).

[3] *The Asymptotic Distribution of Eigenvalues of Partial Differential Operators.* volume 155 of Translations of Mathematical Monographs. American Mathematical Society (1997). 354pp.

Sarnak, P.

[1] *Spectra and eigenfunctions of Laplacians.* CRM Proc. and Lecture Notes, 12:261–273 (1997).

Schenk, D.; Shubin, M. A.

[1] *Asymptotic expansion of the density of states and the spectral function of the Hill operator.* Mat. Sborn., 12(4):474–491 (1985).

Seco, L. A.; Sigal, I. M.; Solovej, J. P.

[1] *Bound of the ionization energy of large atoms*. Commun. Math. Phys., 131:307–315 (1990).

Seeley, R.

[1] *A sharp asymptotic estimate for the eigenvalues of the Laplacian in a domain of* $\boldsymbol{R}^3$. Advances in Math., 102(3):244–264 (1978).

[2] *An estimate near the boundary for the spectral function of the Laplace operator*. Amer. J. Math., 102(3):869–902 (1980).

Shargorodsky E.

[1] *On negative eigenvalues of two-dimensional Schrödinger operators*. Proc. Lond. Math. Soc. (3) 108 (2014), no. 2, 441–483.

Shigekawa, I.

[1] *Eigenvalue problems for the Schrödinger operator with magnetic field on a compact Riemannian manifold*. J. Funct. Anal., 87(75):92–127 (1989).

Shubin, M. A.

[1] *Weyl's theorem for the Schrödinger operator with an almost periodic potential*. Trans. Moscow Math. Soc., 35:103–164 (1976).

[2] *Spectral theory and the index of elliptic operators with almost periodic potential*. Soviet Math. Surv., 34(2):95–135 (1979).

[3] *Pseudodifferential Operators and Spectral Theory*. Nauka (1978). In English: Springer-Verlag (2001).

Shubin, M. A.; Tulovskii, V. A.

[1] *On the asymptotic distribution of eigenvalues of p.d.o. in* $\mathbb{R}^n$. Math. USSR Sbornik, 21:565–573 (1973).

Siedentop, H.; Weikard R.

[1] *On the leading energy correction for the statistical model of the atom: interacting case*. Comm. Math. Phys. 112:471-490 (1987).

[2] *On the leading correction of the Thomas-Fermi model: lower bound*. Invent. Math. 97, 159–193 (1989).

[3] *A new phase space localization technique with application to the sum negative eigenvalues of Schrödinger operators*. Annales Sci. de l'É.N.S., 4e sér, 24(2):215-225 (1991).

Sigal, I. M.

[1] *Lectures on large Coulomb systems*. CRM Proc. and Lecture Note, 8:73–107 (1995).

Simon, B.

[1] *The bound state of weakly coupled Schrödinger operators in one and two dimensions*. Annals of Physics. 97(2):279–288 (1976).

[2] *Nonclassical eigenvalue asymptotics*. J. Funct. Anal., 33:84–98 (1983).

[3] *The Neumann Laplacian of a jelly roll*. Proc. AMS, 114(3):783–785 (1992).

[4] *Comprehensive Course in Analysis. Parts I-IV*. American Mathematical Society (2015).

Sjöstrand, J.

[1] *Analytic singularities and microhyperbolic boundary value problems*. Math. Ann., 254:211–256 (1980).

[2] *On the eigenvalues of a class of hypoelliptic operators. IV*. Ann. Inst. Fourier, 30(2) (1980).

[3] *Singularités analytiques microlocales*. Astérisque, 95 (1982).

[4] *Propagation of analytic singularities for second order Dirichlet problems. I*. Commun. Part. Differ. Equat., 5(1):41–94 (1980).

[5] *Propagation of analytic singularities for second order Dirichlet problems. II*. Commun. Part. Differ. Equat., 5(2):187–207 (1980).

[6] *Propagation of analytic singularities for second order Dirichlet problems. III*. Commun. Part. Differ. Equat., 6(5):499–567 (1981).

Sobolev, A. V.

[1] *On the asymptotic for energy levels of a quantum particle in a homogeneous magnetic field, perturbed by a decreasing electric field. 1*. J. Soviet Math., 35(1):2201–2211 (1986).

[2] *On the asymptotic for energy levels of a quantum particle in a homogeneous magnetic field, perturbed by a decreasing electric field. 2*. Probl. Math. Phys, 11:232–248 (1986).

[3] *Discrete spectrum asymptotics for the Schrödinger operator with electric and homogeneous magnetic field*. Notices LOMI sem., 182:131–141 (1990).

[4] *On the asymptotics of discrete spectrum for the Schrödinger operator in electric and homogeneous magnetic field*. Operator Theory: Advances and Applications, 46:27–31 (1990).

[5] *Weyl asymptotics for the discrete spectrum of the perturbed Hill operator*. Adv. Soviet Math., 7:159–178 (1991).

[6] *The quasiclassical asymptotics of local Riesz means for the Schrödinger operator in a strong homogeneous magnetic field*. Duke Math. J., 74(2):319–429 (1994).

[7] *Quasiclassical asymptotics of local Riesz means for the Schrödinger operator in a moderate magnetic field*. Ann. Inst. H. Poincaré, 62(4):325–359 (1995).

[8] *Discrete Spectrum Asymptotics for the Schrödinger Operator in a Moderate Magnetic Field*. volume 78, pages 357–367. Birkhäuser (1995).

Sobolev, A. V.; Yafaev D. R.

[1] *Phase analysis in the problem of scattering by a radial potential*. Zap. Nauchn. Sem. Leningrad. Otdel. Mat. Inst. Steklov. (LOMI), 147:155–178, 206, 1985.

Sogge, C. D.

[1] *Hangzhou lectures on eigenfunctions of the Laplacian (AM-188)*. Princeton Univ. Press, (2014).

Sogge, C. D.; Zelditch, S.

[1] *Riemannian manifolds with maximal eigenfunction growth*. Duke Math. J., 114(3):387–437 (2002).

Sobolev, S. L.

[1] *Some Applications of Functional Analysis in Mathematical Physics*. volume 90 of Trans. of Math. Monographs. AMS (1991)

Solnyshkin, S. N.

[1] *Asymptotic of the energy of bound states of the Schrödinger operator in the presence of electric and homogeneous magnetic fields*. Select. Math. Sov., 5(3):297–306 (1986).

[2] *On the discrete spectrum of a quantum particle in an electrical and a homogeneous magnetic field*. Algebra i Analiz, 3(6):164–172 (1991).

Solomyak, M. Z.

[1] *Spectral asymptotics of Schrödinger operators with non-regular homogeneous potential*. Math. USSR Sbornik, 55(1):9–38 (1986).

[2] *Piecewise-polynomial approximation of functions from* $H^l((0,1)^d)$, $2l = d$, *and applications to the spectral theory of the Schrödinger operator*. Israel J. Math., 86(1–3):253–275 (1994).

[3] *On the negative discrete spectrum of the operator $-\Delta_N - \alpha V$ for a class of unbounded domains in $\mathbb{R}^d$*. CRM Proceedings and Lecture Notes, Centre de Recherches Mathematiques, 12:283–296 (1997).

[4] *On the discrete spectrum of a class of problems involving the Neumann Laplacian in unbounded domains*. Advances in Mathematics, AMS (volume dedicated to 80-th birthday of S.G.Krein (P. Kuchment and V.Lin, Editors) - in press.

Solomyak, M. Z.; Vulis, L. I.

[1] *Spectral asymptotics for second order degenerating elliptic operators*. Math. USSR Izv., 8(6):1343–1371 (1974).

Solovej, J. P.

[1] *Asymptotic neutrality of diatomic molecules*. Commun. Math. Phys., 130:185–204 (1990).

[2] *A new look at Thomas-Fermi Theory*. Molecular Physics, 114(7-8), 1036–1040 (2016).

[3] *Thomas Fermi type theories (and their relation to exact models)*. In Encyclopedia of Applied and Computational Mathematics, Eds. B. Engquist, Springer-Verlag Berlin Heidelberg, pp. 1471–1475 (2015).

Solovej, J. P.; Sørensen, T. Ø.; Spitzer W. L.

[1] *The relativistic Scott correction for atoms and molecules*. Comm. Pure Appl. Math., 63: 39–118 (2010).

Solovej, J. P.; Spitzer W. L.

[1] *A new coherent states approach to semiclassics which gives Scott's correction*. Commun. Math. Phys. 241, 383–420, (2003).

Sommerfeld, A.

[1] *Die Greensche Funktion der Schwingungsgleichung für ein beliebiges Gebiet*. Physikal. Zeitschr., 11, 1057–1066 (1910).

Tamura, H.

[1] *The asymptotic distribution of eigenvalues of the Laplace operator in an unbounded domain.* Nagoya Math. J. 60:7–33 (1976).

[2] *Asymptotic formulas with sharp remainder estimates for bound states of Schrödinger operators.* J. d'Anal. Math., 40:166–182 (1981).

[3] *Asymptotic formulas with sharp remainder estimates for bound states of Schrödinger operators. II.* J. d'Anal. Math., 41:85–108 (1982).

[4] *Asymptotic formulas with sharp remainder estimates for eigenvalues of elliptic second order operators.* Duke Math. J., 49(1):87–119 (1982).

[5] *Asymptotic formulas with remainder estimates for eigenvalues of Schrödinger operators.* Commun. Part. Differ. Equat., 7(1):1–54 (1982).

[6] *The asymptotic formulas for the number of bound states in the strong coupling limit.* J. Math. Soc. Japan, 36:355–374 (1984).

[7] *Eigenvalue asymptotics below the bottom of essential spectrum for magnetic Schrödinger operators.* In Proc. Conf. on Spectral and Scattering Theory for Differential Operators, Fujisakara-so, 1986, pages 198–205, Tokyo. Seizo Itô (1986).

[8] *Asymptotic distribution of eigenvalues for Schrödinger operators with magnetic fields.* Nagoya Math. J., 105(10):49–69 (1987).

[9] *Asymptotic distribution of eigenvalues for Schrödinger operators with homogeneous magnetics fields.* Osaka J. Math., 25:633–647 (1988).

[10] *Asymptotic distribution of eigenvalues for Schrödinger operators with homogeneous magnetic fields. II.* Osaka J. Math., 26:119–137 (1989).

Taylor, M. E.

[1] *Reflection of singularities of solutions to systems of differential equations.* Comm. Pure Appl. Math. 28(4):457–478,1975.

[2] *Grazing rays and reflection of singularities of solutions to wave equations.* Comm. Pure Appl. Math., 29(1):1–38 (1976).

[3] *Grazing rays and reflection of singularities of solutions to wave equations. II. Systems*. Comm. Pure Appl. Math., 29(5):463–481, (1976).

[4] *Rayleigh waves in linear elasticity as a propagation of singularities phenomenon. Partial differential equations and geometry*. Proc. Conf., Park City, Utah, 1977), Lecture Notes in Pure and Appl. Math., 48:273–291 (1979), Dekker, New York.

[5] *Diffraction effects in the scattering of waves*. Singularities in boundary value problems (Proc. NATO Adv. Study Inst., Maratea, 1980), NATO Adv. Study Inst. Ser. C: Math. Phys. Sci., 65, Reidel, Dordrecht-Boston, Mass., 271–316, 1981. pages 271–316. D. Reidel Publ. Co., Dordrecht, Boston, London (1981).

[6] *Diffraction of waves by cones and polyhedra*. Analytical and numerical approaches to asymptotic problems in analysis Proc. Conf., Univ. Nijmegen, Nijmegen, 1980, 235–248, North-Holland Math. Stud., 47, North-Holland, Amsterdam-New York, 1981.

[7] *Pseudodifferential Operators*. Princeton University Press (1981).

Treves, F.

[1] *Introduction to Pseudo-differential and Fourier Integral Operators*. Vol. 1,2 *Plenum Press* (1982).

Vainberg, B. R.

[1] *A complete asymptotic expansion of the spectral function of second order elliptic operators in $\mathbb{R}^n$*. Math. USSR Sb., 51(1):191–206 (1986).

Van Den Berg, M.

[1] *Dirichlet-Neumann bracketing for horn-shaped regions*. Journal Funct. Anal., 104(1), 110–120 (1992).

[12] *On the spectral counting function for the Dirichlet laplacian*. Journal Funct. Anal., 107(12), 352–361 (1992).

Vassiliev, D. G.

[1] *Binomial asymptotics of spectrum of boundary value problems*. Funct. Anal. Appl., 17(4):309–311 (1983).

[2] *Two-term asymptotics of the spectrum of a boundary value problem under an interior reflection of general form*. Funct. Anal. Appl., 18(4):267–277 (1984).

[3] *Asymptotics of the spectrum of pseudodifferential operators with small parameters*. Math. USSR Sbornik, 49(1):61–72 (1984).

[4] *Two-term asymptotics of the spectrum of a boundary value problem in the case of a piecewise smooth boundary*. Soviet Math. Dokl., 33(1):227–230 (1986).

[5] *Two-term asymptotics of the spectrum of natural frequencies of a thin elastic shell*. Soviet Math. Dokl., 41(1):108–112 (1990).

[6] *One can hear the dimension of a connected fractal in R^2*. In Petkov & Lazarov: Integral Equations and Inverse Problems, 270–273; Longman Academic, Scientific & Technical (1991).

Volovoy, A.

[1] *Improved two-term asymptotics for the eigenvalue distribution function of an elliptic operator on a compact manifold*. Commun. Part. Differ. Equat., 15(11):1509–1563 (1990).

[2] *Verification of the Hamiltonian flow conditions associated with Weyl's conjecture*. Ann. Glob. Anal. Geom., 8(2):127–136 (1990).

Vorobets, Y. B.; Galperin, G. A.; Stepin, A. M.

[1] *Periodic billiard trajectories in polygons.* Uspechi Matem. Nauk, (6):165–166 (1991).

Vugal'ter, S. A.

[1] *On asymptotics of eigenvalues of many-particles Hamiltonians on subspaces of functions of the given symmetry*. Theor. Math. Physics, 83(2):236–246 (1990).

Vugal'ter, S. A.; Zhislin, G. M.

[1] *On the spectrum of Schrödinger operator of multiparticle systems with short-range potentials*. Trans. Moscow Math. Soc., 49:97–114 (1986).

[2] *Asymptotics of the discrete spectrum of Hamiltonians of quantum systems with a homogeneous magnetic field*. Operator Theory: Advances and Applications, 46:33–53 (1990).

[3] *On precise asymptotics of the discrete spectrum of n-particle Schrödinger operator in the spaces with symmetry*. Soviet Math. Dokl., 312(2):339–342 (1990).

[4] *On the asymptotics of the discrete spectrum of the given symmetry for many-particle Hamiltonians*. Trans. Moscow Math. Soc., 54:187–213 (1991). In Russian.

[5] *On the discrete spectrum of the given symmetry for the Schrödinger operator of the n-particle system in the homogeneous magnetic field*. Funct. Anal. Appl., 25(4):83–86 (1991). In Russian, also in Soviet Math. Dokl., 317, no 6, 1365–1369 (1991).

[6] *On the asymptotics of the discrete spectrum of given symmetry of multiparticle Hamiltonians*. (Russian) Trudy Moskov. Mat. Obshch. 54:186–212 (1992).

[7] *Spectral asymptotics of N-particle Schrödinger operators with a homogeneous magnetic field on subspaces with fixed $SO(2)$ symmetry*. (Russian) Algebra i Analiz 5(2):108–125 (1993).

Wakabayashi, S.

[1] *Analytic singularities of solutions of the hyperbolic Cauchy problem*. Japan Acad. Proc. Ser. A. Math. Sci. 59(10):449–452 (1983).

[2] *Singularities of solutions of the Cauchy problem for symmetric hyperbolic systems*. Commun. Part. Differ. Equat., 9(12):1147–1177 (1984).

Wang, X. P.

[1] *Asymptotic behavior of spectral means of pseudo-differential operators. I*. J. Approx. Theor. Appl., 1(2):119–136 (1985).

[2] *Asymptotic behavior of spectral means of pseudo-differential operators. II*. J. Approx. Theor. Appl., 1(3):1–32 (1985).

Weidl, T.

[1] *On the Lieb-Thirring Constants* $L_{\gamma,1}$ *for* $\gamma \geq \frac{1}{2}$. Comm. Math. Phys., 178:135–146 (1996).

[2] *Improved Berezin-Li-Yau inequalities with a remainder term*. In Spectral Theory of Differential Operators, Amer. Math. Soc. Transl. (2) 225:253–263 (2008).

Weinstein, A.

[1] *Asymptotics of the eigenvalues, clusters for Laplacian plus a potential*. Duke Math. J., 44:883–892 (1977).

Weyl, H.

[1] *Über die Asymptotische Verteilung der Eigenwerte*. Nachr. Konigl. Ges. Wiss. Göttingen 110–117 (1911).

[2] *Das asymptotische Verteilungsgesetz linearen partiellen Differentialgleichungen*. Math. Ann., 71:441–479 (1912).

[3] *Über die Abhängigkeit der Eigenschwingungen einer Membran von deren Begrenzung*. J. Für die Angew. Math., 141:1–11 (1912).

[4] *Über die Randwertaufgabe der Strahlungstheorie und asymptotische Spektralgeometrie*. J. Reine Angew. Math., (143):177–202 (1913).

[5] *Das asymptotische Verteilungsgesetz der Eigenschwingungen eines beliebig gestalteten elastischen Körpers*. Rend. Circ. Mat. Palermo. 39:1–49 (1915).

[6] *Quantenmechanik und Gruppentheorie*. Zeitschrift für Physik, 46:1–46 (1927) (see The Theory of Groups and Quantum Mechanics, Dover, 1950, xxiv+422).

[7] *Ramifications, old and new, of the eigenvalue problem*. Bull. Amer. Math. Soc. 56(2):115–139 (1950).

Widom, H.

[1] *A complete symbolic calculus for pseudo-differential operators*. Bull. Sci. Math., Sér. 2, 104:19–63 (1980).

Yngvason, J.

[1] *Thomas-Fermi Theory for Matter in a Magnetic Field as a Limit of Quantum Mechanics*. Lett. Math. Phys. 22:107–117 (1991).

Zaretskaya, M. A.

[1] *On the spectrum of a class of differential operators*. Izvestiya VUZ. Matematika, 27(5):76–78 (1983).

[2] The character of the spectrum of square quantum Hamiltonians, Izvestiya VUZ. Matematika, 29(5):68–69 (1985).

[3] *Joint spectrum of system of commuting quadratic quantum Hamiltonians*. Izvestiya VUZ. Matematika, 31(4):80–82 (1987).

[4] *Spectrum of quadratic operators different from normal*. Izvestiya VUZ. Matematika, 33(11):24–30 (1990).

Zelditch, S.

[1] *Selberg trace formulae, Ψdo's and equidistribution theorems for closed geodesics and Laplace eigenfunctions*. Lect. Notes Math., 1256:467–479 (1987).

Zhislin, G.

[1] *Discussion of the spectrum of Schrodinger operator for systems of many particles*. Tr. Mosk. Mat. Obs., 9, 81–128 (1960).

Zielinski, L.

[1] *Asymptotic distribution of eigenvalues of some elliptic operators with simple remainder estimates*. J. Operator Theory, 39:249–282 (1998).

[2] *Asymptotic distribution of eigenvalues for some elliptic operators with intermediate remainder estimate*. Asymptot. Anal., 17(2):93–120 (1998).

[3] *Asymptotic distribution of eigenvalues for elliptic boundary value problems*. Asymptot. Anal., 16(3–4):181–201 (1998).

[4] *Sharp spectral asymptotics and Weyl formula for elliptic operators with non-smooth coefficients*. Math. Phys. Anal. Geom., 2(3):291–321 (1999).

[5] *Sharp semiclassical estimates for the number of eigenvalues below a degenerate critical level*. Asymptot. Anal., 53(1-2):97–123 (2007).

[6] *Semiclassical Weyl formula for a class of weakly regular elliptic operators*. Math. Phys. Anal. Geom., 9(1):1–21 (2006).

[7] *Sharp semiclassical estimates for the number of eigenvalues below a totally degenerate critical level*. J. Funct. Anal., 248(2):259–302 (2007).

Zworski, M.

[1] *Semiclassical Analysis*. volume 138 of Graduate Studies in Mathematics. AMS, (2012).

Presentations

[1] *Sharp spectral asymptotics for irregular operators*

[2] *Sharp spectral asymptotics for magnetic Schrödinger operator*

[3] *25 years after*

[4] *Spectral asymptotics for 2-dimensional Schrödinger operator with strong degenerating magnetic field*

[5] *Magnetic Schrödinger operator: geometry, classical and quantum dynamics and spectral asymptotics*

[6] *Spectral asymptotics and dynamics*

[7] *Magnetic Schrödinger operator near boundary*

[8] *2D- and 3D-magnetic Schrödinger operator: short loops and pointwise spectral asymptotics*

[9] *100 years of Weyl's law*

[10] *Some open problems, related to spectral theory of PDOs*

[11] *Large atoms and molecules with magnetic field, including self-generated magnetic field (results: old, new, in progress and in perspective)*

[12] *Semiclassical theory with self-generated magnetic field*

[13] *Eigenvalue asymptotics for Dirichlet-to-Neumann operator*

Available at `http://weyl.math.toronto.edu/victor_ivrii/research/talks/`

V. Ivrii, *Microlocal Analysis, Sharp Spectral Asymptotics and Applications II*, https://doi.org/10.1007/978-3-030-30541-3

[14] *Eigenvalue asymptotics for Fractional Laplacians*

[15] *Asymptotics of the ground state energy for relativistic atoms and molecules*

[16] *Etudes in spectral theory*

[17] *Eigenvalue asymptotics for Steklov's problem in the domain with edges*

[18] *Complete semiclassical spectral asymptotics for periodic and almost periodic perturbations of constant operators*

[19] *Complete Spectral Asymptotics for Periodic and Almost Periodic Perturbations of Constant Operators and Bethe-Sommerfeld Conjecture in Semiclassical Settings*

Index

V. Ivrii, *Microlocal Analysis, Sharp Spectral Asymptotics and Applications II*, https://doi.org/10.1007/978-3-030-30541-3

Content of All Volumes

Volume I. Semiclassical Microlocal Analysis and Local and Microlocal Semiclassical Asymptotics

Volume II. Functional Methods and Eigenvalue Asymptotics

Volume III. Magnetic Schrödinger Operator. 1

Volume IV. Magnetic Schrödinger Operator. 2

Volume V. Applications to Quantum Theory and Miscellaneous Problems

Printed in the United States
By Bookmasters